Non-Stationary Electromagnetics

Second Edition

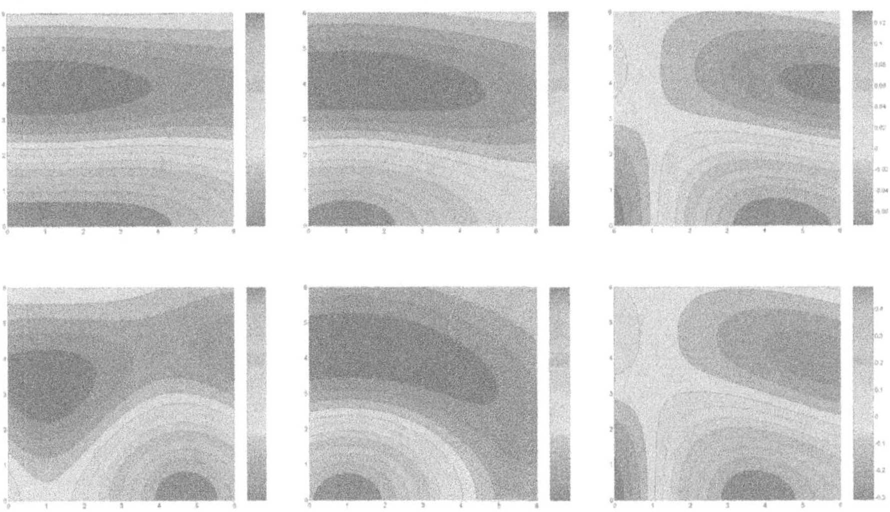

$$\mathbf{E}(\mathbf{x}) = \mathbf{E}_0(\mathbf{x}') + \int d\mathbf{x}' \left\langle \mathbf{x} \middle| \hat{K} \middle| \mathbf{x}' \right\rangle \mathbf{E}(\mathbf{x}')$$

$$\mathbf{E}(\mathbf{x}) = \mathbf{E}_0(\mathbf{x}') + \int d\mathbf{x}' \left\langle \mathbf{x} \middle| \hat{R} \middle| \mathbf{x}' \right\rangle \mathbf{E}_0(\mathbf{x}')$$

Non-Stationary Electromagnetics

An Integral Equations Approach

Second Edition

Alexander Nerukh
Trevor Benson

PAN STANFORD PUBLISHING

Published by

Pan Stanford Publishing Pte. Ltd.
Penthouse Level, Suntec Tower 3
8 Temasek Boulevard
Singapore 038988

Email: editorial@panstanford.com
Web: www.panstanford.com

British Library Cataloguing-in-Publication Data
A catalogue record for this book is available from the British Library.

Non-Stationary Electromagnetics: An Integral Equations Approach (Second Edition)

ISBN 978-981-4774-95-6 (Hardcover)
ISBN 978-0-429-02773-4 (eBook)

Contents

Preface

The present book is a revised version of *Non-Stationary Electromagnetics*, by Alexander Nerukh, Nataliya Sakhnenko, Trevor Benson, and Phillip Sewell. It differs from the earlier volume by its emphasis on an integral equation approach. Consequently, the chapter of the previous book that introduced the differential method for non-stationary problems is omitted. It is planned that such an approach to non-stationary problems will form the basis of another self-contained book.

The book is devoted to the investigations of non-stationary electromagnetic processes. It contains results concerning the non-stationary electromagnetic processes initiated by time variations of material objects. The main idea of the book can be characterized by the phrase "Any change makes a path for other changes" from Niccolo dei Machiavelli (1469–1527). This book offers a good opportunity to introduce the Volterra integral equation method for widespread investigations of electromagnetic phenomena. A systematic presentation of this method in the time domain provides important theoretical results; the explicit mathematical theory is combined with examples of its application in electromagnetic devices in microwaves, optoelectronics, and photonics where time-domain methods provide a powerful modeling tool. Particular consideration is given to electromagnetic transients in time-varying media and their potential applications. The approach is formulated and electromagnetic phenomena are investigated in detail for a hollow metal waveguide, which contains a moving dielectric or plasma-bounded medium, and for dielectric waveguides with time-varying medium inside the core. The content of the present version has been enriched by the consideration of the transformation of Airy pulses, which provides new insights into the preservation of pulses of complex form. Considering the influences of medium changes on electromagnetic fields in optoelectronic devices is very important for the realistic description of such devices. Many electromagnetic phenomena studied in the book may lead to innovative ideas for experimentalists and engineers developing new classes of photonic devices.

This book systematizes and collects almost all pertinent results obtained by the authors since the 1970s. Some of these results were published in Russian, and some are unpublished but may be of interest to the wider electromagnetic community. It is a pleasure to express our sincere gratitude to the people who contributed to obtaining the results during all these years, especially to Peter E. Minko, Oleg N. Rybin, Irina Yu Shavorykina, Fedor V. Fedotov, Phillip Sewell, and Nataliya K. Sakhnenko.

Alexander Nerukh
Kharkov, Ukraine
Trevor Mark Benson
Nottingham, UK
2018

Introduction

Any change in the state of a medium, for example, a change of its material properties or movement of its boundaries, affects the characteristics of an electromagnetic field existing in this medium. This influence is very strong, even in the simplest non-dispersive electromagnetic structures. As there are two temporal processes in this case, medium change and field change, the points of their beginnings acquires principal importance and corresponding mathematical problems becomes initial-boundary value ones. It is evident that a dispersive structure adds new special features to the change of the electromagnetic field state, and can greatly influence transient electromagnetic processes. In practice, waveguides and resonators where the electromagnetic field interacts with matter in bounded space areas constrained by waveguide or resonator walls are very important dispersive structures; the presence of the walls brings a dispersive character to electromagnetic wave propagation in the region considered. The field interaction with a non-stationary medium acquires new features in these conditions. In addition, according to the difference between the phase and the group velocities of the waves conditioned by the dispersion, the importance of taking into account some initial time of the interaction process arises. This importance increases in the case where a medium or its borders move, when the relationship between all three velocities, the phase and the group velocities of the waves and the motion velocity, begins to play a significant role. Investigations of transients in waveguides have a long history, but they have often been concerned with the degradation of pulses in stationary waveguides, principally metallic waveguides.

Maxwell's equations are self-consistent only for electromagnetic fields in a vacuum. In a general medium, the constitutive equations and boundary conditions complicate significantly both the formulation and the solution of electromagnetic problems. Such problems become even more complex when the media are not only inhomogeneous but also time varying. Such a situation can be met when considering the propagation of electromagnetic signals in dielectric or semiconductor waveguides, in particular in the context of modulators, pulsed lasers, and frequency conversion. The proper description and investigation of the physics of these phenomena is motivated by their significant importance to optical communication technology; the interactions between microwave and optical pulses and semiconductor active media in waveguides have, therefore, received considerable attention in recent years. The solution of such electromagnetic problems has demanded accurate time-domain techniques, some variants of which have received widespread attention in the literature, mainly owing to their computational superiority for solving wide-band problems in comparison with frequency-domain methods. Unfortunately, most of these techniques are focused upon numerical calculations and are not suitable for identifying the general features of the phenomena. This is especially true for the important case of understanding the behavior of the guided modes supported by dielectric optical waveguides, a central task in the simulation of integrated optical components.

F. R. Morgenthaler revealed in 1958 that a temporal change in the permittivity of an unbounded medium transforms a primary harmonic plane wave to new secondary ones having different frequencies but the same wavenumber as the primary wave. This general feature is also observed when a plane wave is normally incident onto a plane interface between two media, the permittivity of one of which changes abruptly. However, in this case the spatial structure of waves also becomes more complex. Nevertheless, the monochromatic character of the secondary waves is not disturbed if the medium is non-dissipative. The picture of such phenomena becomes even more complex in the case of the oblique incidence of an electromagnetic wave onto a plane boundary with a time-varying medium. In this case, not only does the structure of the system of monochromatic waves become more complex, but a continuous wave spectrum also appears. All the circumstances just discussed arise in a dielectric waveguide with time-varying media; in this book, a time domain integral equation technique is presented that takes into account, in one formulation, a complex combination of boundary and initial conditions, as well as permitting the medium parameters to change in time. Investigations are made by using the evolution approach developed. The approach is applied to an investigation of the interaction of a guided wave with a medium moving in a rectangular waveguide with perfectly conducting walls. The relativistic movement of a non-dispersive medium and the effects caused by a double-dispersion mechanism, waveguide and plasma dispersions, are considered.

The need to consider the interaction of optical beams with time-varying media is becoming ever more common. Applications, such as the production of terahertz sources are exploiting the phenomena observed in such circumstances and, as data rates increase, designers of switched lasers and modulators and similar devices must confront the consequences of these interactions. There is a significant literature considering the simple case of plane waves interacting with time changes in the parameters of open and semi-open regions. However, to date, the practically important case of time-variant materials in spatially limited and optically confining waveguides has received far less attention. The principal objective of this work is provide a formal, non-numerical, framework within which to investigate this case and it shall be shown that certain general conclusions regarding the nature of the optical field in these circumstances can be demonstrated. This is clearly an important precursor to the detailed numerical analysis of specific configurations in the design of a wide variety of novel devices.

The book is organized as follows. An essential point for elaborating a common approach to the investigation of transient electromagnetic phenomena is the evolutionary character of such phenomena, and an initial moment, when the non-stationarity starts, carries an important meaning. The introduction of the non-stationarity initial moment is dictated in many cases by a necessity to separate the moment of "switching on" the field and the moment of the non-stationarity beginning. Non-stationarity, which starts at some certain moment of time, is accompanied by the appearance of a transient (non- harmonic) field, so-called "transients." These transients can exist for a long time being a significant part of a whole field. However, they fall out from the field of vision of a stationary approach when all periodic processes are assumed to start at the infinite past. The commonly used approximation of the adiabatic "switching on" of a process at the infinite past can easily lead to indefiniteness in the problem formulation because of the irreversibility of the non-stationary phenomenon. Therefore, an investigation of non-stationary electromagnetic phenomena should be based on equations, which include a general representation of the medium parameters, where an inhomogeneity has a time-dependent shape and medium properties inside it. The mathematical technique describing transient electromagnetic phenomena should contain a description of both continuous and abrupt changes of both the field functions and the medium parameters. This technique has also to take into account the correlation between spatial and temporal changes in the media. Such a correlation occurs, for example, when a medium boundary moves in space. In this case, a sharp time jump of the medium parameters occurs at every fixed point passed by the medium boundary.

The theory of generalized functions is an adequate mathematical technique for treating such problems. The generalized functions describe uniformly continuous and discontinuous functions of the field and media parameters. Applying this theory to the classical electromagnetic equations requires a substitution of the generalized derivatives instead of the conventional (classical) derivatives with corresponding modification of Maxwell's equations. The mathematical formulation of a non-stationary electromagnetic problem into a differential equation in the space of generalized functions and then conversion of a differential equation into an integral one is given in Chapter 1. This allows one to include all conditions for the fields on the discontinuity surfaces (boundaries), as well as at the moments of time changes in parameters, directly into the equations. A description is given of the causality of the time-spatial evolution of an electromagnetic field, and a technique is developed for the consideration of problems.

The main phenomena caused by a time changing of an unbounded medium are considered in Chapter 2. It is shown that medium modulation, by a finite chain of a medium permittivity time disturbance, can lead to the appearance of chaotic behavior of some field characteristics that is estimated by the calculation of statistical complexity, Hurst's index and the Lyapunov exponent. The dispersive medium is represented by plasma with an abrupt change of density and by a magnetized plasma in which the magnetization is switched on at some moment of time. A wide variety of problems pronounced by the influence of a medium boundary on electromagnetic transients is considered in Chapter 3. These problems include the normal incidence of a plane electromagnetic onto a plane boundary of a dielectric or plasma, created at some moment of time, and the evolution of an electromagnetic plane wave in the layer of a linear or nonlinear dielectric after its creation. The consideration of an Airy pulse allows one to investigate triple asymmetry in time-spatial structure in a non-stationary environment. The 3D case of a bounded medium is considered in Chapter 4, where the resolvent operators are derived. It allows one to obtain Fresnel's formulas for a plane boundary appearing at some moment of time. By virtue of these operators, a new effect, the focusing of secondary waves by a non-stationary medium plane boundary, is investigated. A more complex medium boundary is represented by a dielectric sphere created at some moment of time and by plasma ignition in a spherical region.

The interaction of electromagnetic waves with a medium moving boundary is investigated in Chapter 5. Resolution of moving boundary "paradoxes" that occur when the number of supposed waves does not correspond to the number of boundary conditions is shown. Peculiarities of the interaction of a wave with a moving boundary whose movement begins at zero moment are investigated also in this chapter. A sharp origin of uniform movement, as well as continuous relativistic uniform accelerated movement when velocity changes from zero to relativistic value, is considered. Investigation of a collapsing dielectric layer reveals energy accumulation in the layer and the generation of electromagnetic pulses.

The influence on a guided wave of a moving medium in a rectangular waveguide with perfectly conducting walls is considered, together with the wave evolution, in Chapter 6. The transformation of the guided wave in a waveguide filled by a uniform dielectric relativistic moving along the waveguide is also investigated. Using these problems as a basis, the possibility to model various phenomena in which electromagnetic waves interact with a boundary of a relativistic moving medium in the presence of waveguide dispersion is shown. The interaction of the electromagnetic field with plasma "flashed" at zero moment of time in the waveguide and then uniformly expanding is considered. In Chapter 6, it is shown that the interplay between phase and group velocities can enhance the interaction of an electromagnetic wave with a plasma bunch moving in a metallic waveguide and make this interaction very effective for small values of the bunch.

All the effects considered in the previous chapters arise in a dielectric or plasma waveguide and resonators with time-varying media; this is considered in some detail in Chapter 7. The Volterra integral equation is formulated and a corresponding resolvent is built. It allows one to investigate the field in a dielectric waveguide with a time jump of the core permittivity in a linear as well as in a nonlinear regime. Two ways for the calculation of field evolution in a dielectric layer (via Brillouin waves or Eigen-waves) are also considered.

Many of these questions were considered in the previous edition of the book, *Non-Stationary Electromagnetics*, but we now feel that the discussion there was overburdened by minor problems that led to an extensive volume. In this revised edition, such secondary questions have been reconsidered or removed. These changes have enabled us to reduce the size of the book by a factor of around two, so refining the focus on key concepts. Chapters 1 and 2 remain practically unchanged; Chapter 1 is devoted to the integral equation method itself and the application of this method to phenomena in unbounded medium is considered in Chapter 2. Chapters 3 and 4 are formed from the material that was covered in one chapter of the previous edition. As can be imagined, it is reasonable to consider consecutively 1D problems and 3D ones, which are a development of the former. Chapter 3 has been enriched by the consideration of the transformation of Airy pulses, which has acquired new insights recently in connection with the preservation of pulses of complex form. A medium restricted by a plane boundary only, but now taking a more complex, spherical, form, is considered in Chapter 4. Chapter 5 is the previous Chapter 4, and its content remains practically unaltered, while Chapter 6 of this revised edition merges the fifth and sixth chapters of the previous edition. Chapter 7 is practically unchanged in comparison with the previous edition. Chapter 8 of the previous edition is removed, as it does not contain the integral equation method and it is planned to use this material as the basis of a future separate edition.

Chapter 1

Basic Relations for an Electromagnetic Field in a Medium with Time-Varying Parameters and/or Moving Boundaries

An essential point for elaborating a common approach to the investigation of transient electromagnetic phenomena is the evolutionary character of such phenomena, and an initial moment, when the non-stationary condition starts, takes an important meaning. The introduction of the non-stationary initial moment is dictated in many cases by a necessity to separate the moment of "switching on" the field and the moment of the non-stationary behavior beginning. The non-stationary state, which starts at some definite moment of time, is accompanied by the appearance of a transient (non-harmonic) field. These so-called "transients" can exist for a long time, being a significant part of the total field. However, they fall out of view in a stationary approach when all periodic processes are assumed to start at the infinite past. It should be noted that the commonly used approximation of an adiabatic "switching on" of a process at the infinite past can easily lead to indefiniteness in the problem formulation because of the irreversibility of the non-stationary phenomenon. Therefore,

Non-Stationary Electromagnetics: An Integral Equations Approach (Second Edition)
Alexander Nerukh and Trevor Benson

ISBN 978-981-4774-95-6 (Hardcover), 978-0-429-02773-4 (eBook)
www.panstanford.com

an investigation of non-stationary electromagnetic phenomena should be based on equations that include a general representation of the medium parameters, where an inhomogeneity has a time-dependent shape and time-dependent medium properties inside it. A mathematical approach to the theory of transient electromagnetic phenomena should contain a description of both continuous and abrupt changes of both the field functions and the medium parameters. This technique also has to take into account the correlation between spatial and temporal changes in the media. Such a correlation occurs, for example, when a medium boundary moves in space. In this case, a sharp time-jump of the medium parameters occurs at every fixed point passed by the medium boundary.

The theory of generalized functions [1–6] is an adequate mathematical technique for treating such problems. The generalized functions describe uniformly continuous and discontinuous functions of the field and media parameters. Applying this theory to the classical electromagnetic equations means a substitution of the generalized derivatives instead of the conventional (classical) derivatives with a corresponding modification of Maxwell's equation.

In this chapter, a non-stationary electromagnetic problem is mathematically formulated as a differential equation in a generalized function space. This allows all conditions for the fields on the discontinuity surfaces (boundaries) as well as parameter time jumps to be included directly into the equations.

1.1 Generalized Wave Equation for an Electromagnetic Field in a Time-Varying Medium

1.1.1 Generalized Derivatives

To use the theory of generalized functions, one must consider the generalized derivatives [1–6] instead of the classic ones, as follows. Assume a vector-function $\mathbf{a}(t, \mathbf{r})$ has a discontinuity on an arbitrary time-varying surface, $S(t)$, and that its jump value is equal to $[\mathbf{a}]_S = \mathbf{a}^+ - \mathbf{a}^-$, where $\mathbf{a}^\pm$ are the magnitudes of the vector function on the positive and negative sides of this surface.

These sides are determined by a normal vector **n**, as shown in Fig. 1.1:

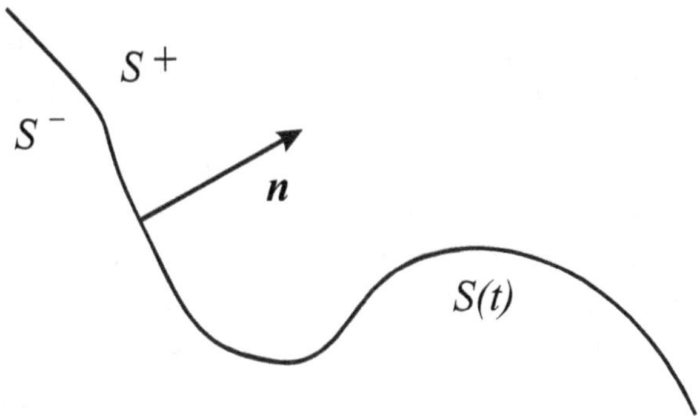

Figure 1.1 The orientation of the normal vector on the discontinuity surface.

The generalized derivatives are defined by the following formulae:

$$\frac{\partial \mathbf{a}}{\partial t} = \left\{\frac{\partial \mathbf{a}}{\partial t}\right\} - [\mathbf{a}]_s u_n \delta(S(t)), \text{ curl}\mathbf{a} = \{\text{curl}\mathbf{a}\} + \mathbf{n} \times [\mathbf{a}]_s \delta(S(t)), \quad (1.1)$$

where the braces mean an ordinary derivative where it exists, $\delta(S)$ is a surface delta-function and u_n is a velocity component normal to a certain surface domain. Here and later, bold characters are used for vectors.

Time jumps of solid medium parameters can be taken into account by the formula

$$\frac{\partial \mathbf{a}}{\partial t} = \left\{\frac{\partial \mathbf{a}}{\partial t}\right\} - [\mathbf{a}]_{t=0} \delta(t), \quad (1.2)$$

where $[\mathbf{a}]_{t=0} = \mathbf{a}(t = +0) - \mathbf{a}(t = -0)$.

Switching to these generalized derivatives allows the conditions for the fields on the surfaces and at the time points to be included directly into Maxwell's equations, where the medium parameters are discontinuous. These conditions are given by the terms in the square brackets in Eqs. (1.1) and (1.2). Comparing to the continuous medium case, the equation form remains unchanged almost everywhere. To show this, let us consider the classical Maxwell's equations in a continuous medium. These equations have the following form in an SI system:

$$\frac{1}{\mu_0}\{\text{curl}\mathbf{B}\} = \varepsilon_0\left\{\frac{\partial \mathbf{E}}{\partial t}\right\} + \left\{\frac{\partial \mathbf{P}}{\partial t}\right\} + \{\text{curl}\mathbf{M}\} + \mathbf{j}, \quad \{\text{curl}\mathbf{E}\} = -\left\{\frac{\partial \mathbf{B}}{\partial t}\right\}, \tag{1.3}$$

where $\mathbf{E}$ is the electric field strength, $\mathbf{B}$ is the magnetic flux density, $\mathbf{P}$ and $\mathbf{M}$ are vectors of medium electric and magnetic polarizations, $\mathbf{j}$ is a conductivity current; $\varepsilon_0 = 10^{-9}/36\pi$ [F·m^{-1}], $\mu_0 = 4\pi\cdot 10^{-7}$ [H·m^{-1}] are the permittivity and permeability of free space, and $\frac{1}{\sqrt{\varepsilon_0\mu_0}} = c = 3 \times 10^8$ [m s^{-1}] is the velocity of light in vacuum. The polarizations, $\mathbf{P}$ and $\mathbf{M}$, and the electric field flux density $\mathbf{D}$ and the magnetic field strength $\mathbf{H}$ are connected in a conventional way:

$$\mathbf{D} = \varepsilon_0\mathbf{E} + \mathbf{P} = \varepsilon_0(1+\hat{\kappa}_\varepsilon)\mathbf{E} = \varepsilon_0\hat{\varepsilon}\mathbf{E},$$

$$\mathbf{H} = \frac{1}{\mu_0}\mathbf{B} - \mathbf{M} = \frac{1}{\mu_0(1+\hat{\kappa}_\varepsilon)}\mathbf{B} = \frac{1}{\mu_0\hat{\mu}}\mathbf{B}. \tag{1.4}$$

Here, $1+\hat{\kappa}_\varepsilon = \hat{\varepsilon}$, $\hat{\kappa}_\varepsilon$ is an operator of the electrical susceptibility and $\hat{\varepsilon}$ is an operator of the relative permittivity of the medium, and, analogously, $1+\hat{\kappa}_\mu = \hat{\mu}$, $\hat{\kappa}_\mu$ is an operator of the magnetic susceptibility and $\hat{\mu}$ is an operator of the relative permeability of the medium. As $\hat{\kappa}_\varepsilon$ and $\hat{\kappa}_\mu$ are assumed to be operators, the relations (1.4) are general ones and they describe all possible media, including dispersive and anisotropic ones.

1.1.2 Initial and Boundary Conditions for Electromagnetic Fields in a Time-Varying Medium

It is convenient to determine the conditions for the field on a discontinuity surface on the basis of Maxwell's equations in an integral form [7–9]. They have an invariant form independent of the way the medium parameters change:

$$\oint_L \mathbf{H}d\mathbf{l} = \frac{d}{dt}\int_S \mathbf{D}d\mathbf{s} + \int_S \mathbf{j}d\mathbf{s},$$
$$\oint_L \mathbf{E}d\mathbf{l} = -\frac{d}{dt}\int_S \mathbf{B}d\mathbf{s},$$
$$\oint_S \mathbf{D}d\mathbf{s} = q,$$
$$\oint_S \mathbf{B}d\mathbf{s} = 0, \tag{1.5}$$

where $\mathbf{j}$ and q are bulk densities of current and charge in a laboratory frame of reference. It follows from Eq. (1.5) that the boundary conditions have the following form for an arbitrarily moving surface [9, 10]:

$$\mathbf{n}\times[\mathbf{H}]_s + u_n\left[\mathbf{D}_{\tan}\right]_s = \gamma^{-1}(\mathbf{n}\times(\mathbf{i}'\times\mathbf{n})),$$
$$\mathbf{n}\times[\mathbf{E}]_s - u_n\left[\mathbf{B}_{\tan}\right]_s = 0,\quad \left[\mathbf{D}_n\right]_s = \sigma',\quad \left[\mathbf{B}_n\right]_s = 0 \tag{1.6}$$

where $\mathbf{D}_{\tan}$, $\mathbf{B}_{\tan}$ are the components of the fields tangential to the surface, $\gamma^{-1} = \sqrt{1-\beta_n^2}$ is the relativistic factor, $\beta_n = u_n/c$, and $\mathbf{i}'$ and σ' are the surface current and charge densities in the intrinsic frame of reference of the moving surface domain, respectively.

The electric field flux density, as well as the magnetic induction, remains continuous at time jumps of the medium features. This follows from the classical Maxwell's Eq. (1.3) together with the limiting value of the spatial derivatives of the field. Indeed, integrating the relation $\mathrm{curl}\left(\frac{1}{\mu_0}\mathbf{B}-\mathbf{M}\right) = \frac{\partial \mathbf{D}}{\partial t}$ yields the following:

$$\mathbf{D}(t=+0) - \mathbf{D}(t=-0) = \lim_{\Delta t\to 0}\int_{-\Delta t}^{\Delta t} \mathrm{curl}\left(\frac{1}{\mu_0}\mathbf{B}-\mathbf{M}\right)dt = 0. \tag{1.7}$$

Analogously, the equation $\mathrm{curl}\mathbf{E} = -\frac{\partial \mathbf{B}}{\partial t}$ leads to

$$\mathbf{B}(t=+0) - \mathbf{B}(t=-0) = -\lim_{\alpha\to 0}\int_{-\alpha}^{\alpha} \mathrm{curl}\mathbf{E}dt = 0. \tag{1.8}$$

Note that in the general case the time derivatives of the electric flux density and the magnetic induction do not remain continuous. Indeed, the jump of electric flux density differs from zero in the general case, obeying

$$\left[\frac{\partial \mathbf{D}}{\partial t}\right]_{t=0} = \lim_{\alpha \to 0} \int_{-\alpha}^{\alpha} \text{curl} \frac{\partial}{\partial t}\left(\frac{1}{\mu_0}\mathbf{B} - \mathbf{M}\right) dt = \left[\text{curl}\left(\frac{1}{\mu_0}\mathbf{B} - \mathbf{M}\right)\right]_{t=0} \neq 0.$$

Similarly, for the magnetic induction we have

$$\left[\frac{\partial \mathbf{B}}{\partial t}\right]_{t=0} = -[\text{curl}\mathbf{E}]_{t=0} \neq 0.$$

The electric flux density derivative remains continuous only when the mediums' magnetic features are continuous. Analogously, the magnetic induction is continuous in the case when the medium electric features are continuous. If this is not the case then the initial condition for the magnetic induction derivative can be derived as follows. The equation $\text{curl}\,\mathbf{E} = -\frac{\partial \mathbf{B}}{\partial t}$ in the case when the permittivity changes in time, $\mathbf{E}(t) = \frac{\mathbf{D}(t)}{\varepsilon(t)}$, leads to

$$\begin{aligned}
\frac{\partial \mathbf{B}}{\partial t}(t=+0) - \frac{\partial \mathbf{B}}{\partial t}(t=-0) &= -\lim_{\alpha \to 0} \int_{-\alpha}^{\alpha} \text{curl} \frac{\partial}{\partial t} \mathbf{E} dt \\
&= -\lim_{\alpha \to 0} \text{curl}\left[\frac{\mathbf{D}(t+\alpha)}{\varepsilon(t+\alpha)} - \frac{\mathbf{D}(t-\alpha)}{\varepsilon(t-\alpha)}\right] \\
&= -\lim_{\alpha \to 0}\left[\frac{1}{\varepsilon(t+\alpha)}\text{curl}\mathbf{D}(t+\alpha) - \frac{1}{\varepsilon(t-\alpha)}\text{curl}\mathbf{D}(t-\alpha)\right] \\
&= -\frac{1}{\varepsilon^+}\text{curl}\mathbf{D}(t) + \frac{1}{\varepsilon^-}\text{curl}\mathbf{D}(t) = -\left(\frac{1}{\varepsilon^+} - \frac{1}{\varepsilon^-}\right)\text{curl}\mathbf{D}(t) \\
&= -\left(\frac{1}{\varepsilon^+} - \frac{1}{\varepsilon^-}\right)\text{curl}\varepsilon^-\mathbf{E}^{(-)}(t) \\
&= -\left(\frac{\varepsilon^-}{\varepsilon^+} - 1\right)\text{curl}\mathbf{E}^{(-)}(t)
\end{aligned} \tag{1.9}$$

It gives the initial condition for the magnetic induction derivative as

$$\left[\frac{\partial \mathbf{B}}{\partial t}\right]_{t=0} = -\left(\frac{\varepsilon^-}{\varepsilon^+} - 1\right)\text{curl}\mathbf{E}^{(-)}(t). \tag{1.10}$$

Initial conditions for electromagnetic fields in a dispersive medium such a plasma have another form. Indeed, continuity of $\mathbf{D}_+ - \mathbf{D}_- = 0$ follows directly from Maxwell's equations as well as $\mathbf{B}_+ - \mathbf{B}_- = 0$ or $\mathbf{H}_+ - \mathbf{H}_- = 0$ if the magnetic properties do not change. The equation $\frac{\partial \mathbf{D}}{\partial t} = \varepsilon_0 \frac{\partial \mathbf{E}}{\partial t} + \mathbf{j}$ yields

$$\begin{aligned}[\mathbf{E}]_{t=0} &= \lim_{\Delta t \to 0} \int_{-\Delta t}^{\Delta t} \frac{\partial \mathbf{E}}{\partial t} dt = \lim_{\Delta t \to 0} \int_{-\Delta t}^{\Delta t} \left(\frac{\partial \mathbf{D}}{\partial t} - \mathbf{j}\right) dt = [\mathbf{D}]_{t=0} - \lim_{\Delta t \to 0} \int_{-\Delta t}^{\Delta t} \mathbf{j}\, dt \\ &= \lim_{\Delta t \to 0} \Delta t\, \overline{\mathbf{j}} = 0,\end{aligned} \tag{1.11}$$

where $\overline{\mathbf{j}}$ is a jump of current at zero moment of time. It gives the continuity for the elect ric field $\mathbf{E}_+ - \mathbf{E}_- = 0$ as well as for the polarization $\mathbf{P}_+ - \mathbf{P}_- = 0$ as $\mathbf{P} = \mathbf{D} - \varepsilon_0 \mathbf{E}$. Continuity of the electric flux density derivative $\left[\frac{\partial \mathbf{D}}{\partial t}\right]_{t=0} = 0$ provides continuity of the electric field derivative $\frac{\partial \mathbf{E}_+}{\partial t} - \frac{\partial \mathbf{E}_-}{\partial t} = 0$ as

$$\begin{aligned}\left[\frac{\partial \mathbf{E}}{\partial t}\right]_{t=0} &= \lim_{\Delta t \to 0} \int_{-\Delta t}^{\Delta t} \left\{\text{curl}\left(\frac{1}{\mu_0}\mathbf{B} - \mathbf{M}\right) - \mathbf{j}\right\} dt \\ &= \left[\text{curl}\left(\frac{1}{\mu_0}\mathbf{B} - \mathbf{M}\right)\right]_{t=0} - \lim_{\Delta t \to 0} \int_{-\Delta t}^{\Delta t} \text{curl}\mathbf{j}\, dt \\ &= -\lim_{\Delta t \to 0} \Delta t\, \overline{\text{curl}\mathbf{j}} = 0,\end{aligned} \tag{1.12}$$

Continuity of the derivative of the magnetic induction follows from the equation

$$\begin{aligned}\frac{\partial \mathbf{B}}{\partial t}(t = +0) - \frac{\partial \mathbf{B}}{\partial t}(t = -0) &= -\lim_{\Delta t \to 0} \int_{-\Delta t}^{\Delta t} \text{curl} \frac{\partial}{\partial t} \mathbf{E}\, dt \\ &= -\frac{1}{\varepsilon_0} \lim_{\Delta t \to 0} \int_{-\Delta t}^{\Delta t} \text{curl}\left[\frac{\partial \mathbf{D}}{\partial t} - \mathbf{j}\right] dt \\ &= -\frac{1}{\varepsilon_0} \lim_{\Delta t \to 0} \text{curl}[\mathbf{D}(t + \Delta t - \mathbf{D}(t - \Delta t)] - \frac{1}{\varepsilon_0} \lim_{\Delta t \to 0} \overline{\text{curl}\mathbf{j}}\, \Delta t = 0,\end{aligned} \tag{1.13}$$

1.1.3 Maxwell's Equations in a Generalized Derivatives Representation

Returning to the differential form of Maxwell's Eq. (1.3), we see that the second equation in (1.3) is not changed when the classical derivatives are replaced by the generalized ones, while the first equation gains an additional term determined by the surface current $\mathbf{i} = \nu_n\sigma' + \gamma^{-1}(\mathbf{n} \times (\mathbf{i}' \times \mathbf{n}))$ in the laboratory frame of reference:

$$\text{curl}\mathbf{B} = \mu_0\left(\varepsilon_0\frac{\partial \mathbf{E}}{\partial t} + \frac{\partial \mathbf{P}}{\partial t}\right) + \mu_0(\text{curl}\mathbf{M} + \mathbf{j} + \mathbf{i}\delta(S(t))),$$

$$\text{curl}\mathbf{E} = -\frac{\partial \mathbf{B}}{\partial t}. \tag{1.14}$$

In these equations, all the derivatives are generalized; therefore they readily contain the boundary conditions for the fields on the discontinuity surfaces [11] that distinguish them from Eq. (1.3). The discontinuity surface $S(t)$ restricts the region $V(t)$ and is moving in the general case. Equation (1.14) forms the basis for further description and investigation of electromagnetic phenomena in time-varying inhomogeneous media.

Merging the two equations results in a wave equation in a generalized derivatives representation. This equation describes an electromagnetic field in a medium whose parameters can vary arbitrarily in time as well as in space (this variation includes the arbitrarily moving surface as well) [12]:

$$\frac{1}{\mu_0}\text{curlcurl}\mathbf{E} + \varepsilon_0\frac{\partial^2 \mathbf{E}}{\partial t^2} = -\frac{\partial^2 \mathbf{P}}{\partial t^2} - \text{curl}\frac{\partial \mathbf{M}}{\partial t} - \frac{\partial(\mathbf{j} + \mathbf{i}\delta(S))}{\partial t} \tag{1.15}$$

Using the characteristic function χ, which is equal to unity inside $V(t)$ and equal to zero outside this region, we can define the generalized functions ***P***, ***M*** and ***j*** that describe the medium electromagnetic representation in the whole space:

$$\mathbf{P} = \chi(\mathbf{P}_1 - \mathbf{P}_{\text{ex}}) + \mathbf{P}_{\text{ex}},$$

$$\mathbf{M} = \chi(\mathbf{M}_1 - \mathbf{M}_{\text{ex}}) + \mathbf{M}_{\text{ex}},$$

$$\mathbf{j} = \chi\mathbf{j}_1 + \mathbf{j}_{\text{extr}}, \tag{1.16}$$

where the values with the index "1" are defined inside the region $V(t)$ and those with the index "ex" are defined outside of it. This external region is further referred to as a "background medium." $\mathbf{j}_{\text{extr}}$ is a current describing extrinsic sources of the field (see Fig. 1.2).

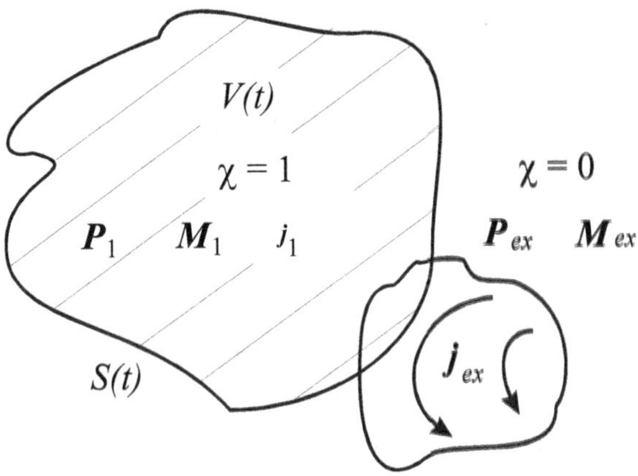

Figure 1.2 The medium description and the arrangement of an inhomogeneity and field sources in the general problem formulation.

Substituting (1.16) into (1.15) yields

$$\frac{1}{\mu_0}\text{curlcurl}\mathbf{E} + \text{curl}\frac{\partial \mathbf{M}_{\text{ex}}}{\partial t} + \frac{\partial^2}{\partial t^2}(\varepsilon_0\mathbf{E} + \mathbf{P}_{\text{ex}})$$

$$= -\frac{\partial^2}{\partial t^2}\chi(\mathbf{P}_1 - \mathbf{P}_{\text{ex}}) - \text{curl}\frac{\partial}{\partial t}\chi(\mathbf{M}_1 - \mathbf{M}_{\text{ex}}) - \frac{\partial(\chi\mathbf{j}_1 + \mathbf{i}\delta(S))}{\partial t} - \frac{\partial \mathbf{j}_{\text{extr}}}{\partial t}. \tag{1.17}$$

According to the main idea of the approach originated by Khizhnyak [13], the left-hand side of this equation has the form as in the background, and the right-hand side is distinct from zero inside the region $V(t)$ only. This equation allows consideration of the electromagnetic problem for an arbitrary inhomogeneity placed in various backgrounds. We will consider two such cases: a non-dispersive background, and plasma as an example of a dispersive medium.

1.1.4 Generalized Wave Equation for the Case of a Non-Dispersive Background

First, we consider the generalized wave equation in the non-dispersive background that is described by the relative permittivity and permeability ε and μ, respectively. In this case

$$\mathbf{P}_{ex} = \varepsilon_0(\varepsilon - 1)\mathbf{E}, \quad \mathbf{M}_{ex} = \frac{1}{\mu_0}(1 - \mu^{-1})\mathbf{B}. \tag{1.18}$$

If these operators commute with the operators $\frac{\partial}{\partial t}$ and curl then we have for the background polarizations in (1.17)

$$\begin{aligned}\frac{\partial^2 \mathbf{P}_{ex}}{\partial t^2} + \mathrm{curl}\frac{\partial \mathbf{M}_{ex}}{\partial t} &= \frac{\partial^2}{\partial t^2}\varepsilon_0(\varepsilon - 1)\mathbf{E} + \mathrm{curl}\frac{\partial}{\partial t}\frac{1}{\mu_0}(1 - \mu^{-1})\mathbf{B} \\ &= \frac{\partial^2}{\partial t^2}\varepsilon_0(\varepsilon - 1)\mathbf{E} - \mathrm{curl}\frac{1}{\mu_0}(1 - \mu^{-1})\mathrm{curl}\mathbf{E},\end{aligned} \tag{1.19}$$

Now the generalized wave equation takes the form

$$\begin{aligned}\mathrm{curlcurl}\mathbf{E} + \varepsilon_0\mu_0\frac{\partial^2}{\partial t^2}\varepsilon\mu\mathbf{E} = -\mu_0\mu\Bigg[&\frac{\partial^2}{\partial t^2}\chi(\mathbf{P}_1 - \mathbf{P}_{ex}) \\ &+ \mathrm{curl}\frac{\partial}{\partial t}\chi(\mathbf{M}_1 - \mathbf{M}_{ex}) + \frac{\partial(\chi\mathbf{j}_1 + \mathbf{i}\delta(S))}{\partial t} + \frac{\partial \mathbf{j}_{extr}}{\partial t}\Bigg]\end{aligned} \tag{1.20}$$

where all parameters of the region $V(t)$ are collected on the right-hand side of this equation. Introducing the shorthand notation

$$\begin{aligned}\mathbf{Q}_1 = &-\frac{1}{v^2}\frac{\partial^2}{\partial t^2}\chi\frac{1}{\varepsilon_0\varepsilon}(\mathbf{P}_1 - \mathbf{P}_{ex}) - curl\frac{\partial}{\partial t}\chi\mu_0\mu(\mathbf{M}_1 - \mathbf{M}_{ex}) \\ &-\frac{\partial}{\partial t}\mu_0\mu(\chi\mathbf{j}_1 + \mathbf{i}\delta(S))\end{aligned} \tag{1.21}$$

Eq. (1.20) can be re-written as follows:

$$\mathrm{curlcurl}\mathbf{E} + \frac{1}{v^2}\frac{\partial^2}{\partial t^2}\mathbf{E} = \mathbf{Q}_1 - \mu_0\mu\frac{\partial \mathbf{j}_{extr}}{\partial t}. \tag{1.22}$$

The coefficient $v = c/\sqrt{\varepsilon\mu} = 1/\sqrt{\varepsilon_0\mu_0\varepsilon\mu}$ in (1.22) is the wave phase velocity in the background.

1.1.5 Generalized Wave Equation for the Case of a Dispersive Background

Equation (1.22) describes the electromagnetic field in the case of an object placed in a uniform non-dispersive medium. Let us now consider the case of a dispersive background, the simplest and most applicable of which is a cold isotropic plasma. It is known that such a plasma is described by the constitutive relations

$$\mathbf{P}_{ex}(t,\mathbf{r}) = \varepsilon_0 \int_{-\infty}^{t} \omega_e^2 (t-t')\mathbf{E}(t',\mathbf{r})dt', \quad \mathbf{M}_{ex} = \mathbf{j}_{ex} = 0, \tag{1.23}$$

where $\omega_e = \sqrt{Ne^2/m\varepsilon_0}$ is a plasma frequency, N is the density of electrons, and e and m are, respectively the electron charge and mass.

With these relations, the first equation in (1.3) can be represented in the form valid for the region outside of the object $V(t)$:

$$\frac{1}{\mu_0}\mathrm{curl}\mathbf{B}_{ex} - \varepsilon_0 \frac{\partial \mathbf{E}_{ex}}{\partial t} - \varepsilon_0 \int_{-\infty}^{t} \omega_e^2 \mathbf{E}_{ex} dt' = 0 \tag{1.24}$$

Inside the object, this equation has the right-hand side

$$\frac{1}{\mu_0}\mathrm{curl}\mathbf{B}_{in} - \varepsilon_0 \frac{\partial \mathbf{E}_{in}}{\partial t} - \varepsilon_0 \int_{-\infty}^{t} \omega_e^2 \mathbf{E}_{in} dt'$$
$$= \frac{\partial}{\partial t}\left(\mathbf{P}_{in} - \varepsilon_0 \int_{-\infty}^{t} \omega_e^2 (t-t')\mathbf{E}_{in} dt'\right) + (\mathrm{curl}\mathbf{M}_{in} + \mathbf{j}_{in} + \mathbf{i}\delta(S)), \tag{1.25}$$

Similarly to the previous case of a non-dispersive medium, we can also introduce the discontinuous functions

$$\mathbf{P} = \chi\mathbf{P}_{in} + (1-\chi)\varepsilon_0 \int_{-\infty}^{t} \omega_e^2 (t-t')\mathbf{E}(t',\mathbf{r})dt', \ \mathbf{M} = \chi\mathbf{M}_{in}, \ \mathbf{j} = \chi\mathbf{j}_{in} \tag{1.26}$$

and expand Eq. (1.26) over the whole space. This gives the following equation defined over the whole space considered:

$$\frac{1}{\mu_0}\text{curl}\mathbf{B} - \varepsilon_0 \frac{\partial \mathbf{E}}{\partial t} - \varepsilon_0 \int_{-\infty}^{t} \omega_e^2 \mathbf{E} dt'$$

$$= \frac{\partial}{\partial t}\left(\mathbf{P} - \varepsilon_0 \int_{-\infty}^{t} \omega_e^2 (t-t')\mathbf{E} dt'\right) + (\text{curl}\mathbf{M} + \mathbf{j} + \mathbf{i}\delta(S)) \quad (1.27)$$

Combining this equation with Maxwell's second equation yields the inhomogeneous Klein–Gordon's equation

$$\text{curlcurl}\mathbf{E} + \frac{1}{c^2}\frac{\partial^2}{\partial t^2}\mathbf{E} + \frac{\omega_e^2}{c^2}\mathbf{E} = \mathbf{Q}_{1,p} - \mu_0 \frac{\partial \mathbf{j}_{\text{extr}}}{\partial t}, \quad (1.28)$$

where

$$\mathbf{Q}_{1,p} = -\mu_0 \left[\frac{\partial^2}{\partial t^2}\left(\mathbf{P} - \varepsilon_0 \int_{-\infty}^{t} \omega_e^2 (t-t')\mathbf{E} dt'\right) + \text{curl}\frac{\partial}{\partial t}\mathbf{M} + \frac{\partial(\mathbf{j}_1 + \mathbf{i}\delta(S))}{\partial t}\right].$$

$$(1.29)$$

Equations (1.22) and (1.28) are defined on the whole time axis and over the whole of space and describe the electromagnetic field completely, even though they contain only the electric field. The magnetic field **B** can be determined from the second equation in (1.3) according to $\mathbf{B} = -\int_{-\infty}^{t} curl\mathbf{E}dt$. The magnetic field is defined uniquely, if the condition of zero value limit is fulfilled for the electromagnetic field when $t \to -\infty$.

The right-hand sides of Eqs. (1.22) and (1.28) depend on the extrinsic current, $\mathbf{j}_{\text{extr}}$, and on the presence of the inhomogeneity, which is described by the function χ in $\mathbf{Q}_1$. If both these components are absent, then the right-hand side is equal to zero, and the equation describes a free-space electromagnetic field $\mathbf{E}_0$ in the background without sources. This field is a general solution to a homogeneous equation in the non-dispersive medium

$$\text{curlcurl}\mathbf{E}_0' + \frac{1}{v^2}\frac{\partial^2}{\partial t^2}\mathbf{E}_0' = 0, \quad (1.30)$$

or in plasma

$$\text{curlcurl}\mathbf{E}' + \frac{1}{c^2}\frac{\partial^2}{\partial t^2}\mathbf{E}' + \frac{\omega_e^2}{c^2}\mathbf{E}' = 0. \quad (1.31)$$

It is natural to assume that the right-hand sides of the inhomogeneous equations differ from zero on the time interval that is restricted in the past by the moment $-t_\infty$, where t_∞ can be assumed as some great value. This means that field sources start operating after the moment $-t_\infty$. This assumption allows the right-hand sides of (1.22) and (1.28) to be considered as finite functions, and application of the theory of generalized functions to this equation. According to this theory the formal solutions to the equations for a time-varying and moving medium (1.22) or (1.28) can be written as a convolution of the fundamental solution to this equation with the right-hand side of the latter.

1.2 Fundamental Solutions to Maxwell's Equations (Green's Functions)

1.2.1 The Non-Dispersive Background

The fundamental solutions (Green's functions) to Eqs. (1.22) or (1.28) are determined ambiguously because an arbitrary solution of the homogeneous equation can be added. This arbitrary solution is given by a regular part of Green's function and the convolution with it gives the solution to the corresponding homogeneous equation, i.e., the field $\mathbf{E}_0'$ in the background. This field is a general solution of the homogeneous equation and it must be added to obtain a general solution of the inhomogeneous equation. The part of the solution, which is determined by the right-hand side of these equations, is given by a singular part of the fundamental solution G_0. The latter contains all the information about the field in the region $V(t)$. The singular part of the fundamental solution must satisfy an equation similar to (1.22) but with the right-hand side replaced by delta-function:

$$\mathrm{curlcurl}G_0 + \frac{1}{v^2}\frac{\partial^2}{\partial t^2}G_0 = I\delta(\mathbf{x}), \tag{1.32}$$

where I is the unity tensor. To solve Eq. (1.32), we define Green's function as a tensor with elements ξ_{ij}, for which the equation has the form

$$\frac{\partial^2 \xi_{kj}}{\partial x_i \partial x_k} - \frac{\partial^2 \xi_{kj}}{\partial x_k^2} + \frac{1}{v^2}\frac{\partial^2 \xi_{ij}}{\partial t^2} = \delta_{ij}\delta(t-t')\delta(\mathbf{r}-\mathbf{r}'). \tag{1.33}$$

Using a common approach [12–14], the solution to this equation will be sought in the form of the inverse Fourier–Laplace transform:

$$\xi_{ij}(t,\mathbf{r}) = \frac{1}{(2\pi)^4 i}\int d\mathbf{k} \int_{\alpha-i\infty}^{\alpha+i\infty} dp g_{ij}(p,\mathbf{k}) e^{i\mathbf{kr}+pt}. \tag{1.34}$$

Here, and subsequently, i, if not an index, is the imaginary unit, $i=\sqrt{-1}$.

Substituting (1.34) in (1.33) yields a tensor equation for the elements g_{ij}

$$-k_i k_n g_{nj} + k^2 g_{ij} + \frac{p^2}{v^2} g_{ij} = \delta_{ij}. \tag{1.35}$$

An assumption $g_{ij} = g_1 k_i k_j + g_2 \delta_{ij}$ allows one to find

$$g_{ij} = v^2 \frac{v^2 k_i k_j + p^2 \delta_{ij}}{p^2(p^2+v^2k^2)}. \tag{1.36}$$

Calculation of the inverse transform gives the elements of Green's function in the spatial-time representation

$$\xi_{ij} = -\frac{v^2}{4\pi}\left(\frac{\partial^2}{\partial x_i \partial x_\kappa} - \frac{1}{v^2}\frac{\partial^2}{\partial t^2}\delta_{ik}\right)\tilde{\tilde{f}}\delta_{kj}, \tag{1.37}$$

where

$$\tilde{\tilde{f}} = \int_{-i\infty}^{i\infty} \frac{dp}{2\pi i}\frac{e^{p(t-t')}}{p^2}\frac{e^{-\frac{p}{v}|\mathbf{r}-\mathbf{r}'|}}{|\mathbf{r}-\mathbf{r}'|}. \tag{1.38}$$

It is easy to see after substitution of (1.37) into (1.33) that a function $f = \frac{\partial^2}{\partial t^2}\tilde{\tilde{f}}$ satisfies the scalar equation

$$-\frac{\partial^2 f}{\partial x_k^2}+\frac{1}{v^2}\frac{\partial^2 f}{\partial t^2}=4\pi\delta(t-t')\delta(\mathbf{r}-\mathbf{r}'). \tag{1.39}$$

In addition to Eq. (1.39), the function f must satisfy some conditions on the boundary of the region being considered. In the case of an unbounded background, such a condition is the radiation condition at infinity. The solution to Eq. (1.38) satisfying this condition takes the form [5]

$$f=\frac{1}{|\mathbf{r}-\mathbf{r}'|}\delta\left(t-t'-\frac{1}{v}|\mathbf{r}-\mathbf{r}'|\right). \tag{1.40}$$

Double integration of this expression gives

$$\tilde{\tilde{f}}=\int_{-\infty}^{t}d\tau\int_{-\infty}^{\tau}f(\tau'-t')d\tau'=\frac{t-t'-\frac{|\mathbf{r}-\mathbf{r}'|}{v}}{|\mathbf{r}-\mathbf{r}'|}\theta\left(t-t'-\frac{|\mathbf{r}-\mathbf{r}'|}{v}\right) \tag{1.41}$$

where θ is the Heaviside unit function. The same result is obtained by calculation of the integral (1.38) [15].

1.2.2 The Dispersive Background

The fundamental solution to Eq. (1.28) corresponding to the plasma background [15] must satisfy the equation

$$\frac{\partial^2 \xi_{lj}}{\partial x_i \partial x_k}-\frac{\partial^2 \xi_{ij}}{\partial x_k^2}+\frac{1}{c^2}\left(\frac{\partial^2}{\partial t^2}+\omega_e^2\right)\xi_{ij}=\delta_{ij}\delta(t-t')\delta(\mathbf{r}-\mathbf{r}'). \tag{1.42}$$

Substituting (1.34) into this equation, we obtain the tensor equation for the elements g_{ij}

$$-k_i k_n g_{nj}+k^2 g_{ij}+\frac{1}{c^2}(p^2+\omega_e^2)g_{ij}=\delta_{ij}, \tag{1.43}$$

whose solution is equal to

$$g_{ij}=c^2\frac{c^2 k_i k_j+(p^2+\omega_e^2)\delta_{ij}}{(p^2+\omega_e^2)(p^2+c^2k^2+\omega_e^2)}. \tag{1.44}$$

Integration of this expression by virtue of the formula (1.34), and using tabulated integrals [16], gives the solution to (1.42) which is the fundamental solution (Green's function) to the wave equation for the case of a plasma background

$$(\xi_p)_{ij} = -\frac{c^2}{4\pi}\left(\frac{\partial^2}{\partial x_i \partial x_\kappa} - \frac{1}{c^2}\left(\frac{\partial^2}{\partial t^2} + \omega_e^2\right)\delta_{ik}\right)\tilde{\tilde{f}}_e \delta_{kj}, \tag{1.45}$$

where $$\tilde{\tilde{f}}_p = \frac{\theta\left(t - t' - \frac{|\mathbf{r}-\mathbf{r}'|}{v}\right)}{|\mathbf{r}-\mathbf{r}'|} \int_{\frac{1}{c}|\mathbf{r}-\mathbf{r}'|}^{t-t'} J_0(\omega_e \sqrt{(t-t')^2 - u^2})du. \tag{1.46}$$

and $J_0(.)$ is the Bessel function of the zeroth order.

For more compact notation of the formulae obtained, we introduce differential operators $\hat{D}$, $\hat{D}_e$:

$$\hat{D} = \left(\text{graddiv} - \frac{1}{v^2}\frac{\partial^2}{\partial t^2}\right), \quad \hat{D}_e = \text{graddiv} - \frac{1}{c^2}\left(\frac{\partial^2}{\partial t^2} + \omega_e^2\right). \tag{1.47}$$

Then the fundamental solutions to Eqs. (1.22) or (1.28) can be written in the forms

$$\hat{G}_0 = -\frac{v^2}{4\pi}\hat{D}\tilde{\tilde{f}}, \quad \hat{G}_{0p} = -\frac{c^2}{4\pi}\hat{D}_e\tilde{\tilde{f}}_p, \tag{1.48}$$

where the functions $\tilde{\tilde{f}}$ and $\tilde{\tilde{f}}_p$ are determined by the formulae (1.41) and (1.46).

It follows from these formulae that Green's functions take the form of the retarded potential. They allow one to write the formal solutions to the differential Eqs. (1.22) or (1.28); this will be done in the next section.

1.2.3 A Rectangular Waveguide with Perfectly Conducting Walls

To solve electromagnetic problems in a waveguide, as well as for transformation of the differential Eq. (1.22) into an integral one, it is necessary to know a fundamental solution to this differential equation. This solution is a dyadic Green's function and in the case of a magneto-dielectric isotropic medium in a waveguide it must satisfy Eq. (1.33). For a rectangular metal waveguide, Green's function can be represented in the form analogous to (1.37)

$$G_{ij} = -\frac{v^2}{4\pi}\left(\frac{\partial^2}{\partial x_i \partial x_k} - \frac{1}{v^2}\frac{\partial^2}{\partial t^2}\delta_{ik}\right)\tilde{\tilde{f}}_{kj}^{E}, \tag{1.49}$$

where $v^2 = c^2 (\varepsilon\mu)^{-1}$ is the velocity of light in the medium with the relative permittivity ε and relative permeability μ. To make it easier to find the function f_{ij}^E we, in a way analogous to the case of an unbounded medium, consider a new tensor function $f_{ij}^E = \frac{\partial^2}{\partial t^2}\tilde{\tilde{f}}_{ij}^E$, which satisfies the simpler equation

$$-\frac{\partial^2}{\partial x_k^2} f_{ij}^E + \frac{1}{v^2}\frac{\partial^2}{\partial t^2} f_{ij}^E = 4\pi\delta_{ij}\delta(t-t')\delta(\vec{r}-\vec{r}'). \tag{1.50}$$

It is seen from this equation that this new tensor function is diagonal. In addition to Eq. (1.50) this function must satisfy boundary conditions at the waveguide walls. These conditions are followed by setting the tangential electric field components to zero, that means equating to zero the corresponding diagonal components f_{ij}^E. To provide these conditions the fundamental solutions are sought in the form of an expansion in terms of waveguide electric functions [7] which for a rectangular waveguide take the forms

$$\left.\begin{aligned} f_{xx}^{E,mn} &= \sin\kappa_m y \sin\kappa_n z \\ f_{yy}^{E,mn} &= \cos\kappa_m y \sin\kappa_n z \\ f_{zz}^{E,mn} &= \sin\kappa_m y \cos\kappa_n z \end{aligned}\right\}; \tag{1.51}$$

where $\kappa_m = \pi m/a$, $\kappa_n = \pi n/b$ and the axes are directed as shown in Fig. 1.3. Further, the coordinates x, y, z will be numbered as 1, 2, 3, respectively.

Let us expand the function f_{ij}^E in terms of the functions of the system (1.51):

$$f_{ij}^{E} = \sum_{m,n} g_{mn}(t-t', x-x') f_{ij}^{E,mn}(y,z) f_{ij}^{E,mn}(y',z')\delta_{ij}. \tag{1.52}$$

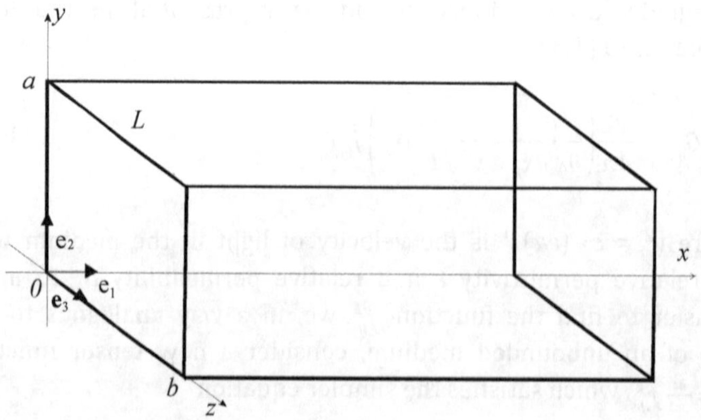

Figure 1.3 A rectangular metal waveguide.

The coefficients in this expansion follow from Eq. (1.50) and they are equal to

$$g_{mn} = \frac{8\pi v}{ab} J_0(\kappa_{mn}\sqrt{v^2(t-t')^2-(x-x')^2})\theta\left(t-t'-\frac{|x-x'|}{v}\right), \tag{1.53}$$

where $\kappa_{mn}^2 = \kappa_m^2 + \kappa_n^2$. The elements f_{ij}^{E} of the tensor Green's function can then be given in the form of the series

$$\tilde{\tilde{f}}_{ij}^{E} = \frac{8\pi v}{ab}\delta_{ij}\sum_{m,n}\tilde{\tilde{\Gamma}}_{mn} f_{ij}^{E,mn} f_{ij}'^{E,mn} \tag{1.54}$$

with coefficients $\tilde{\tilde{\Gamma}}_{mn} = \int_{-\infty}^{t} d\tau \int_{-\infty}^{\tau} d\tau' \Gamma_{mn}(\tau'-t', x-x')$, $\quad$ (1.55)

where $\Gamma_{mn}(t,x) = J_0\left(\omega_{mn}\sqrt{t^2-\frac{x^2}{v^2}}\right)\theta\left(t-\frac{|x|}{v}\right)$, $\quad$ (1.56)

$\omega_{mn} = v\kappa_{mn}$ and the prime in $f_{ij}'^{E,mn}$ denotes the primed arguments, $f_{ij}'^{E,mn} = f_{ij}'^{E,mn}(y',z')$.

1.3 Causal Time-Spatial Interpretation of Electromagnetic Field Interaction with Time-Varying Objects

According to the approach first formulated by Khizhnyak [13], the stationary boundary value problems can be represented as integral equations. These equations follow from the inhomogeneous wave equation and boundary conditions by virtue of Green's function. The integral equations are more physically clear compared to the corresponding differential equations and allow the boundary conditions to be considered more naturally. In this section, this approach is generalized for initial and boundary value problems in the time domain. The integral equations for such problems are a convolution in space of generalized functions, being the Volterra integral equation of the second kind for transient problems. Its type determines the choice of solution method, and in some cases, it significantly simplifies the construction of an algorithm for the problem solution.

In this section, we show that the introduction of the initial moment of the non-stationary behavior yields a splitting of the integral equation into a chain of evolutionary connecting equations, which describe three stages of the interaction of the electromagnetic field with the non-stationary medium. This chain determines the three stages of the problem solution.

1.3.1 The Volterra Integral Equation for the Electromagnetic Field in a Non-Dispersive Background

The condition on the time history for the right-hand side of Eq. (1.22), which has been discussed at the end of Section 1.1.4, means this function is finite in 4D space-time (t, $\mathbf{r}$). This allows one to write a general solution to this equation as a convolution of its fundamental solution G_0 with the right-hand side as follows:

$$\mathbf{E} = \mathbf{E}_0' + G_0 * \left(\mathbf{Q}_1 - \mu_0 \mu \frac{\partial \mathbf{j}_{\text{extr}}}{\partial t} \right)$$

$$= \mathbf{E}_0' + \int_{-\infty}^{\infty} dt' \int_{\infty} d\mathbf{r}' G_0(\mathbf{x} - \mathbf{x}') \left(\mathbf{Q}_1(\mathbf{x}') - \mu_0 \mu \frac{\partial \mathbf{j}_{\text{extr}}(\mathbf{x}')}{\partial t'} \right). \tag{1.57}$$

Here $\mathbf{E}_0'$ is a general solution of the homogeneous Eq. (1.30). The symbol (*) designates the convolution,

$$(a * b)(\mathbf{x}) = \int_{-\infty}^{\infty} dt' \int_{\infty} d\mathbf{r}' a(\mathbf{x} - \mathbf{x}') b(\mathbf{x}'),$$

and $\mathbf{x} = (t, \mathbf{r})$ is a 4D spatial-time vector. Integration is performed over the whole 4D space $-\infty < t' < \infty$, $-\infty < x', y', z' < \infty$. The finiteness of the vector-function $\mathbf{Q}_1$ defined by Eq. (1.38) affords the existence of the convolution [5].

The term in the convolution which is determined by the extrinsic current should be added to the general solution of the homogeneous equation. This sum gives the field in the background medium without the inhomogeneity $V(t)$, i.e., an undisturbed, or primary, or initial field

$$\mathbf{E}_0 = \mathbf{E}_0' - G_0 * \mu_0 \mu \frac{\partial \mathbf{j}_{\text{extr}}}{\partial t}. \tag{1.58}$$

The existence of the convolution ensured by the finiteness of the function $\boldsymbol{Q}_1$ allows one to transfer the differentiation with respect to time from this function to G_0 [4, 5]. It results in transforming Eq. (1.57) into the form

$$\mathbf{E} = \mathbf{E}_0 - \frac{1}{v^2} \frac{\partial^2 G_0}{\partial t^2} * \chi \frac{1}{\varepsilon_0 \varepsilon} (\mathbf{P}_1 - \mathbf{P}_{\text{ex}}) - \frac{\partial G_0}{\partial t} * \{\text{curl} \chi \mu_0 \mu (\mathbf{M}_1 - \mathbf{M}_{\text{ex}}) + \mu_0 \mu (\chi \mathbf{j}_1 + \mathbf{i}\delta(S))\}, \tag{1.59}$$

where the field $\mathbf{E}_0$ is defined by formula (1.58).

The key point of the non-stationary problem is its dual initial and boundary character that marks out some moment of time t_0 before which the state of the field is assumed to be known. It is natural to formulate the problem in such a way that t_0 is assumed to coincide with the beginning of the non-stationary (transient) process. This moment is further considered as one that coincides with zero moment, $t_0 = 0$.

According to the causality principle, which is expressed by the Heaviside unit function in Eq. (1.41), the function G_0 is equal to zero if $t' > t$. Therefore, the integration with respect to time in the convolution (1.59) is executed over the interval $(-\infty, t)$, where t is the observation moment. If the observation moment precedes the initial moment, $t < t_0$, then the integration is carried out over the whole interval $(-\infty, t]$ in the expression (1.59). If the observation occurs after the initial moment, $t > t_0$, then the integration time interval in Eq. (1.59) should be divided into two parts: before and after t_0. Consequently, the expression (1.59) is then rewritten as:

$$\mathbf{E} = \mathbf{F} - \frac{1}{v^2}\frac{\partial^2 G_0}{\partial t^2} * \theta(t)\chi\frac{1}{\varepsilon_0\varepsilon}(\mathbf{P}_1 - \mathbf{P}_{ex})$$
$$-\frac{\partial G_0}{\partial t} * \theta(t)\left\{curl\chi\mu_0\mu(\mathbf{M}_1 - \mathbf{M}_{ex}) + \mu_0\mu(\chi\mathbf{j}_1 + \mathbf{i}\delta(S(t)))\right\}. \quad (1.60)$$

The free term $\boldsymbol{F}$ in Eq. (1.60) contains the undisturbed (initial) field $\mathbf{E}_0$ as well as the whole prehistory of the field interaction with the region $V(t)$ until the initial moment t_0. If the medium possesses inertia then terms conditioned by this property ought to be also included into the free term. These terms are relict parts $\mathbf{P}_{rel}$, $\mathbf{M}_{rel}$,$\mathbf{j}_{rel}$ caused by the medium state before the zero moment and by the jump change of this state at t_0. Therefore, the free term is defined finally as

$$\mathbf{F} = \mathbf{E}_0 - \frac{1}{v^2}\frac{\partial^2 G_0}{\partial t^2} * \theta(-t)\chi\frac{1}{\varepsilon_0\varepsilon}(\mathbf{P}_1 - \mathbf{P}_{ex})$$
$$-\frac{\partial G_0}{\partial t} * \theta(-t)\{\mathrm{curl}\chi\mu_0\mu(\mathbf{M}_1 - \mathbf{M}_{ex}) + \mu_0\mu(\chi\mathbf{j}_1 + \mathbf{i}\delta(S))\}$$
$$-\frac{1}{v^2}\frac{\partial^2 G_0}{\partial t^2} * \theta(t)\chi\frac{1}{\varepsilon_0\varepsilon}(\mathbf{P}_{rel} - \mathbf{P}_{ex})$$
$$-\frac{\partial G_0}{\partial t} * \theta(t)\{\mathrm{curl}\chi\mu_0\mu(\mathbf{M}_{rel} - \mathbf{M}_{ex}) + \mu_0\mu(\chi\mathbf{j}_{rel} + \mathbf{i}_{rel}\delta(S))\}. \quad (1.61)$$

Let us represent the explicit expression for the convolution in Eqs. (1.60) and (1.61) as integral operators. To do this, differentiate explicitly the function G_0 using the notations $\frac{\partial^2\tilde{\tilde{f}}(\mathbf{x}-\mathbf{x}')}{\partial t^2} = f(\mathbf{x}-\mathbf{x}')$ and $\frac{\partial\tilde{\tilde{f}}\mathbf{x}-\mathbf{x}'}{\partial t} = \tilde{f}(\mathbf{x}-\mathbf{x}')$, as follows:

$$\frac{1}{v^2}\frac{\partial^2 G_0}{\partial t^2} = -\frac{1}{4\pi}\hat{D}f(\mathbf{x}-\mathbf{x}'),$$

$$\mathrm{curl}\frac{\partial G_0}{\partial t} = \frac{1}{4\pi}\mathrm{curl}\frac{\partial}{\partial t}f(\mathbf{x}-\mathbf{x}'),$$

$$\frac{\partial G_0}{\partial t} = -\frac{v^2}{4\pi}\hat{D}\tilde{f}(\mathbf{x}-\mathbf{x}') \tag{1.62}$$

where f, $\tilde{f}$, and $\hat{D}$ are following from Eq. (1.40) and Eq. (1.41), and Eq. (1.47), denoted by

$$f(\mathbf{x}-\mathbf{x}') = \frac{1}{|\mathbf{r}-\mathbf{r}'|}\delta\left(t-t'-\frac{|\mathbf{r}-\mathbf{r}'|}{v}\right),$$

$$\tilde{f}(\mathbf{x}-\mathbf{x}') = \frac{1}{|\mathbf{r}-\mathbf{r}'|}\theta\left(t-t'-\frac{|\mathbf{r}-\mathbf{r}'|}{v}\right)$$

$$\hat{D} = \nabla\nabla - \frac{1}{v^2}\frac{\partial^2}{\partial t^2} \tag{1.63}$$

After substituting (1.62) into (1.60), and writing the explicit formula for the convolution, the following is obtained:

$$\mathbf{E} = \mathbf{F} + \int d\mathbf{x}' \left\{ \langle \mathbf{x} | \Gamma_e | \mathbf{x}' \rangle \chi \frac{1}{\varepsilon_0 \varepsilon}(\mathbf{P}_1 - \mathbf{P}_{ex}) - \langle \mathbf{x} | \Gamma_b | \mathbf{x}' \rangle \chi \mu_0 \mu (\mathbf{M}_1 - \mathbf{M}_{ex}) \right.$$
$$\left. + \langle \mathbf{x} | \Gamma_j | \mathbf{x}' \rangle \mu_0 \mu (\chi \mathbf{j}_1 + \mathbf{i}\delta(S)) \right\}, \tag{1.64}$$

where the Dirac bracket symbols are used [18, 19] for the kernels of the integral equation. These kernels follow from Eqs. (1.62) and (1.63), and have the form

$$\langle \mathbf{x} | \Gamma_e | \mathbf{x}' \rangle = \frac{1}{4\pi}\left(\nabla\nabla - \frac{1}{v^2}\frac{\partial^2}{\partial t^2}\hat{I}\right) f(\mathbf{x}-\mathbf{x}'),$$

$$\langle \mathbf{x} | \Gamma_b | \mathbf{x}' \rangle = \frac{1}{4\pi}\operatorname{curl}\frac{\partial}{\partial t} f(\mathbf{x}-\mathbf{x}'),$$

$$\langle \mathbf{x} | \Gamma_j | \mathbf{x}' \rangle = \frac{v^2}{4\pi}\left(\nabla\nabla - \frac{1}{v^2}\frac{\partial^2}{\partial t^2}\hat{I}\right) \tilde{f}(\mathbf{x}-\mathbf{x}'). \tag{1.65}$$

Here $\hat{I}$ is the unit matrix (operator) and ∇ is the nabla operator. The integral $\int d\mathbf{x}'$ in Eq. (1.64) and further on designates the integration over the 4D half-space $t' \geq 0$:

$$\int d\mathbf{x}' = \int_0^\infty dt' \int_\infty d\mathbf{r}'.$$

The explicit expression for the free term is obtained from Eq. (1.61):

$$\mathbf{F}=\mathbf{E}_0+\int_{-\infty}^{0}dt'\int_{\infty}d\mathbf{r}'\Big\{\langle\mathbf{x}|\Gamma_e|\mathbf{x}'\rangle\chi\frac{1}{\varepsilon_0\varepsilon}(\mathbf{P}_1-\mathbf{P}_{ex})$$
$$-\langle\mathbf{x}|\Gamma_b|\mathbf{x}'\rangle\mathrm{curl}\chi\mu_0\mu(\mathbf{M}_1-\mathbf{M}_{ex})+\langle\mathbf{x}|\Gamma_j|\mathbf{x}'\rangle\mu_0\mu(\chi\mathbf{j}_1+\mathbf{i}\delta(S))\Big\}$$
$$+\int_{0}^{\infty}dt'\int_{\infty}d\mathbf{r}'\Big\{\langle\mathbf{x}|\Gamma_e|\mathbf{x}'\rangle\chi\frac{1}{\varepsilon_0\varepsilon}(\mathbf{P}_{\mathrm{rel}}-\mathbf{P}_{ex})$$
$$-\langle\mathbf{x}|\Gamma_b|\mathbf{x}'\rangle\mathrm{curl}\chi\mu_0\mu(\mathbf{M}_{\mathrm{rel}}-\mathbf{M}_{ex})+\langle\mathbf{x}|\Gamma_j|\mathbf{x}'\rangle\mu_0\mu\chi\mathbf{j}_{\mathrm{rel}}\Big\} \quad (1.66)$$

Note that the first integral in Eq. (1.66) is equal to zero when the region $V(t)$ has come into being at the zero moment of time or all the changes of the medium have already occurred at this zero moment. In this case the free term **F** is the sum of the primary field, which existed in the background until zero moment, and the relict field caused by the jump change of the medium state. The relation (1.61) is then the evolutionary relation, which connects the field after zero moment with the previous field.

Equation (1.61) can be written in an operator representation as follows:

$$\mathbf{E}=\mathbf{F}+\hat{K}\mathbf{E}, \quad (1.67)$$

where the operator $\hat{K}$ is

$$\hat{K}=\int d\mathbf{x}'\Big\{\langle\mathbf{x}|\Gamma_e|\mathbf{x}'\rangle\chi\frac{1}{\varepsilon_0\varepsilon}(\hat{P}_1-\hat{P}_{ex})$$
$$-\langle\mathbf{x}|\Gamma_b|\mathbf{x}'\rangle\chi\mu_0\mu(\hat{M}_1-\hat{M}_{ex})+\langle\mathbf{x}|\Gamma_j|\mathbf{x}'\rangle\mu_0\mu(\chi\hat{j}_1+\hat{i}\delta(S))\Big\} \quad (1.68)$$

and the operators $\hat{P}_1,\hat{M}_1,\hat{j}_1,\hat{i}$ correspond to the functions $\mathbf{P}_1(\mathbf{E})$, $\mathbf{M}_1(\mathbf{E})$, $\mathbf{j}_1(\mathbf{E})$ and $\mathbf{i}(\mathbf{E})$.

The operator $\hat{K}$ is the integral Volterra operator, because its kernel is integrable over the 4D space and it is supported in the domain in the past which is the light cone with a vertex at the observation point $(t, \mathbf{r})$. These features of the operator $\hat{K}$ allow one to give a clear causal-geometric interpretation to Eqs. (1.61) and (1.67).

1.3.2 Influence of a Dispersive Background on the Integral Equation Form

For a dispersive background, reasoning similar to the above gives the equation

$$\mathbf{E} = \mathbf{E}_0 - \hat{G}_{0p} * \mu_0 \left[\frac{\partial^2}{\partial t^2} \chi \left(\mathbf{P}_{\text{in}} - \varepsilon_0 \int_{-\infty}^{t} \omega_e^2 (t - t') \mathbf{E} dt' \right) \right.$$

$$\left. + \operatorname{curl} \frac{\partial}{\partial t} \chi \mathbf{M}_{\text{in}} + \frac{\partial (\chi \mathbf{j}_{\text{in}} + \mathbf{i}\delta(S))}{\partial t} \right], \tag{1.69}$$

where the free term is determined analogously to Eq. (1.58) as

$$\mathbf{E}_0 = \mathbf{E}_0' - \hat{G}_{0p} * \mu_0 \frac{\partial \mathbf{j}_{\text{extr}}}{\partial t} \tag{1.70}$$

and $\hat{G}_{0p}$ is as in Eq. (1.48).

An explicit expression for the integral equation follows from Eq. (1.69) after a transfer of the differentiation operations from the second multipliers to Green's function

$$\mathbf{E} = \mathbf{E}_0 + \int_{-\infty}^{\infty} d\mathbf{x}' \left\{ \langle \mathbf{x} | \Gamma_{ep} | \mathbf{x}' \rangle \chi \left[\frac{1}{\varepsilon_0} \mathbf{P}_{\text{in}} - \int_{-\infty}^{t'} \omega_e^2 (t' - t'') \mathbf{E} dt'' \right] - \right.$$

$$\left. - \langle \mathbf{x} | \Gamma_{bp} | \mathbf{x}' \rangle \chi \mu_0 \mathbf{M}_{\text{in}} + \langle \mathbf{x} | \Gamma_{jp} | \mathbf{x}' \rangle \mu_0 \left[\chi \mathbf{j}_{\text{in}} + \mathbf{i}\delta(s(t')) \right] \right\}. \tag{1.71}$$

Here the equation kernels are defined by the expressions following from Eqs. (1.46–1.48)

$$\langle \mathbf{x} | \Gamma_{ep} | \mathbf{x}' \rangle = \frac{1}{4\pi} \left[\nabla\nabla - \frac{1}{c^2} \left(\frac{\partial^2}{\partial t^2} + \omega_e^2 \right) \right] f(\mathbf{x} - \mathbf{x}'),$$

$$\langle \mathbf{x} | \Gamma_{bp} | \mathbf{x}' \rangle = \frac{1}{4\pi} \operatorname{curl} \left(\frac{\partial^2}{\partial t^2} + \omega_e^2 \right) \tilde{f}(\mathbf{x} - \mathbf{x}'),$$

$$\langle \mathbf{x} | \Gamma_{jp} | \mathbf{x}' \rangle = \frac{1}{4\pi} \left[\nabla\nabla - \frac{1}{c^2} \left(\frac{\partial^2}{\partial t^2} + \omega_e^2 \right) \right] \tilde{f}(\mathbf{x} - \mathbf{x}'), \tag{1.72}$$

where

$$f_p(\mathbf{x}-\mathbf{x}') = \frac{1}{|\mathbf{r}-\mathbf{r}'|}\delta\left(t-t'-\frac{|\mathbf{r}-\mathbf{r}'|}{c}\right)$$

$$+\frac{1}{|\mathbf{r}-\mathbf{r}'|}\theta\left(t-t'-\frac{|\mathbf{r}-\mathbf{r}'|}{c}\right)\cdot\frac{\partial^2}{\partial t^2}\int_{|\mathbf{r}-\mathbf{r}'|/c}^{t-t'} J_0(\omega_e\sqrt{(t-t')^2-u^2})du, \tag{1.73}$$

$$\tilde{f}_p(\mathbf{x}-\mathbf{x}') = \frac{1}{|\mathbf{r}-\mathbf{r}'|}\theta\left(t-t'-\frac{|\mathbf{r}-\mathbf{r}'|}{c}\right)\cdot\frac{\partial}{\partial t}\int_{|\mathbf{r}-\mathbf{r}'|/c}^{t-t'} J_0(\omega_e\sqrt{(t-t')^2-u^2})du. \tag{1.74}$$

After zero moment, the field satisfies the integral equation for the non-stationary problem

$$\mathbf{E} = \mathbf{E}_0 + \int d\mathbf{x}'\left\{\langle\mathbf{x}|\Gamma_{ep}|\mathbf{x}'\rangle\chi\left[\frac{1}{\varepsilon_0}\mathbf{P} - \int_0^{t'}\omega_e^2(t'-t'')\mathbf{E}dt''\right]\right.$$

$$\left.-\langle\mathbf{x}|\Gamma_{bp}|\mathbf{x}'\rangle\chi\mu_0\mathbf{M} + \langle\mathbf{x}|\Gamma_{jp}|\mathbf{x}'\rangle\mu_0\left[\chi\mathbf{j}+\mathbf{i}\delta(s(t'))\right]\right\}. \tag{1.75}$$

The free term determines the "field prehistory" and is given by the expression:

$$\mathbf{F} = \mathbf{E}_0 + \mathbf{F}_0 + \mathbf{F}_1,$$

where there is the relict part

$$\mathbf{F}_0 = \int d\mathbf{x}'\left\{\langle\mathbf{x}|\Gamma_{ep}|\mathbf{x}'\rangle\chi\left[\frac{1}{\varepsilon_0}\mathbf{P}_{\text{rel}} - \int_{-\infty}^{0}\omega_e^2(t'-t'')\,\mathbf{E}dt''\right]\right.$$

$$\left.-\langle\mathbf{x}|\Gamma_{bp}|\mathbf{x}'\rangle\chi\mu_0\mathbf{M}_{\text{rel}} + \langle\mathbf{x}|\Gamma_{jp}|\mathbf{x}'\rangle\mu_0\left[\chi\mathbf{j}_{\text{rel}}+\mathbf{i}_{\text{rel}}\delta(s(t'))\right]\right\}. \tag{1.76}$$

and the part determined by the prehistory before zero moment

$$\mathbf{F}_1 = \int_{-\infty}^{0} dt'\int d\mathbf{r}'\left\{\langle\mathbf{x}|\Gamma_{\text{ep}}|\mathbf{x}'\rangle\chi\left[\frac{1}{\varepsilon_0}\mathbf{P}_1 - \int_{-\infty}^{t'}\omega_e^2(t'-t'')\mathbf{E}_1 dt''\right]\right.$$

$$\left.-\langle\mathbf{x}|\Gamma_{bp}|\mathbf{x}'\rangle\chi\mu_0\mathbf{M} + \langle\mathbf{x}|\Gamma_{jp}|\mathbf{x}'\rangle\mu_0\left[\chi\mathbf{j}+\mathbf{i}\delta(s(t'))\right]\right\}. \tag{1.77}$$

Equation (1.75) can also be written in operator form, as Eq. (1.67).

1.3.3 Spatial-Temporal Interpretation of the Volterra Integral Equation

The meaning of the relations (1.64) and (1.75) depends essentially on the location of the observation point **x** defined in the frame of time-spatial reference $(t, \mathbf{r})$. If this point **x** is located in the time-spatial region, corresponding to the region $V(t)$, i.e., inside the time-varying inhomogeneity, then the expressions (1.64) and (1.75) are integral equations. Their kernels are defined in the intersection of the light cone $t - t' - |\mathbf{r} - \mathbf{r}'|/v$ and the region given by the characteristic function $\chi(\mathbf{x})$. The light cone is the support domain of the generalized function in the kernels of all the operators (1.65) and (1.72) and the cone vertex is located at the observation point, **x**. Figure 1.4 illustrates the projection of this intersection onto the coordinate plane (t, x).

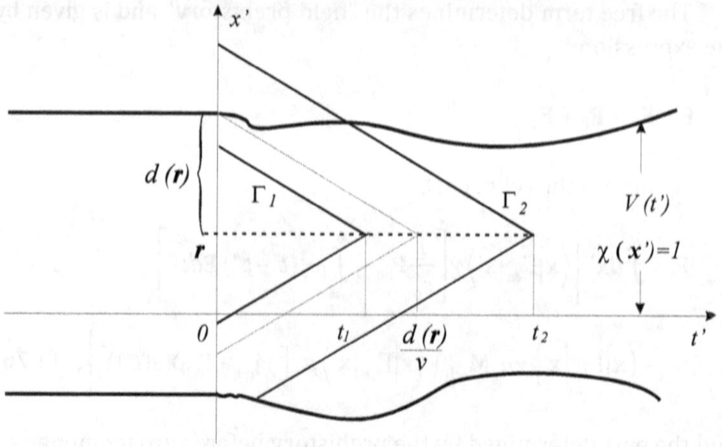

Figure 1.4 The projection of the intersection of the light cone with the time-spatial lines of the boundaries of the region $V(t)$.

This intersection takes qualitatively different forms on the intervals $0 \le t \le d(\mathbf{r})/v$ and $t \ge d(\mathbf{r})/v$. Here, $d(\mathbf{r})$ denotes the shortest distance between the observation point **r** and the boundary of $V(t)$ at the moment $t = 0$. On the interval

$0 \leq t \leq d(\mathbf{r})/v$, the part of the light cone corresponding to the half-axes $t' \geq 0$, (Γ_1 in Fig. 1.4), is enclosed in the region $\chi(\mathbf{x}) = 1$. It means that the characteristic function $\chi(\mathbf{x})$ in Eq. (1.68) is identical to unity, i.e., there is no influence of the region $V(t)$ boundaries on the field on this time interval. This is a natural consequence of a time delay, which is caused by the finiteness of the light velocity. Formally, it means that the electromagnetic field on this time interval interacts with the medium in the region $V(t)$ as with an unbounded medium. Therefore, the problem for the electromagnetic field is reduced to the initial problem in the unbounded medium over this time interval.

On the second time interval, $t \geq d(\mathbf{r})/v$, a part of the light cone, Γ_2 in Fig. 1.4, passes out the boundary of the region determined by the characteristic function $\chi(\mathbf{x})$, and the problem here becomes the boundary value problem. Its solution can be constructed by using the solution over the previous time interval, corresponding to the unbounded medium. This solution should contain all the information about features of the medium varying in the region $V(t)$.

1.3.4 Three Stages of Development of Electromagnetic Transients in a Bounded Medium with Time-Varying Parameters

We analyze in more detail Eq. (1.64) for the case of a non-stationary object in a non-dispersive background. A change of the state of this object at the moment $t = 0$ leads to division of Eq. (1.64) onto a chain of evolutionary coupled equations.

Three time intervals can be separated in the evolution process. When the observation moment of time, $t < 0$, then the field satisfies Eq. (1.64), in which the integration is performed over the infinite intersection $T^{(-)}$ of the light cone $t - t' - |\mathbf{r} - \mathbf{r}'|/v$ and the region restricted by the characteristic function $\chi(\mathbf{x})$, so the integral over the half-axis $t' > 0$ is absent. The solution to this equation gives the initial field for the non-stationary problem.

After the zero moment of time a part of the integration region restricted by the hyperplane $t' = 0$ appears. In this part of the region, the object is in the new state. Possible intersections of the light cone $t - t' - \frac{1}{v}|\vec{r} - \vec{r}'| \geq 0$ restricted by this hyperplane with the region restricted by the surface $\chi(\mathbf{x}) = 1$ can come to two regions T_1 and T_2.

In the region T_1 the intersection bounded by Γ_1 in Fig. 1.4 is enclosed inside the surface $\chi(\mathbf{x}) = 1$. It means that this whole intersection belongs to the region of the event space corresponding to the object $V(t)$. Maximum intersection is achieved at $t' = 0$ and it is a sphere $|\mathbf{r} - \mathbf{r}'| \leq vt$ if the condition $vt \leq d(\vec{r})$ is satisfied. This condition gives the equation of the cone that corresponds to the point $\mathbf{r}$ and does not intersect with the surface of the event space of the object for $t' > 0$:

$$\frac{d(\mathbf{r})}{v} - t' - \frac{|\mathbf{r} - \mathbf{r}'|}{v} \geq 0 \tag{1.78}$$

For all the observation points $\mathbf{r}$ belonging to the whole region $V(0)$ the inequality (1.78) determines a set of all points $\tilde{f}_e(\vec{x}) = \frac{1}{r}\theta\left(t - \frac{r}{c}\right) \cdot \frac{\partial}{\partial t}\int_{r/c}^{t} J_0(\omega_e\sqrt{t^2 - u^2})du$ of 4D space, in which there is no influence of a boundary. It is evident that for boundary points on the surface $S(0)$ the distance $d(\mathbf{r}) = 0$ and the inequality (1.78) produces a cone with zero height. The cone with maximum height will correspond to the point that is the most distant from all boundary points.

This reasoning holds true provided that it can be assumed that the velocity of the movement of a boundary of the region $V(t)$ does not exceed the value of the velocity v that is the slope of the time-spatial lines of the boundary, in other words that the slope of a tangent plane to $S(t)$ is less than the slope of the generators of the light cone. In the opposite case this circumstance must be taken into account when constructing the region T_1.

As examples, we consider the half-space and the sphere as diffraction objects. In the case of the half-space, $\theta(x) = 1$, the minimum distance is equal to $d = x$ that yields the region $vt < x$, which is free from the boundary influence. For given observation point x, the cone that does not intersect with the surface $\chi(\mathbf{x}) = 1$ is determined by inequality

$$\frac{x}{v} - t' - \frac{|x - x'|}{v} > 0 \tag{1.79}$$

This inequality and the condition x > 0 define the region T_1: $x' > vt'$.

In the case of a sphere with radius a centered at the origin of the coordinate system the minimum distance is equal to $d = a - r$. Therefore, the region T_1 will be determined by the 4D cone $a - vt' > \mathbf{r} + |\mathbf{r} - \mathbf{r}'|$.

Within the region T_1 the characteristic function in Eq. (1.64) for the internal field is equal identically to unity

$$\mathbf{E}^{(1)} = \mathbf{F}^{(1)} + \int d\mathbf{x}' \left\{ \langle \mathbf{x}|\Gamma_e|\mathbf{x}'\rangle \frac{1}{\varepsilon_0 \varepsilon}(\mathbf{P}_1 - \mathbf{P}_{ex}) - \langle \mathbf{x}|\Gamma_b|\mathbf{x}'\rangle \mu_0 \mu(\mathbf{M}_1 - \mathbf{M}_{ex}) + \langle \mathbf{x}|\Gamma_j|\mathbf{x}'\rangle \mu_0 \mu \mathbf{j}_1 \right\}. \quad (1.80)$$

So, in the time interval $t < d(\mathbf{r})/v$ there is no influence of the boundary of the diffraction object and the equation takes into account a pure change of medium properties only. If one extends this equation over the whole 4D half-space then this equation will describe the field in an infinite medium with the same properties as the medium within the diffraction object.

Beginning from the moment $t = d(\mathbf{r})/v$ the past light cone will intersect the boundary of the 4D region $\chi(\mathbf{x}) = 1$ and a formation of the internal field will be influenced by the boundary of the region $V(t)$. In 4D space-time it corresponds to passing into the region T_2. The equation for the field in this region transforms to the complete form

$$\mathbf{E} = \mathbf{F} + \int d\mathbf{x}' \left\{ \langle \mathbf{x}|\Gamma_e|\mathbf{x}'\rangle \chi \frac{1}{\varepsilon_0 \varepsilon}(\mathbf{P}_1 - \mathbf{P}_{ex}) - \langle \mathbf{x}|\Gamma_b|\mathbf{x}'\rangle \chi \mu_0 \mu(\mathbf{M}_1 - \mathbf{M}_{ex}) + \langle \mathbf{x}|\Gamma_j|\mathbf{x}'\rangle \mu_0 \mu(\chi \mathbf{j}_1 + \mathbf{i}\delta(S)) \right\}. \quad (1.81)$$

The chain of equations obtained reveals the following evolution of the field interaction with the diffraction object. Before zero moment of time the primary field $\mathbf{E}_0$ generates the internal field $\mathbf{E}^{(-)}$ that is described by Eq. (1.64) for $t < 0$. After a change of the object state at zero moment of time the internal field in the region $T_1 = \{\mathbf{x}: 0 \le t \le d(\mathbf{r})/v\}$ is described by Eq. (1.80), which does not contain the characteristic function. It means that the object $V(t)$ boundary has no influence in the region T_1 and the primary field does not explicitly take part in the creation of the internal field. To see this one can extend Eq. (1.59) from the region $T^{(-)}$ into the region T_1 and transform the expression (1.66) taking this extension into account. Then the free term of Eq. (1.80) can be written in the form

$$\mathbf{F}^{(1)} = \mathbf{E}^{(-)} - \int d\mathbf{x}' \left\{ \langle \mathbf{x}|\Gamma_e|\mathbf{x}'\rangle \frac{1}{\varepsilon_0 \varepsilon}(\mathbf{P}_1 - \mathbf{P}_{\text{ex}}) \right.$$

$$\left. - \langle \mathbf{x}|\Gamma_b|\mathbf{x}'\rangle \mu_0 \mu (\mathbf{M}_1 - \mathbf{M}_{\text{ex}}) + \langle \mathbf{x}|\Gamma_j|\mathbf{x}'\rangle \mu_0 \mu \mathbf{j}_1 \right\}. \quad (1.82)$$

It follows from (1.80) and (1.82) that the field $\mathbf{E}^{(1)}$ is created directly by the field $\mathbf{E}^{(-)}$ that existed inside the diffraction object until the change of its state and implicitly by the field $\mathbf{E}_0$, which contains in the medium "memory" in the form of the polarizations and the currents.

Equation (1.81) shows that the field in the region $T_3 = \{\mathbf{x}: t \ge d(\mathbf{r})/v\}$ is created by the field $\mathbf{E}^{(-)}$, the primary field $\mathbf{E}_0$ passing through the boundary and the field $\mathbf{E}^{(1)}$.

If the observation point is located beyond the event space region corresponding to $V(t)$, as shown in Fig. 1.5, then Eq. (1.64) is the quadrature formula for calculating an external field via the previously obtained internal field.

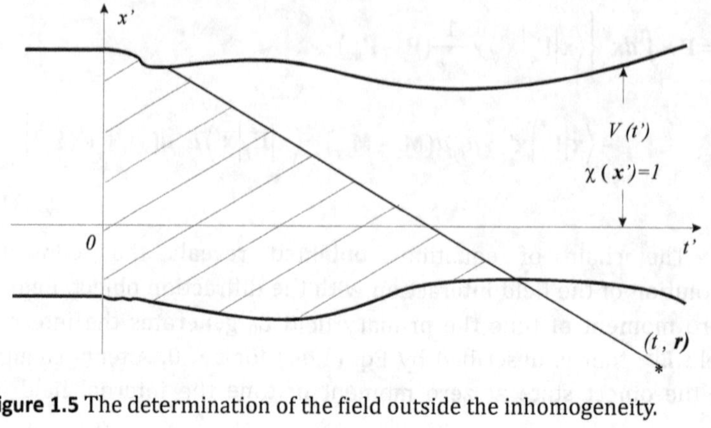

Figure 1.5 The determination of the field outside the inhomogeneity.

In this case, two time intervals are distinguished. When $t < l(\mathbf{r})/v$ where $l(\mathbf{r})$ is a distance from an observation point $\mathbf{r}$ to the nearest point of the region $V(t)$ at the moment $t = 0$ then the external field is determined by the undisturbed state of the diffraction object by virtue of Eq. (1.64), into which the field $\mathbf{E}^{(-)}$ must be substituted. When $t \geq l(\mathbf{r})/v$ the external field is influenced by the new state of the diffraction object and is calculated using the formula (1.81).

1.3.5 Three Stages of Solution of a Non-Stationary Problem

In this evolution chain the expressions describing the field until zero moment of time must be considered as undisturbed ones and the expressions (1.80) and (1.81) must be considered as ones that take into account the disturbance made by a change of a medium state and boundaries of the diffraction object. The undisturbed problem can be considered as a stationary one and all non-stationary behavior can be referred to the time interval $t \geq 0$ including a change of the character of the incident field. Then a key problem in investigations of non-stationary phenomena can be defined as the problem of seeking the internal field.

The solution to this problem is broken up into three stages. In the first stage in the region $T^{(-)}$ one seeks the solution to the undisturbed problem that can be taken as a simple one enough to use known solutions to problems investigated earlier.

In the second stage in the region T_1, one seeks the solution to the disturbed problem but without any influence from the boundaries of the diffraction object, i.e., in this stage the interaction of the electromagnetic field determined by the free term $\mathbf{F}^{(1)}$ with an unbounded non-stationary medium is considered. In this case, the field is disturbed by the medium change only. Therefore, the problem is reduced to the initial Cauchy problem for the field change in an unbounded medium. However, one must understand that some solutions to this problem can be lost as a consequence of an extension of the equation from the region T_1 on the whole 4D half-space. Such solutions satisfy the homogeneous integral equation and will be considered later.

In the third stage one seeks in the region T_2 the solution that takes into account boundaries effects. The solutions found in the second stage can be used as a zero approximation that takes into account the change of the medium properties.

1.4 The Resolvent Method for Solving the Integral Equation

1.4.1 Operator Form of an Integral Equation Representation

For carrying-out analytical operations it is convenient to use the Dirac bracket notations that we now consider in more detail according to the books [18] and [19]. In these notations, the i-th coordinate of a vector $\mathbf{F}(t, \mathbf{r}) = |\mathbf{F}\rangle$ is written in the form $F_i(t, \mathbf{r}) = \langle\mathbf{x}|\mathbf{F}\rangle_i$ where $\mathbf{x} = (t, \mathbf{r})$ is a time-space vector. A matrix element of a kernel of an integral operator $\hat{K}$ is written in the form $\left\langle \mathbf{x}\middle|\hat{K}\middle|\mathbf{x}'\right\rangle_{ij}$, and the result of an operation on the vector that is the vector $\hat{K}\mathbf{F}$ is determined by the formula

$$\left\langle \mathbf{x} \mid \hat{K}\mathbf{F}\right\rangle_i = \int d\mathbf{x}' \left\langle \mathbf{x}\middle|\hat{K}\middle|\mathbf{x}'\right\rangle_{in} \langle \mathbf{x}' \mid \mathbf{F}\rangle_n, \tag{1.83}$$

where the integral means $\int d\mathbf{x} = \int_0^\infty dt \int_\infty d\mathbf{r}$ as above.

The matrix element of the product of the operators is evaluated by the formula

$$\left\langle \mathbf{x}\middle|\hat{K}\hat{R}\middle|\mathbf{x}'\right\rangle_{ij} = \int d\mathbf{x}'' \left\langle \mathbf{x}\middle|\hat{K}\middle|\mathbf{x}''\right\rangle_{in} \left\langle \mathbf{x}''\middle|\hat{R}\middle|\mathbf{x}'\right\rangle_{nj}. \tag{1.84}$$

Further, an impulse representation (Fourier–Laplace representation) of the vectors and operators is used widely side-by-side with the coordinate representation. The transition to the impulse representation in a rectangular system of coordinates is performed by virtue of the transformation functions

$$\langle \mathbf{x} \| \mathbf{p}\rangle_{ij} = \delta_{ij} e^{pt + i\mathbf{k}\mathbf{r}}, \quad \langle \mathbf{p} \| \mathbf{x}\rangle_{ij} = \delta_{ij} e^{-pt - i\mathbf{k}\mathbf{r}}, \tag{1.85}$$

where $\mathbf{p} = (p, \mathbf{k})$, p is a complex variable of the Laplace transformation and $\mathbf{k}$ is a real variable of the 3D Fourier transformation. The transformation functions are orthonormal in the coordinate as well as in the impulse representations:

$$\langle \mathbf{x} \,||\, \mathbf{x}' \rangle_{ij} = \int d\mathbf{p} \langle \mathbf{x} \,||\, \mathbf{p} \rangle_{in} \langle \mathbf{p} \,||\, \mathbf{x}' \rangle_{nj} = \delta_{ij} \int d\mathbf{p} e^{pt + i\mathbf{kr}} e^{-pt' - i\mathbf{kr}'}$$
$$= \delta_{ij} \delta(t - t') \delta(\mathbf{r} - \mathbf{r}'),$$

$$\langle \mathbf{p} \,||\, \mathbf{p}' \rangle_{ij} = \int d\mathbf{x} \langle \mathbf{p} \,||\, \mathbf{x} \rangle_{in} \langle \mathbf{x} \,||\, \mathbf{p}' \rangle_{nj} = \delta_{ij} \int d\mathbf{p} e^{-pt - i\mathbf{kr}} e^{pt' + i\mathbf{kr}'} \tag{1.86}$$
$$= \delta_{ij} \Delta(\mathbf{p} - \mathbf{p}'),$$

Here and below the notations used are

$$\Delta(\mathbf{p} - \mathbf{p}') = \frac{(2\pi)^3 \delta(\mathbf{k} - \mathbf{k}')}{p - p'}, \text{ where } \mathrm{Re}(p - p') > 0; \tag{1.87}$$

the integral notation is $\int d\mathbf{p} = \int_{\eta - i\infty}^{\eta + i\infty} \frac{dp}{2\pi i} \int_{\infty} \frac{d\mathbf{k}}{(2\pi)^3}$, where η is a constant determined by the analytical properties of the integrand.

The transition to the impulse (Fourier–Laplace) representation of the vector $\mathbf{F}(t, \mathbf{r}) = |\mathbf{F}\rangle$ is given by virtue of the transformation functions

$$\langle \mathbf{p} \,|\, \mathbf{F} \rangle_i = \int d\mathbf{x} \langle \mathbf{p} \,||\, \mathbf{x} \rangle_{in} \langle \mathbf{x} \,|\, \mathbf{F} \rangle_n = \int d\mathbf{x} e^{-pt - i\mathbf{kr}} \langle \mathbf{x} \,|\, \mathbf{F} \rangle_i. \tag{1.88}$$

The inverse transition is implemented by the inverse Fourier–Laplace transformation:

$$\langle \mathbf{x} \,|\, \mathbf{F} \rangle_i = \int d\mathbf{p} \langle \mathbf{x} \,||\, \mathbf{p} \rangle_{in} \langle \mathbf{p} \,|\, \mathbf{F} \rangle_n = \int d\mathbf{p} e^{pt + i\mathbf{kr}} \langle \mathbf{p} \,|\, \mathbf{F} \rangle_i. \tag{1.89}$$

The impulse representation for the matrix elements is found by virtue of the double transformation with respect to each time-spatial vector

$$\left\langle \mathbf{p} \middle| \hat{K} \middle| \mathbf{p}' \right\rangle_{ij} = \int d\mathbf{x} \langle \mathbf{p} \,||\, \mathbf{x} \rangle_{im} \hat{K}_{mn} \langle \mathbf{x} \,||\, \mathbf{p}' \rangle_{nj}$$
$$= \int d\mathbf{x} \int d\mathbf{x}' \langle \mathbf{p} \,||\, \mathbf{x} \rangle_{im} \left\langle \mathbf{x} \middle| \hat{K} \middle| \mathbf{x}' \right\rangle_{mn} \langle \mathbf{x}' \,||\, \mathbf{p}' \rangle_{nj}. \tag{1.90}$$

The transition to the coordinate representation of the matrix elements is given by the double Fourier–Laplace transformation:

$$\left\langle \mathbf{x}\middle|\hat{K}\middle|\mathbf{x}'\right\rangle_{ij} = \int d\mathbf{p}\int d\mathbf{p}'\langle \mathbf{x}\,||\,\mathbf{p}\rangle_{im}\left\langle \mathbf{p}\middle|\hat{K}\middle|\mathbf{p}'\right\rangle_{mn}\langle \mathbf{p}'\,||\,\mathbf{x}'\rangle_{nj}. \tag{1.91}$$

The matrix elements of a product of operators are defined as a conventional matrix product in the coordinate as well as in the impulse representation

$$\left\langle \mathbf{p}\middle|\hat{K}\hat{R}\middle|\mathbf{p}'\right\rangle_{ij} = \int d\mathbf{p}''\langle \mathbf{p}\,|\,\hat{K}\,|\,\mathbf{p}''\rangle_{im}\langle \mathbf{p}''\,|\,\hat{R}\,|\,\mathbf{p}'\rangle_{mj}. \tag{1.92}$$

The relations given above are true for the general case of functions of four variables. The number of the spatial variables can be reduced in advance to two or one variables in many problems owing to the presence some symmetry. In this case, the relations are simplified and this simplification can be made in the general form by virtue of the operation of coordinate descending [5]. We show this by an example of the descending with respect to two coordinates, which are called lateral ones.

The result of the operation on the vector $\mathbf{E} = \mathbf{E}(t, x) = \mathbf{E}(\mathbf{x}_{||})$, that depends on one, longitudinal, coordinate only is written in the form

$$\hat{K}\mathbf{E} = \int_0^\infty dt'\int_{-\infty}^\infty dx'\mathbf{E}(t',x')\int d\mathbf{r}'_\perp\left\langle \mathbf{x}\middle|\hat{K}\middle|\mathbf{x}'_{||},\mathbf{r}'_\perp\right\rangle, \tag{1.93}$$

where $\mathbf{x}_{||} = (t, x)$ and $\mathbf{r} = (y, z)$ is the lateral vector.

Integration over the lateral coordinates gives

$$\begin{aligned}\int d\mathbf{r}'_\perp\left\langle \mathbf{x}\middle|\hat{K}\middle|\mathbf{x}'_{||},\mathbf{r}'_\perp\right\rangle &= \int d\mathbf{r}'_\perp\int d\mathbf{p}\int d\mathbf{p}' e^{\mathbf{p}_{||}\mathbf{x}_{||}+i\mathbf{k}_\perp\mathbf{r}_\perp}\left\langle \mathbf{p}\middle|\hat{K}\middle|\mathbf{p}'_{||},\mathbf{k}'_\perp\right\rangle e^{-\mathbf{p}'_{||}\mathbf{x}'_{||}-i\mathbf{k}'_\perp\mathbf{r}'_\perp}\\ &= \int d\mathbf{p}\int d\mathbf{p}' e^{\mathbf{p}_{||}\mathbf{x}_{||}+i\mathbf{k}_\perp\mathbf{r}_\perp}\left\langle \mathbf{p}\middle|\hat{K}\middle|\mathbf{p}'_{||},\mathbf{k}'_\perp\right\rangle e^{-\mathbf{p}'_{||}\mathbf{x}'_{||}}(2\pi)^2\delta(\mathbf{k}'_\perp)\\ &= \int d\mathbf{p}\int d\mathbf{p}'_{||} e^{\mathbf{p}_{||}\mathbf{x}_{||}+i\mathbf{k}_\perp\mathbf{r}_\perp}\left\langle \mathbf{p}\middle|\hat{K}\middle|\mathbf{p}'_{||},\mathbf{k}'_\perp=0\right\rangle e^{-\mathbf{p}'_{||}\mathbf{x}'_{||}}\end{aligned} \tag{1.94}$$

Using Eqs. (1.86), (1.88), (1.93) and (1.94) we obtain a simple result in the impulse representation

$$\langle \mathbf{p}\,|\,\mathbf{E}\rangle = \int d\mathbf{p}'_{||}\left\langle \mathbf{p}\middle|\hat{K}\middle|\mathbf{p}'_{||},\mathbf{k}'_\perp=0\right\rangle\langle \mathbf{p}'_{||}\,|\,\mathbf{E}\rangle, \tag{1.95}$$

where it is seen that the inner lateral variables of the Fourier transformation can be put equal to zero in the kernel. The vector

$$\left\langle \mathbf{p}'_{||} \,|\, \mathbf{E} \right\rangle = \int d\mathbf{x}'_{||} \left\langle \mathbf{p}'_{||} \,||\, \mathbf{x}'_{||} \right\rangle \left\langle \mathbf{x}'_{||} \,|\, \mathbf{E} \right\rangle = \int d\mathbf{x}'_{||} e^{-p't'-i\mathbf{k}'\mathbf{r}'} \left\langle \mathbf{x}'_{||} \,|\, \mathbf{E} \right\rangle$$

is the Fourier–Laplace transformation of the vector $\langle \mathbf{x}_{||} | \mathbf{E} | \rangle$ with respect to the variables t, x.

If the kernel depends on the lateral coordinates in the form of a difference,

$$\left\langle \mathbf{x} \middle| \hat{K} \middle| \mathbf{x}' \right\rangle = \left\langle \mathbf{x}_{||} \middle| \hat{K} \middle| \mathbf{x}'_{||}, \mathbf{r}_{\perp} - \mathbf{r}'_{\perp} \right\rangle,$$

then

$$\begin{aligned} \left\langle \mathbf{p} \middle| \hat{K} \middle| \mathbf{p}' \right\rangle &= \int d\mathbf{x}_{||} \int d\mathbf{x}'_{||} e^{-\mathbf{p}_{||}\mathbf{x}_{||} + \mathbf{p}'_{||}\mathbf{x}'_{||}} \int_{\infty} d\mathbf{R}_{\perp} e^{-i\mathbf{k}_{\perp}\mathbf{R}_{\perp}} \left\langle \mathbf{x}_{||} \middle| \hat{K} \middle| \mathbf{x}'_{||}, \mathbf{R}_{\perp} \right\rangle (2\pi)^2 \delta(\mathbf{k}_{\perp} - \mathbf{k}'_{\perp}) \\ &= \left\langle \mathbf{p}_{||}, \mathbf{k}_{\perp} \middle| \hat{K} \middle| \mathbf{p}'_{||}, \mathbf{k}'_{\perp} \right\rangle (2\pi)^2 \delta(\mathbf{k}'_{\perp}) (2\pi)^2 \delta(\mathbf{k}_{\perp} - \mathbf{k}'_{\perp}) \end{aligned} \tag{1.96}$$

where the result (1.94) is used.

To implement explicitly the operation of coordinate descending in the kernels of integral equations the following formulae will be useful:

$$\int_{\infty} d\mathbf{r}'_{\perp} \frac{\theta\left(t - \frac{1}{v}\sqrt{x^2 + (\mathbf{r}_{\perp} - \mathbf{r}'_{\perp})^2}\right)}{\sqrt{x^2 + (\mathbf{r}_{\perp} - \mathbf{r}'_{\perp})^2}} \theta(t) = 2\pi(vt - |x|)\theta(vt - |x|)\theta(t) \tag{1.97}$$

and

$$\begin{aligned} \int_{\infty} d\mathbf{r}'_{\perp} \frac{\delta\left(t - \frac{1}{v}\sqrt{x^2 + (\mathbf{r}_{\perp} - \mathbf{r}'_{\perp})^2}\right)}{\sqrt{x^2 + (\mathbf{r}_{\perp} - \mathbf{r}'_{\perp})^2}} \theta(t) &= \int_{\infty} d\mathbf{R}_{\perp} \frac{\delta\left(t - \frac{1}{v}\sqrt{x^2 + R_{\perp}^2}\right)}{\sqrt{x^2 + R_{\perp}^2}} \theta(t) \\ &= \int_0^{\infty} dR_{\perp} \frac{2\pi R_{\perp} v^2 t}{\sqrt{x^2 + R_{\perp}^2}\sqrt{v^2t^2 - x^2}} \delta(R_{\perp} - \sqrt{v^2t^2 - x^2})\theta(v^2t^2 - x^2)\theta(t) \\ &= 2\pi v \theta(v^2t^2 - x^2)\theta(t) = 2\pi v \theta(vt - |x|)\theta(t) \end{aligned} \tag{1.98}$$

1.4.2 Kernels of the Integral Equations for Typical Media

To show the implementation of the resolvent method an explicit expression for the integral equation kernel is needed. We first obtain the kernel and then the resolvent for problems where the object is in a non-dispersive background. The kernel operators $\hat{K}$ in Eqs. (1.64) and (1.67) can be written as a sum of products of the propagators (1.65) and medium operators $\hat{V}_a$

$$\hat{K} = \hat{\Gamma}_{\mathrm{e}}\hat{V}_{\mathrm{e}} + \hat{\Gamma}_{\mathrm{b}}\hat{V}_{\mathrm{b}} + \hat{\Gamma}_{\mathrm{j}}\hat{V}_{\mathrm{j}}. \tag{1.99}$$

The medium operators are determined by the operators of the constitutive relations, $\hat{P}$, $\hat{M}$, $\hat{j}$, $\hat{i}$, and the shape of object is given by the characteristic function χ

$$\hat{V}_e = \chi\frac{1}{\varepsilon_0\varepsilon}(\hat{P}_1 - \hat{P}_{\mathrm{ex}}),\ \hat{V}_b = \chi\mu_0\mu(\hat{M}_1 - \hat{M}_{\mathrm{ex}}),\ \hat{V}_j = \chi\mu_0\mu\hat{j}_1. \tag{1.100}$$

The impulse (Fourier–Laplace) representations of these operators are derived in the next sections.

(a) *Propagator operators in the case of a non-dispersive background*
Matrix elements of the propagator $\hat{\Gamma}_e$ in the impulse representations background follow from (1.65)

$$\begin{aligned}\left\langle \mathbf{p}\middle|\hat{\Gamma}_e\middle|\mathbf{p}'\right\rangle_{ij} &= \int d\mathbf{x}\int d\mathbf{x}'\langle\mathbf{p}\|\mathbf{x}\rangle_{im}\left\langle\mathbf{x}\middle|\hat{\Gamma}_e\middle|\mathbf{x}'\right\rangle_{mn}\langle\mathbf{x}'\|\mathbf{p}'\rangle_{nj}\\ &= \int d\mathbf{x}\int d\mathbf{x}'\langle\mathbf{p}\|\mathbf{x}\rangle_{im}\frac{1}{4\pi}\left(\frac{\partial^2}{\partial x_m\partial x_n} - \frac{1}{v^2}\frac{\partial^2}{\partial t^2}\delta_{mn}\right)f(\mathbf{x}-\mathbf{x}')\langle\mathbf{x}'\|\mathbf{p}'\rangle_{nj},\end{aligned} \tag{1.101}$$

In (1.101) all the evaluations can be made analytically. Indeed, the differentiation in the internal integral can be taken out of the spatial integral because the latter is the convolution,

$$\int_0^\infty dt'\frac{1}{4\pi}\left(\frac{\partial^2}{\partial x_m\partial x_n} - \frac{1}{v^2}\frac{\partial^2}{\partial t^2}\delta_{mn}\right)\int_\infty d\mathbf{r}' f(\mathbf{x}-\mathbf{x}')e^{p't'+i\mathbf{k}'\mathbf{r}'}. \tag{1.102}$$

The integral with respect to $\mathbf{r}'$ can be easily calculated as

$$\int_{\infty} d\mathbf{r}' f(\mathbf{x}-\mathbf{x}')e^{i\mathbf{k}'\mathbf{r}'} = \int_{\infty} d\mathbf{r}' \frac{\delta\left(t-t'-\dfrac{|\mathbf{r}-\mathbf{r}'|}{v}\right)}{|\mathbf{r}-\mathbf{r}'|} e^{i\mathbf{k}'\mathbf{r}'}$$
$$= 4\pi e^{i\mathbf{k}'\mathbf{r}} \int_0^{\infty} R\delta\left(t-t'-\frac{R}{v}\right)\frac{\sin(k'R)}{k'R} dR$$
$$= 4\pi v e^{i\mathbf{k}'\mathbf{r}} \frac{\sin(vk'(t-t'))}{k'}\theta(t-t'). \qquad (1.103)$$

The external integration in (1.101), after differentiation with respect to spatial coordinates x_m and time t and integration with respect to t' in (1.102), yields the final result,

$$\left\langle \mathbf{p}\left|\hat{\Gamma}_e\right|\mathbf{p}'\right\rangle_{ij} = -\frac{v^2 k_i k_j + p^2\delta_{ij}}{p^2 + v^2k^2}\Delta(\mathbf{p}-\mathbf{p}'), \qquad (1.104)$$

where $\Delta(\mathbf{p}-\mathbf{p}') = \frac{(2\pi)^3\delta(\mathbf{k}-\mathbf{k}')}{p-p'}$, and Re $(p - p') > 0$.

Impulse representations for the operators $\hat{\Gamma}_b$ and $\hat{\Gamma}_j$ follow from (1.65) analogously:

$$\left\langle \mathbf{p}\left|\hat{\Gamma}_b\right|\mathbf{p}'\right\rangle_{ij} = iv^2 e_{inj} k_n \frac{p}{p^2 + v^2k^2}\Delta(\mathbf{p}-\mathbf{p}'),$$
$$\left\langle \mathbf{p}\left|\hat{\Gamma}_j\right|\mathbf{p}'\right\rangle_{ij} = -v^2\frac{v^2 k_i k_j + p^2\delta_{ij}}{p(p^2 + v^2k^2)}\Delta(\mathbf{p}-\mathbf{p}') \qquad (1.105)$$

where e_{inj} are the elements of the perfectly anti-symmetric tensor. These elements are equal to zero if any two of the indices are equal to each other. If the indices form an even or odd transposition then e_{inj} are equal to +1, respectively.

(b) *Medium operators for typical media*

We consider here medium operators for an unbounded medium because taking into account the shape of an object demands a special approach in every case. The medium operators of the electric kind are described by the integral operator taking into account the causality principle:

$$\mathbf{P}_1(t,\mathbf{r}) = \hat{P}_1\mathbf{E}(t,\mathbf{r}), \quad \hat{P}_1 = \varepsilon_0 \int_0^t dt' \hat{\alpha}_{\text{eff}}(t-t'), \tag{1.106}$$

where $$\hat{\alpha}_{\text{eff}}(t,t') = \hat{\alpha}(t-t') + \frac{1}{\varepsilon_0}\theta(t-t')\int_0^{t'} dt'' \hat{\sigma}(t'-t''), \tag{1.107}$$

$\hat{\alpha}(t - t')$ is a susceptibility tensor of a dispersive dielectric medium and a conductivity tensor $\hat{\sigma}(t - t')$ describes the dissipation of the medium. The impulse representation of the operator is equal to

$$\left\langle \mathbf{p} \middle| \hat{P}_1 \middle| \mathbf{p}' \right\rangle_{ij} = \varepsilon_0 \alpha_{eff}(p)\delta_{ij}\Delta(\mathbf{p}-\mathbf{p}') \tag{1.108}$$

$$\alpha_{\text{eff}}(p) = \alpha(p) + \frac{1}{\varepsilon_0 p}\sigma(p),$$

$$\alpha(p) = \int_0^\infty \hat{\alpha}(t')e^{-pt'}dt',$$

$$\sigma(p) = \int_0^\infty \sigma(t')e^{-pt'}dt' \tag{1.109}$$

The formula (1.108), gives the impulse representation for the medium electric operator for the case when the background medium is described by the operator $\hat{P}_{\text{ex}} = \varepsilon_0(\varepsilon - 1)$ before zero moment:

$$\left\langle \mathbf{p} \middle| \hat{V}_e \middle| \mathbf{p}' \right\rangle = s(p)\Delta(\mathbf{p}-\mathbf{p}'), \tag{1.110}$$

where $$s(p) = \frac{1}{\varepsilon}\left[\alpha(p) + \frac{\sigma(p)}{\varepsilon_0 p} - \varepsilon + 1\right]. \tag{1.111}$$

For example, $s(p) = \frac{\varepsilon_1 - \varepsilon}{\varepsilon} + \frac{\sigma(p)}{\varepsilon_0 \varepsilon p}$ for a dissipative dielectric, and $s(p) = \frac{\omega_e^2}{p^2}$ for a cold isotropic plasma with electron plasma frequency equal to ω_e. For magnetized cold plasma placed in a static magnetic field directed along the z-axis

$$s(p) = \omega_e^2 s_B, \tag{1.112}$$

$$\text{where } s_B = \begin{pmatrix} \frac{1}{p^2+\omega_B^2} & \frac{-\omega_B}{p(p^2+\omega_B^2)} & 0 \\ \frac{\omega_B}{p(p^2+\omega_B^2)} & \frac{1}{p^2+\omega_B^2} & 0 \\ 0 & 0 & \frac{1}{p^2} \end{pmatrix}, \tag{1.113}$$

$\omega_B = \frac{eB_0}{m}$ is the electron gyrofrequency, and B_0 is the value of the magnetic field.

For magnetic media the magnetic polarization is represented as a sum of two terms, $\mathbf{M}_{rel} + \mathbf{M}_1$, because of $\mathbf{B} = -\int_{-\infty}^{t} curl\mathbf{E}dt$, i.e., the magnetic vector depends on the whole history of the electric field evolution. Here, $\mathbf{M}_{rel}$ is a relict component which is determined by the history of the electric field evolution until zero moment, namely, by the value $\mathbf{B}(0) = -\int_{-\infty}^{0} \text{curl}\mathbf{E}dt$.

The magnetic medium operator in the case of the background described by $\hat{M}_{ex} = \frac{1}{\mu_0}(1-\mu^{-1})$ is an integral-differential operator

$$\hat{M}_1 - \hat{M}_{ex} = \frac{1}{\mu_0\mu}\text{curl}\int_0^t dt'\left(\frac{\mu}{\mu_1(t-t')} - 1\right) \tag{1.114}$$

where μ_1 is the permeability after zero moment. Then the impulse representation (1.110) has the form

$$\left\langle \mathbf{p} \middle| \hat{V}_b \middle| \mathbf{p}' \right\rangle_{mn} = i\left(\mu(p) - \frac{1}{p}\right) e_{mjn} k_j \Delta(\mathbf{p} - \mathbf{p}'), \tag{1.115}$$

where $\mu(p) = \int_0^\infty \frac{\mu}{\mu_1(t)} e^{-pt} dt$ and μ_{mjn} are the elements of the perfectly anti-symmetric tensor as in (1.105).

The current medium operator for a simple conductor with a finite conductivity σ_1, for which the current operator is described by the constant

$$\hat{j}_1 = \sigma_1, \tag{1.116}$$

is equal to

$$\hat{V}_j = \chi\mu_0\mu\sigma_1, \tag{1.117}$$

or in impulse representation

$$\left\langle \mathbf{p}\left|\hat{V}_j\right|\mathbf{p}'\right\rangle_{ij} = \mu_0\mu\sigma_1\delta_{ij}\Delta(\mathbf{p}-\mathbf{p}'). \tag{1.118}$$

(c) *Kernels of the integral equations*

The products of the propagators and the medium operators yield, according to the formula (1.92), expressions for the complete integral equation kernels. The impulse representation of the kernel of the electric kind has the form

$$\left\langle \mathbf{p}\left|\hat{K}_e\right|\mathbf{p}'\right\rangle_{ij} = -s(p)\frac{v^2k_ik_j + p^2\delta_{ij}}{p^2+v^2k^2}\Delta(\mathbf{p}-\mathbf{p}'), \tag{1.119}$$

where $s(p)$ is given by (1.111) and $\mathrm{Re}(p - p') > 0$. Note that the kernel for the simple conductor can be obtained from the kernel for the dissipative dielectric by assuming $\varepsilon_1 = \varepsilon$.

The coordinate form of the kernel can be derived by the following integration:

$$\left\langle \mathbf{x}\left|\hat{K}_e\right|\mathbf{x}'\right\rangle_{ij}$$

$$= \int_{-i\infty}^{i\infty}\frac{dp}{2\pi i}e^{pt}\int_{\infty}\frac{d\mathbf{k}}{(2\pi)^3}e^{i\mathbf{kr}}s(p)\frac{v^2k_ik_j+p^2\delta_{ij}}{p^2+v^2k^2}\int_{-i\infty}^{i\infty}\frac{dp'}{2\pi i}e^{-p't'}\int_{\infty}\frac{d\mathbf{k}'}{(2\pi)^3}e^{-i\mathbf{k}'\mathbf{r}'}\frac{(2\pi)^3\delta(\mathbf{k}-\mathbf{k}')}{p-p'}$$

$$= \int_{-i\infty}^{i\infty}\frac{dp}{2\pi i}e^{p(t-t')}s(p)\left(-\frac{\partial^2}{\partial x_i\partial x_j}+\frac{1}{v^2}\frac{\partial^2}{\partial t^2}\delta_{ij}\right)\int_{\infty}\frac{d\mathbf{k}}{(2\pi)^3}\frac{e^{i\mathbf{k}(\mathbf{r}-\mathbf{r}')}}{p^2+v^2k^2}$$

The latter integral in this formula is equal to

$$\int_{\infty}\frac{d\mathbf{k}}{(2\pi)^3}\frac{e^{i\mathbf{k}(\mathbf{r}-\mathbf{r}')}}{p^2+v^2k^2} = 2\pi\int_0^{\infty}\frac{dk}{(2\pi)^3}\frac{k^2}{p^2+v^2k^2}\int_0^{\pi}d\theta\sin\theta e^{ik|\mathbf{r}-\mathbf{r}'|\cos\theta} = \frac{1}{4\pi v^2|\mathbf{r}-\mathbf{r}'|}e^{-\frac{p}{v}|\mathbf{r}-\mathbf{r}'|}.$$

So, finally we obtain

$$\left\langle \mathbf{x}\left|\hat{K}_e\right|\mathbf{x}'\right\rangle_{ij} = -\left(\frac{\partial^2}{\partial x_i \partial x_j} - \frac{1}{v^2}\frac{\partial^2}{\partial t^2}\delta_{ij}\right)\frac{1}{4\pi|\mathbf{r}-\mathbf{r}'|}\int_{-i\infty}^{i\infty}\frac{dp}{2\pi i}\xi(p)e^{p(t-t'-|\mathbf{r}-\mathbf{r}'|/v)},$$

(1.120)

where Re $p > 0$.

In the case of the cold isotropic plasma created in a medium with $\varepsilon = 1$, the coefficient s in (1.119) is equal to $s(p) = -\frac{\omega_e^2}{p^2}$. When the plasma is placed in an external static magnetic field directed along the z-axis, then the kernel is more complicated,

$$\left\langle \mathbf{p}\left|\hat{K}_{eB}\right|\mathbf{p}'\right\rangle_{ij} = \omega_e^2\frac{v^2k_ik_n + p^2\delta_{ln}}{p^2+v^2k^2}(\hat{s}_{eB})_{nj}\Delta(\mathbf{p}-\mathbf{p}'), \tag{1.121}$$

where the matrix elements $(\hat{s}_{eB})_{ij}$ are determined by (1.113).

For the kernel of the magnetic kind we have

$$\left\langle \mathbf{p}\left|\hat{K}_b\right|\mathbf{p}'\right\rangle_{ij} = v^2(p\mu(p)-1)\frac{k_ik_j - k^2\delta_{ij}}{p^2+v^2k^2}\Delta(\mathbf{p}-\mathbf{p}'), \tag{1.122}$$

This expression was obtained using the relation $\varepsilon_{ins}\varepsilon_{smj} = \delta_{in}\delta_{mj} - \delta_{nm} - \delta_{nm}\delta_{ij}$ [14].

The coordinate form of this kernel is

$$\left\langle \mathbf{x}\left|\hat{K}_b\right|\mathbf{x}'\right\rangle_{ij} = -\left(\frac{\partial^2}{\partial x_i \partial x_j} - \Delta\delta_{ij}\right)\frac{1}{4\pi|\mathbf{r}-\mathbf{r}'|}\int_{-i\infty}^{i\infty}\frac{dp}{2\pi i}(p\mu(p)-1)e^{p(t-t'-|\mathbf{r}-\mathbf{r}'|)}.$$

(1.123)

Note that the symbol $\frac{\partial^2}{\partial x_i \partial x_j} - \Delta\delta_{ij}$, where $\Delta = \nabla^2$, used is equivalent to graddiv – Δ = curlcurl in the vector form.

1.4.3 The Resolvent Method

(a) *The equation for the resolvent*

If the observation point is located inside the region $V(t)$ then Eq. (1.67) is the Volterra integral equation of the second kind. Therefore its solution can be found by means of a resolvent $\hat{R}$ as follows [22–24]:

$$\mathbf{E} = \mathbf{F} + \hat{R}\mathbf{F}, \tag{1.124}$$

where

$$\hat{R} = \int d\mathbf{x}' \langle \mathbf{x} | \hat{R} | \mathbf{x}' \rangle. \tag{1.125}$$

The resolvent obtained as a result of a successive approximation is represented by the convergent series

$$\hat{R} = \sum_{n=1}^{\infty} \hat{K}^n = (\hat{I} - \hat{K})^{-1} \hat{K} = \hat{K}(\hat{I} - \hat{K})^{-1}, \tag{1.126}$$

where $\hat{I}$ is an identity operator. From this formula the operator equations for the resolvent follow

$$\hat{R} - \hat{K}\hat{R} = \hat{K}, \tag{1.127}$$

$$\hat{R} - \hat{R}\hat{K} = \hat{K}. \tag{1.128}$$

We will use the first form of these equations further. Substituting the field which was found by the formula (1.124) into (1.67) and supposing that the observation point is beyond the region $V(t)$, an expression for the electric field outside this region is as follows:

$$\mathbf{E} = \mathbf{F}_{\text{ex}} + \hat{K}_{\text{ex}} \mathbf{E}_{in} = \mathbf{F}_{\text{ex}} + \hat{K}_{\text{ex}} (\mathbf{F}_{in} + \hat{R}\mathbf{F}_{in}) = \mathbf{F}_{\text{ex}} + \hat{N}\mathbf{F}_{in}, \tag{1.129}$$

where the indices "in" and "ex" designate the observation point belonging to the internal or external parts of the region $V(t)$, respectively. A reflection operator is introduced in (1.129) as

$$\hat{N} = \hat{K}_{ex}(\hat{I} + \hat{R}). \tag{1.130}$$

Thus, the problem of finding the electric field is reduced to the construction of the resolvent operator $\hat{R}$. One of the ways to do this is to solve Eq. (1.127). This equation for the matrix elements of the resolvent has the same form in different representations:
In the coordinate representation:

$$\langle \mathbf{x} | \hat{R} | \mathbf{x}' \rangle_{ij} = \langle \mathbf{x} | \hat{K} | \mathbf{x}' \rangle_{ij} + \int d\mathbf{x}'' \langle \mathbf{x} | \hat{K} | \mathbf{x}'' \rangle_{in} \langle \mathbf{x}'' | \hat{R} | \mathbf{x}' \rangle_{nj}; \tag{1.131}$$

In the impulse (Fourier–Laplace) representation:

$$\left\langle \mathbf{p}\middle|\hat{R}\middle|\mathbf{p}'\right\rangle_{ij} = \left\langle \mathbf{p}\middle|\hat{K}\middle|\mathbf{p}'\right\rangle_{ij} + \int d\mathbf{p}''\left\langle \mathbf{p}\middle|\hat{K}\middle|\mathbf{p}''\right\rangle_{in}\left\langle \mathbf{p}''\middle|\hat{R}\middle|\mathbf{p}'\right\rangle_{nj}. \qquad (1.132)$$

The matrix elements in these two representations are connected by relations (1.90) and (1.91).

(b) *The resolvents for an unbounded medium with a jump change in its properties*

If Eq. (1.132) for the resolvent has the kernel of the form (1.119), then the resolvent can be found as

$$\left\langle \mathbf{p}\middle|\hat{R}\middle|\mathbf{p}'\right\rangle_{ij} = A(p,k)k_ik_j + B(p,k)\delta_{ij}. \qquad (1.133)$$

Substitution of this expression and the kernel (1.119) into Eq. (1.132) yields

$$A(p,k)k_ik_j + B(p,k)\delta_{ij}$$
$$= s(p)\frac{v^2k_ik_j + p^2\delta_{ij}}{p^2+v^2k^2}\Delta(\mathbf{p}-\mathbf{p}') + \int_{\eta-i\infty}^{\eta+i\infty}\frac{dp''}{2\pi i}$$
$$s(p)\frac{\left[(v^2k^2+p^2)A(p'',k)+v^2B(p'',k)\right]k_ik_j + p^2B(p'',k)\delta_{ij}}{p^2+v^2k^2}\frac{1}{p-p''}$$
$$(1.134)$$

To find the functions A and B assume that they have the same analytical features as the kernel in Eq. (1.134). This assumption will be verified easily *a posteriori*, using the result obtained. It means that the integration line in (1.134) has to pass in parallel with the imaginary axis of p in such a way that the conditions $\mathrm{Re}p' > \mathrm{Re}p'' > \mathrm{Re}p'$ are fulfilled (see Fig. 1.6). Singular points of the integrand with respect to the variable p'' are on the imaginary axis and at the points p and p'. The integrating contour will be closed as shown in Fig. 1.6. Then only the pole of the function $\Delta(\mathbf{p} - \mathbf{p}')$ at the point p contributes to the integral.

Integration over the contour shown in Fig. 1.6 gives a simple system of equations with respect to A and B. Its solution gives the impulse representation for the resolvent

$$\left\langle \mathbf{p}\middle|\hat{R}\middle|\mathbf{p}'\right\rangle_{ij} = \frac{s(p)}{1-s(p)}\frac{v^2k_ik_j + [1-s(p)]p^2\delta_{ij}}{[1-s(p)]p^2+v^2k^2}\Delta(\mathbf{p}-\mathbf{p}'). \qquad (1.135)$$

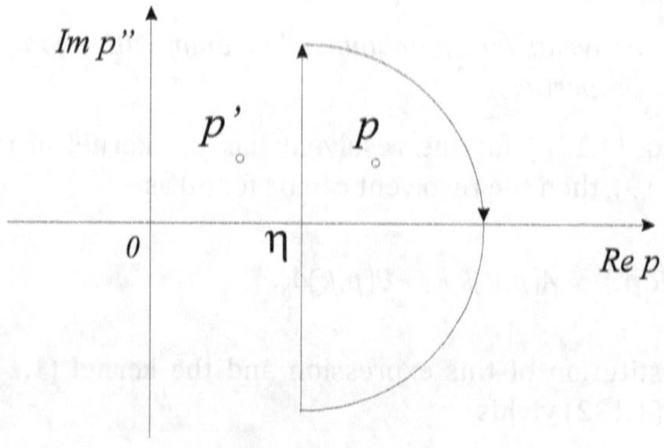

Figure 1.6 The integrating contour for Eq. (1.134).

The specific expressions for the resolvent can be derived from Eq. (1.135). The substitution $s(p)=\frac{\varepsilon-\varepsilon_1}{\varepsilon}-\frac{\overline{\sigma}_1}{\varepsilon p}$ leads to the impulse representation of the resolvent for a dissipative dielectric as follows:

$$\left\langle \mathbf{p}\middle|\hat{R}_{ej}\middle|\mathbf{p}'\right\rangle_{ij} = \frac{p(\varepsilon-\varepsilon_1)-\overline{\sigma}_1}{\varepsilon_1 p+\overline{\sigma}_1}\frac{v_1^2k_ik_j + p\left(p+\dfrac{\overline{\sigma}_1}{\varepsilon_1}\right)\delta_{ij}}{p^2+p\dfrac{\overline{\sigma}_1}{\varepsilon_1}+v_1^2k^2}\Delta(\mathbf{p}-\mathbf{p}'), \quad (1.136)$$

where $v_1^2=\frac{c^2}{\varepsilon_1}=\frac{1}{\varepsilon_0\mu_0\varepsilon_1}$ and $\overline{\sigma}_1=\sigma_1/\varepsilon_0$. The coordinate representation can then be found by the inverse Fourier–Laplace transformation,

$$\left\langle \mathbf{x}\middle|\hat{R}_j\middle|\mathbf{x}'\right\rangle_{ij} = \frac{1}{4\pi}\left\{\frac{\partial^2}{\partial x_i\partial x_j}-\frac{1}{v_1^{\,2}}\delta_{ij}\left(\frac{\partial^2}{\partial t^2}+\frac{\overline{\sigma}_1}{\varepsilon_1}\frac{\partial}{\partial t}\right)\right\}\Phi(\tau,R), \qquad (1.137)$$

where $t = t - t'$, $R = |\mathbf{r}-\mathbf{r}'|$ and

$$\Phi(\tau,R) = 4\pi v_1^2 \int d\mathbf{p} \frac{(\varepsilon_1 - \varepsilon)p + \bar{\sigma}_1}{\varepsilon_1 p + \bar{\sigma}_1} \frac{e^{p\tau + i\mathbf{kR}}}{p^2 + p\frac{\bar{\sigma}_1}{\varepsilon_1} + v_1^2 k^2}$$

$$= \frac{1}{R} \int_{\eta - i\infty}^{\eta + i\infty} \frac{dp}{2\pi i} \frac{(\varepsilon_1 - \varepsilon)p + \bar{\sigma}_1}{\varepsilon_1 p + \bar{\sigma}_1} e^{p\tau - \frac{R}{v_1}\sqrt{p^2 + \frac{\bar{\sigma}_1}{\varepsilon_1}p}}. \qquad (1.138)$$

The integral in (1.138) is calculated along the line for which $\eta > \frac{\bar{\sigma}_1}{\varepsilon_1}$. One can easily make certain that the function $\Phi(t,\ \mathbf{r})$ satisfies the equality

$$\left(\Delta - \frac{1}{v_1^2}\frac{\partial}{\partial t^2} - \frac{\bar{\sigma}_1}{v_1^2}\frac{\partial}{\partial t}\right)\Phi(\tau,\mathbf{R}) = 4\pi\left[\frac{\varepsilon - \varepsilon_1}{\varepsilon_1}\delta(\tau) - \frac{\varepsilon\bar{\sigma}_1}{\varepsilon_1^2} e^{-\frac{\bar{\sigma}_1}{\varepsilon_1}\tau}\theta(\tau)\right]\delta(\mathbf{R}).$$

$$(1.139)$$

Equations (1.136) and (1.138) yield the following specific expressions in special cases.

An ideal dielectric corresponds to $\bar{\sigma}_1 = 0$. Then

$$\left\langle \mathbf{p}\middle|\hat{R}_e\middle|\mathbf{p}'\right\rangle_{ij} = -\frac{\varepsilon_1 - \varepsilon}{\varepsilon_1}\frac{v_1^2 k_i k_j + p^2\delta_{ij}}{p^2 + v_1^2 k^2}\Delta(\mathbf{p} - \mathbf{p}'), \qquad (1.140)$$

where $v_1^2 = \frac{c^2}{\varepsilon_1} = \frac{1}{\varepsilon_0\mu_0\varepsilon_1}$.

The kernel of the resolvent in the coordinate representation is

$$\left\langle \mathbf{x}\middle|\hat{R}_e\middle|\mathbf{x}'\right\rangle_{ij} = \frac{\varepsilon_1 - \varepsilon}{4\pi\varepsilon_1}\left(\frac{\partial^2}{\partial x_i \partial x_j} - \frac{1}{v_1^2}\delta_{ij}\frac{\partial^2}{\partial t^2}\right)\frac{\delta(t - t' - |\mathbf{r} - \mathbf{r}'|/v_1)}{|\mathbf{r} - \mathbf{r}'|}. \qquad (1.141)$$

A perfect conductor ($\varepsilon = \varepsilon_1$) has such a resolvent

$$\left\langle \mathbf{p}\middle|\hat{R}_j\middle|\mathbf{p}'\right\rangle_{ij} = -\bar{\sigma}_1 \frac{v^2 k_i k_j + p\left(p + \frac{\bar{\sigma}_1}{\varepsilon}\right)\delta_{ij}}{(\varepsilon p + \bar{\sigma}_1)\left(p^2 + p\frac{\bar{\sigma}_1}{\varepsilon} + v^2 k^2\right)}\Delta(\mathbf{p} - \mathbf{p}'). \qquad (1.142)$$

In the coordinate representation it has the form

$$\left\langle \mathbf{x}\left|\hat{R}_j\right|\mathbf{x}'\right\rangle_{ij} = \frac{1}{4\pi}\left\{\frac{\partial^2}{\partial x_i \partial x_j} - \frac{1}{v^2}\delta_{ij}\left(\frac{\partial^2}{\partial t^2} + \frac{\overline{\sigma}_1}{\varepsilon}\frac{\partial}{\partial t}\right)\right\}\tilde{\Phi}(\tau,R), \qquad (1.143)$$

where $$\tilde{\Phi}(\tau,R) = \frac{\overline{\sigma}_1}{\varepsilon R} e^{-\frac{\overline{\sigma}_1 \tau}{2\varepsilon}} \int_{\zeta - i\infty}^{\zeta + i\infty} \frac{dq}{2\pi i} \frac{e^{q\tau}}{q + \overline{\sigma}_1/(2\varepsilon)} e^{-\frac{R}{v}\sqrt{q^2 - \overline{\sigma}_1{}^2/(4\varepsilon^2)}}, \zeta > \frac{\overline{\sigma}_1}{2\varepsilon}, \qquad (1.144)$$

or

$$\tilde{\Phi}(\tau,R) = \frac{\overline{\sigma}_1}{R\varepsilon v} e^{-\frac{\overline{\sigma}_1 \tau}{2\varepsilon}} \left(\frac{\partial}{\partial \tau} - \frac{\overline{\sigma}_1}{2\varepsilon}\right)\left[\theta(v\tau - R)\int_R^{v\tau} I_0\left(\frac{\overline{\sigma}_1}{2\varepsilon}\sqrt{\tau^2 - \frac{R^2}{v^2}}\right) dR\right], \qquad (1.145)$$

where I_0 is the modified Bessel function.

In the case of a plasma medium $s(p) = -\frac{\omega_e^2}{p^2}$ the resolvent follows from Eq. (1.135)

$$\left\langle \mathbf{p}\left|\hat{R}_{ep}\right|\mathbf{p}'\right\rangle_{ij} = -\frac{\omega_e^2}{p^2 + c^2k^2 + \omega_e{}^2}\left(c^2\frac{k_i k_j}{p^2 + \omega_e^2} + \delta_{ij}\right)\Delta(\mathbf{p} - \mathbf{p}'). \qquad (1.146)$$

The coordinate representation can be written in one of the following two forms:

$$\left\langle \mathbf{x}\left|\hat{R}_{ep}\right|\mathbf{x}'\right\rangle_{ij} = \frac{\omega_e{}^2}{4\pi}\left\{\frac{\partial^2}{\partial x_i \partial x_j} - \frac{1}{c^2}\delta_{ij}\left(\frac{\partial^2}{\partial t^2} + \omega_e{}^2\right)\right\}\Psi(\tau,R), \qquad (1.147)$$

where

$$\Psi(\tau, R) = \frac{1}{R}\int_{\alpha - i\infty}^{\alpha + i\infty} \frac{dp}{2\pi i} \frac{\exp\left(p\tau - \frac{R}{c}\sqrt{p^2 + \omega_e^2}\right)}{p^2 + \omega_e^2}$$

$$= \frac{1}{R}\int_0^{\tau} J_0(\omega_e\sqrt{\tau^2 - u^2})\theta\left(\tau - \frac{R}{c}\right) du, \qquad (1.148)$$

or

$$\langle \mathbf{x}|\hat{R}_{ep}|\mathbf{x}'\rangle_{ij} = \frac{\omega_e^2}{4\pi c^2}\left\{\frac{\partial^2}{\partial x_i \partial x_j}\frac{c^2}{R}\int_0^{\tau} J_0(\omega_e\sqrt{\tau^2 - u^2})\theta\left(\tau - \frac{R}{c}\right)du\right.$$

$$\left.-\delta_{ij}\left(\frac{\delta\left(\tau - \frac{R}{c}\right)}{R} - \frac{\omega_e^2}{\sqrt{c^2\tau^2 - R^2}}J_1\left(\omega_e\sqrt{\tau^2 - \left(\frac{R}{c}\right)^2}\right)\theta\left(\tau - \frac{R}{c}\right)\right)\right\},$$

(1.149)

where $J_{0,1}$ are Bessel functions.

More complex relations are found in the case of an anisotropic medium (magnetized plasma). Considering the case when the external magnetic field is directed along the z-axis, the kernel of the integral equation is described by the expression (1.121) which can be rewritten in the form

$$\langle \mathbf{p}|\hat{K}_{eB}|\mathbf{p}'\rangle_{ij} = \frac{\omega_e^2}{p^2 + v^2k^2}M_{ij}(\mathbf{p})\Delta(\mathbf{p} - \mathbf{p}'), \tag{1.150}$$

where $M_{ij}(\mathbf{p}) = (c^2k_ik_n + p^2\delta_{in})(s_{eB})_{nj}$ and the matrix $(s_{eB})_{nj}$ is determined by (1.113). Substituting it into Eq. (1.132), under the assumption that the resolvent has the same analytical properties as the kernel, yields the matrix equation

$$\langle \mathbf{p}|\hat{R}_{eB}|\mathbf{p}'\rangle_{ij} = -\frac{\omega_e^2}{p^2 + c^2k^2}\left(M_{in}(\mathbf{p})\langle \mathbf{p}|\hat{R}_{eB}|\mathbf{p}'\rangle_{nj} + M_{ij}(\mathbf{p})\Delta(\mathbf{p} - \mathbf{p}')\right),$$

(1.151)

the solution to which gives the resolvent

$$\langle \mathbf{p}|\hat{R}_{eB}|\mathbf{p}'\rangle = \frac{-\omega_e^2c^2}{G(p,\mathbf{k})}\varphi_e^2\left(p^2T_1 + p\Omega T_2 + \frac{p^2}{c^2}T_3 + \frac{p^2}{c^2}(p^2 + \omega_e^2)\hat{I}\right)\Delta(\mathbf{p} - \mathbf{p}').$$

(1.152)

Here $\varphi_e^2 = p^2 + c^2k^2 + \omega_e^2$, and $\hat{I}$ is the identity matrix. The T_j in Eq. (1.152) matrices are equal to

$$T_1 = \begin{pmatrix} k_1^2 & k_1k_2 & \left(1+\dfrac{\Omega^2\varphi^2}{p^2\varphi_e^2}\right)k_1k_3 \\ k_2k_1 & k_2^2 & \left(1+\dfrac{\Omega^2\varphi^2}{p^2\varphi_e^2}\right)k_2k_3 \\ k_3k_1 & k_3k_2 & \left(1+\dfrac{\Omega^2\varphi^2}{p^2\varphi_e^2}\right)k_3^2 \end{pmatrix},$$

$$T_2 = \begin{pmatrix} -k_1k_2 & k_2^2+\dfrac{\omega_e^2}{\varphi_e^2}k_3^2 & -\dfrac{\omega_e^2}{\varphi_e^2}k_2k_3 \\ -k_2^2-\dfrac{\omega_e^2}{\varphi_e^2}k_3^2 & k_2^2 & \dfrac{\omega_e^2}{\varphi_e^2}k_1k_3 \\ -\dfrac{\varphi^2}{\varphi_e^2}k_3k_2 & \dfrac{\varphi^2}{\varphi_e^2}k_3k_1 & 0 \end{pmatrix},$$

$$\text{and } T_3 = \begin{pmatrix} 0 & p\Omega & 0 \\ -p\Omega & 0 & 0 \\ 0 & 0 & \Omega^2\varphi^2/\varphi_e^2 \end{pmatrix}, \qquad (1.153)$$

where $\varphi^2 = p^2 + c^2k^2$.

The gyromagnetic polynomial in (1.152) has the following representation:

$$G(p,\mathbf{k}) = p^2\varphi_e^4(p^2+\omega_e^2)+\Omega^2\varphi^2(p^2\varphi_e^4+\omega_e^2c^2k_3^2). \qquad (1.154)$$

A magnetic kernel (1.122) for the case of constant permeability μ_1 yields the impulse representation for the resolvent

$$\langle\mathbf{p}|\hat{R}_b|\mathbf{p}'\rangle_{ij} = -v_1^2\frac{\mu_1-\mu}{\mu}\frac{k_ik_j-k^2\delta_{ij}}{p^2+v_1^2k^2}\Delta(\mathbf{p}-\mathbf{p}'), \qquad (1.155)$$

where $v_1^2 = \frac{c^2}{\mu_1} = \frac{1}{\varepsilon_0\mu_0\mu_1}$. In the coordinate representation, it takes the form

$$\langle\mathbf{x}|\hat{R}_m|\mathbf{x}'\rangle_{ij} = \frac{\mu_1-\mu}{4\pi\mu}\left(\frac{\partial^2}{\partial x_i\partial x_j}-\delta_{ij}\frac{\partial^2}{\partial x_n^2}\right)\frac{\delta(t-t'-|\mathbf{r}-\mathbf{r}'|/v_1)}{|\mathbf{r}-\mathbf{r}'|} \qquad (1.156)$$

The expressions obtained show the validation of the assumption mentioned above on the distribution of the singularities of the impulse representations of the resolvent.

The resolvents constructed make it possible to obtain analytically the electromagnetic field transformation caused by the time jump of the medium parameters.

References

1. Sobolev, S. L. (1950) Applications of functional analysis in mathematical physics, Leningrad [English translation: *Amer. Math. Soc. Transl., Math. Mono*, 7 (1963)].
2. Adams, R. A. (1975) *Sobolev Spaces*, Academic Press, New York, 268 p.
3. Schwartz, L. (1966) *Theorie Des Distributions*, Herman, Paris.
4. Gelfand, I. M., and Shilov, G. E. (1964) *Generalized Functions*, vol. 1, Academic Press, New York, 423 p.
5. Vladimirov, V. S. (1967) *Equations of Mathematical Physics* (in Russian), Nauka, Moscow, 437 p.
6. Kecs, W., and Teodorescu, P. P. (1975) *Introducere in Teoria Distributilor cu Aplicatii in Tehnica*, Editura Tehnica, Bucuresti, 520 p.
7. Nikolskiy, V. V., and Nikolskaya, T. N. (1989) *Electrodynamics and Propagation of Radiowaves* (in Russian), Nauka, Moscow, 544 p.
8. Kong, J. (1990) *Electromagnetic Field Theory*, John Wiley & Sons, New York, 704 p.
9. Fedorchenko, A. M. (1988) *Theoretical Physics. Classical Electrodynamics* (in Russian), Vyscha shkola, Kiev, 280 p.
10. Sedov, L. I. (1973) *Mechanics of Solid Medium*, vol. 1 (in Russian), Nauka, Moscow, 536 p.
11. Born, M., and Wolf, E. (1964) *Principles of Optics*, Pergamon Press, Oxford, 856 p.
12. Bolotovskiy B. M., and Stolyarov, S. N. (1974) Modern state of electrodynamics of moving media, *Uspekhi Fizicheskikh Nauk*, **114**(4), 489–529.
13. Khizhnyak, N. A. (1958) Green's function of Maxwell's equations for inhomogeneous media, Zhurnal tekhnicheskoy fiziki (in Russian), *Soviet Physics – Technical Physics*, **28**(7), 1592–1609.
14. Ginzburg, V. L., and Rukhadze, A. A. (1975) *Waves in Magnetoactive Plasma* (in Russian), Nauka, Moscow, 256 p.

15. Nerukh A. G., and Khizhnyak, N. A. Integral Maxwell's equations in problems of wave scattering on moving media, Zhurnal tekhnicheskoy fiziki (in Russian), *Soviet Physics – Technical Physics*, 1973, **43**(6), 1113–1120.

16. Bateman, H., and Erdely, A. (1954) *Tables of Integral Transforms*, McGraw-Hill, New York, 344 p.

17. Mitra, R., and Lee, S. W. (1971) *Analytic Techniques in the Theory of Guided Waves*, The Macmillan Company, New York, 324 p.

18. Balescu, R. (1963) *Statistical Mechanics of Charged Particles*, Interscience Publishers, London, 516 p.

19. Davydov, A. S. (1973) *Quantum Mechanics* (in Russian), Nauka, Moscow, 704 p.

20. Korn, G., and Korn, T. (1961) *Mathematical Handbook for Scientists and Engineers*, McGraw-Hill, New York, 720 p.

21. Zabreiko, P. P., Koshelev, A. I., Krasnoselskiy, M. A., et al. (1968) *Integral Equations* (in Russian), Nauka, Moscow, 448 p.

22. Muntz, G. (1934) *Integral Equations. Part I: Linear Volterra Equations* (in Russian), GTTI, Leningrad, Moscow, 330 p.

23. Mikhlin, S. G. (1959) *Lectures on Linear Integral Equations* (in Russian), Moscow, Fizmatgiz, 232 p.

24. Volterra, V. (1930) *Theory of Functionals and Integral and Integro-Differential Equations* (in Russian), London, 226 p.

Chapter 2

Transformation of an Electromagnetic Field in an Unbounded Medium with Time-Varying Parameters

The first stage of a non-stationary electromagnetic problem is realized in some parametric phenomena in active media, for example in a time modulation of medium parameters. If these parameters change slowly in comparison with an electromagnetic field then investigations can be made via adiabatic approximations without transients that yield a slow change of the results. These results can be predicted relatively easily, but they are of little interest. Transients become apparent in the other extreme when the medium parameters change rapidly with time. The main features of the field transformation in this case can be revealed by considering a simple law for the time variation of medium parameters because exact solutions of the problem can be constructed.

One possible way to temporally adjust medium parameters, so as to control an electromagnetic field, is by varying the medium parameters in a semiconductor material by an extrinsic electrical field or in a magnetic by an extrinsic magnetic field. Using a finite number of extraneous actions, a complex transformation

Non-Stationary Electromagnetics: An Integral Equations Approach (Second Edition)
Alexander Nerukh and Trevor Benson

ISBN 978-981-4774-95-6 (Hardcover), 978-0-429-02773-4 (eBook)
www.panstanford.com

of an electromagnetic wave transformation can be studying by rather simple methods. In this chapter, the variation of medium parameters by a finite packet of periodically repeated actions is considered. An exact solution for the field transformation is derived and new features arising from the appearance of dynamical chaotic elements are revealed. The influence of an extrinsic time varying magnetic field on electromagnetic wave structure is shown.

2.1 Transformation of an Electromagnetic Wave in a Non-Dispersive Time-Varying Medium

Pure effects in the transformation of an electromagnetic field by a temporal medium change can be revealed in a model of a non-dispersive medium with sharp changes of its parameters. To avoid redundant sophistication we assume that the medium is infinite in extent and that changes in parameters are jump-like. Such a situation is idealized but it admits a simple solution of a problem that allows one to obtain all basic results. Without loss of generality, we assume also that the background medium has a relative permittivity ε and no conductivity until the zero moment of time. After that moment, the medium goes simultaneously to a new state which is characterized by a permittivity e_1 and a conductivity σ_1. According to Eq. 1.67, the field after this zero moment of time satisfies the equation

$$\mathbf{E} = \mathbf{F} + \hat{K}\mathbf{E}, \tag{2.1}$$

where $\hat{K}$ is the integral operator (1.68) and $\mathbf{F}$ is the free term that takes into account the previous state of the electromagnetic field and in the case considered is equal to

$$\mathbf{E}_0 = \mathbf{E}_0' - G_0 * \mu_0\mu \frac{\partial \mathbf{j}_{\text{extr}}}{\partial t}. \tag{2.2}$$

Here, $\mathbf{E}_0'$ is the field that existed in the background without sources (this field satisfies the homogeneous wave equation in the background), and $\mathbf{j}_{\text{extr}}$ is the current of an extrinsic source. A solution to Eq. (2.1), made by the resolvent method (Section 1.4.3) is given by virtue of the resolvent operator

$$\mathbf{E} = \mathbf{F} + \int d\mathbf{x}' \left\langle \mathbf{x} \middle| \hat{R}_e \middle| \mathbf{x}' \right\rangle \langle \mathbf{x}' | \mathbf{F} \rangle . \tag{2.3}$$

2.1.1 The Resolvent for an Unbounded Medium with a Jump Change in Its Properties

Let us find the resolvent satisfying the Eq. (1.132) with the kernel (1.119). The resolvent can be found in the form

$$\left\langle \mathbf{p} \middle| \hat{R} \middle| \mathbf{p}' \right\rangle_{ij} = A(p,k)k_i k_j + B(p,k)\delta_{ij}. \tag{2.4}$$

Substitution of this expression and the kernel (1.119) into the Eq. (1.132) yields

$$\begin{aligned} &A(p,k)k_i k_j + B(p,k)\delta_{ij} \\ &\quad = s(p)\frac{v^2 k_i k_j + p^2\delta_{ij}}{p^2 + v^2 k^2}\Delta(\mathbf{p}-\mathbf{p}') + \int_{\eta - i\infty}^{\eta + i\infty} \frac{dp''}{2\pi i} s(p) \\ &\qquad \frac{\left[(v^2k^2 + p^2)A(p'',k) + v^2 B(p'',k)\right]k_i k_j + p^2 B(p'',k)\delta_{ij}}{p^2 + v^2k^2}\frac{1}{p-p''}. \end{aligned} \tag{2.5}$$

To find the functions A and B assume that they have the same analytical features as the kernel in (1.134). This assumption will be verified easily *a posteriori*, using the result obtained. It means that the integration line in (1.134) has to pass in parallel with the imaginary axis of p in such a way that the conditions $\mathrm{Re}\,p > \mathrm{Re}\,p'' > \mathrm{Re}\,p'$ are fulfilled, see Fig. 1.6. Singular points of the integrand with respect to the variable p'' are on the imaginary axis and at the points p and p''. The integrating contour will be closed as shown in Fig. 1.6. Then only the pole of the function $\Delta(\mathbf{p}-\mathbf{p}')$ at the point p contributes into the integral.

Integration over the contour shown in Fig. 1.6 gives a simple system of equations with respect to A and B. Its solution gives the impulse representation for the resolvent

$$\left\langle \mathbf{p} \middle| \hat{R} \middle| \mathbf{p}' \right\rangle_{ij} = \frac{s(p)}{1-s(p)}\frac{v^2 k_i k_j + [1-s(p)]p^2\delta_{ij}}{[1-s(p)]p^2 + v^2k^2}\Delta(\mathbf{p}-\mathbf{p}'). \tag{2.6}$$

The specific expressions for the resolvent can be derived from Eq. (2.6). The substitution $s(p) = \frac{\varepsilon - \varepsilon_1}{\varepsilon} - \frac{\sigma_1}{\varepsilon p}$ leads to the impulse representation of the resolvent for a dissipative dielectric as follows:

$$\left\langle \mathbf{p}\left|\hat{R}_{ej}\right|\mathbf{p}'\right\rangle_{ij} = \frac{p(\varepsilon-\varepsilon_1)-\bar{\sigma}_1}{\varepsilon_1 p+\bar{\sigma}_1}\frac{v_1^2 k_i k_j + p\left(p+\frac{\bar{\sigma}_1}{\varepsilon_1}\right)\delta_{ij}}{p^2+p\frac{\bar{\sigma}_1}{\varepsilon_1}+v_1^2 k^2}\Delta(\mathbf{p}-\mathbf{p}'), \quad (2.7)$$

where $v_1^2 = \frac{c^2}{\varepsilon_1} = \frac{1}{\varepsilon_0\mu_0\varepsilon_1}$ and $\bar{\sigma}_1 = \sigma_1/\varepsilon_0$. The coordinate representation can then be found by the inverse Fourier–Laplace transformation,

$$\left\langle \mathbf{x}\left|\hat{R}_j\right|\mathbf{x}'\right\rangle_{ij} = \frac{1}{4\pi}\left\{\frac{\partial^2}{\partial x_i \partial x_j} - \frac{1}{v_1^2}\delta_{ij}\left(\frac{\partial^2}{\partial t^2}+\frac{\bar{\sigma}_1}{\varepsilon_1}\frac{\partial}{\partial t}\right)\right\}\Phi(\tau,R), \quad (2.8)$$

where $\tau = t - t'$, $R = |\mathbf{r} - \mathbf{r}'|$ and

$$\Phi(\tau,R) = \frac{1}{R}\int_{\eta-i\infty}^{\eta+i\infty}\frac{dp}{2\pi i}\frac{(\varepsilon_1-\varepsilon)p+\bar{\sigma}_1}{\varepsilon_1 p+\bar{\sigma}_1}e^{p\tau-\frac{R}{v_1}\sqrt{p^2+\frac{\bar{\sigma}_1}{\varepsilon_1}p}}. \quad (2.9)$$

The integral in (2.9) is calculated along the line for which $\eta > \frac{\bar{\sigma}_1}{\varepsilon_1}$. One can easily make certain that the function $\Phi(t, \mathbf{r})$ satisfies the equality

$$\left(\Delta - \frac{1}{v_1^2}\frac{\partial}{\partial t^2} - \frac{\bar{\sigma}_1}{v_1{}^2}\frac{\partial}{\partial t}\right)\Phi(\tau,\mathbf{R}) = 4\pi\left[\frac{\varepsilon-\varepsilon_1}{\varepsilon_1}\delta(\tau) - \frac{\varepsilon\bar{\sigma}_1}{\varepsilon_1{}^2}e^{-\frac{\bar{\sigma}_1}{\varepsilon_1}\tau}\theta(\tau)\right]\delta(\mathbf{R}). \quad (2.10)$$

Equations (2.7) and (2.8) yield the following specific expressions in special cases.

An ideal dielectric corresponds to $\bar{\sigma}_1 = 0$. Then

$$\left\langle \mathbf{p}\left|\hat{R}_e\right|\mathbf{p}'\right\rangle_{ij} = -\frac{\varepsilon_1-\varepsilon}{\varepsilon_1}\frac{v_1^2 k_i k_j + p^2\delta_{ij}}{p^2+v_1^2 k^2}\Delta(\mathbf{p}-\mathbf{p}'), \quad (2.11)$$

where $v_1^2 = \frac{c^2}{\varepsilon_1} = \frac{1}{\varepsilon_0\mu_0\varepsilon_1}$.

The kernel of the resolvent in the coordinate representation is

$$\left\langle \mathbf{x}\left|\hat{R}_e\right|\mathbf{x}'\right\rangle_{ij} = \frac{\varepsilon_1-\varepsilon}{4\pi\varepsilon_1}\left(\frac{\partial^2}{\partial x_i \partial x_j} - \frac{1}{v_1^2}\delta_{ij}\frac{\partial^2}{\partial t^2}\right)\frac{\delta(t-t'-|\mathbf{r}-\mathbf{r}'|/v_1)}{|\mathbf{r}-\mathbf{r}'|}. \quad (2.12)$$

A perfect conductor ($\varepsilon = \varepsilon_1$) has such a resolvent

$$\left\langle \mathbf{p}\left|\hat{R}_j\right|\mathbf{p}'\right\rangle_{ij} = -\bar{\sigma}_1 \frac{v^2 k_i k_j + p\left(p+\dfrac{\bar{\sigma}_1}{\varepsilon}\right)\delta_{ij}}{(\varepsilon p+\bar{\sigma}_1)\left(p^2+p\dfrac{\bar{\sigma}_1}{\varepsilon}+v^2k^2\right)}\Delta(\mathbf{p}-\mathbf{p}'). \quad (2.13)$$

In the coordinate representation, it has the form

$$\left\langle \mathbf{x}\left|\hat{R}_j\right|\mathbf{x}'\right\rangle_{ij} = \frac{1}{4\pi}\left\{\frac{\partial^2}{\partial x_i \partial x_j} - \frac{1}{v^2}\delta_{ij}\left(\frac{\partial^2}{\partial t^2}+\frac{\bar{\sigma}_1}{\varepsilon}\frac{\partial}{\partial t}\right)\right\}\tilde{\Phi}(\tau,R). \quad (2.14)$$

Here,

$$\begin{aligned}\tilde{\Phi}(\tau,R) &= \frac{\bar{\sigma}_1}{\varepsilon R}e^{-\frac{\bar{\sigma}_1\tau}{2\varepsilon}}\int_{\zeta-i\infty}^{\zeta+i\infty}\frac{dq}{2\pi i}\frac{e^{q\tau}}{q+\bar{\sigma}_1/(2\varepsilon)}e^{-\frac{R}{v}\sqrt{q^2-\bar{\sigma}_1^2/(4\varepsilon^2)}} \\ &= \frac{\bar{\sigma}_1}{R\varepsilon v}e^{-\frac{\bar{\sigma}_1\tau}{2\varepsilon}}\left(\frac{\partial}{\partial\tau}-\frac{\bar{\sigma}_1}{2\varepsilon}\right)\left[\theta(v\tau-R)\int_R^{v\tau} I_0\left(\frac{\bar{\sigma}_1}{2\varepsilon}\sqrt{\tau^2-\frac{R^2}{v^2}}\right)dR\right],\end{aligned} \quad (2.15)$$

where $\zeta > \frac{\bar{\sigma}_1}{2\varepsilon}$ and I_0 is the modified Bessel function.

2.1.2 Splitting of a Plane Harmonic Wave by a Time Jump in Medium Parameters

First, we consider the simplest situation where an initial electromagnetic plane harmonic wave $\mathbf{E}_0(t, \mathbf{r}) = \mathbf{E}_0 \exp[i(\omega t - \mathbf{k}_0\mathbf{r})]$ exists in an unbounded homogeneous non-magnetic non-dispersive medium with the relative permittivity ε. The wave vector $\mathbf{k}_0$ has the magnitude $k_0 = \omega/v$, where $v = c/\sqrt{\varepsilon}$ is a wave phase velocity. This field corresponds to the term $\mathbf{E}_0'$ in the formula (2.2). Let us consider transformation of this wave when the medium goes to a new state characterized by the complex permittivity $\xi(p) = \varepsilon_1 + \varepsilon\sigma_1/(\varepsilon_0 p)$. The problem can be simplified without loss of generality if we assume that the initial field depends on one coordinate x only and the direction of the vector $\mathbf{E}_0$ is perpendicular to this axis. In this case, the resolvent operator is reduced to a scalar one after integrating with respect to the lateral coordinates

$$\int_{\infty} d\mathbf{r}'_{\perp} \frac{\delta\left(t-t'-\frac{1}{v}\sqrt{(x-x')^2+(\mathbf{r}_{\perp}-\mathbf{r}'_{\perp})^2}\right)}{\sqrt{(x-x')^2+(\mathbf{r}_{\perp}-\mathbf{r}'_{\perp})^2}} = 2\pi v\theta\left(t-t'-\frac{|x-x'|}{v}\right) \quad (2.16)$$

and it takes the form

$$\left\langle \mathbf{x}\left|\hat{R}_e\right|\mathbf{x}'\right\rangle = \int_{\eta-i\infty}^{\eta+i\infty} \frac{dp}{2\pi i} \int_{-\infty}^{\infty} \frac{dk}{2\pi} \frac{(\varepsilon-\xi(p))p^2}{p^2\xi(p)+\varepsilon v^2 k^2} e^{p(t-t')+ik(x-x')}, \quad (2.17)$$

which gives the transformed field

$$\begin{aligned} \mathbf{E}_1(t,x) &= \mathbf{E}_0 e^{i(\omega t-k_0 x)} + \mathbf{E}_0 \int_0^{\infty} dt' \int_{-\infty}^{\infty} dx' \left\langle \mathbf{x}\left|\hat{R}_e\right|\mathbf{x}'\right\rangle e^{i(\omega t'-k_0 x')} \\ &= \mathbf{E}_0 e^{i(\omega t-k_0 x)} + \mathbf{E}_0 \int_{\eta-i\infty}^{\eta+i\infty} \frac{dp}{2\pi i} \frac{\varepsilon-\xi(p)}{p-i\omega} \frac{p^2 e^{pt-ik_0 x}}{p^2\xi(p)+\varepsilon v^2 k_0^2} \end{aligned} \quad (2.18)$$

To calculate this integral let us specify the model of the medium assuming that ε_1 and σ_1 do not depend on the variable p, that is, $\xi(p) = \varepsilon_1 + \sigma_1/(\varepsilon_0 p)$. In this case, the resolvent after integration over k is equal to

$$\begin{aligned} \left\langle \mathbf{x}\,|\,\hat{R}_e\,|\,\mathbf{x}'\right\rangle &= R(t-t', x-x') \\ &= \frac{1}{2v_1} \int_{\eta'-i\infty}^{\eta'+i\infty} \frac{dq}{2\pi i} \frac{(q-b/2)[a^2(q-b/2)-(q+b/2)]}{\sqrt{q^2-(b/2)^2}} e^{(q-b/2)(t-t')-\frac{|x-x'|}{v_1}\sqrt{q^2-(b/2)^2}}, \end{aligned}$$

(2.19)

Here, the contracted designations $a = \sqrt{\varepsilon/\varepsilon_1}$ and $b = \sigma_1/(\varepsilon_0\varepsilon_1)$ are introduced to simplify the expressions. The square root in

Eq. (2.19) is taken such that $\mathrm{Re}\sqrt{q^2 - b^2/4} > 0$ and the integral does not depend on η, if $\eta > b$.

The expression for the field (2.18) takes the form

$$\mathbf{E}_1(t,x) = \mathbf{E}_0 e^{i(\omega t - k_0 x)} - \mathbf{E}_0 e^{-ik_0 x}\left(\frac{\partial^2}{\partial t^2} + \frac{\sigma_1}{\varepsilon_0\varepsilon_1}\frac{\partial}{\partial t}\right)\int_{\eta-i\infty}^{\eta+i\infty}\frac{dp}{2\pi i}$$

$$= \frac{(\varepsilon_1 - \varepsilon)p + \sigma_1/\varepsilon_0}{(\varepsilon_1 p + \sigma_1/\varepsilon_0)(p - i\omega)}\,\frac{e^{pt}}{p^2 + p\sigma_1/(\varepsilon_0\varepsilon_1) + v_1^2 k_0^2}, \tag{2.20}$$

where $v_1 = c/\sqrt{\varepsilon_1}$. The integrand in Eq. (2.18) has four poles: $p_1 = -\sigma_1/(\varepsilon_0\varepsilon_1)$, $p_2 = i\omega$ and $p_{3,4} = -\sigma_1/(2\varepsilon_0\varepsilon_1) \pm \sqrt{(\sigma_1/(2\varepsilon_0\varepsilon_1))^2 - \varepsilon\omega^2/\varepsilon_1}$.

The first pole gives zero contribution to the resultant field. The second pole gives a term, which compensates the incident field in accordance with the Evald–Oseen theorem [1]

$$\mathbf{E}_1(t,\mathbf{r}) = \mathbf{E}_0(t,x) - \mathbf{E}_0[e^{i\omega t} + A(t)]e^{-ik_0 x}. \tag{2.21}$$

The last two poles give two opposite propagating waves

$$A(t) = \frac{\varepsilon}{2\varepsilon_1}\left(\left(1 + \frac{\omega + i\sigma_1/(2\varepsilon_0\varepsilon_1)}{\bar{\omega}_1}\right)e^{(-\sigma_1/(2\varepsilon_0\varepsilon_1) + i\bar{\omega}_1)t}\right.$$

$$\left. + \left(1 - \frac{\omega + i\sigma_1/(2\varepsilon_0\varepsilon_1)}{\bar{\omega}_1}\right)e^{(-\sigma_1/(2\varepsilon_0\varepsilon_1) - i\bar{\omega}_1)t}\right) \tag{2.22}$$

with the new frequency

$$\bar{\omega}_1 = \sqrt{\omega_1^2 - \sigma_1^2(2\varepsilon_0\varepsilon_1)^2} \quad \omega_1 = v_1\omega/v \tag{2.23}$$

and the same wave vector as in the initial wave. These waves propagate in opposite directions and have the transformed frequencies, given by Eq. (2.23). Splitting of a wave when medium parameters change in a jump manner with time, in such a way as to change its frequency and conserve wave number, was revealed for the first time by Morgenthaler [2] in 1958, and further by other authors [3–6].

The waves (Eq. (2.1)) have a monochromatic character, with a new frequency ω_1, in the case of an ideal dielectric ($\sigma_1 = 0$)

$$A(t) = \frac{v_1}{2v}\left(\frac{v_1 + v}{v} e^{i\omega_1 t} + \frac{v_1 - v}{v} e^{-i\omega_1 t}\right) \tag{2.24}$$

when only the permittivity has a jump (from ε up to ε_1, Fig. 2.1a). In a lossy medium, the waves have an oscillatory character with a damping effect only in the case when $\omega > \sigma_1\varepsilon_1/(2\varepsilon_0\varepsilon^2)$. If the conductivity is strong, $\omega < \sigma_1\varepsilon_1/(2\varepsilon_0\varepsilon^2)$, the damping obtains a qualitatively different character, because the sign of the expression under the square root in Eq. (2.23) becomes negative. It leads to a disappearance of the time oscillations of the field.

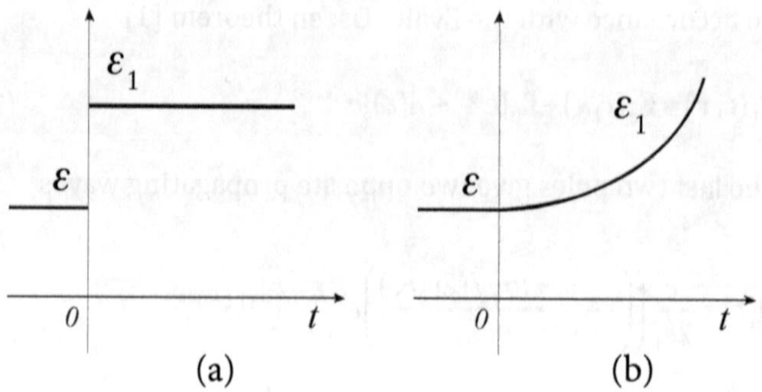

Figure 2.1 The laws of changing permittivity.

The wave splitting occurs not only for a plane harmonic wave. Let the initial field be a pulse described by the function $E_0(t, x) = f(x \pm vt)$ before the zero moment of time. Using the resolvent and integrating via the formula (2.3), we obtain the result of the resolvent action

$$\int_0^{\infty} dt' \int_{-\infty}^{\infty} dx' R(t-t', x-x') f(x' \pm vt') =$$

$$-f(x \pm vt) + \frac{a}{2} e^{-bt} \{(a \mp 1) f(x - v_1 t) + (a \pm 1) f(x + v_1 t)\} + \Phi(t,x) \tag{2.25}$$

The result of the time jump in parameters consists of four terms. The first one cancels the primary wave, in accordance with Evald–Oseen theorem [1]. The others yield the superposition of the new waves that travel in opposite directions with a new phase velocity and these waves consist of a discrete spectrum, the two waves which appear in brackets in Eq. (2.25), and the continuous spectrum

$$\Phi(t,x)=\frac{ab}{2}e^{-bt}\int_0^t\left\{\pm[f(x+v_1\sqrt{t^2-z^2})-f(x-v_1\sqrt{t^2-z^2})]I_1(bz)\right.$$
$$\left.+a[f(x+v_1\sqrt{t^2-z^2})+f(x-v_1\sqrt{t^2-z^2})]\frac{tI_1(bz)-zI_0(bz)}{\sqrt{t^2-z^2}}\right\}dz, \tag{2.26}$$

where $I_n(z)$ are the modified Bessel functions of the first kind. Finally, the expression for the field is

$$E(t,x)=\frac{a}{2}e^{-bt}\left[(a\mp 1)f(x-v_1t)+(a\pm 1)f(x+v_1t)\right]+\Phi(t,x) \tag{2.27}$$

All waves of these spectra may be thought of as a splitting of the primary wave. The continuous superposition $\Phi(t, x)$ is formed owing to the wave dispersion caused by the change of the medium conductivity ($b \neq 0$).

In the case when the primary field is a plane harmonic wave, $f = \exp[-ik_0(x \pm vt)]$, $k_0 - \omega/v$, the continuous superposition can be calculated explicitly as

$$\Phi(t,x)=-\frac{a}{2}e^{-bt-ikx}\left[(a\mp 1)e^{i\omega_1 t}+(a\pm 1)e^{-i\omega_1 t}\right]$$
$$+\frac{a^2}{2}e^{-bt-ikx}\left[\left(1+\frac{\mp\omega+ib}{\sqrt{\omega_1^2-b^2}}\right)e^{it\sqrt{\omega_1^2-b^2}}+\left(1-\frac{\mp\omega+ib}{\sqrt{\omega_1^2-b^2}}\right)e^{-it\sqrt{\omega_1^2-b^2}}\right], \tag{2.28}$$

where $\omega_1 = a\omega$ is the new frequency.

It follows from Eq. (2.21) that the split waves emerging from the continuous superposition cancel the first two waves in Eq. (2.28) and replace them by the new two waves with new frequencies and amplitudes:

$$E(t,x)=\frac{a^2}{2}e^{-bt-ikx}\left[\left(1+\frac{\mp\omega+ib}{\sqrt{\omega_1^2-b^2}}\right)e^{it\sqrt{\omega_1^2-b^2}}+\left(1-\frac{\mp\omega+ib}{\sqrt{\omega_1^2-b^2}}\right)e^{-it\sqrt{\omega_1^2-b^2}}\right] \tag{2.29}$$

that agree with Eq. (2.22), taking into account the contracted designations $a=\sqrt{\varepsilon/\varepsilon_1}$ and $b=\sigma_1/(\varepsilon_0\varepsilon_1)$

The wave splitting is a phenomenon that is common for any change in time of medium parameters, not only for a sharp jump. It follows from the existence of two characteristic families of hyperbolic equations and, generally, from the solution to such an equation with constant coefficients in the form of two d'Alembert's waves propagating in opposite directions. It can be illustrated by the example of when the permittivity changes according to the law $\varepsilon(t)=\varepsilon(1+gt)^4, g>0$ (Fig. 2.1b) [7]. The wave equation in this case

$$\frac{\partial^2}{\partial t^2}\left(\frac{\varepsilon(t)}{\varepsilon}E\right)+(vk_0)^2E=0 \tag{2.30}$$

has an exact analytic solution which describes the evolution of the plane wave $E=e^{ik_0(vt-x)}$ when the permittivity begins to change:

$$E(t,x)=\frac{1}{(1+gt)^3}\left[\left(1+i\frac{g\sqrt{\varepsilon}}{2ck_0}\right)e^{i\frac{k_0vt}{1+gt}}-\frac{g\sqrt{\varepsilon}}{2ck_0}e^{-i\frac{k_0vt}{1+gt}}\right]e^{-ik_0x} \tag{2.31}$$

The initial wave also splits into two waves propagating in opposite directions despite the fact that they have a non-monochromatic character. The effect of the incident wave splitting into two waves takes place even for a slow change of the permittivity ($g \ll 1$).

2.1.3 Transformation of Extrinsic Source Radiation by a Jump in Medium Parameters

For consideration of a 3D+T problem, let us assume that an initial field is generated by a source given by an extrinsic current $\mathbf{j}_{extr}$. This field is determined by the formula (2.2) in which $\mathbf{E}_0$ should be omitted. Substituting the exact expression for Green's function (1.37) into Eq. (2.2) yields the radiation field of this source

$$\mathbf{E}_0 = \frac{v^2\mu_0\mu}{4\pi}\int_{\infty} d\mathbf{x}'\hat{D}\tilde{\tilde{f}}(\mathbf{x}-\mathbf{x}')\frac{\partial \mathbf{j}_{\text{extr}}}{\partial t'}, \tag{2.32}$$

where $\tilde{\tilde{f}}$ is from Eq. 1.41 and the integral $\int_{\infty} d\mathbf{x}'$ is an integral over the whole four-dimensional space-time. Using the property of the convolution, the differentiation can be transferred in this formula from the factor $\mathbf{j}_{\text{extr}}$ to the factor $\tilde{\tilde{f}}$. Representing the operator $\hat{D}$ in the form $\hat{D} = \text{curlcurl} + \Delta - \frac{1}{v^2}\frac{\partial^2}{\partial t^2}$, one can use the equations for the function $\tilde{f} = \frac{\theta(\tau - R/v)}{R}$, where the contracted designations $\tau = t - t'$ и $R = |\mathbf{r} - \mathbf{r}'|$ are used,

$$\Delta\tilde{f}(\mathbf{x}-\mathbf{x}') - \frac{1}{v^2}\frac{\partial^2 \tilde{f}(\mathbf{x}-\mathbf{x}')}{\partial t^2} = -4\pi\theta(t-t')\delta(\mathbf{r}-\mathbf{r}'), \tag{2.33}$$

and for the function $f(\mathbf{x}-\mathbf{x}') = \frac{\partial}{\partial t}\tilde{f}(\mathbf{x}-\mathbf{x}') = \frac{\delta(\tau - R/v)}{R}$,

$$\Delta f(\mathbf{x}-\mathbf{x}') - \frac{1}{v^2}\frac{\partial f(\mathbf{x}-\mathbf{x}')}{\partial t^2} = -4\pi\delta(t-t')\delta(\mathbf{r}-\mathbf{r}'). \tag{2.34}$$

Considering these relations, the formula for calculation of the primary field is changed to

$$\mathbf{E}_0 = -v^2\mu_0\mu\int_{-\infty}^{t} \mathbf{j}_{\text{extr}}(t',\mathbf{r})dt'' + \frac{v^2\mu_0\mu}{4\pi}\text{curlcurl}\int_{\infty} d\mathbf{x}'\tilde{f}(\mathbf{x}-\mathbf{x}')\mathbf{j}_{\text{extr}}(\mathbf{x}') \tag{2.35}$$

The field transformation, which is a result of the medium state jump, is found by means of the same resolvent (2.4). In a coordinate representation the expression for the resolvent can be written in the form of a sum of two terms

$$\left\langle \mathbf{x}\middle|\hat{R}_j\middle|\mathbf{x}'\right\rangle = \text{curlcurl}\Phi_1(\tau,R) + \delta(\mathbf{R})\Phi_2(\tau) \tag{2.36}$$

where

$$\Phi_1(\tau,R) = \varepsilon v^2\int d\mathbf{p}\frac{\varepsilon - \xi(p)}{\xi(p)}\frac{e^{p\tau + i\mathbf{kR}}}{p^2\xi(p) + \varepsilon v^2 k^2}$$
$$= \frac{1}{4\pi R}\int_{\eta - i\infty}^{\eta + i\infty}\frac{dp}{2\pi i}\frac{\varepsilon - \xi(p)}{\xi(p)}e^{p\tau - \frac{R}{\sqrt{\varepsilon v}}\sqrt{p^2\xi(p)}} \tag{2.37}$$

$$\Phi_2(\tau) = \int_{\eta-i\infty}^{\eta+i\infty} \frac{dp}{2\pi i} \frac{\varepsilon - \xi(p)}{\xi(p)} e^{p\tau}. \tag{2.38}$$

Due to this, the transformed field is determined by the formula

$$\mathbf{E}_1 = \mathbf{E}_0 + \text{curlcurl}\int d\mathbf{x}'\Phi_1(t-t',\mathbf{r}-\mathbf{r}')\mathbf{E}_0(t',\mathbf{r}') + \int_0^\infty \Phi_2(t-t')\mathbf{E}_0(t',\mathbf{r})dt'. \tag{2.39}$$

The first integral in Eq. (2.39) can be simplified by introducing the expression (2.32) instead of $\mathbf{E}_0$ and using the property of the convolution differentiation as well as the well-known relation curl grad $f = 0$:

$$\text{curlcurl}\int d\mathbf{x}'\Phi_1(\mathbf{x}-\mathbf{x}')\frac{v^2\mu_0\mu}{4\pi}\left(\nabla'\nabla' - \frac{1}{v^2}\frac{\partial^2}{\partial t'^2}\right)\int_\infty d\mathbf{x}'' f^{**}(\mathbf{x}'-\mathbf{x}'')\frac{\partial \mathbf{j}_{\text{extr}}}{\partial t''}$$

$$= -\frac{\mu_0\mu}{4\pi}\text{curlcurl}\int d\mathbf{x}'\Phi_1(\mathbf{x}-\mathbf{x}')\int_\infty d\mathbf{x}'' f(\mathbf{x}'-\mathbf{x}'')\frac{\partial \mathbf{j}_{\text{extr}}}{\partial t''} \tag{2.40}$$

Then, substituting the formula (2.35) into the second integral in Eq. (2.39), the following can be obtained:

$$\mathbf{E}_1 = \mathbf{E}_0 - v^2\mu_0\mu\int_{-\infty}^{\infty} dt''\mathbf{j}_{\text{extr}}(t'',\mathbf{r})\int_0^\infty dt'\Phi_2(t-t')\theta(t'-t'')$$

$$+\frac{v^2\mu_0\mu}{4\pi}\text{curlcurl}\int_\infty d\mathbf{x}''\mathbf{j}_{\text{extr}}(\mathbf{x}'')\int_0^\infty dt'\Phi_2(t-t')f^*(t'-t'',\mathbf{r}-\mathbf{r}'')$$

$$-\frac{\mu_0\mu}{4\pi}\text{curlcurl}\int_\infty d\mathbf{x}''\frac{\partial \mathbf{j}_{\text{extr}}}{\partial t''}\int d\mathbf{x}'\Phi_1(\mathbf{x}-\mathbf{x}')f(\mathbf{x}'-\mathbf{x}'') \tag{2.41}$$

In the last term, the internal integral can be solved explicitly by substituting the expression (2.37) instead of Φ_1:

$$\Psi(\mathbf{x},\mathbf{x}') = \int d\mathbf{x}'\Phi_1(\mathbf{x}-\mathbf{x}')f(\mathbf{x}'-\mathbf{x}'')$$

$$= \frac{1}{4\pi}\int_{\eta-i\infty}^{\eta+i\infty}\frac{dp}{2\pi i}\frac{\varepsilon-\xi(p)}{\xi(p)}\int d\mathbf{x}'\frac{e^{p\tau-|\mathbf{r}-\mathbf{r}'|\alpha(p)}}{|\mathbf{r}-\mathbf{r}'|}\frac{\delta(t'-t''-|\mathbf{r}'-\mathbf{r}''|/v)}{|\mathbf{r}'-\mathbf{r}''|}, \tag{2.42}$$

where

$$\alpha(p) = \frac{1}{v}\sqrt{p^2\xi(p)/\varepsilon}. \tag{2.43}$$

For executing this integral, change the variables, introducing new ones, according to Fig. 2.2.

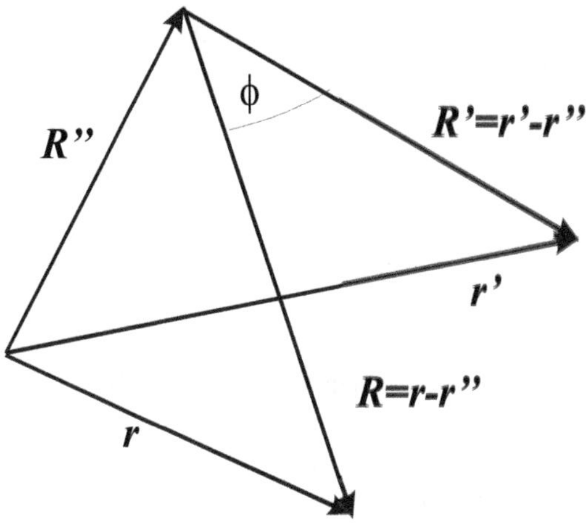

Figure 2.2 Changing the variables in the integral (2.42).

Then the internal integral in Eq. (2.42) is transformed into

$$\begin{aligned} I(t,t'',R,p) &= \int d\mathbf{x}' \frac{e^{p\tau - |\mathbf{r}-\mathbf{r}'|\alpha(p)}}{|\mathbf{r}-\mathbf{r}'|} \frac{\delta(t'-t''-|\mathbf{r}'-\mathbf{r}''|/v)}{\left||\mathbf{r}'-\mathbf{r}''|\right|} \\ &= \int d\mathbf{r}' \frac{e^{p(t-t''-|\mathbf{r}'-\mathbf{r}''|/v) - |\mathbf{r}-\mathbf{r}'|\alpha(p)}}{|\mathbf{r}-\mathbf{r}'|} \frac{\theta(t''+|\mathbf{r}'-\mathbf{r}''|/v)}{|\mathbf{r}'-\mathbf{r}''|} \\ &= \int d\mathbf{R}' \frac{e^{p(t-t''-R'/v) - |\mathbf{R}-\mathbf{R}'|\alpha(p)}}{|\mathbf{R}-\mathbf{R}'|} \frac{\theta(t''+R'/v)}{R'} \\ &= 2\pi \int_0^{\infty} dR' \theta(t''+R'/v) R' \int_{-1}^{1} dz' \frac{e^{p(t-t''-R'/v) - h(R,R',z')\alpha(p)}}{h(R,R',z')}, \end{aligned} \tag{2.44}$$

where $h(R,R',z') = \sqrt{R^2 + R'^2 - 2RR'z'}$. Changing the variables $R' = R'$, $h = h(R, R', z')$ yields the integral over the region shown in Fig. 2.3:

$$I(t,t'',R,p) = 2\pi \iint_S dR' dh \theta(t''+R'/v) \frac{e^{p(t-t''-R/v) - h\alpha(p)}}{R} \tag{2.45}$$

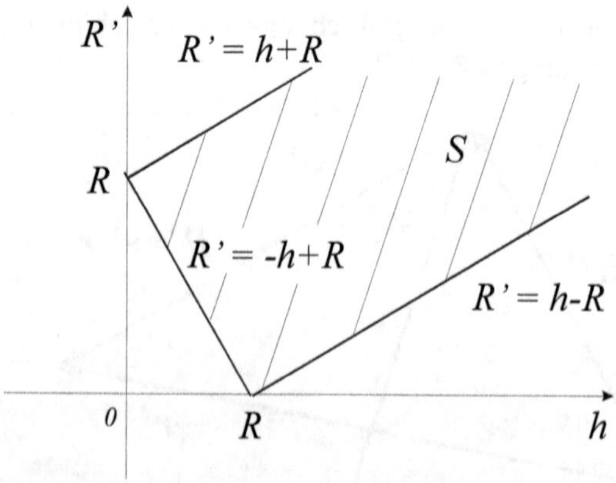

Figure 2.3 The region of integration in the integral (2.45).

The result of integration in Eq. (2.45) depends on the ratio between R and t'' as follows:

(a) $-vt'' > R$

$$I(t,t'',R,p) = -\frac{v}{\alpha(p)}\frac{e^{pt+v\alpha(p)t''}}{p+v\alpha(p)}\left[e^{-\alpha(p)R} - e^{\alpha(p)R}\right]$$

(b) $0 < -vt'' < R$

$$I(t,t'',R,p) = \frac{v}{\alpha(p)}\frac{e^{pt}}{p+v\alpha(p)}$$

$$\times\left[-e^{\alpha(p)(vt''-R)} - \frac{2v\alpha(p)}{p-v\alpha(p)}e^{-pt''-\frac{p}{v}R} + \frac{p+v\alpha(p)}{p-v\alpha(p)}e^{-\alpha(p)(vt''+R)}\right]$$

(c) $0 > -vt''$

$$I(t,t'',R,p) = \frac{2v^2}{p^2 - v^2\alpha^2(p)}e^{p(t-t'')}\left[e^{-\alpha(p)R} - e^{-pR/v}\right]$$

Integration of the function Φ_2 in Eq. (2.41) with respect to the variable t'' yields

$$\int_0^\infty dt'\,\Phi_2(t-t')f(t'-t'',R) = \frac{1}{R}\left[\theta\left(t''+\frac{R}{v}\right)\int_{\eta-i\infty}^{\eta+i\infty}\frac{dp}{2\pi i}\frac{\varepsilon-\xi(p)}{p\xi(p)}e^{p(t-t''-R/v)}\right.$$

$$\left. +\,\theta\left(-t''-\frac{R}{v}\right)\int_{\eta-i\infty}^{\eta+i\infty}\frac{dp}{2\pi i}\frac{\varepsilon-\xi(p)}{p\xi(p)}e^{pt}\right].$$

Further calculation requires choosing the medium model. For an ideal dielectric ($\xi = p/e_1$) the function $I(t, t'', R, p)$ is easily integrated over p. Executing the integral containing the function $\Psi(x, x')$ yields

$$\int_{-\infty}^{\infty} dt'' \frac{\partial \mathbf{j}_{\text{extr}}}{\partial t''} \Psi(\mathbf{x}, \mathbf{x}')$$

$$= -\frac{1}{2R}\left\{ v_1(v_1 - v) \int_{\tau^-}^{-\frac{v_1}{v}\tau_1^+} \mathbf{j}_{\text{extr}} dt'' + \theta(R - v_1 t) v_1(v_1 + v) \int_{\tau^-}^{\frac{v_1}{v}\tau_1^-} \mathbf{j}_{\text{extr}} dt'' \right.$$

$$\left. + \theta(R - v_1 t)\left[v_1(v_1 + v) \int_{\tau^-}^{\tau_1^-} \mathbf{j}_{\text{extr}} dt'' - v_1(v_1 - v) \int_{\tau_1^-}^{-\frac{v_1}{v}\tau_1^-} \mathbf{j}_{\text{extr}} dt'' \right]\right\}. \quad (2.46)$$

Here $\tau^- = t - \dfrac{R}{v}$, $\tau_1^\pm = t \pm \dfrac{R}{v_1}$ and $R = |\mathbf{r} - \mathbf{r}'|$.

The integrals containing the function Φ_2 are equal to

$$\int_{-\infty}^{\infty} dt'' \mathbf{j}_{\text{extr}} \int_0^{\infty} dt' \Phi_2(t - t') f(t' - t'', R) = \frac{\varepsilon - \varepsilon_1}{\varepsilon_1} \int_{-\infty}^{t-R/v} \mathbf{j}_{\text{extr}} dt'' \quad (2.47)$$

$$\int_{-\infty}^{\infty} dt'' \mathbf{j}_{\text{extr}} \int_0^{\infty} dt' \Phi_2(t - t') \theta(t' - t'') = \frac{\varepsilon - \varepsilon_1}{\varepsilon_1} \int_{-\infty}^{t} \mathbf{j}_{\text{extr}} dt''. \quad (2.48)$$

Substituting the obtained expressions (2.46), (2.47) and (2.48) into (2.41), the transformed field as the result of the jump of the dielectric permittivity is found in the following form:

$$\mathbf{E}_1 = \mathbf{E}_0$$

$$-v^2 \mu_0 \mu \frac{\varepsilon - \varepsilon_1}{\varepsilon_1} \int_{-\infty}^{t} \mathbf{j}_{\text{extr}}(t', \mathbf{r}) dt' + \frac{v^2 \mu_0 \mu}{4\pi} \frac{\varepsilon - \varepsilon_1}{\varepsilon_1} \text{rotrot} \int_{\infty} d\mathbf{r}' \frac{1}{R} \int_{-\infty}^{\tau^-} \mathbf{j}_{\text{extr}}(\mathbf{x}') dt'$$

$$+ \frac{\mu_0 \mu}{8\pi} \text{curlcurl} \int_{\infty} d\mathbf{r}' \frac{1}{R}\left\{ v_1(v_1 - v) \int_{\tau^-}^{-\frac{v_1}{v}\tau_1^+} \mathbf{j}_{\text{extr}} dt'' + \theta(R - v_1 t) v_1(v_1 + v) \int_{\tau^-}^{\frac{v_1}{v}\tau_1^-} \mathbf{j}_{\text{extr}} dt'' \right.$$

$$\left. + \theta(v_1 t - R)\left[v_1(v_1 + v) \int_{\tau^-}^{\tau_1^-} \mathbf{j}_{\text{extr}} dt'' - v_1(v_1 - v) \int_{\tau_1^-}^{-\frac{v_1}{v}\tau_1^-} \mathbf{j}_{\text{extr}} dt'' \right]\right\} \quad (2.49)$$

The limits of the integral in this expression can be considered as indicators of the initial and final states of the field, which, in their turn, are indicators of the transitions caused by the dielectric permittivity time jump. Indeed, the first integral in the brackets corresponds to the transition $\tau^- \to -\frac{v_1}{v}\tau_1^+$. This transition describes the creation of the wave which is converging to the source point. This wave is created in the whole space as a transformation of the initial field at each point of the non-stationary space with a change of the time scale determined by the relation of the phase velocities. The second integral corresponding to the transition $\tau^- \to \frac{v_1}{v}\tau_1^-$ describes the transformation of the incident divergent wave into a divergent wave with the same change of the time scale. As this wave is formed from the incident wave, and it is divergent, it can exist only outside the sphere $R - v_1 t = 0$. The integrals in the square brackets determine the field inside this sphere. The first term in the square brackets that conforms to the transition $\tau^- \to \tau_1^-$ is a field created by a source in the new medium after its changing. The second term, which conforms to the transition $\tau_1^- \to -\frac{v_1}{v}\tau_1^-$ is a field formed as a result of the converging wave passing through the source point.

Re-writing all the integrals in Eq. (2.49) as

$$\mathbf{I}(\xi, \mathbf{r}') = \int_{-\infty}^{\xi} \mathbf{j}_{\text{extr}}(t', \mathbf{r}')dt' \tag{2.50}$$

and using the expression for the incident field, the transformed field can be represented in another form

$$\mathbf{E}_1 = -v_1^2\mu_0\mu \int_{-\infty}^{t} \mathbf{j}_{\text{extr}}(t', \mathbf{r})dt' + \frac{v_1^2\mu_0\mu}{4\pi}\text{curlcurl}\int_{\infty} d\mathbf{r}'\frac{1}{R}\theta(v_1 t - R)\mathbf{I}(\tau_1^-, \mathbf{r}')$$
$$+\frac{v_1^2\mu_0\mu}{8\pi}\text{curlcurl}\int_{\infty} d\mathbf{r}'\frac{1}{R}\left[\frac{v_1 - v}{v_1}\mathbf{I}\left(-\frac{v_1}{v}\tau_1^+, \mathbf{r}'\right)\right.$$
$$\left. +\theta(R - v_1 t)\frac{v_1 + v}{v_1}\mathbf{I}\left(\frac{v_1}{v}\tau_1^-, \mathbf{r}'\right) - \theta(v_1 t - R)\frac{v_1 - v}{v_1}\mathbf{I}\left(-\frac{v_1}{v}\tau_1^-, \mathbf{r}'\right)\right], \tag{2.51}$$

where the structure of all the terms clearly demonstrates the transformations noted above.

The radiated field of the source point is split into two waves with opposite propagation directions, similarly to the case of the plane wave incidence. However, the spatial picture of the distribution of the secondary fields looks much more complex (see Fig. 2.4).

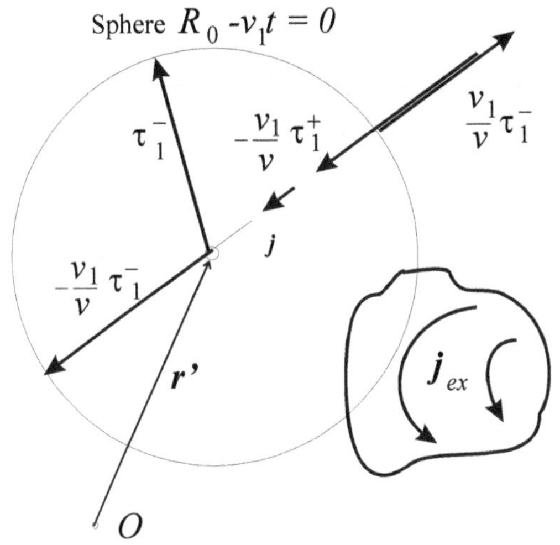

Figure 2.4 Spatial distribution of the wave originated due to a current source in the case of an ideal dielectric permittivity jump.

The first two terms in Eq. (2.51), being similar to those in the formula (2.35), represent the field created by the source in the medium with new properties. This field is concentrated inside the sphere extending with the phase velocity v_1. The other term represents the field that is a result of the primary wave transformation associated with the dielectric permittivity jump. The first two components of this term are the results of the primary field splitting into the forward wave (divergent from the source) and the backward (convergent to the source) one. The forward wave differs from zero outside the sphere, but the backward one exists in the whole space. Passing through the source point, the backward wave is transformed into the divergent wave existing inside the sphere with radius $v_1 t$. As was shown for the first time in Ref. 3, the convergent wave is focused in the source point so that for a harmonic source this wave amplitude tends to infinity. This result is a consequence of the idealized problem, assuming unlimited power of the mechanism which changes the dielectric permittivity in a jump manner simultaneously in the whole space.

Note also, that after a dielectric permittivity jump in the case of a monochromatic source the electromagnetic field is also transformed into monochromatic waves.

2.2 Evolution of a Harmonic Wave in a Medium Modulated by Repetitive Identical Pulses

2.2.1 Electromagnetic Waves in a Linear Modulated Medium

Splitting of the wave under a time jump of medium parameters gains new features when such jumps consequently repeat many times. To investigate this process we consider the transformation of an electromagnetic wave in which electric field is directed along $\hat{y}$ axis $\mathbf{E}(t, x) = (0, E_0(t, x), 0)$ in an unbounded dielectric dissipative medium whose permittivity $\varepsilon(t)$ and conductivity $\sigma(t)$ change from the zero moment of time according to the law of a finite packet of N rectangular pulses under the action of external forces [8, 9]. In this change the permittivity and the conductivity of the medium take constant magnitudes ε_n and σ_n on the disturbance intervals $(n - 1)T < t < T_1 + (n - 1)T$ $n = 1, \dots N$ and they have constant magnitudes ε and $\sigma = 0$ on the inactive intervals $T_1 + (n - 1)T < t < nT$ $n = 1, \dots N$:

$$\varepsilon(t) = \varepsilon + (\varepsilon_1 - \varepsilon)\sum_{k=1}^{N}\{\theta(t-(k-1)T) - \theta(t-T_1-(k-1)T)\}$$

$$\sigma(t) = \sigma_1\sum_{k=1}^{N}\{\theta(t-(k-1)T) - \theta(t-T_1-(k-1)T)\} \tag{2.52}$$

Here, $\theta(t)$ is the Heaviside unit function, T is the duration of the period of the parameters change, and T_1 is the duration of the disturbance interval. The values $(\varepsilon_1 - \varepsilon)$ and σ_1 do not have any limitations, nor T_1.

The problem of determining the field will be solved by the integral equation method developed in this book according to which the electric field is calculated via the resolvent by virtue of the formula (2.3) in Section 2.1.1. In the case considered the electric field has only one component and depends only on one coordinate. This simplifies the equations and on the disturbance interval of the n-th period the expression for the field has the form

$$E_n(t,x) = F_{n-1}(t,x) + \int_{(n-1)T}^{\infty} dt' \int_{-\infty}^{\infty} dx' \, R(t-t', x-x') F_{n-1}(t', x') \quad n \geq 1, \quad (2.53)$$

where R is the problem resolvent (2.19). The free term of Eq. (2.53) is determined by the field in the previous periods

$$F_n(t,x) = F_0(t,x) + \sum_{k=0}^{n-1} \int_{kT}^{kT+T_1} dt' \int_{-\infty}^{\infty} dx' \, K(t-t', x-x') E_{k+1}(t', x') \qquad (2.54)$$

where

$$K(t,t',x,x') = -\frac{1}{2va^2}\left\{2b + (1-a^2)\frac{\partial}{\partial t}\right\}\delta(v(t-t') - |x-x'|) \qquad (2.55)$$

and $\delta(t)$ is the Dirac delta-function.

The free term with zero index is $F_0(t, x) = E_0(t, x)$ because the change of the medium parameters begins at the zero moment of time.

One can easily obtain from Eq. (2.12) that $R(t - t', x - x') = 0$ on the inactive interval of the n-th period of the medium parameters change, if one takes into account that $\varepsilon(t) = \varepsilon$ and $\sigma(t) = 0$ ($a = 1$, $b = 0$) on these intervals. Consequently, on these intervals the field equals the free term of the equation, $E_n = F_n$, where F_n is determined by the formula (2.54). An example of the time intervals used for the field definition for the case of a packet consisting of three pulses, $N = 3$, is shown in Fig. 2.5.

The kernel of Eq. (2.55) is invariant with respect to the period of the medium parameter change; therefore, the field structure will be repeated after each period. The sole jump of the medium parameters transforms the initial plane harmonic wave $f = \exp[-ik_0(x + vt)$, $k_0 = \omega/v]$ into the split waves (Eq. (2.29)):

$$E(t,x) = \frac{a^2}{2} e^{-bt-ikx}\left[\left(1 + \frac{\mp\omega + ib}{\sqrt{\Omega^2 - b^2}}\right) e^{it\sqrt{\Omega^2-b^2}} + \left(1 - \frac{\mp\omega + ib}{\sqrt{\Omega^2 - b^2}}\right) e^{-it\sqrt{\Omega^2-b^2}}\right],$$

(2.56)

where $a = \sqrt{\varepsilon/\varepsilon_1}$, $b = \sigma_1(\varepsilon_0\varepsilon_1)$ and $\Omega = a\omega$ is the new frequency.

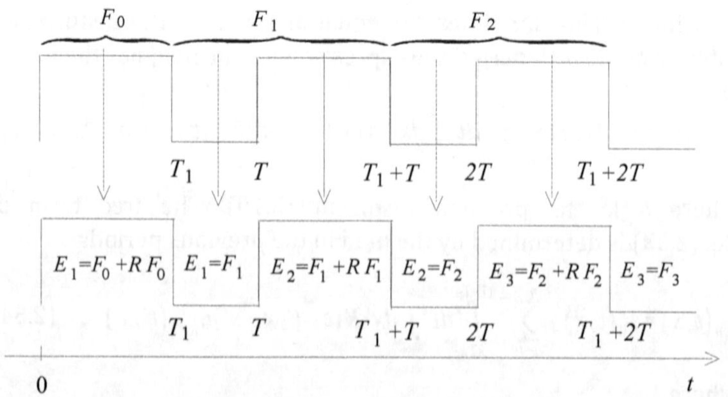

Figure 2.5 The time intervals for the field definition.

In the case of N periods of the medium parameter variations according to the law (2.52) we assume the initial wave $E_0(t, x) = \exp[i(\omega t - kx)]$ travels in the positive x direction. In the disturbance interval of the first period the field, according to Eq. (2.22), consists of two waves (direct and inverse ones) $E_1 = \exp(-bt - ikx)[C_1 \exp(i\Omega t) + D_1 \exp(-i\Omega t)]$ with amplitudes

$$C_1 = \frac{a^2}{2\Omega}(\Omega + \omega + ib) \quad D_1 = \frac{a^2}{2\Omega}(\Omega - \omega - ib) \tag{2.57}$$

In the remaining interval of the first period the field is calculated by the formula (2.54). It also consists of two waves that take the frequency of the primary wave

$$F_1 = \exp(-ik)[A_1 \exp(i\omega t) + B_1 \exp(-i\omega t)]$$

and where the amplitudes are given by

$$A_1 = \frac{1}{2\Omega} e^{-(b+i\omega)T_1}\left[i\omega(a^2+1)\sin\Omega T_1 + 2\Omega\cos\Omega T_1\right]$$

$$B_1 = \frac{-1}{2\Omega} e^{-(b-i\omega)T_1}\left[i\omega(a^2-1) + 2b\right]\sin\Omega T_1, \tag{2.58}$$

where $\Omega = a\omega$, is the same as in Eq. (2.56).

Shifting successively the time origin from one period to the next we find that the field in each disturbance interval of the n-th period consists of two waves, direct and inverse ones,

$$E_n = \exp(-bt - ikx)[C_n \exp(i\Omega t) + Dn \exp(-i\Omega t)]$$

with amplitudes

$$C_n = \frac{a^2}{2\Omega} e^{(n-1)bT - i(n-1)\Omega T} [c_{11} A_{n-1} e^{i(n-1)\omega T} + c_{12} B_{n-1} e^{-i(n-1)\omega T}]$$

$$D_n = \frac{a^2}{2\Omega} e^{(n-1)bT + i(n-1)\Omega T} [c_{21} A_{n-1} e^{i(n-1)\omega T} + c_{22} B_{n-1} e^{-i(n-1)\omega T}], \quad n > 1, \quad (2.59)$$

where

$c_{11} = \Omega + \omega + ib$, $c_{12} = \Omega - \omega + ib$, $c_{21} = \Omega - w - ib$ and $c_{22} = \Omega + \omega - ib$.

In the remaining period of inactivity the field

$$F_n = \exp(-ikx)\left[A_n \exp(i\omega t) + B_n \exp(-i\omega t)\right]$$

also consists of two waves; these possess the frequencies of the primary wave, but have new amplitudes that satisfy the recurrent relations

$$A_n = \frac{1}{2\Omega} e^{-(b+i\omega)T_1} \left[l_{11} A_{n-1} + l_{12} B_{n-1} e^{-i2(n-1)\omega T}\right]$$

$$B_n = \frac{-1}{2\Omega} e^{-(b-i\omega)T_1} \left[l_{21} A_{n-1} e^{i2(n-1)\omega T} + l_{22} B_{n-1}\right], \quad n > 1, \qquad (2.60)$$

where

$$l_{11} = i\omega(a^2+1)\sin\Omega T_1 + 2\Omega\cos\Omega T_1 \quad l_{12} = i[\omega(a^2-1)+2ib]\sin\Omega T_1$$

$$l_{21} = i[\omega(a^2-1)-2ib]\sin\Omega T_1 \quad l_{22} = i\omega(a^2+1)\sin\Omega T_1 - 2\Omega\cos\Omega T_1$$

These expressions represent the exact solution of the problem, which is correct for arbitrary magnitudes of the amplitude and arbitrary durations of the medium parameter variations.

So, the influence of one cycle of the medium modulation upon the electromagnetic wave leads to splitting of the latter into two new waves, with new frequencies and amplitudes and traveling in opposite directions with a new phase velocity. The next cycles of medium modulation make the picture of the field very complex. Nevertheless, in the case of a harmonic plane wave the transformed field preserves the structure of the first transformation, so that it consists of two harmonic plane waves with residual wave number, bouncing frequency and amplitudes changing in a complex way from one modulation cycle to another. The appearance of backward traveling waves is an inevitable result of the medium time change.

In the following treatment, it is convenient to use the reduced amplitudes $a_n = A_n \exp(in\tilde{T})$, $b_n = B_n \exp(-in\tilde{T})$, $c_n = C_n \exp(i(n-1)q\tilde{T})$ and $d_n = D_n \exp(-i(n-1)q\tilde{T})$ where the transition to the dimensionless parameters $\omega T \to \tilde{T}$, $\omega T_1 \to \tilde{T}_1$ is made. The recurrence relation for the amplitudes of the secondary waves follows from Eqs. (2.59) and (2.60). The amplitudes on the disturbance intervals of the n-th cycle are determined by the waves amplitudes on the previous $(n-1)$-th "inactivity" interval via the matrix formula

$$\begin{pmatrix} c_n \\ d_n \end{pmatrix} = \frac{a^2}{2q} e^{(n-1)s\tilde{T}} Q \begin{pmatrix} a_n \\ b_n \end{pmatrix} \quad Q = \begin{pmatrix} \overline{c}_{11} & \overline{c}_{12} \\ \overline{c}_{21} & \overline{c}_{22} \end{pmatrix}, \tag{2.61}$$

where $\overline{c}_{11} = \overline{c}_{22}^* = q + 1 + is$, $\overline{c}_{12} = \overline{c}_{22}^* = q - 1 + is$, c^* is conjugate to c, $a_1 = m$, $b_1 = -h$ and

$$m = \frac{2q\cos(q\tilde{T}_1) + i(a^2+1)\sin(q\tilde{T}_1)}{2q} \exp[-s\tilde{T}_1 + i(\tilde{T} - \tilde{T}_1)] \tag{2.62}$$

$$h = i\frac{1}{2q}(a^2 - 1 - i2s)\sin(q\tilde{T}_1)\exp[-s\tilde{T}_1 - i(\tilde{T} - \tilde{T}_1)].$$

Here $a = \sqrt{\varepsilon/\varepsilon_1}$, $b = \sigma/\varepsilon_0\varepsilon_1$, $q = \sqrt{a^2 - s^2}$, $\overline{q} = i\sqrt{s^2 - a^2}$ and $s = b/\omega$.

The recursion relations for the amplitudes of the secondary waves on inactivity intervals are given by the matrix equation:

$$\begin{pmatrix} a_n \\ b_n \end{pmatrix} = e^{-s\tilde{T}_1} \mathrm{P} \begin{pmatrix} a_{n-1} \\ b_{n-1} \end{pmatrix}, \quad \mathrm{P} = e^{s\tilde{T}_1} \begin{pmatrix} m & -h^* \\ -h & m^* \end{pmatrix} \tag{2.63}$$

with det $P = 1$. The formulae (2.63) also determine the amplitudes of the transformed fields after n complete cycles for $n \geq 2$.

The relation between the backward and forward secondary wave amplitudes is given by the following ratios:

On the disturbance intervals

$$w_N = \frac{d_N}{c_N} = \frac{\{p_2 + (p_1 - p_2)r_{N-1}\}\overline{c}_{21} + p_1 p_2 \overline{c}_{22}}{\{p_2 + (p_1 - p_2)r_{N-1}\}\overline{c}_{11} + p_1 p_2 \overline{c}_{12}} \tag{2.64}$$

On the inactivity intervals

$$p_N = \frac{b_N}{a_N} = \frac{p_1 p_2}{p_2 + (p_1 - p_2)r_N}, \quad N \geq 2 \tag{2.65}$$

where $p_1 = -\frac{h}{m}$ and $p_2 = -\frac{h(m+m^*)}{hh^*+m^2}$. The variable r_N is determined by a real sequence

$$r_{N+1} = \frac{4u^2}{4u^2 - r_N}, \; r_1 = 0 \tag{2.66}$$

The behavior of this variable determines the behavior of the amplitude ratios w_N and p_N and they are all determined by the magnitude of the generalized parameter

$$u = \cos(q\tilde{T}_1)\cos\tilde{T}_2 - \frac{a^2+1}{2q}\sin(q\tilde{T}_1)\sin\tilde{T}_2, \tag{2.67}$$

where $\tilde{T}_2 = \tilde{T} - \tilde{T}_1$.

After N cycles of medium parameter changes, according to the rectangular pulse law, the primary wave transforms into two waves propagating in opposite directions and having the same frequency as the primary wave. The amplitudes of these waves are given by the following formula derived from the relation (2.63)

$$\begin{pmatrix} a_N \\ b_N \end{pmatrix} = e^{-(N-1)s\tilde{T}_1}\hat{\mathrm{P}}^{N-1}\begin{pmatrix} a_1 \\ b_1 \end{pmatrix}, \tag{2.68}$$

where (see [8])

$$\hat{\mathrm{P}}^n = U_{n-1}(u)\hat{\mathrm{P}} - U_{n-2}(u)\hat{\mathrm{I}} \tag{2.69}$$

$U_n(u) = \frac{\sin((n+1)\arccos u)}{\sqrt{1-u^2}}$ is the Chebyshev n-th order polynomial of the second kind, $\hat{\mathrm{I}}$ is the unit matrix, and u is the parameter given in Eq. (2.67).

The behavior of the amplitudes of the secondary waves differs qualitatively with the number of pulses N depending on the value of the generalized parameter u with respect to 1. If this parameter is less than 1, then the behavior of the Chebyshev polynomials has an oscillating character and the amplitudes of both the direct and the reverse waves decrease as the exponent $\exp(-Ns\tilde{T}_1)$ owing to the medium dissipation only. If the parameter u is greater than 1, then the argument of the sine in the Chebyshev polynomials become imaginary, $\arccos u = \pm i\ln(u+\sqrt{u^2-1})$, and the asymptotic variation of the amplitudes of these waves is described by the exponent $\exp\{-N\gamma\}$ with $\gamma = s\tilde{T}_1 \pm \ln(u+\sqrt{u^2-1})$. The amplification of the secondary waves is possible only if $\gamma < 0$. This condition cannot be fulfilled in the case when only the medium conductivity is varying but the permittivity remains constant, $a \equiv 1$. In this case the generalized parameter (2.67) is equal to $u = \cos(q\tilde{T}_1)\cos\tilde{T}_2 - \frac{1}{q}\sin(q\tilde{T}_1)\sin\tilde{T}_2$. For $s = 0$ ($q = 1$), this parameter is not greater than unity as $u = \cos(\tilde{T}_1 + \tilde{T}_2)$. If $s \to 1$, that is $q \to 0$, the parameter $u \to (\cos\tilde{T}_2 - \tilde{T}_1\sin\tilde{T}_2)$ and its magnitude is of the order of $\tilde{T}_1$, if $\tilde{T}_1 > 1$. However, $\gamma \approx s\tilde{T}_1 \pm \ln 2u \approx s\tilde{T}_1 \pm \ln 2\tilde{T}_1 > 0$ in this case, meaning that the amplification condition is not satisfied. If $s >> 1$, then $u = [\cosh(\bar{q}\tilde{T}_1)\cos\tilde{T}_2 - \bar{q}^{-1}\sinh(\bar{q}\tilde{T}_1)\sin\tilde{T}_2] \sim \exp(\bar{q}\tilde{T}_1)$ and also $\gamma \approx s\tilde{T}_1 \pm \bar{q}\tilde{T}_1 > 0$. Thus the implementation of the amplification condition is possible only if $a > 1$.

2.2.2 "Intermittency" in Electromagnetic Wave Transients in a Time-Varying Linear Medium

The relation of the parameter u with respect to 1 determines another feature of the secondary waves, namely, their regular or irregular behavior with an increase in the number of cycles N of the medium parameter variations. To reveal this, consider the successive behavior of the amplitudes with N

$$\begin{cases} \dfrac{a_{N+1}}{a_N} = m - \dfrac{h^*}{z_N} \\ \dfrac{b_{N+1}}{b_N} = -hz_N + m^* \end{cases}, \tag{2.70}$$

which is expressed through the sequence of the complex variable z_N

$$z_N = \frac{a_N}{b_N} \quad N \geq 2\tilde{T}_1 = -\frac{m}{h} \tag{2.71}$$

This sequence changes according to the recursion relation following from Eq. (2.63)

$$z_N + 1 = -\frac{h^* - mz_N}{m^* - hz_N} = -\frac{h^*}{h}\frac{1 - gz_N}{g^* - z_N} \tag{2.72}$$

where $g = \frac{m}{h^*} = -\frac{2q\cot(q\tau_1) + i(a^2+1)}{2s - i(a^2-1)}$

The relation (2.72) represents a conformal mapping, which maps a unit circle into another unit circle. If $|g| > 1$, the circle interior is mapped onto a circle interior and a circle exterior is mapped onto a circle exterior. If $|z_1| \geq 1$ then $|z_N| \geq 1$ consequently, and the inverse wave amplitude does not exceed the amplitude of direct wave. Equality of these amplitudes is possible only in the case where $z_{N+1} = z_N = \bar{z}$. In this case, the formula (2.72) gives $\bar{z} = -(m - m^* \pm \sqrt{(m+m^*)^2 - 4\Delta})/2h$, where $\Delta = \exp(-2s\tilde{T}_1)$. Indeed, setting $z_{N+1} = z_N = z$ in Eq. (2.72) we arrive at the equation $hz^2 + (m-m^*)\bar{z} - h^* = 0$, from which the expression for $\bar{z}$ follows.

The equality of the wave amplitudes provides that $|\bar{z}| = 1$. Since the relation (2.72) maps a unit circle into a unit circle, then $|z| = 1$ follows from $|\bar{z}| = 1$. The definition $z_1 = -m/h$ by Eq. (2.71) gives such expressions:

$$|z_1|^2 = \begin{cases} \dfrac{4q^2\cot^2(q\tilde{T}_1) + (a^2+1)^2}{-4q^2 + (a^2+1)^2}, & \text{if } q^2 > 0 \\[2ex] \dfrac{4\bar{q}^2\coth^2(\bar{q}\tilde{T}_1) + (a^2+1)^2}{4\bar{q}^2 + (a^2+1)^2}, & \text{if } q^2 < 0 \end{cases}$$

It leads to $|z_1|^2 = 1$ if either $q^2\cot^2(q\tilde{T}_1) = -q^2$, for $q^2 > 0$; or $\bar{q}^2\coth^2(\bar{q}\tilde{T}_1) = \bar{q}^2$, for $q^2 < 0$.

If the absolute value of $\bar{z}$ is equal to 1 then the modulus of z_1 should be unity, $|z_1| = 1$. However, this is possible only when $q^2(\cot^2(q\tilde{T}_1) + 1) = 0$ or $\bar{q}^2(\coth^2(\bar{q}\tilde{T}_1) - 1) = 0$. Since the modulus $|z_1|^2 \to 1 + \frac{4}{\tau_1^2(a^2+1)^2} > 1$ with $q \to 0$ $(\bar{q} \to 0)$, the equality $|z_1| = 1$ is possible only if $\coth^2(\bar{q}\tilde{T}_1) = 1$, that is when $(\bar{q}\tilde{T}_1) \to \infty$. Consequently, if the values of the parameters are finite, the modulus of z_N does not remain constant when varying the number of pulses, and so the equality of the amplitudes of the direct and inverse waves can only be approached asymptotically.

The analysis of the complex value z_N sequence becomes simpler if it is expressed in terms of the real sequence r_N given in Eq. (2.66)

$$z_N = z_1 + (z_2 - z_1)r_N \tag{2.73}$$

The characters of the behavior of r_N and the ratio z_N, as well as the amplitudes a_N and b_N themselves, are determined by the magnitude of the generalized parameter u. If $u^2 > 1$, then the mapping (2.66) approaches to the stable stationary point $r^* = 2u(u - \sqrt{u^2 - 1})$ monotonically, as shown in the Lamerey diagram of the mapping (2.66) in Fig. 2.6a [9]. The absolute values of the amplitudes increase monotonically if $\gamma < 0$ (see Fig. 2.7a), and decrease monotonically starting from a certain number N if $\gamma > 0$ (see Fig. 2.7b). They coalesce asymptotically when $N \to \infty$.

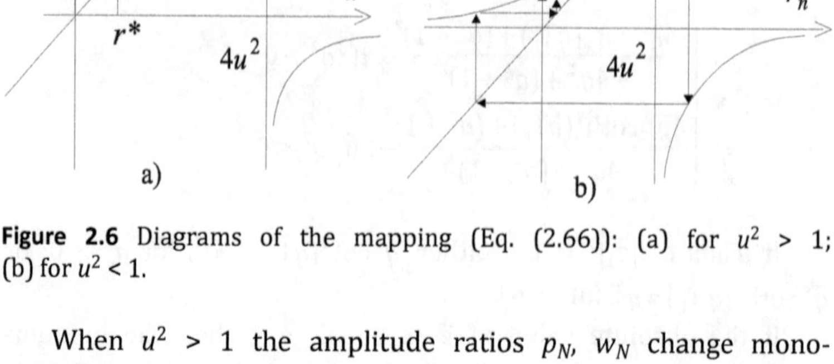

Figure 2.6 Diagrams of the mapping (Eq. (2.66)): (a) for $u^2 > 1$; (b) for $u^2 < 1$.

When $u^2 > 1$ the amplitude ratios p_N, w_N change monotonically with the number of modulation cycles, as shown in Fig. 2.8, for $u = 1.0002$.

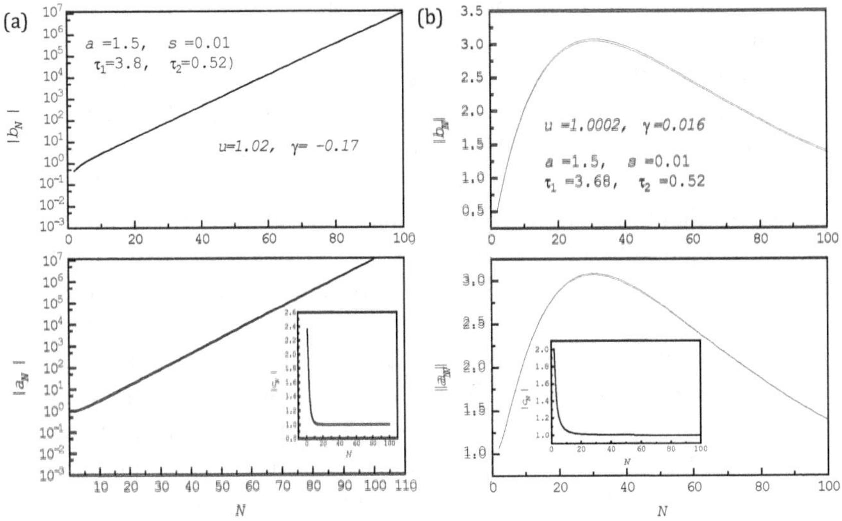

Figure 2.7 Monotonic amplification (a) and fading of the secondary wave amplitudes (b) (the insets are for the ratio z_N).

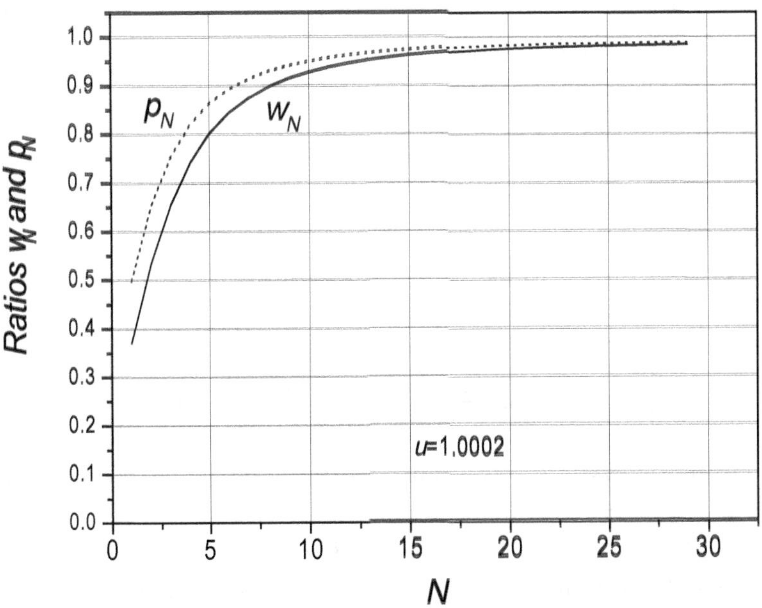

Figure 2.8 Relations between direct and inverse secondary wave amplitudes for the parameters of modulation $a = 1.5$, $s = 0.01$, $\tilde{T}_1 = 3.68$ and $\tilde{T}_2 = 0.52$.

The behavior of r_N has a non-monotonic, and even irregular, character if $u^2 < 1$ (Fig. 2.6b). The monotonic behavior of the amplitude ratios breaks and these ratios both have a minimum after some number of cycles, as shown in Fig. 2.9, which corresponds to $u = 0.98988$. In both cases, the ratios of the amplitudes on the disturbance and inactivity intervals have similar character. The case when the behavior of the amplitudes has a quasi-chaotic character is shown in Fig. 2.10. Here $u = 0.18978$ and the ratio of the amplitudes on the disturbance intervals does not correlate with the ratio on the inactivity ones. Small variations in the number of modulation cycles lead to considerable changes in the ratio value. So, the generalized parameter u is responsible for the character of behavior of the relation between the forward and backward wave amplitudes and such a relation can become incoherent for small values of this parameter.

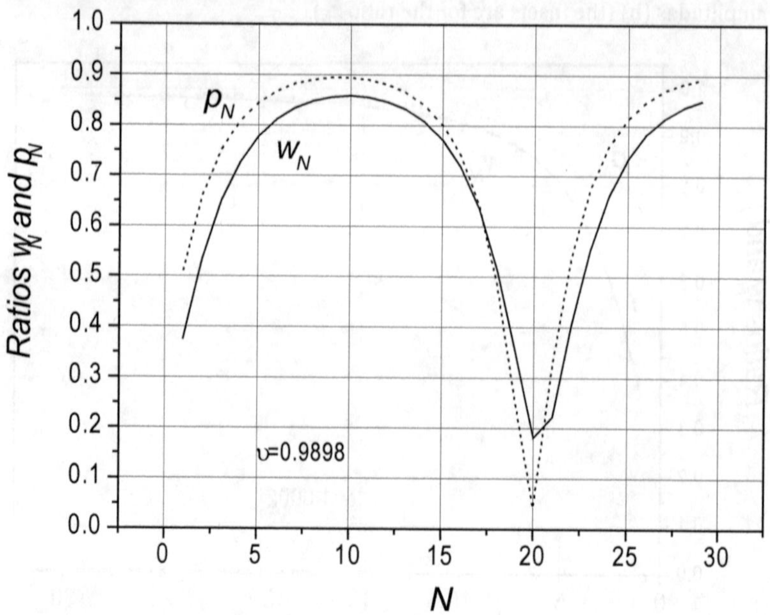

Figure 2.9 Relations between direct and inverse secondary wave amplitudes for the parameters of modulation $a = 1.5$, $s = 0.01$, $\tilde{T}_1 = 3.65$ and $\tilde{T}_2 = 0.52$.

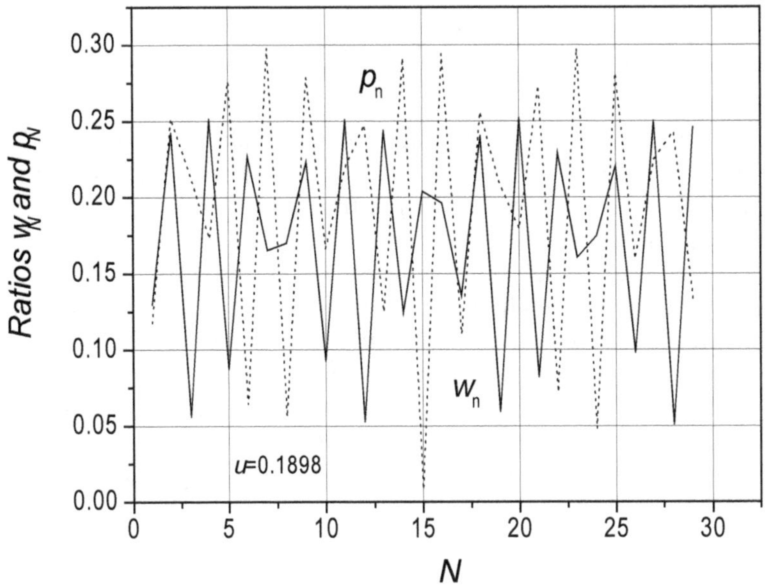

Figure 2.10 Relations between direct and inverse secondary wave amplitudes for the parameters of modulation a = 1.5, s = 0.01, $\tilde{T}_1$ = 0.55 and $\tilde{T}_2$ = 0.52.

If $u^2 < 1$, then the sequence r_N has an irregular character. Indeed, as can be seen in Fig. 2.6b, there are great intervals of consequent N where r_N changes almost regularly. These intervals alternate with the relatively short ones where a strongly irregular behavior of r_N takes place. This phenomenon is known as "intermittence" [10–12]. The presence of the "intermittence" can be proved by Hurst's method [10] according to which the time series of r_i is characterized by the Hurst exponent H in the asymptotic behavior of the function

$$\frac{R_n}{S_n} \sim n^H, \tag{2.74}$$

where

$$R_n = \max_{1\le k\le n} X(k,n) - \min_{1\le k\le n} X(k,n),\ X(k,n) = \sum_{i=1}^{k}(r_i - \langle r \rangle_n) \tag{2.75}$$

$$<r>_n = \frac{1}{n}\sum_{i=1}^{n} r_i,\ S_n = \sqrt{\frac{1}{n}\sum_{i=1}^{n}(r_i - <r>_n)^2} \tag{2.76}$$

For white noise (i.e. a completely uncorrelated signal) Hurst's index $H = 0.5$. The values $H > 0.5$ ($H < 0.5$) are associated with the long-range correlation when a time series exhibit persistence (anti-persistence).

The analysis of the sequence r_N in the case considered shows that the amplitude ratio variations are associated with the non-correlated process when $u^2 < 1$. The deviation of u^2 from 1 does not sufficiently influence the irregularity of the amplitude ratio behavior. The index H is not stable only for a small deviation of u^2 from 1. When $u \approx 0.98988$ the index H decreases from $H > 0.5$ to $H < 0.5$. The behavior of the amplitude and the index H are illustrated in Fig. 2.11.

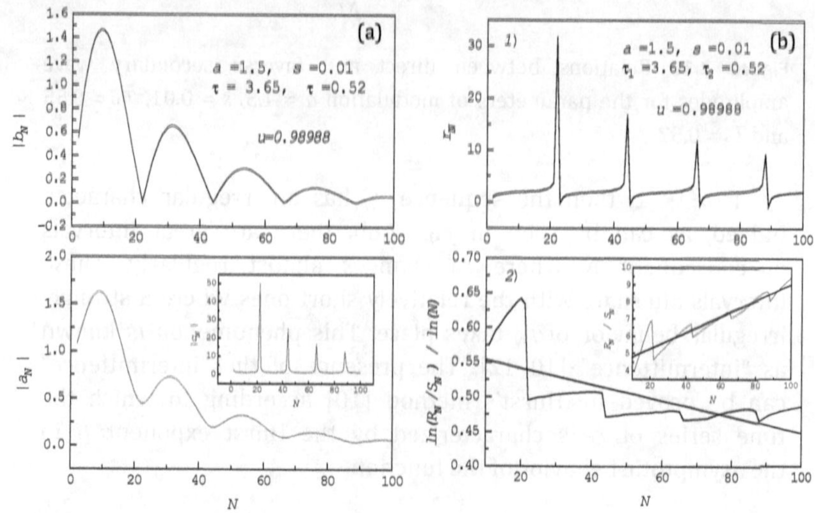

Figure 2.11 "Intermittence" of the secondary waves for $u \approx 0.98988$: (a) change of the inverse and direct wave amplitudes with the pulse number; (b) change of the sequence r_N (1) and Hurst's index (2).

For $u \approx 0.77536$ the index $H \approx 0.46$ and remains stable, as can be seen in Fig. 2.12. The character of the amplitude behavior changes slowly with a further decrease of $|u|$, as shown in Fig. 2.13.

All the cases considered above were investigated with a variation of only one parameter, namely, the duration of the disturbance interval of the medium state. Nevertheless, it is possible to adjust the generalized parameter u via a variation of the other parameters of the system. It should be noted that in the case considered, the "intermittence" is observed in a linear electromagnetic wave propagation problem.

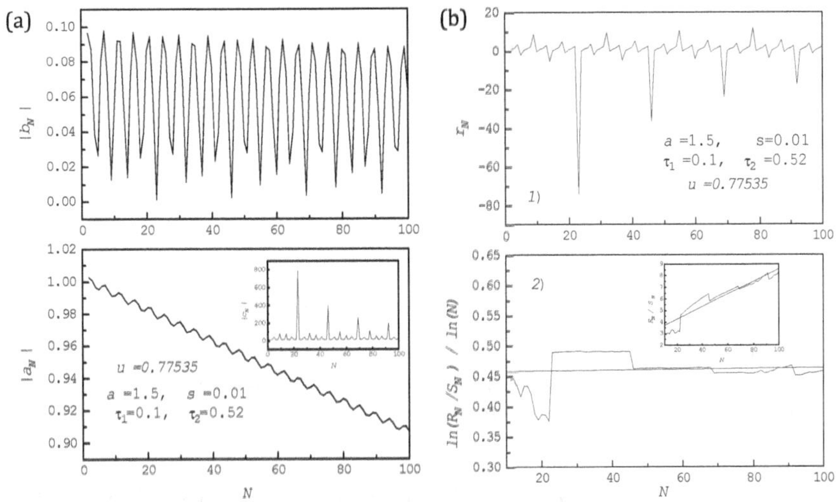

Figure 2.12 "Intermittence" of the secondary waves when $u \approx 0.77535$: (a) change of the inverse and direct wave amplitudes with the pulse number; (b) change of the sequence r_N (1) and Hurst's index (2).

A conclusion from these results follows from the exact solution of the problem: the field after its transformation in the medium with parameters disturbed by the finite packet of pulses that passes consists of two waves propagating in opposite directions. The frequencies of these waves coincide with the frequency of the primary wave. Their amplitudes have complicated dependencies on medium parameters and on the pulse duration and number. We have succeeded in finding a generalized parameter which determines the increase and decrease of the secondary waves as well as their irregular behavior. It is shown that parametric amplification is impossible if the medium conductivity varies alone in time. It is also shown that the inverse wave amplitude does not exceed the direct wave amplitude, and that they can both become mutually equal only in an asymptotic manner.

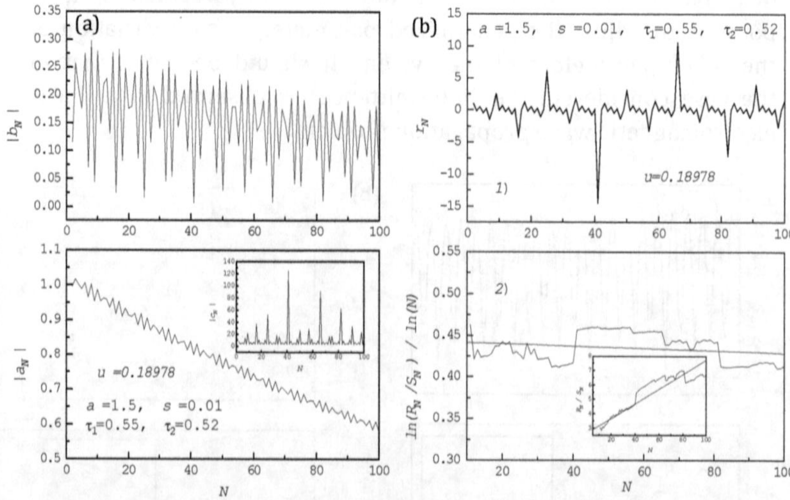

Figure 2.13 "Intermittence" of the secondary waves when u = 0.18978: (a) change of the inverse and direct wave amplitudes with the pulse number; (b) change of the sequence r_N (1) and Hurst's index (2).

2.3 Wave Chaotic Behavior Generated by Linear Time-Varying Systems

2.3.1 Complexity of the Signals

It was shown in the previous section that a simple electromagnetic signal in a time-varying medium can acquire very complex behavior. In order to estimate how complex the signals are we can calculate a "finite statistical complexity" measure of the signals. This approach to estimating the complexity of dynamical process rests on such well-known theories as Kolmogorov–Chaitin algorithmic complexity [13] and Shannon entropy [14]. The formalism is called "computational mechanics" and was originated in the works by Crutchfield and others [15–19]. The measure of complexity shows how much information is stored in the signal. It also indicates how much information is needed to predict the next value of the signal if we know all the values up to some moment in time. Intuitively, two limiting cases have zero complexity in this framework: if a signal has constant value at all times or if the signal is completely random. In both cases, no information exists about the previous evolution which is needed to predict the signal at later times; it is either constant or random regardless of the previous values. All intermediate cases have a finite, non-zero, value of complexity.

In "computational mechanics" a continuum signal is converted into a sequence of symbols from a pre-defined alphabet. Then, the "symbolic dynamics" are analyzed such that the number of dynamical "patterns" is extracted. The diversity of these patterns and their interrelations define the resulting complexity. The procedure of "symbolization" is illustrated in Fig. 2.14. The signal is considered as a continuous function represented by the dots at the discrete experimental data points. The alphabet is constructed by partitioning and, for example, for k = 3 consists of three symbols: $\{s_0, s_1, s_2\}$. The resulting symbolic sequence is shown in the bottom row in Fig. 2.14.

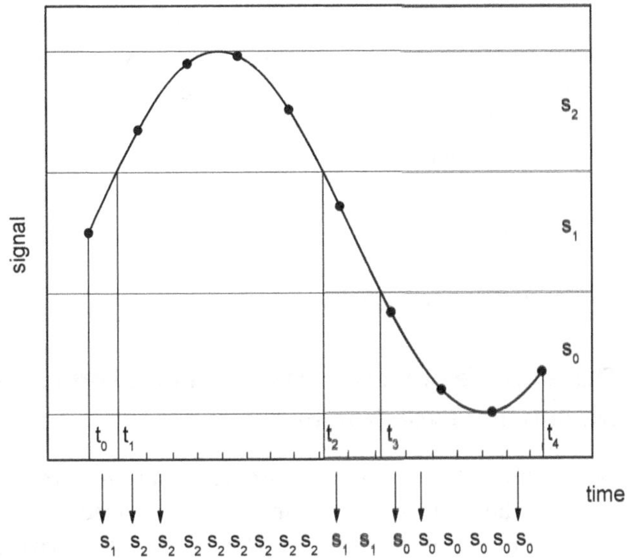

Figure 2.14 Symbolization of a continuous signal. Only discrete data points (the dots) representing the continuous signal (the solid curve) are available.

The procedure of symbolization is as follows [15]: (i) find the intersection points of a signal with the partition lines (e.g., t_1, t_2, t_3 in Fig. 2.14), taking the first and the last points of the signal as well (t_0, t_4); (ii) find the smallest interval in time Δt (equals to the length of $[t_1, t_1]$ for this example); (iii) for each interval find the number of symbols produced by this interval by dividing its length by Δt; (iv) form the final sequence by choosing the symbol from the partition space where the signal falls between the intersection points (Fig. 2.14).

It is important to consider the intersection points t_i as time interval boundaries and not the data points themselves. Otherwise, if the data points do not fall in the points of natural periodicity an artificial randomness is introduced into the final symbolic sequence. Also, it is our belief that generating a sequence of repeating symbols, like the ones on the $[t_1, t_1]$ interval, preserves more information from the original signal.

The algorithm for computing the finite statistical complexity follows the method described in Ref. [17]. It consists of considering the symbolic sub-sequences of a finite length and analyzing the "past" and "future" parts of them. The probabilities of various "futures" are calculated as the occurrence frequencies. Different sub-sequences form the dynamical "states" of the system and the time evolution is described as transitions between these states with some probabilities. The finite statistical complexity is calculated by the formula

$$C = -\sum_i P_i \log_2 P_i \tag{2.77}$$

where P_i is a probability of each dynamical state.

2.3.2 Dependence of Electromagnetic Pulse Form on a Modulation Cycle Number

When the conductivity of the medium is equal to zero exact expressions for the transformed field can be derived explicitly for an arbitrary initial signal $E_0(t, x) = F_0(t, x) = f(x - vt)$ by virtue of the resolvent (2.19), which in this case takes a simple form

$$\langle \mathbf{x}|R|\mathbf{x}'\rangle = R(t-t', x-x') = -\frac{1-a^2}{2}\frac{\partial}{\partial t}\delta(v_1(t-t') - |x-x'|) \tag{2.78}$$

Using this resolvent, one can obtain via Eq. (2.54) of the previous section the transformed field on the disturbance interval of the n-th cycle, $n > 1$. This field consists of two, forward and backward, waves

$$E_n(t,x) = E_n^{(+)}(t - x/v_1) + E_n^{(-)}(t + x/v_1) \tag{2.79}$$

that are determined by the free term on the previous cycle:

$$E_n^{(+)}(t - x/v_1) = \frac{a}{2}\Bigg[(a+1)F_{n-1}^{(+)}\left(\frac{t - x/v_1}{a} - (a-1)(n-1)T\right)$$
$$+ (a-1)F_{n-1}^{(-)}\left(-\frac{t - x/v_1}{a} - (a+1)(n-1)T\right)\Bigg]$$

$$E_n^{(-)}(t + x/v_1) = \frac{a}{2}\Bigg[(a-1)F_{n-1}^{(+)}\left(-\frac{t + x/v_1}{a} - (a+1)(n-1)T\right)$$
$$+ (a+1)F_{n-1}^{(-)}\left(\frac{t + x/v_1}{a} + (a-1)(n-1)T\right)\Bigg] \tag{2.80}$$

On the inactivity interval of the n-th cycle the field is equal to the free term which is calculated by virtue of Eq. (2.55)

$$E_n = F_n(t,x) = F_n^{(+)}(t - x/v) + F_n^{(-)}(t + x/v), \tag{2.81}$$

where

$$F_n^{(+)}(t - x/v) = F_0(t - x/v)$$
$$-\frac{1}{2a^2}\sum_{k=0}^{n-1}\Bigg[(a+1)\left\{E_{k+1}^{(+)}\left(\frac{t - x/v}{a} + \frac{a-1}{a}(kT + T_1)\right) - E_{k+1}^{(+)}\left(\frac{t - x/v}{a} + \frac{a-1}{a}kT\right)\right\}$$
$$-(a-1)\left\{E_{k+1}^{(-)}\left(-\frac{t - x/v}{a} + \frac{a+1}{a}(kT + T_1)\right) - E_{k+1}^{(-)}\left(-\frac{t - x/v}{a} + \frac{a+1}{a}kT\right)\right\}\Bigg]$$

$$F_n^{(-)}(t + x/v)$$
$$= \frac{-1}{2a^2}\sum_{k=0}^{n-1}\Bigg[(a-1)\left\{E_{k+1}^{(+)}\left(-\frac{t + x/v}{a} + \frac{a+1}{a}(kT + T_1)\right) - E_{k+1}^{(+)}\left(-\frac{t + x/v}{a} + \frac{a+1}{a}kT\right)\right\}$$
$$-(a+1)\left\{E_{k+1}^{(-)}\left(\frac{t + x/v}{a} + \frac{a-1}{a}(kT + T_1)\right) - E_{k+1}^{(-)}\left(\frac{t + x/v}{a} + \frac{a-1}{a}kT\right)\right\}\Bigg]$$

(2.82)

2.3.3 Wave Chaotic Behavior Generated by Linear Time-Varying Systems

Parametric phenomena in active media have been attracting much attention for a long time in connection with the transformation of electromagnetic waves by the time variation of the medium parameters. In the previous section, a linear parametric problem for the electromagnetic field in the homogeneous dielectric medium whose permittivity is modulated in time beginning from some moment is investigated. In such a system with distributed parameters, the main features of the wave transformation by the non-stationary medium are revealed by changing the medium parameters according to a simple law that allows one to construct an exact solution of the problem.

The electromagnetic wave transformation in a medium whose parameters undergo changes in the form of a finite packet of periodic rectangular pulses reveals the possibility of irregular behavior. Here we estimate the regularity of the transformation by three characteristics, Hurst's index [10], the signal complexity [15] and the Lyapunov exponent [11, 12].

Such a model allows one to carry out an exact investigation of the process. If the initial field is a plane monochromatic wave and the modulation is in the form of rectangular pulses then the problem has an exact analytical solution obtained in Section 2.2.1. This solution describes the temporal process of the wave transformation. Since under the modulation the permittivity changes abruptly from an initial value to the new one and then back, the transformation process consists of a progressive repetition of the known effect of each wave splitting into a pair of waves at each jump of the permittivity. These waves propagate in opposite directions and their frequencies change periodically by a jump, also from the initial value to the new one and back. The wave amplitudes undergo sophisticated transformations as time (the number of the permittivity jumps) increases. However, they are the solutions to the linear electrodynamic wave equations with variable coefficients. Since the essential feature in this process is the wave splitting (the wave reflection in time), it is interesting to observe the change in the relationship between the amplitudes of the forward and backward waves in the transformation process. These relations are obtained exactly in an analytical form.

There are long intervals in the sequence of the modulation periods where the sequence r_N changes almost regularly. After this interval, the relatively short intervals of strong irregular behavior of r_N take place. Larger deviations of u^2 from 1 lead to more irregular behavior of r_N. This phenomenon can be termed "quasi-intermittency." A similar behavior takes place for the transformed field. If $u^2 > 1$ it has the regular character with time as well as the sequence r_N (Fig. 2.15a). Otherwise, if $u^2 < 1$, the sequence as well as the field have irregular behavior (Fig. 2.15b).

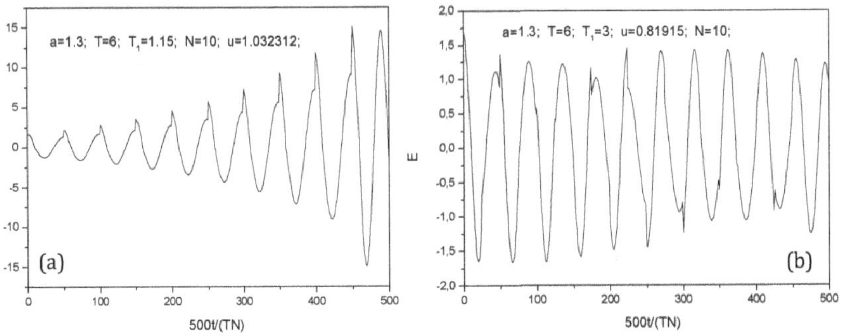

Figure 2.15 The behavior of the transformed field with time: (a) the parametric amplification; (b) the irregular changing.

The presence of the quasi-intermittency is confirmed by Hurst's method described in Section 2.1.4. The chaotic character of r_N is demonstrated by the calculation of Hurst's index. It can be seen from Fig. 2.16 that the decrease of the generalized parameter leads to a situation when Hurst's index takes magnitudes corresponding to white noise, $H > 0.5$.

The r_N behavior can be also characterized by a complexity measure [15–18]. This measure of complexity shows how much information is stored in the signal and how much information is needed to predict the next value of the signal if we know all the values up to some moment in time. In two limiting cases, when a signal has constant value at all times and when the signal is completely random, the complexity is equal to zero in this framework because in both cases no information is available from the previous evolution to predict the signal. All intermediate cases have a finite, non-zero, value of the complexity. The dependence of this measure on the modulation period shows a correlation between the complexity and the generalized parameter u (Fig. 2.17). The electromagnetic signal is regular and its complexity drops to zero when the absolute value of the generalized parameter u (the dash-dot sine-like line) becomes greater than 1.

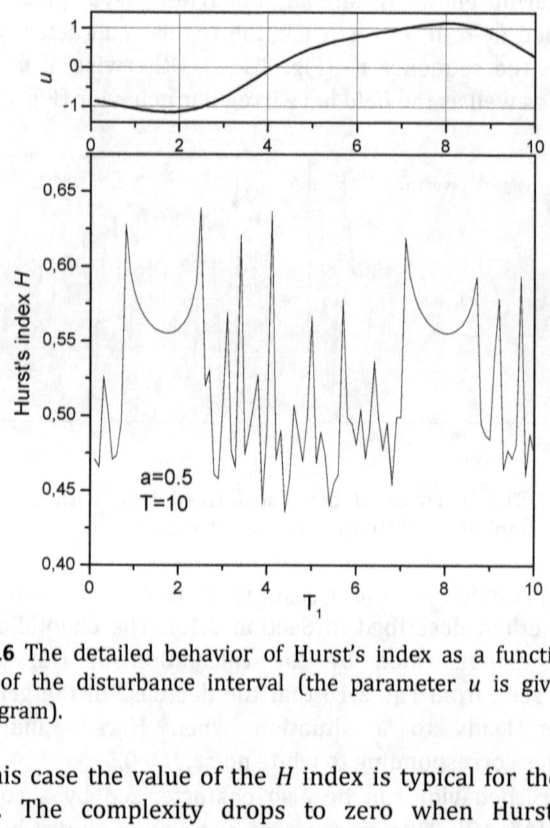

Figure 2.16 The detailed behavior of Hurst's index as a function of the duration of the disturbance interval (the parameter u is given in the upper diagram).

In this case the value of the H index is typical for the regular behavior. The complexity drops to zero when Hurst's index deviates notably from the value of 0.5 (on average) that corresponds to regular behavior of the signal (Fig. 2.17). Thus, a correlation exists between Hurst's index and the complexity of the signal, and it corresponds to the behavior of these two characteristics and the sequence r_N. Both characteristics correlate with the generalized parameter u. This is true for both cases when the medium becomes more, $a < 1$, or less, $a > 1$, dense on the disturbance intervals.

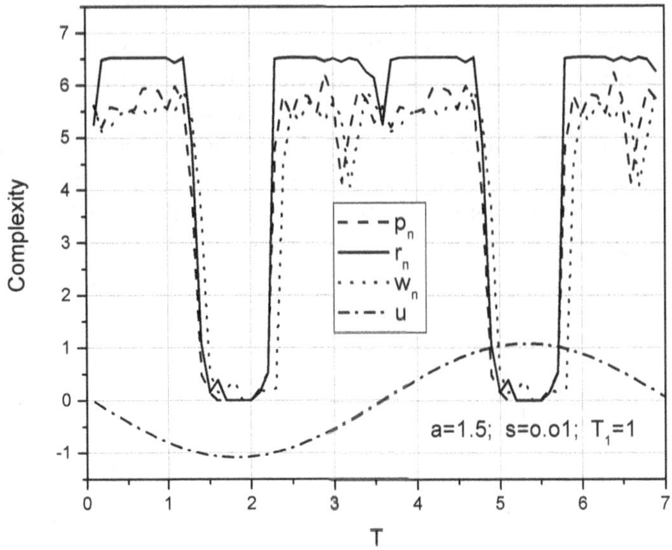

Figure 2.17 The behavior of the complexity as a function of the modulation periods (the parameter u is shown by the dash-dot sine-like line).

The behavior of r_n can be quantified by the Lyapunov exponent. Let us derive the estimation of this exponent as in Ref. 12. For this, we consider two nearby trajectories r_n and $r_n' = r_n + \tilde{r}_n$ of recurrence mapping $r_{n+1} = f(r_n)$ given by the formula (2.66). Using Taylor's series expansion of this formula one can derive

$$\tilde{r}_n = \tilde{r}_1 / (u^{2(n-1)} + B_1 u^{2(n-2)} + \dots + B_{n-1}), \tag{2.83}$$

where B_i are some coefficients that do not depend on u.

The Lyapunov exponent is determined by the evolution of the small disturbance in a linear approximation as

$$\Lambda = \overline{\lim_{n\to\infty}} \frac{1}{n} \ln \|\tilde{r}_n\| \tag{2.84}$$

Substitution of $\tilde{r}_n$ from Eq. 2.12 gives, at least for $u^2 > 1$,

$$\Lambda \approx \overline{\lim_{n\to\infty}} \frac{1}{n} \ln \|\tilde{r}_1 / u^{2(n-1)}\| = \overline{\lim_{n\to\infty}} \left(\ln \|\tilde{r}_1\| / n - \frac{n-1}{n} \ln \|u^2\| \right) = -\ln \|u^2\| \tag{2.85}$$

Therefore, the Lyapunov exponent is negative, if $u^2 > 1$. If one assumes that the estimation is true for $u^2 \le 1$ (which is not evident), then the Lyapunov exponent becomes positive. This analysis is confirmed by the direct calculation of the Lyapunov exponent by the formula (2.84), according to which its magnitudes become negative beginning from the generalized parameter value of $u \approx 0.5$.

Figure 2.18 shows that the Lyapunov exponent becomes positive for $u^2 \leq 0.6$. In the region $\sim 0.6 < u < 1$, there is a set of intermittent intervals of chaotic and non-chaotic behavior of the Lyapunov exponent. For $u > 1$, the Lyapunov exponents become strictly negative.

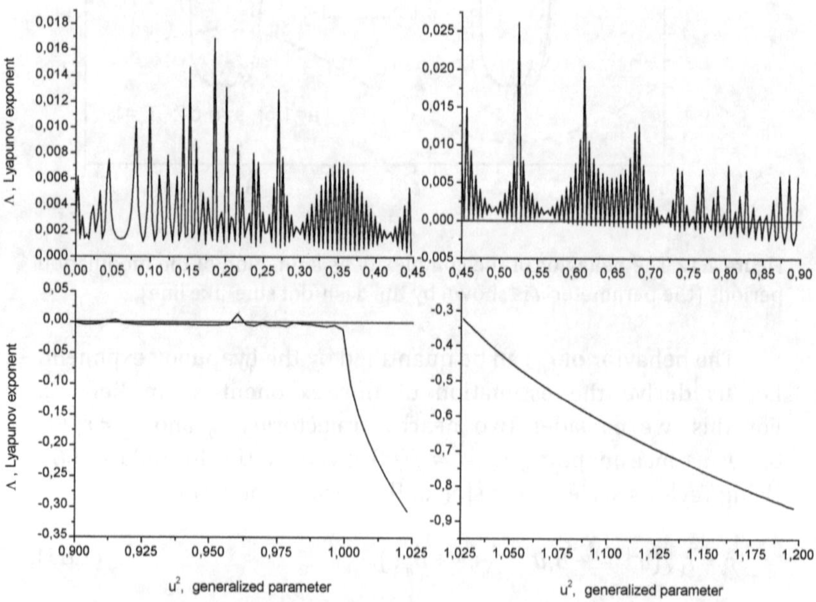

Figure 2.18 Dependence of the Lyapunov exponent on the generalized parameter.

So, it is shown that these linear systems and some of their characteristics can have chaotic behavior. It is worth noting that these types of models are often used in the description of various physical processes. One can conclude that the change of the dependent variables can transform the initially linear equations into non-linear ones that can have chaotic solutions. The chaotic behavior may reflect also the fact that the new variables themselves satisfy non-linear equations.

The chaotic dynamics generated by linear systems allow one to reveal unknown features of such systems that may be useful for their better and more complete understanding. Moreover, the methods of statistical physics can be used for the investigation of the system characteristics in the regime of dynamical chaos.

2.4 Electromagnetic Wave in Isotropic Plasma with Step-Wise Change of Plasma Density

One of the reasons for the non-stationary nature of a dispersive medium can be a time variation of plasma density. The electrodynamic processes in such a plasma were examined in many papers, for example [32–40]. However, these studies were restricted either to the one-dimensional case, or to a scalar problem, or to the initial (undisturbed) field taking some special form. This is why it is interesting to examine the non-stationary behavior from a unified point of view based on the evolutionary approach.

The general formulation of the problem is that a plasma filling an object $V(t)$ changes its parameters, for example plasma frequency $\omega_e(t)$, determined by plasma density, starting at the zero moment of time, or by the value of an extrinsic magnetic field, that is characterized by a gyrotropic frequency Ω_B. Within the integral equation approach it is not difficult to examine two kinds of medium inside the object $V(t)$. One kind matches the case when there is no background medium inside the object, that is, inside there is pure plasma, and the other matches the case when inside the object there is a mix of plasma with the background medium.

The electromagnetic field after zero moment is described by the equation

$$E = F + \int d\mathbf{x}' \left\langle \mathbf{x} \middle| \hat{K} \middle| \mathbf{x}' \right\rangle E(\mathbf{x}'), \tag{2.86}$$

where $\left\langle \mathbf{x} \,|\, \hat{K} \,|\, \mathbf{x}' \right\rangle = \left\langle \mathbf{x} \,|\, \hat{\Gamma} \frac{1}{\varepsilon\varepsilon_0} (\hat{\mathrm{P}}_e - \hat{\mathrm{P}}_{\mathrm{ex}}) \,|\, \mathbf{x}' \right\rangle$.

In this section, the initial-boundary-value problem is investigated for the case of a dispersive medium, namely the problem with a plasma whose plasma density experiences jump changes in time such that its plasma frequency changes from ω_{e1} to ω_{e2}. We use an equation corresponding to the plasma background examined in Chapter 1. In this case the propagation operator in impulse representation is defined by the expression:

$$\left\langle \mathbf{p} \middle| \hat{\Gamma}_e \middle| \mathbf{p}' \right\rangle_{ij} = -4\pi \frac{c^2 k_i k_j + (p^2 + \omega_e^2)\delta_{ij}}{p^2 + c^2 k^2 + \omega_e^2} \frac{p^2}{p^2 + \omega_e^2} \Delta(\mathbf{p} - \mathbf{p}'), \tag{2.87}$$

where ω_e is a background plasma density in the stationary plasma and $\Delta(\mathbf{p} - \mathbf{p}') = \frac{2\pi\delta(\mathbf{k}-\mathbf{k}')}{p-p'}$. A polarization vector of plasma is defined by the operator

$$\frac{1}{\varepsilon\varepsilon_0}(\mathbf{P} - \mathbf{P}_{\text{ex}}) = (\omega_{e2}^2 - \omega_e^2) \int_0^t dt'(t - t')\mathbf{E}(t', \mathbf{r}). \tag{2.88}$$

Here $\varepsilon = 1$ and ω_{e2} is a new plasma frequency. The matrix elements for the equation kernel for the intermediate stage of the evolution are given by

$$\left\langle \mathbf{p} \middle| \hat{\mathrm{K}}_2 \middle| \mathbf{p}' \right\rangle_{ij} = -\frac{c^2 k_i k_j + (p^2 + \omega_e^2)\delta_{ij}}{p^2 + c^2 k^2 + \omega_e^2} \frac{\omega_{e2}^2 - \omega_e^2}{p^2 + \omega_{e2}^2} \Delta(\mathbf{p} - \mathbf{p}'), \tag{2.89}$$

where $\mathrm{Re}(p - p') > 0$. The solution to the equation for the resolvent with this kernel results in

$$\left\langle \mathbf{p} \middle| \hat{R}_2 \middle| \mathbf{p}' \right\rangle_{ij} = -\frac{c^2 k_i k_j + (p^2 + \omega_e^2)\delta_{ij}}{p^2 + c^2 k^2 + \omega_e^2} \frac{\omega_{e2}^2 - \omega_{e1}^2}{p^2 + \omega_{e2}^2} \Delta(\mathbf{p} - \mathbf{p}'), \tag{2.90}$$

where $\mathrm{Re}(p - p') > 0$.

To construct the third stage resolvent we confine the investigation to the one-dimensional scalar case. This means that in the formulae (2.89), (2.90) we discard the tensor part $k_i k_j$ and carry out descending with respect to two transverse spatial coordinates. It gives the kernel of the second stage as

$$\left\langle \mathbf{p} \middle| \hat{\mathrm{K}}_2 \middle| \mathbf{p}' \right\rangle_{ij} = -\frac{\omega_{e2}^2 - \omega_{e1}^2}{p^2 + c^2 k^2 + \omega_e^2} \Delta(\mathbf{p} - \mathbf{p}'), \tag{2.91}$$

$$\left\langle \mathbf{x}\middle|\hat{\mathrm{K}}_2\middle|\mathbf{x}'\right\rangle_{ij} = -\frac{\omega_{e2}^2-\omega_{e1}^2}{2c}J_0\left(\omega_e\sqrt{(t-t')^2-\frac{(x-x')^2}{c^2}}\right)\theta\left(t-t'-\frac{|x-x'|}{c}\right),$$

(2.92)

and the second stage resolvent as

$$\left\langle \mathbf{p}\middle|\hat{R}_2\middle|\mathbf{p}'\right\rangle_{ij} = -\frac{\omega_{e2}^2-\omega_{e1}^2}{p^2+c^2k^2+\omega_{e2}^2}\frac{2\pi\delta(k-k')}{p-p'}, \tag{2.93}$$

$$\left\langle \mathbf{x}\middle|\hat{R}_2\middle|\mathbf{x}'\right\rangle = -\frac{\omega_{e2}^2-\omega_{e1}^2}{2c}J_0\left(\omega_{e2}\sqrt{(t-t')^2-\frac{(x-x')^2}{c^2}}\right)\theta\left(t-t'-\frac{|x-x'|}{c}\right),$$

(2.94)

Let us first examine a simple case when after the zero moment of time the plasma frequency becomes equal to ω_{e2}, that is, that at the zero moment a plasma frequency jump takes place. Then the polarization of the plasma after the jump is described by the formula

$$\mathbf{P} = \varepsilon_0\omega_{e2}^2\int_0^t (t-t')\mathbf{E}(t',\mathbf{r})dt' \tag{2.95}$$

Then matrix elements of the integral equation kernel $\hat{K}_2$ are

$$\left\langle \mathbf{p}\middle|\hat{\mathrm{K}}_2\middle|\mathbf{p}'\right\rangle_{ij} = -\frac{\omega_{e2}^2}{p^2}\frac{v^2k_ik^j+p^2\delta^{ij}}{p^2+v^2k^2}\Delta(\mathbf{p}-\mathbf{p}') \tag{2.96}$$

This determines the equation for the resolvent

$$\left\langle \mathbf{p}\middle|\hat{R}_2\middle|\mathbf{p}'\right\rangle = \left\langle \mathbf{p}\middle|\hat{K}_2\middle|\mathbf{p}'\right\rangle + \int d\mathbf{p}''\left\langle \mathbf{p}\middle|\hat{K}_2\middle|\mathbf{p}''\right\rangle\left\langle \mathbf{p}''\middle|\hat{R}_2\middle|\mathbf{p}'\right\rangle, \tag{2.97}$$

which after integration by $\mathbf{k}''$ takes the form

$$\left\langle \mathbf{p}\middle|\hat{R}_2\middle|\mathbf{p}'\right\rangle_{ij} = \left\langle \mathbf{p}\middle|\hat{K}_2\middle|\mathbf{p}'\right\rangle - \frac{\omega_{e2}^2}{p^2}\frac{v^2k_ik_m+p^2\delta_{im}}{p^2+v^2k^2}\left\langle \mathbf{p}\middle|\hat{R}_2\middle|\mathbf{p}'\right\rangle_{mj}. \tag{2.98}$$

If we represent the solution to this equation as a tensor

$$\left\langle \mathbf{p}\left|\hat{R}_2\right|\mathbf{p}'\right\rangle_{ij} = \left\langle \mathbf{p}\left|Q\right|\mathbf{p}'\right\rangle k_i k_j + \left\langle \mathbf{p}\left|S\right|\mathbf{p}'\right\rangle \delta_{ij}, \tag{2.99}$$

where the coefficients Q and S do not have singularities on the right side of the integration line then we obtain an algebraic system of equations for these coefficients. Solution of this system gives

$$\left\langle \mathbf{p}\left|\hat{R}_2\right|\mathbf{p}'\right\rangle_{ij} = -\frac{\omega_{e2}^2}{p^2 + v_2^2 k^2 + \omega_{e2}^2}\left(v_2^2 \frac{k_i k_j}{p^2 + \omega_{e2}^2} + \delta_{ij}\right)\Delta(\mathbf{p}-\mathbf{p}'), \tag{2.100}$$

where $v_2^2 = c^2/\mu$ and Re$p >$ Rep'.

Making the reverse transformation and a descent with respect to two transverse coordinates we obtain the coordinate representation of the resolvents:

$$\left\langle \mathbf{x}\left|\hat{R}_2^{(2)}\right|\mathbf{x}'\right\rangle_{ij} = -\frac{\omega_{e2}}{2v}\delta_{ij} J_0\left(\omega_{e2}\sqrt{\tau^2 - \frac{\left|x-x'\right|^2}{v^2}}\right)\theta\left(\tau - \frac{\left|x-x'\right|}{v}\right) \tag{2.101}$$

2.5 Plane Wave in Gyrotropic Plasma with "Switching On" of a Magnetizing Field

Switching on of an external magnetic field which acts on a plasma changes the type of electromagnetic oscillations inside the plasma, that is, it changes the plasma's properties [26–28, 34–36]. The transformation of electromagnetic waves in plasma with a slow time-varying external magnetic field has been considered in the adiabatic approximation. However, the adiabatic conditions also hold in the case of an abrupt magnetic field jump, when the period of oscillations is much longer than the duration of the jump. For instance, in free electron lasers [41], an electronic beam passes through a domain of change of a magnetic field polarity within $\Delta t \approx 10^{-8}$–10^{-10} s. Therefore, in the microwave band, $\omega \approx 10^7 \div 10^9$ rad s^{-1} the condition $\omega\Delta t << 1$ of an abrupt jump can be fulfilled. Naturally, the external magnetic field becomes inhomogeneous when it changes in time. Indeed, if the dimension of the domain considered is a, then the duration of the transient magnetic field is $\Delta t \approx a/u$, where u is the propagation velocity of the magnetic field disturbance. Therefore, the condition for the abrupt jump is $a << u/\omega$. The magnetic field in its turn can be considered homogenous if $a\kappa << 1$, where κ is a spatial scale of the field. As $\kappa > \omega_m/u$, where ω_m is the frequency of the magnetic field oscillations, we obtain the following conditions for the time change of the homogenous magnetic field to be considered abrupt:

$$a << \frac{u}{\omega} \qquad a << \frac{u}{\omega_m}$$

The fulfillment of these conditions is determined both by the type of electromagnetic oscillations considered and by the external magnetic field orientation that can be directed along one of the numerous branches of the plasma oscillations [42]. These conditions can be satisfied at least locally.

2.5.1 Basic Equations

Let us consider an electromagnetic field in plasma when the external magnetic field and the plasma density vary in time starting from the zero moment. The field is described by the operator equation

$$\mathbf{E} = \mathbf{F} + \hat{K}\mathbf{E} \tag{2.102}$$

with an operator $\hat{K} = \hat{\Gamma}\hat{V}$ where the propagator $\hat{\Gamma}$ is defined as

$$\hat{\Gamma} = \int_0^t dt' \int_\infty dt' \left\langle \mathbf{x} | \hat{\Gamma} | \mathbf{x}' \right\rangle \qquad \left\langle \mathbf{x} | \hat{\Gamma} | \mathbf{x}' \right\rangle = \left(\nabla\nabla - \frac{1}{c^2}\frac{\partial^2}{\partial t^2} \right) \frac{\delta\left(t - t' - \frac{1}{c}|\mathbf{r} - \mathbf{r}'| \right)}{|\mathbf{r} - \mathbf{r}'|}$$

and the medium operator $\hat{V}$ is given by the constitutive relation

$$\hat{V} = \varepsilon_0 \omega_e^2 \int_0^t dt' \hat{\alpha}(t - t') \tag{2.103}$$

$$\hat{\alpha}(t) = \frac{1}{\Omega} \begin{pmatrix} \sin\Omega t & 1 - \cos\Omega t & 0 \\ -1 + \cos\Omega t & \sin\Omega t & 0 \\ 0 & 0 & \Omega t \end{pmatrix}. \tag{2.104}$$

Here $\Omega = \omega_B = eB_0(mc)^{-1}$ is a gyrotropic frequency, and it is assumed that the magnetic field $\mathbf{B}_0$ is directed along the z axis.

The free term in Eq. (2.102) consists of two items

$$\mathbf{F}(t,\mathbf{r}) = \mathbf{F}^{(0)}(t,\mathbf{r}) + \int_0^\infty dt' \int_0^\infty d\mathbf{r}' \langle \mathbf{x} | \hat{\Gamma} | \mathbf{x}' \rangle \mathbf{P}_r(t',\mathbf{r}'), \tag{2.105}$$

where the first item

$$\mathbf{F}^{(0)}(t,\mathbf{r}) = \int_{-\infty}^0 dt' \int_\infty d\mathbf{r}' \langle \mathbf{x} | \hat{\Gamma} | \mathbf{x}' \rangle \frac{\omega_e^2}{4\pi} \int_{-\infty}^t dt''(t'-t'')E(t'',\mathbf{r}') \tag{2.106}$$

takes into account only the pre-history of the electromagnetic field interaction with the plasma (i.e., until the beginning of the magnetic field change), and $\mathbf{P}_r$ is the residual polarization.

A solution to Eq. 2.102 is given by the resolvent $\hat{R}$, $\mathbf{E} = \mathbf{F} + \hat{R}\mathbf{F}$, which satisfies the corresponding equation $\hat{R} = \hat{K} + \hat{K}\hat{R}$. An integral equation for the matrix elements of the resolvent operator follows from this operator equation and has the form

$$\langle \mathbf{x} | \hat{R} | \mathbf{x}' \rangle_{ij} = \langle \mathbf{x} | \hat{K} | \mathbf{x}' \rangle_{ij} + \int_0^\infty dt'' \int_\infty d\mathbf{r}'' \langle \mathbf{x} | \hat{K} | \mathbf{x}'' \rangle_{in} \langle \mathbf{x}'' | \hat{R} | \mathbf{x}' \rangle_{nj}, \tag{2.107}$$

where $\langle \mathbf{x} | \hat{K} | \mathbf{x}' \rangle_{ij} = \int_0^\infty dt'' \int_\infty d\mathbf{r}'' \langle \mathbf{x} | \hat{\Gamma} | \mathbf{x}'' \rangle_{in} \langle \mathbf{x}'' | \hat{V} | \mathbf{x}' \rangle_{nj}$ (2.108)

$$\langle \mathbf{x} | \hat{\Gamma} | \mathbf{x}' \rangle_{in} = \left(\frac{\partial^2}{\partial x_i \partial x_n} - \frac{1}{c}\frac{\partial^2}{\partial t^2}\delta_{in} \right) \frac{\delta\left(t - t' - \frac{1}{c}|\mathbf{r}-\mathbf{r}'|\right)}{|\mathbf{r}-\mathbf{r}'|}. \tag{2.109}$$

δ_{ij} is the Kroneker index and

$$\langle \mathbf{x} | \hat{V} | \mathbf{x}' \rangle_{\text{in}} = \varepsilon_0 \omega_e^2 \delta(\mathbf{r}-\mathbf{r}')\hat{\alpha}(t-t')\delta_{\text{in}} \tag{2.110}$$

2.5.2 The Resolvent for the Integral Equation

It is convenient to obtain a solution to Eq. 2.107 in terms of the impulse representations

$$\left\langle \mathbf{p}\left|\hat{R}\right|\mathbf{p}\right\rangle_{ij} = \int_0^\infty dt \int_0^\infty dt' \int_\infty d\mathbf{r} \int_\infty d\mathbf{r}' e^{-pt-i\mathbf{kr}} \left\langle \mathbf{x}\left|\hat{R}\right|\mathbf{x}'\right\rangle_{ij} e^{p't'+i\mathbf{k}'\mathbf{r}'} \quad (2.111)$$

using the formulae for operators in the impulse representation

$$\left\langle \mathbf{P}\,|\,\hat{\Gamma}\,|\,\mathbf{P}'\right\rangle_{ij} = -4\pi \frac{c^2 k_i k_j + p^2 \delta_{ij}}{p^2 + c^2 k^2} \Delta(\mathbf{p}-\mathbf{p}') \quad (2.112)$$

$$\left\langle \mathbf{p}\,|\,\hat{V}\,|\,\mathbf{p}'\right\rangle_{ij} = \varepsilon_0 \omega_p^2 \alpha_{ij} \Delta(\mathbf{p}-\mathbf{p}')$$

$$\left\langle \mathbf{p}\,|\,\hat{K}\,|\,\mathbf{p}'\right\rangle_{ij} = \int_{\zeta-i\infty}^{\zeta-i\infty} \frac{dp''}{2\pi i} \int_\infty \frac{dp''}{(2\pi)^3} \left\langle \mathbf{p}\,|\,\hat{\Gamma}\,|\,\mathbf{p}''\right\rangle_{im} \left\langle \mathbf{p}''\,|\,\hat{V}\,|\,\mathbf{p}'\right\rangle_{mj},$$

where $\hat{\alpha}_{ij}$ are the elements of the matrix

$$\hat{\alpha} = \frac{1}{p^2(p^2+\Omega^2)} \begin{pmatrix} p^2 & p\Omega & 0 \\ -p\Omega & p^2 & 0 \\ 0 & 0 & p^2+\Omega^2 \end{pmatrix}.$$

In these formulae, $\mathbf{p} = (p, \mathbf{k})$ where p is a complex variable, $\mathbf{k}$ is a real vector, $\Delta(\mathbf{p}-\mathbf{p}') = (2\pi)^3 \frac{\delta(\mathbf{k}-\mathbf{k}')}{p-p'}$ and the external magnetic field is assumed to be directed along the z axis $\mathbf{b} = \mathbf{B}_0/B_0 = \{0, 0, 1\}$.

After manipulations we obtain the resolvent

$$\left\langle \mathbf{x}\,|\,R^{(1)}\,|\,\mathbf{x}'\right\rangle = \int_{-i\infty}^{i\infty} \frac{dp}{2\pi i} \int_\infty \frac{d\mathbf{k}}{(2\pi)^3} T(p,\mathbf{k})\, e^{p(t-t')+i\mathbf{k}(\mathbf{r}-\mathbf{r}')}, \quad (2.113)$$

where the matrix T has the following form:

$$T(p,\mathbf{k}) = \frac{-\omega_e^2 c^2}{G(p,\mathbf{k})} \varphi_e^2 (p^2 T_1 + p\Omega T_2 + \frac{p^2}{c^2} T_3 + \frac{p^2}{c^2}(p^2+\omega_e^2)\hat{I}) \quad (2.114)$$

Here $\varphi_e^2 = p^2 + c^2k^2 + \omega_e^2$ and I is the identity matrix.

The T_j matrices are equal to

$$T_1 = \begin{pmatrix} k_1^2 & k_1 k_2 & \left(1+\dfrac{\Omega^2\varphi^2}{p^2\varphi_e^2}\right)k_1 k_3 \\ k_2 k_1 & k_2^2 & \left(1+\dfrac{\Omega^2\varphi^2}{p^2\varphi_e^2}\right)k_2 k_3 \\ k_3 k_1 & k_3 k_2 & \left(1+\dfrac{\Omega^2\varphi^2}{p^2\varphi_e^2}\right)k_3^2 \end{pmatrix} \tag{2.115}$$

$$T_2 = \begin{pmatrix} -k_1 k_2 & k_2^2 + \dfrac{\omega_e^2}{\varphi_e^2}k_3^2 & -\dfrac{\omega_e^2}{\varphi_e^2}k_2 k_3 \\ -k_2^2 - \dfrac{\omega_e^2}{\varphi_e^2}k_3^2 & k_2^2 & \dfrac{\omega_e^2}{\varphi_e^2}k_1 k_3 \\ -\dfrac{\varphi^2}{\varphi_e^2}k_3 k_2 & \dfrac{\varphi^2}{\varphi_e^2}k_3 k_1 & 0 \end{pmatrix}$$

$$T_3 = \begin{pmatrix} 0 & p\,\Omega & 0 \\ -p\,\Omega & 0 & 0 \\ 0 & 0 & \Omega^2\varphi^2/\varphi_p^2 \end{pmatrix},$$

where $\varphi^2 = p^2 + c^2k^2$.

The gyromagnetic polynomial in Eq. 2.114 has the following representation:

$$G(p, \mathbf{k}) = p^2\varphi_e^4(p^2 + \omega_e^2) + \Omega^2\varphi^2(p^2\varphi_e^4 + \omega_e^2c^2k_3^2) \tag{2.116}$$

2.5.3 The Case of the Arbitrary Time-Varying Magnetic Field Approximation

An arbitrary time-varying magnetic field can be approximated by a sequence of magnetic field jumps. In this case there is a need to advance the initial point of time sequentially from step to step. After each step, the transformed electromagnetic field is determined by the constructed resolvent in which the value Ω should be replaced by a new one corresponding to a new value of the external magnetic field, that is, Ω_{n-1} should be replaced by Ω_n when passing from the $(n - 1)$-th step to the n-th one. Clearly, there is a need to take into account the change in the equation free term to which the integral over the previous time interval should be added after each step. Let us consider a situation when the external magnetic field is absent until the zero moment and then from this moment it varies with time in a step-by-step manner, being constant between jumps. After the n-th jump in the external magnetic field the polarization vector $\mathbf{P}^{(n)}$ is determined by the following relations:

$$\mathbf{P}^{(n)}(t) = \mathbf{P}_r^{(n)}(t) + \varepsilon_0\omega_e^2 \int_{t_n}^{t} \hat{\alpha}^{(n)}(t-t')\mathbf{E}^{(n)}(t')dt' \tag{2.117}$$

$$\mathbf{P}_r^{(n)}(t) = \mathbf{P}^{(n-1)}(t_n) + \hat{\alpha}^{(n)}(t-t_n)\frac{d\mathbf{P}^{(n-1)}(t_n)}{dt} \quad \mathbf{P}_r^{(0)} = 0, \tag{2.118}$$

where t_n is the moment of the n-th jump in the magnetic field and $\hat{\alpha}^{(n)}$ is the corresponding susceptibility tensor.

Taking into account the expressions for the electric polarization in Eqs. (2.117) and (2.118), the equation for the electric intensity of the electromagnetic field, after the n-th jump in the external magnetic field, can be represented as the Volterra integral equation of the second kind:

$$\mathbf{E}^{(n)}(t,\mathbf{r}) = \mathbf{F}^{(n)}(t,\mathbf{r}) + \int_{t_n}^{t} dt' \int_{\infty} d\mathbf{r}' \left\langle \mathbf{x} | \hat{\Gamma} | \mathbf{x}' \right\rangle \varepsilon_0\omega_e^2 \int_{t_n}^{t'} \hat{\alpha}^{(n)}(t-t'')\mathbf{E}^{(n)}(t'',\mathbf{r}')dt'' \tag{2.119}$$

The free term of this equation is equal to

$$\mathbf{F}^{(n)}(t,\mathbf{r}) = \mathbf{F}^{(0)}(t,\mathbf{r}) + \sum_{k+1}^{n-1} \int_{t_k}^{t_{k+1}} dt' \int_{\infty} d\mathbf{r}' \left\langle \mathbf{x} | \hat{\Gamma} | \mathbf{x}' \right\rangle \mathbf{P}_r^{(k)}(t',\mathbf{r}')$$
$$+ \int_{t_n}^{t} dt' \int_{\infty} d\mathbf{r}' \left\langle \mathbf{x} | \hat{\Gamma} | \mathbf{x}' \right\rangle P_r^{(n)}(t',\mathbf{r}') \tag{2.120}$$

The first term in Eq. (2.120) takes into account only the pre-history of the electromagnetic field interaction with the plasma until the beginning of the change in the magnetic field and is determined by Eq. (2.105); the other terms describe after-effects caused by the external magnetic field jumps.

To solve Eq. (2.119), for $\mathbf{E}^{(n)}$, the resolvent method is used. With the aid of the resolvent, the formula for the field intensity in the magneto-active plasma after the n-th jump in the magnetic field can be expressed as

$$\mathbf{E}^{(n)}(t,\mathbf{r})=\mathbf{F}^{(n)}(t,\mathbf{r})+\int_{0}^{t}dt'\int_{\infty}d\mathbf{r}'\ \langle\mathbf{x}|R^{(n)}|\mathbf{x}'\rangle\mathbf{F}^{(n)}(t',\mathbf{r}') \qquad (2.121)$$

As has been mentioned above, the expression for the resolvent for the sequence of jumps is the same as the resolvent for a single jump in the external magnetic field if the Larmor frequency is changed from step to step.

2.5.4 The Transformation of a Plane Wave by a Jump of a Magnetic Field

Consider the initial field as that of an eigen-wave of the plasma, that is, a plane wave: $\mathbf{E}_0(t, \mathbf{r}) = \mathbf{E}_0 \exp[i(\omega t - \mathbf{sr})]$, where $s^2 = c^2(\omega^2 - \omega_e^2)$. Suppose that it exists until the external magnetic field in the plasma is switched on. Let us also consider that the transformations of the wave are made for a single jump in the magnetic field. Substituting the expression for the initial field in Eqs. 2.117 and 2.118 and then substituting the obtained expressions for $\mathbf{P}_r$ and $\mathbf{P}$ in Eq. 2.120 yields the free term as follows:

$$\mathbf{F}^{(1)}(t,\mathbf{r})=\int_{-i\infty}^{i\infty}\frac{dp}{2\pi i}\Phi(p)e^{pt-i\mathbf{sr}}, \qquad (2.122)$$

where

$$\Phi(p)=\frac{i}{\omega(p^2+c^2s^2)}[(c^2s^2-i\omega p)\mathbf{E}_0+p^2\omega_e^2\hat{\alpha}\mathbf{E}_0+c^2\omega_e^2\mathbf{s}(\mathbf{s}\hat{\alpha}\mathbf{E}_0)]$$

Substituting Eq. 1.122 into 2.121 yields the expression for the transformed electric field when the external magnetic field is arbitrarily orientated:

$$\mathbf{E}^{(1)}(t,\mathbf{r})=\int_{-i\infty}^{i\infty}\frac{dp}{2\pi i}(\hat{I}+T(p,\mathbf{s}))\Phi(p)e^{pt-i\mathbf{sr}} \qquad (2.123)$$

Formula (2.123) is a general expression which determines the transformed electromagnetic field for an arbitrary orientation of the external magnetic field. It can be simplified if the orientation of the external magnetic field is properly specified. Let us investigate two special cases of the mutual orientation of the magnetic field $\mathbf{B}_0$ and the wave vector $\mathbf{s}$.

Let us consider the specific cases of the orientation of the external magnetic field and the primary wave. At first let the external magnetic field $\mathbf{B}_0$ be orientated along the propagation direction of the primary wave, that is, $\mathbf{b}||\mathbf{s} = \{0, 0, s\}$ and $\mathbf{E}_0 = \{E_0, 0, 0\}$. In such a case, the gyromagnetic polynomial can be simplified as

$$G(p,\mathbf{s}) = (p^2 + \omega_e^2)H(p), \tag{2.124}$$

where $H(p) = p^2(p^2 + \omega^2)^2 + \Omega^2 (p^2 + c^2s^2)^2$. The electromagnetic field is defined by the following expression:

$$\mathbf{E}^{(1)}(t,\mathbf{r}) = \int_{-i\infty}^{i\infty} \frac{dp}{2\pi i} \frac{e^{pt-i\mathbf{sr}}}{\omega H(p)} \begin{Bmatrix} \omega(p+i\omega)Q^{(1)}(p) - i\omega_e^2\Omega^2(p^2 + c^2s^2) \\ -i\omega_e{}^2\Omega p^2(p+i\omega) \\ 0 \end{Bmatrix}, \tag{2.125}$$

where $Q^{(1)}(p) = p^4 + (\omega^2 + \Omega^2)p^2 + \Omega^2c^2s^2$.

The analysis of Eq. 2.125 gives us a possibility to conclude that the transformed electric field consists of three pairs of waves with the frequencies p_k, which are the roots of the polynomial $H(p)$ (see Fig. 2.19). Waves in each pair remain transverse, preserve the wave vector $\mathbf{s}$, and propagate in opposite directions (see Fig. 2.20) [43].

In the second case, the external magnetic field is switched on normally to the wave vector, that is, $\mathbf{b} \perp \mathbf{s} = \{s, 0, 0\}$, $\mathbf{E}_0 = \{0, E_2, E_3\}$. The transformed electric field will be determined by the expression

$$\mathbf{E}^{(1)}(t,\mathbf{r}) = \mathbf{E}_{||}^{(1)} \exp[i(\omega t - \mathbf{sr})] + \sum_{m=0}^{1}\sum_{k=1}^{2} \mathbf{E}_{\perp mk}^{(1)} \frac{\exp((-1)^m i p_k t - i\mathbf{sr})}{\omega\ Q^{(2)}((-1)^m i\ p_k)}, \tag{2.126}$$

where $Q^{(2)}(p) = p^4 + p^2(\omega^2 + \omega_e^2 + \Omega^2) + \omega^2\omega_e^2 + \Omega^2c^2s^2$. The first term in Eq. 2.126 $\mathbf{E}_{||}^{(1)} = \{0, 0, E_3\}$ is directed along the magnetic field and is the same as the primary wave (see Fig. 2.21), whereas the second one is perpendicular to the external magnetic field

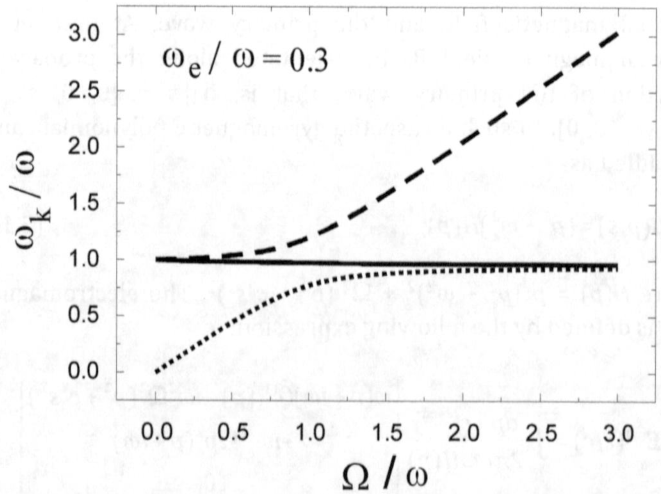

Figure 2.19 New frequencies in the case when the external magnetic field is orientated along the propagation direction of the primary wave (**b**||**s**).

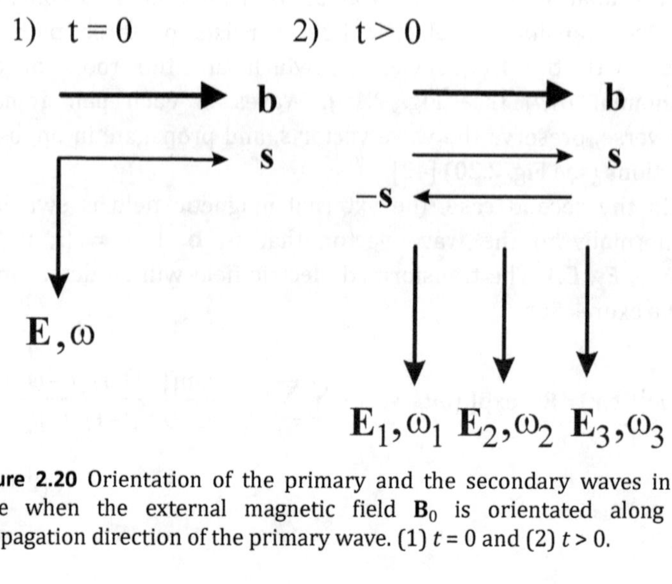

Figure 2.20 Orientation of the primary and the secondary waves in the case when the external magnetic field $\mathbf{B}_0$ is orientated along the propagation direction of the primary wave. (1) $t = 0$ and (2) $t > 0$.

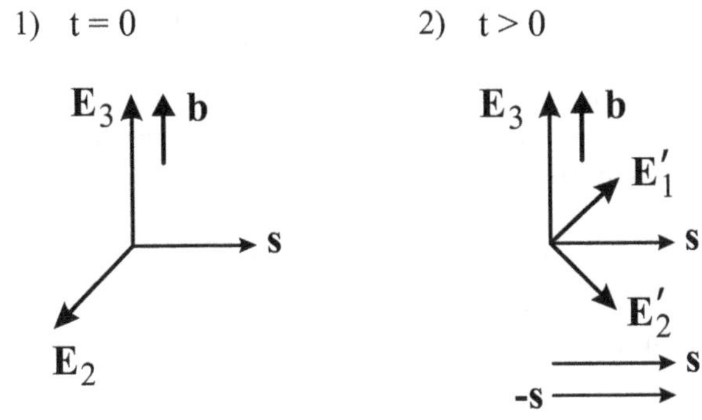

Figure 2.21 Orientation of the primary and the secondary waves in the case when the external magnetic field $\mathbf{B}_0$ is orientated perpendicular to the direction of the primary wave propagation. (1) $t = 0$ and (2) $t > 0$.

$$\mathbf{E}^{(1)}_{\perp mk} = \begin{Bmatrix} -\Omega\omega_p^2((-1)^m p_k + \omega) \\ i\omega((-1)^m p_k + \omega)(-p_k^2 + \omega_p^2 + \Omega^2) - i\Omega^2\omega_p^2 \\ 0 \end{Bmatrix} E_2 \tag{2.127}$$

The new frequencies are defined by the expression

$$\omega_k^2 = \frac{1}{2}(\omega^2 + \omega_p^2 + \Omega^2 + (-1)^{k-1}\sqrt{(\omega^2 - \omega_p^2 - \Omega^2)^2 + 4\omega_p^2\Omega^2}) \tag{2.128}$$

and are presented in Fig. 2.22.

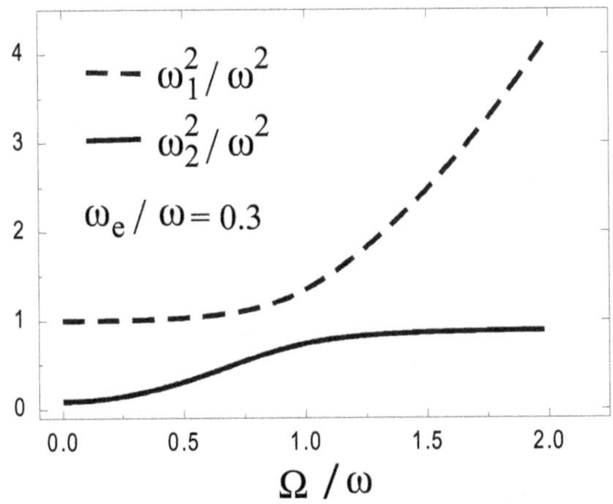

Figure 2.22 The new frequencies in the case when the external magnetic field is normal to the direction of the primary wave $\mathbf{b} \perp \mathbf{E}_0$.

In this case the component of the primary electric vector which is parallel to **b** is not changed. The component normal to **b** forms two pairs of waves with the frequencies $\pm\omega_k$ (Eq. 2.128). These waves have both longitudinal and transversal components and correspond to the fast and the slow extraordinary waves, respectively.

It should be pointed out that in all cases an abrupt change in the external magnetic field transfers a line spectrum of a primary electromagnetic field to the line spectrum of the transformed field.

2.5.5 The Transformation of the Plasma Oscillations

Substituting **s** = 0 and $\omega = \omega_e$ in Eq. 2.125 yields the transformation of plasma oscillations when the external magnetic field is switched on at the zero moment and its vector **b** is normal to the electric field in the plasma oscillations. The latter transforms into two elliptically polarized oscillations:

$$\mathbf{E}^{(1)}(t)=\sum_{k=1}^{2}\frac{E_0}{2(\omega_e^2-\omega_k^2)+\Omega^2}\left\{(\omega_e^2-\omega_k^2+\Omega^2)\cos\omega_k t,\quad \frac{\omega_e^2\Omega}{\omega_k}\sin\omega_k t,\quad 0\right\},\tag{2.129}$$

where $\omega_k^2=\frac{1}{2}(2\omega_e^2+\Omega^2+(-1)^{k-1}\Omega\sqrt{4\omega_e^2+\Omega^2})$. The dependence of the temporal behavior of the plasma oscillation field on the external magnetic field is shown in Fig. 2.23.

The new frequencies of the plasma oscillations are presented in Fig. 2.24.

In the case of a weak magnetic field, $\Omega \ll \omega_e$, both types of oscillations are almost circularly polarized and have the frequencies $\omega_k^2 \approx \omega_e^2 \pm \omega_e\Omega$, which are slightly different from each other

$$\mathbf{E}^{(1)}(t)=\sum_{k=1}^{2}\frac{E_0\omega_e}{\Omega-(-1)^{k-1}2\omega_e}\left\{\left[\frac{\Omega}{\omega_e}-(-1)^{k-1}\right]\cos\omega_k t,\ \left[1-(-1)^{k-1}\frac{\Omega}{2\omega_e}\right]\sin\omega_k t,\ 0\right\}\tag{2.130}$$

In the case of a strong magnetic field, $\Omega \gg \omega_e$, the oscillation with the frequency $\omega_1 \approx \Omega$ will be a circularly polarized cyclotron oscillation with a small amplitude

$$\mathbf{E}_1(t) = -E_0 \frac{\omega_e^2}{\Omega^2}\{\cos\Omega t, \sin\Omega t, 0\}. \tag{2.131}$$

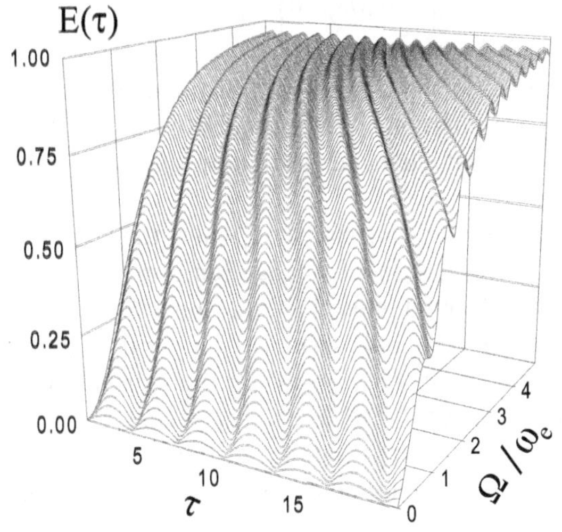

Figure 2.23 The time dependence of the plasma oscillation field when the external magnetic field is normal to the electric field in the plasma oscillation, $\mathbf{b} \perp \mathbf{E}_0$.

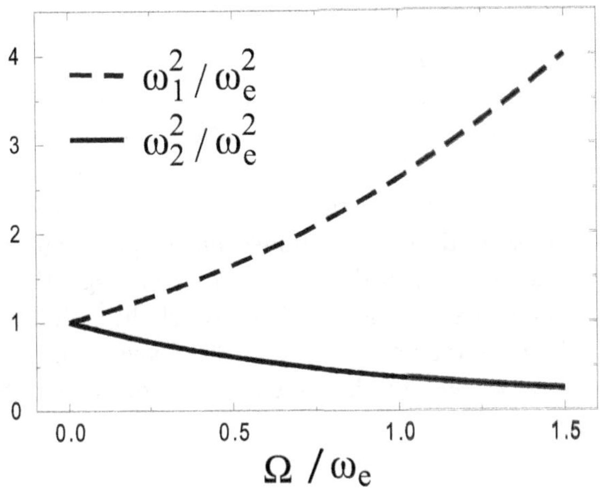

Figure 2.24 New frequencies of the plasma oscillations when the external magnetic field is normal to the electric field in the plasma oscillations $\mathbf{b} \perp \mathbf{E}_0$.

The other oscillation will have the frequency $\omega_2^2 \approx 4\omega_e^4/\Omega_1^2$ and elliptical polarization:

$$\mathbf{E}_2(t) = E_0 \left\{ \cos\omega_2 t, \frac{1}{2}\sin\omega_2 t, 0 \right\} \tag{2.132}$$

When the external magnetic field is parallel to the electric field of the plasma oscillations, that is $\mathbf{b} \parallel \mathbf{E}_0 = \{E_0, 0, 0\}$, the plasma oscillations are not changed, as follows from Eqs. (2.126) and (2.128).

To conclude, one can say that when the external magnetic field is orientated along the propagation direction of the primary wave, the transformed electric field consists of three pairs of waves. Waves in each pair remain transverse and propagate in opposite directions. In the case when the external magnetic field is switched on normally to the wave vector, the component of the primary electric vector parallel to the external magnetic field is not changed, and the component normal to the external magnetic field forms two pairs of waves which have both longitudinal and transverse components. The transformation of waves in the case of a sequence of arbitrary magnetic field jumps is given by the sequential transfer of the zero moment of time; the field transformed under the influence of magnetization will have a discrete spectrum.

Switching of a magnetic field that is normal to the electric field in plasma oscillations transforms the oscillations into two elliptically polarized oscillations with different frequencies. In the case of a strong magnetic field, the first of these oscillations has almost a circular polarization and a frequency close to the cyclotron one, and the second has an almost linear polarization and a frequency close to zero. In the case of a weak magnetic field, the oscillations have frequencies which are slightly different, and both are almost circularly polarized. The switching on of the magnetic field parallel to the electric field does not affect the electric field.

References

1. Born, M., and Wolf, E. (1964) *Principles of Optics*, Pergamon Press, Oxford, 856 p.

2. Morgenthaler, F. R. (1958) Velocity modulation of electromagnetic waves, *IRE Trans. Microwave Theory Techn.*, **MTT-6**(4), 167–172.
3. Felsen, L. B., and Whitman, G. M. (1970) Wave propagation in time-varying media, *IEEE Trans. Antennas Propag.*, **AP-18**(2), 242–253.
4. Fante, R. L. (1971) Transmission of electromagnetic waves into time-varying media, *IEEE Trans. Antennas Propag.*, **AP-193**, 417–424.
5. Ostrovsky, L. A., and Stepanov, N. S. (1971) Nonresonance parametric phenomena in distributed systems, *Radiophysics Quantum Electron.* (English transl.), **14**, 489–529 (in Russian pagination).
6. Stolyarov, S. N. (1983) Resonance transform of waves in periodic nonstationary medium, *Radiophys. Quantum Electron.* (English transl.), **26,** 514–516 (in Russian pagination).
7. Averkov, S. I., and Boldin, V. P. (1980) Waves in nondispersive nonstationary inhomogeneous media, *Radiophys. Quantum Electron.* (English transl.), **23**, 1060–1066 (in Russian pagination).
8. Nerukh, A. G. (1999) Intermittency of electromagnetic waves in a regular time-varying medium, *J. Phys. D Appl. Phys.*, **32**, 2006–2013.
9. Ruzhytska, N. N., Nerukh, A. G., and Nerukh, D. A. (2003) Accurate modelling of pulse transformation by adjustable-in-time medium parameters, *Opt. Quantum Electron.*, **35**(4), 347–364.
10. Hurst, H. E., Black, R. P., and Simaika, Y. M. (1965) *Long-Term Storage: An Experimental Study*, Constable, London.
11. Lichtenberg, A. J., and Lieberman, M. A. (1983) Regular and Stochastic Motion, Springer-Verlag, New York, 499 p.
12. Kuznetsov, S. P. (2000) *Dynamical Chaos* (in Russian), Moscow.
13. Li, M., and Vitanyi, P. M. B. (1993) *An Introduction to Kolmogorov Complexity and its Applications*, Springer-Verlag, New York.
14. Shannon, C. E. (1948) A mathematic theory of communication, *Bell System Tech. J.*, **27**, 379.
15. Crutchfield, J. P., and Young, K. (1989), Inferring statistical complexity, *Phys. Rev. Lett.*, **63,** 105–108.
16. Crutchfield, J. P. (1994) The calculi of emergence: Computation, dynamics and induction, *Phys. D*, **75,** 11.
17. Nerukh, D., Karvounis, G., and Glen, R. C. (2002) Complexity of classical dynamics of molecular systems. Part I: Methodology, *J. Chem. Phys.*, **117**(21), 9611–9617.
18. Perry, N., and Binder, P.-M. (1999) Finite statistical complexity for soft systems, *Phys. Rev. E*, **60**, 459.

19. Cover, T. M., and Thomas, J. A. (1991) *Elements of Information Theory,* John Wiley & Sons, Inc.
20. Loskutov, Yu, A., and Mikhailov, A. S. (1990) *Introduction into Synergetics* (in Russian), "Nauka" Main Editorial Board for Literature on Physics and Mathematics, Moscow, 272 p.
21. Akhromeyeva, T. S., Kurdyumov, S. P., Malinetskii, G. G., and Samarskii, A. A. (1992) *Chaos and Dissipative Structures in Reaction-Diffusion Systems* (in Russian), "Nauka" Main Editorial Board for Literature on Physics and Mathematics, Moscow, 544 p.
22. Ostrovskiy, L. A. (1971) Some general relations for waves on moving boundary between two media, (in Russian), *Zhurnal Eksperimentalnoy I Teoreticheskoy Fiziki*, 61, **2**(8), 551–561.
23. Ostrovskiy, L. A., and Stepanov, N. S. (1971) Non-resonance parametric phenomena in distributive systems (in Russian), *Izv. Vuz. Radiofizika*, **14**(4), 489–529.
24. Ostrovskiy, L. A. (1975) Some "paradoxes" of moving boundaries in electrodynamics (in Russian), *Uspekhi Fizicheskikh Nauk*, **116**(2), 313–326.
25. Silin, V. P., and Rukhadze, A. A. (1961) *Electromagnetic Features of Plasma and Plasma-like Media*, GosAtomIzdat, Moscow, 280 p.
26. Ginzburg, V. L., and Rukhadze, A. A. (1975) *Waves in Magnito-Active Plasma* (in Russian), Nauka, Moscow, 255 p.
27. Stepanov, N. S. (1976) About permittivity of non-stationary plasma (in Russian), *Izv. Vuz. Radiofizika*, **19**(7), 960–968.
28. Ginsburg, V. L. (1967) *Electromagnetic Wave Propagation in Plasma* (in Russian), Nauka, Moscow.
29. Pauli, W. (1958) *Theory of Relativity*, Pergamon Press.
30. Tolmen, R. C. (1969) *Relativity, Thermodynamics and Cosmology*, Clarendon Press, Oxford.
31. Tamm, I. E. (1924) Electrodynamics of anisotropic medium in special relativity theory (in Russian), *J. Russian Phys. Chem. Soc.*, **56**(2–3), 248–262.
32. Kalluri, D. K. (1999) *Electromagnetics of Complex Media: Frequency Shifting by a Transient Magnetoplasma Medium*, CRC Press, Boca Raton, 292 p.
33. Huang, T. T., Lee, J. H., Kalluri, D. K., and Groves, K. M. (1998) Wave propagation in transient plasma: Development of Green's function, *IEEE Trans. Plasma Sci.*, **26**(1), 19–25.

34. Kalluri, D. K., and Goteti, V. R. (1996) Frequency upshifting with power intensification of a whistler wave by a collapsing plasma medium, *J. Appl. Phys.*, **79**(8), 3875–3899.

35. Kalluri, D. K. (1993) Frequency shifting using magnetoplasma medium: Flash ionization, *IEEE Trans. Plasma Sci.*, **21**(1), 77–81.

36. Stanic, B. V., and Draganovich, I. N. (1996) Electromagnetic waves in a suddenly created magnetized Lorentz medium (transverse propagation), *IEEE Trans. Antennas Propagation*, **44**(10), 1394–1398.

37. Bakunov, M. I., and Zhukov, S. N. (1998) Trapping of an electromagnetic wave by the boundary of a suddenly created plasma half-space, *Proceedings of the International Conference on Mathematical Methods in Electromagnetic Theory (MMET'98)*, vol. 1, Kharkov, 340–342.

38. Stepanov, N. S. (1967) About adiabatic transform of electromagnetic waves spectrum in plasma (in Russian), *Sov. Phys. JETP*, **53**, 6(1), 2186–2193.

39. Borisov, V. V. (1987) *Nonsteady Electromagnetic Waves* (in Russian), Leningrad University Publishing House, Leningrad, 240 p.

40. Nerukh, A. G. (1984) Scattering of electromagnetic waves by a plasma half-space after its movement beginning (in Russian), *Zhurnal Technicheskoy Fiziki*, **54**(2), 216–226.

41. Marshall, T. (1985) *Free-Electron Lasers*, Macmillan Publ., London.

42. Akhiezer, A. I. (ed) (1974) *Plasma Electrodynamics* (in Russian), Nauka, Moscow.

43. Yemelyanov, K. M. (2001) *Method of Evolution Recursion in Investigations of Electromagnetic Fields in Media with Time-Varying Parameters*, PhD thesis, National University of Radio Electronics, Kharkov.

Chapter 3

Influence of Medium Plane Boundaries on Electromagnetic Transients

The next most complex stage of a transient electromagnetic problem occurs when the influence of medium boundaries must be taken into account. Such influence is of significant importance in quantum electronics and optical communications technology because of optical pulse generation, amplification, compression, reshaping and filtering, as well as four-wave mixing [1–8]. All of these studies concern ultrafast electromagnetic transients in complex media, and their solution in the time domain is the subject of much research.

As a rule, an exact consideration of electromagnetic processes in media with arbitrary time-varying parameters is possible only in special cases. Numerical methods, on the other hand, can be used to investigate such problems in the general case. The finite difference time domain (FDTD) method has received wide recognition [9] for solving the differential problem. FDTD methods are extremely popular because of their relative simplicity and clarity, and because they lend themselves to elegant numerical algorithms. Indeed, it is difficult to identify areas in electromagnetic theory where FDTD-based methods have not been applied. However, this method is a purely numerical one

Non-Stationary Electromagnetics: An Integral Equations Approach (Second Edition)
Alexander Nerukh and Trevor Benson

ISBN 978-981-4774-95-6 (Hardcover), 978-0-429-02773-4 (eBook)
www.panstanford.com

and, as such, it does not allow an analytical investigation of the problem. Furthermore, the classical Yee FDTD meshing scheme is limited by the accumulation of phase errors for electrically large domains and late-time analysis [10].

An alternative approach is to solve the problem using integral equations, in particular the integral Volterra equations, which are equivalent to Maxwell's equations in the time domain and initial and boundary conditions [11, 12]. The most attractive feature of this approach is that the Volterra integral equation of the second kind allows the construction of a solution by a convergent sequence of approximations. It permits analytical as well as numerical investigations, and in some cases of the latter, this method has been found to give better accuracy and stability than FDTD [13]. To investigate truly non-stationary phenomena, we must consider initial value problems. A Volterra integral equation allows one to take into account the initial conditions in a natural way, and the numerical investigation of initial problems has been implemented for an unbounded medium in [14, 15].

In the previous chapter, the influence of a time change in an unbounded medium on the electromagnetic field was considered. A more realistic problem must account for the fact that a real non-stationary medium is restricted by boundaries, which themselves can be non-stationary. This can also influence non-stationary electromagnetic processes. In this chapter, such an influence is considered in a simple problem when plane boundaries emerge abruptly at some moment of time.

The influence of a medium boundary on an electromagnetic field is taken into account explicitly in the third stage of the interaction evolution described in Sections 1.3.3–1.3.5, when the characteristic function χ of the diffraction object explicitly appears in the kernel of Eq. (1.64) or (1.75). Therefore, this equation is the most general one. Indeed, it is defined in the whole region considered (i.e., the defined domain of the field) and takes into account features of the medium inside the diffraction object as well as its boundaries. This equation is equivalent to an initial-boundary-value problem for Maxwell's equations. Such a problem has an unique solution if initial values for the electromagnetic vectors **E** and **B** are given in a region restricted by a surface S at zero moment of time ($t = 0$) and boundary values for these vectors tangent components are given on this surface S during the entire time interval from the moment $t = 0$ to $t = t_1$ [16].

A solution to this problem depends on the specific geometry of the medium boundary. This dependence is revealed as a modification of the resolvent if the problem is solved by virtue of the integral equation. To reveal the main features of this modification in comparison with an unbounded medium we consider at first a plane medium boundary as the simplest one. The non-stationary nature of this problem is introduced through time changes in the medium permittivity in one of two half-spaces separated by a plane boundary. Moreover, such a problem is a key one in the reflection/transmission of electromagnetic waves by/through a boundary between two media. The solution to this problem can be used for the analysis of more complicated problems.

In the first part of this section, the influence of a medium boundary on the field is considered for a half-closed region when the plane boundary emerges abruptly at some moment of time. If the electric field is parallel to the boundary, then the problem is a 1D scalar one.

3.1 A Resolvent for an Initial-Boundary-Value 1D Problem in a Dielectric

To reveal the main features of a scheme of resolvent construction, we consider the simplest 1D problem for an electromagnetic field in a dielectric half-space $x \geq 0$ where the permittivity changes from the background value ε to the new one ε_1. The medium in this half-space, as well as in the background, is supposed to be non-magnetic ($\mu = 1$), non-dispersive and lossless. The characteristic function in this problem is defined as $\chi(\mathbf{x}) = \theta(x)$. Assume that the electromagnetic fields do not depend on the lateral coordinates y and z and that they only have a component transverse to the x-axis, which is perpendicular to the non-stationary medium boundary $x = 0$. Such a condition is fulfilled for electromagnetic waves falling onto the interface along the normal to the boundary. In this case, one can implement the operation of coordinate descent [17] in the integral Eq. (1.64) integrating with respect to the coordinates y and z by the formula

$$\int_{-\infty}^{\infty}\int dy'dz' \frac{\delta(\tau - |\mathbf{r}-\mathbf{r}'|/v)}{|\mathbf{r}-\mathbf{r}'|} = 2\pi v\theta(\tau)\theta(v\tau - |x-x'|), \tag{3.1}$$

where $\tau = t - t'$. Taking into account also that $\nabla(\nabla\mathbf{E}) = 0$ in this case, we obtain the scalar equation for the electric field

$$E = F + \int d\mathbf{x}' \left\langle \mathbf{x} \middle| \hat{K} \middle| \mathbf{x}' \right\rangle E(\mathbf{x}'). \tag{3.2}$$

The kernel in this equation is defined by

$$\left\langle \mathbf{x} \,|\, \hat{K} \,|\, \mathbf{x}' \right\rangle = -\frac{\varepsilon_1 - \varepsilon}{2\varepsilon v}\theta(x)\left\langle \mathbf{x} \,|\, \hat{\Gamma} \,|\, \mathbf{x}' \right\rangle \theta(x'), \tag{3.3}$$

where $\left\langle \mathbf{x} \,|\, \hat{\Gamma} \,|\, \mathbf{x}' \right\rangle = \frac{\partial^2}{\partial t^2}\theta\left(\tau - \frac{x-x'}{v}\right)$ is the propagator. The integral denotes integration with respect to $\mathbf{x} = (t, x)$ over the half-plane $0 \le t < \infty, -\infty < x < \infty$.

If the multipliers $\chi(\mathbf{x}) = \theta(x)$ are omitted, we obtain the problem in an unbounded medium which has the resolvent

$$\left\langle \mathbf{x} \middle| \hat{R}^{(1)} \middle| \mathbf{x}' \right\rangle = -\frac{\varepsilon_1 - \varepsilon}{2\varepsilon v}\frac{v_1}{v}\frac{\partial^2}{\partial t^2}\theta\left(t - t' - \frac{|x-x'|}{v_1}\right). \tag{3.4}$$

To construct a resolvent for a problem with a boundary, we analyze the expression (3.4) from the information point of view. Indeed, the single differentiation in gives the form

$$\left\langle \mathbf{x} \middle| \hat{R}^{(1)} \middle| \mathbf{x}' \right\rangle = -\frac{\varepsilon_1 - \varepsilon}{2\varepsilon v}\frac{v_1}{v}\frac{\partial}{\partial t}\delta\left(t - t' - \frac{|x-x'|}{v_1}\right), \tag{3.5}$$

which shows that straight lines which generate the resolvent light cone are the characteristics along which information about the field state propagate. This information forms the field at the observation point $\mathbf{x} = (t, x)$. If the characteristic intersects with the world line of the medium boundary, the line $x' = 0$ in Fig. 3.1, then the former reflects from this world line. A slope of the reflected line is determined by the medium features only, not the boundary, and is equal to the slope of the characteristic.

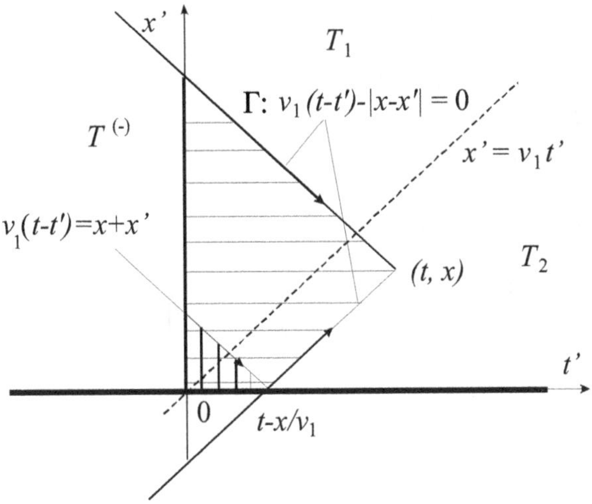

Figure 3.1 Reflection of the light cone from the boundary world line.

Taking into account a boundary directly by multiplying the resolvent (3.5) by characteristics functions χ gives the following expression for the resolvent in an impulse representation

$$\left\langle \mathbf{p}\left|\chi\hat{R}^{(1)}\chi\right|\mathbf{p}'\right\rangle = \varepsilon\frac{v_1^2}{v}\frac{1}{p-p'}\left\{\frac{2p^2}{p^2+v_1^2k^2}\frac{i}{k-k'-io} + \frac{v_1}{v}\frac{p}{p-iv_1k}\frac{1}{p-iv_1k'}\right\} \tag{3.6}$$

But this expression does not satisfy the equation for the resolvent

$$\left\langle \mathbf{p}\left|\hat{R}\right|\mathbf{p}'\right\rangle = \left\langle \mathbf{p}\left|\hat{K}\right|\mathbf{p}'\right\rangle + \int d\mathbf{p}_1 \left\langle \mathbf{p}\left|\hat{K}\right|\mathbf{p}_1\right\rangle\left\langle \mathbf{p}_1\left|\hat{R}\right|\mathbf{p}'\right\rangle, \tag{3.7}$$

where the kernel is given by the impulse representation of (3.3)

$$\left\langle \mathbf{p}\left|\hat{K}\right|\mathbf{p}'\right\rangle = -\frac{\varepsilon_1-\varepsilon}{2\varepsilon v}\left\langle \mathbf{p}\left|\chi\hat{\Gamma}\chi\right|\mathbf{p}'\right\rangle$$
$$= \frac{\varepsilon_1-\varepsilon}{2\varepsilon v}\left(\frac{2vp^2}{p^2+v^2k^2}\frac{1}{p-p'}\frac{i}{k-k'-io} + \frac{v^2p}{(p-ivk)(p-ivk')}\frac{1}{p-p'}\right) \tag{3.8}$$

and o is an infinitesimal quantity ensuring the existence of a Fourier transform. It is worth noting that the form of Eq. (3.7) does not depend on the dimension of the problem, and it will be used for the 3D problem.

The assumption that a contribution of the reflected characteristic is equal to R leads to the following form for the resolvent for a problem with a boundary

$$\langle \mathbf{x}|\hat{R}|\mathbf{x}'\rangle = \theta(x)\left\{\langle \mathbf{x}\,|\,\hat{R}^{(1)}\,|\,\mathbf{x}'\rangle - \frac{\varepsilon_1-\varepsilon}{2\varepsilon v}\frac{v_1}{v}\frac{\partial}{\partial t}R\frac{\partial}{\partial t}\theta\left(t-t'+\frac{x+x'}{v_1}\right)\right\}\theta(x') \tag{3.9}$$

Assuming that R is constant, we find from (3.7) and (3.9)

$$\langle \mathbf{p}|\hat{R}|\mathbf{p}'\rangle = \frac{\varepsilon_1-\varepsilon}{2\varepsilon v}\frac{v_1^2}{v}\frac{1}{p-p'}\left\{\frac{2p^2}{p^2+v_1^2k^2}\frac{i}{k-k'-io}\right.$$
$$\left. +\frac{v_1}{v}p\left[\frac{1}{p-iv_1k}-\frac{R}{p+iv_1k}\right]\frac{1}{p-iv_1k'}\right\}, \tag{3.10}$$

where $\mathrm{Re}(p-p')>0$.

It is worth noting that this expression differs from (3.8) by the term proportional to R only. This condition will be used below for construction of the resolvent in more complex cases.

Substituting the expressions (3.8) and (3.10) into Eq. (3.7) we find that it is satisfied if

$$R = \frac{v-v_1}{v+v_1}. \tag{3.11}$$

The expression obtained for the coefficient R coincides with the reflectivity for a plane wave falling on the interface from inside the non-stationary media with permittivity ε_1. So the integral operator for the bounded medium is characterized by the presence of the reflectivity in its kernel that is typical for such problems.

The coefficient R found determines completely the resolvent $\hat{R}$ in the form of expression (3.9). It allows one to construct the solution to the diffraction problem on the third stage of the interaction when the boundary effects are completely taken into account. Thus, the algorithm for the solution of the diffraction problem along the whole process of the interaction evolution is entirely complete. It is worth noting that the resolvent $\hat{R}$ can be used for construction of the resolvent $\hat{R}^{(-)}$ for the first stage of the interaction corresponding to the period before zero moment of time. To do this one may make the corresponding substitution of the medium parameters and prolong the historical light cone of the resolvent onto the whole negative time half-axis.

It is worth making some remarks concerning the resolvents $\hat{R}^{(1)}$, $\hat{R}$ and the domain of the field definition. Equation (1.80) is defined only in some part of the defined domain of the field and the resolvent (3.4) gives a unique solution to (1.80) only in the case when the resolvent light cone does not intersect the surface $\chi(\mathbf{x}) = 1$. If this is not the case, the resolvent (3.4) gives only a partial solution corresponding to a free term of the equation. The general solution in this case is ambiguous because of the existence of non-trivial solutions to the homogeneous equation, which has an integral-differential character in the general case. It can be shown that a unique solution to the problem is given by the resolvent (3.10) in the general case.

3.2 Electromagnetic Field in a Half-Restricted Time-Varying Medium

3.2.1 Transformation of a Plane Wave by a Jump Change of a Dielectric Permittivity

We consider the influence of synchronous pulse variations of the medium permittivity $\varepsilon(t)$ and conductivity $\sigma(t)$ on a plane harmonic electromagnetic wave [19]. The medium variation starts at zero moment of time in the half-space $x \geq 0$. The medium before the zero moment of time is non-magnetic and lossless and its permittivity is equal to ε_0. Temporal variations of the medium parameters in the half-space $x \geq 0$ are allowed to take the form of rectangular pulses of duration τ, being described by the following expression:

$$\varepsilon(t) = \varepsilon_0 \left[\theta(-t) + \theta(t-\tau)\right] + \varepsilon_1 \left[\theta(t) - \theta(t-\tau)\right],$$
$$\sigma(t) = \sigma_1 \left[\theta(t) - \theta(t-\tau)\right], \tag{3.12}$$

where ε_1 and σ_1 are the permittivity and conductivity in the half-space $x \geq 0$ at the time interval $t \in [0, t]$ and $\theta(t)$ is the Heaviside unit function. The medium in the half-space $x < 0$ remains invariable (see Fig. 3.2).

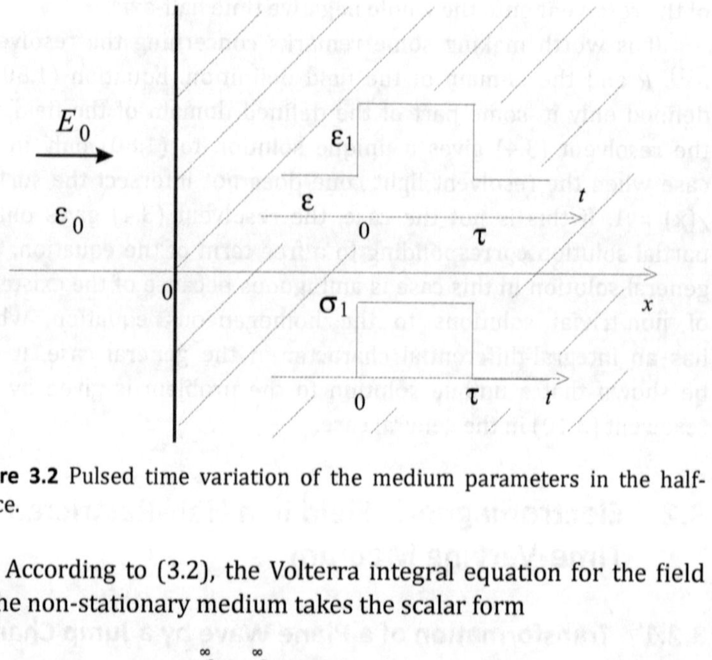

Figure 3.2 Pulsed time variation of the medium parameters in the half-space.

According to (3.2), the Volterra integral equation for the field in the non-stationary medium takes the scalar form

$$E(t,x) = E_0(t,x) + \int_0^\infty dt' \int_0^\infty dx' K(t,t',x,x') E(t',x), \tag{3.13}$$

Here $E_0(t, x)$ is an initial field in the undisturbed medium and (t, t', x, x') is the kernel

$$K(t,t',x,x') = -\frac{1}{\bar{a}^2(t)} \left\{ \bar{\sigma}(t) + \frac{1}{2}(1 - \bar{a}^2(t)) \frac{\partial}{\partial t} \right\}$$
$$\times \delta(v_0(t-t') - |x - x'|), \tag{3.14}$$

where $\bar{a}(t) = \sqrt{\varepsilon_0/\varepsilon(t)}$, $v_0 = c/\sqrt{\varepsilon_0}$, $\bar{\sigma}(t) = 2\pi\sigma(t)/\varepsilon_1$ and c is the light velocity in vacuum.

The solution of Eq. (3.13) in the region $x \geq 0$ is expressed after the resolvent $R(t, t', x, x')$ by the integral

$$E(t,x) = E_0(t,x) + \int_0^\infty dt' \int_0^\infty dx' R(t,t',x,x') E_0(t',x'). \tag{3.15}$$

For the case when $\varepsilon(t) = \varepsilon_1 = const$ and $\sigma(t) = \sigma_1 = const$ the resolvent can be found from Eq. (3.7), reducing them to 1D problem. According to (3.9) the resolvent consists of two terms

$$R(t,t',x,x') = R_1(t,t',x,x') + R_2(t,t',x,x'), \tag{3.16}$$

which can be derived from the inverse Laplace transformation

$$R_n(t,t',x,x') = e^{-\bar{\sigma}_1(t-t')} \int_{\eta - i\infty}^{\eta + i\infty} \frac{dp}{2\pi i} S_n(p,t,t',x,x') e^{p(t-t')}. \tag{3.17}$$

Here $n = 1, 2$ and

$$S_1(p) = \frac{a}{2v_0}\left(a^2 \frac{p - \bar{\sigma}_1}{p + \bar{\sigma}_1} - 1\right)\sqrt{p^2 - \bar{\sigma}_1^{\,2}} \times \exp\left\{-pt' - \frac{|x - x'|}{v_1}\sqrt{p^2 - \bar{\sigma}_1^{\,2}}\right\} \tag{3.18}$$

$$S_2(p) = \frac{a}{2v_0}\left(a\sqrt{\frac{p - \bar{\sigma}_1}{p + \bar{\sigma}_1}} - 1\right)^2 \sqrt{p^2 - \bar{\sigma}_1^{\,2}} \times \exp\left\{-pt' - \frac{x + x'}{v_1}\sqrt{p^2 - \bar{\sigma}_1^{\,2}}\right\}, \tag{3.19}$$

where $a = \sqrt{\varepsilon_0/\varepsilon_1}$, $\eta > \bar{\sigma}_1$, $\bar{\sigma}_1 = 2\pi\sigma_1/\varepsilon_1$, $v_1 = \sqrt{c/\varepsilon_1}$, and $\mathrm{Re}\sqrt{p^2 - \bar{\sigma}_1^{\,2}} > 0$.

The term S_1 in (3.17) is the resolvent of the unbounded problem, and the term S_2 takes into account the influence of the boundary, which is created at the zero time moment dividing the stationary and non-stationary half-spaces. The expression (3.15) with the expressions (3.16–3.19) determines the field in the non-stationary medium after the jump of the medium parameters to $\bar{\sigma}_1$ and ε_1 in the whole time interval $t \in [0, t]$, in which the medium is disturbed.

In the region $x < 0$, the field is determined by the same relation (3.13) through the found field (3.15).

Let us analyze the field in the excited medium in the disturbance time interval. If the initial field is a plane monochromatic wave $E_0(t, x) = \exp\{i(\omega t - kx)\}$, $k = \omega/v_0$ then the field in the time interval $t \in [0, t]$ consists of two terms. The first one is the result of integration of the first part of the resolvent R_1 and it represents the transformed waves which coincide with the waves in the unbounded non-stationary medium [20–25]:

$$E_1(t,x) = e^{-\bar{\sigma}_1 t - kx}[B_1^+ e^{i\Omega t} + B_1^- e^{-i\Omega t}] \tag{3.20}$$

These waves have the amplitudes $B_1^{\pm} = -a^2 \frac{(a^2-1)\omega^2 \pm 2\bar{\sigma}_1(i\Omega \mp \bar{\sigma}_1)}{2i\Omega[i\Omega \mp (\bar{\sigma}_1 + i\omega)]}$, the new frequency $\Omega = \sqrt{a^2\omega^2 - \bar{\sigma}_1^2}$ and the same wavenumber as the initial wave.

The appearance of these waves corresponds to the general law, i.e., the splitting of the primary plane wave, the wave 1 in Fig. 3.3, (here and further the wave names coincide with their amplitude coefficients) into two secondary plane waves caused by the medium variation in time. One of these waves (the first term in the expression (3.20) indicated as B_1^+ in Fig. 3.3) is the direct one, which propagates in the direction of the initial wave $E_0(t, x)$. The inverse wave (the second term in (3.20) indicated as B_1^- in Fig. 3.3) propagates backward to the boundary $x = 0$ created between two media. These waves have the same absolute values of the wavenumber as that of the initial wave, although they have a new frequency Ω. If the conductivity is large, $\bar{\sigma}_1 > kv_1$, the frequency becomes imaginary and the transformed waves become aperiodic. In any case, both waves decrease exponentially in time owing to the appearance of the conductivity σ_1 induced in the region $x \geq 0$. Consequently, the expression (3.20) describes exactly the whole transformed field in the region $x \geq 0$ immediately after the medium excitation that is shown in Fig. 3.3.

Further evolution of the electromagnetic field is caused by the influence of the emerged boundary between half-spaces and it is entirely described by the integral

$$E_2(t,x) = E_1(t,x) - a\int_{\alpha-i\infty}^{\alpha+i\infty}\frac{dp}{2\pi i}\,\frac{p-\bar{\sigma}_1+i\omega}{p-\bar{\sigma}_1-i\omega}$$

$$\times\frac{a(p-\bar{\sigma}_1)-\sqrt{p^2-\bar{\sigma}_1^2}}{p^2-\bar{\sigma}_1^2+k^2v_1^2}\,e^{(p-\bar{\sigma}_1)t-\frac{x}{v_1}\sqrt{p^2-\bar{\sigma}_1^2}}, \tag{3.21}$$

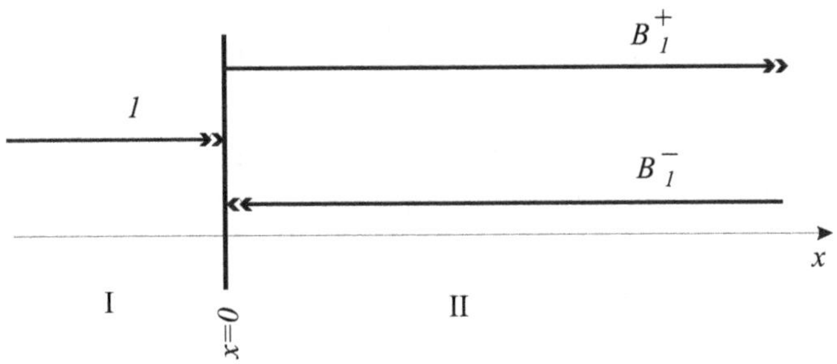

Figure 3.3 The waves at the moment of the interface creation.

where $\mathrm{Re}\sqrt{p^2-\bar{\sigma}_1^2}>0$. The transient is described by the integral term in (3.21) and differs from zero only in the stripe $x \in [0, v, t]$ that represents the boundary influence region. The poles of the integrand in (3.21) describe the steady-state waves for $t \to \infty$ because the integrals along the cuts asymptotically tend to zero. The whole transient influence on the field is out of sight in this approximation, and the essential information is lost if the medium excitation is short. It is interesting to consider the field evolution in all stages of the medium variation. To reveal features of this evolution, consider two cases: $x \approx v_1 t$ $(x < v_1 t)$, i.e., the field near the moving front of the boundary influence, and $x << v_1 t$ making possible the late-time approximation.

The expression (3.21) can be rewritten in the following form:

$$E_2(t,x) = E_1(t,x) + A_2 e(t,x,-\bar{\sigma}_1-i\omega) + B_2^+ e(t,x,-i\Omega) + B_2^- e(t,x,i\Omega)$$

$$-\theta\left(t-\frac{x}{v_1}\right)aI_0\left(\bar{\sigma}_1\sqrt{t^2-\left(\frac{x}{v_1}\right)^2}\right) + A_3 f(t,x,-\bar{\sigma}_1-i\omega)$$

$$+B_3^+ f(t,x,-i\Omega) + B_3^- f(t,x,i\Omega), \tag{3.22}$$

where $I_0(t)$ is the modified Bessel function of the first kind and the constants are defined as

$$A_2 = -a^2 \frac{2\omega}{(1-a^2)\omega - 2i\bar{\sigma}_1}, \quad B_2^{\pm} = \pm a^2 \frac{i\Omega \mp \bar{\sigma}_1}{2i\Omega} \frac{i\Omega \mp (\bar{\sigma}_1 - i\omega)}{i\Omega \mp (\bar{\sigma}_1 + i\omega)},$$

$$A_3 = -a^4 \frac{2i\omega^2}{(1-a^2)\omega - 2i\bar{\sigma}_1}, \quad B_3^{\pm} = \pm a^4 \frac{\omega^2}{2i\Omega} \frac{i\Omega \mp (\bar{\sigma}_1 - i\omega)}{i\Omega \mp (\bar{\sigma}_1 + i\omega)}.$$

The functions $e(t, x, \beta)$ and $f(t, x, \beta)$ in (3.22) differ from zero in the region $x < v_1 t$ and are defined by the expressions [26–28]:

$$\left.\begin{aligned} e(t,x,\beta) &= \int_{\alpha-i\infty}^{\alpha+i\infty} \frac{dp}{2\pi i} \frac{1}{p+\beta} e^{pt - \frac{x}{v_1}\sqrt{p^2 - \bar{\sigma}_1^2}}, \\ f(t,x,\beta) &= -\int_{\alpha-i\infty}^{\alpha+i\infty} \frac{dp}{2\pi i} \frac{1}{(p+\beta)\sqrt{p^2 - \bar{\sigma}_1^2}} e^{pt - \frac{x}{v_1}\sqrt{p^2 - \bar{\sigma}_1^2}} \end{aligned}\right\} \tag{3.23}$$

Applying the convolution theorem [29] to these functions, the expression (3.22) can be rewritten as follows:

$$\begin{aligned} E_2(t,x) = {} & E_1(t,x) + \theta(t - x/v_1) e^{-\bar{\sigma}_1 t} \Big[A_2 e^{(\bar{\sigma}_1 + i\omega)(t - x/v_1)} + B_2^+ e^{i\Omega(t - x/v_1)} \\ & + B_2^- e^{-i\Omega(t - x/v_1)} - a I_0\big(\bar{\sigma}_1 \sqrt{t^2 - (x/v_1)^2}\big) + \frac{\bar{\sigma}_1 x I_1\big(\bar{\sigma}_1 \sqrt{t^2 - (x/v_1)^2}\big)}{v_1 \sqrt{t^2 - (x/v_1)^2}} \\ & \int_{x/v_1}^{t} d\xi \Big(A_2 e^{(\bar{\sigma}_1 + i\omega)(t-\xi)} + B_2^+ e^{i\Omega(t-\xi)} + B_2^- e^{-i\Omega(t-\xi)} \Big) - I_0\big(\bar{\sigma}_1 \sqrt{t^2 - (x/v_1)^2}\big) \\ & \int_{x/v_1}^{t} d\xi \Big(A_3 e^{(\bar{\sigma}_1 + i\omega)(t-\xi)} + B_3^+ e^{i\Omega(t-\xi)} + B_3^- e^{-i\Omega(t-\xi)} \Big) \Big] \end{aligned} \tag{3.24}$$

The formula (3.24) completely describes the electromagnetic field distribution in the half-space of the medium excitement in the whole time interval $t \in [0, \tau]$.

In the region close to the moving front of the boundary influence, $x \approx v_1 t$, the field can be approximated by the following expression, which is accurate up to the first order of the magnitude $\bar{\sigma}_1(t - x/v_1)$:

$$E_2(t,x) \approx E_1(t,x) + e^{-\bar{\sigma}_1 x/v_1} F_1(x) e^{i(\omega t - kx)}$$
$$+ e^{-\bar{\sigma}_1 t}[F_2(x) e^{i\Omega(t-x/v_1)} + F_3(x) e^{-i\Omega(t-x/v_1)} + F_4(x)], \qquad (3.25)$$

where

$$F_1(x) = 2a\omega \frac{\bar{\sigma}_1 + i(1-a)\omega + \bar{\sigma}_1^2 x/v_1}{((1-a^2)\omega - 2i\bar{\sigma}_1)(\bar{\sigma}_1 + i\omega)}$$

$$F_2(x) = -\frac{a^2}{2i\Omega} \frac{\bar{\sigma}_1 - i(\omega+\Omega)}{\bar{\sigma}_1 + i(\omega-\Omega)} \left(\bar{\sigma}_1 - i\Omega + i\frac{a\omega^2}{\Omega} + i\frac{(\bar{\sigma}_1 - i\Omega)\bar{\sigma}_1^{\,2}}{\Omega} x/v_1 \right),$$

$$F_3(x) = -B_4 \left(\bar{\sigma}_1 + i\Omega + i\frac{a\omega^2}{\Omega} + i\frac{(\bar{\sigma}_1 - i\Omega)\bar{\sigma}_1^{\,2}}{\Omega} x/v_1 \right),$$

$$B_4 = -ia^2 2\Omega \frac{[\bar{\sigma}_1 - i(\omega-\Omega)]^2}{\bar{\sigma}_1 + i(\omega+\Omega)}$$

and

$$F_4(x) = -a + \frac{a}{(1-a^2)\omega - 2i\bar{\sigma}_1} \left(a^2(1+a^2)\frac{\omega^3}{\Omega^2} - 2ia^2 \frac{\omega^2}{\bar{\sigma}_1 + i\omega} \right.$$
$$\left. - \frac{\bar{\sigma}_1^2}{2}\left[\frac{2\omega}{\bar{\sigma}_1 + i\omega} + i\frac{(1+a^2)\omega + 2i\bar{\sigma}_1}{\Omega} \right] x/v_1 \right).$$

As it can be seen from the expression (3.25), the parameter jumps in the bounded medium lead to the appearance of a spectrum of waves with inhomogeneous amplitudes near the front of the boundary influence. The first wave, proportional to $F_1(x)$, represents the ordinary transmitted wave. The wave proportional to $F_2(x)$ is a "quasi-antipode" of the direct wave B_1^+ contained in the field E_1 which is described by (3.20). The phase of this wave is equal to $\Omega t - kx\sqrt{1-a^2\left(\frac{\bar{\sigma}_1}{\omega}\right)^2}$ and its amplitude varies in such a way that this wave recombines with the wave B_1^+ asymptotically. Note that the continuous spectrum does not appear in the case of a non-dissipative medium and the recombination of the waves occurs immediately. The third wave, proportional to $F_3(x)$, is the result of the wave B_1^- reflection from the boundary created between the two media. It follows from (3.20) and (3.25) that the permittivity jump in the half-space causes a field discontinuity at the moving plane $x = v_1 t$ of the boundary influence:

$$E_2(x/v_1+0,x)-E_2(x/v_1-0,x)=a(a-1)\exp(-\bar{\sigma}_1 t).$$

The tendency of the electromagnetic field evolution to attain the steady-state regime is described by the late-time approximation of the expression (3.22). To obtain this approximation the representation of the integrals $e(t, x, \beta)$ and $f(t, x, \beta)$ using the Lipschitz–Hankel function [26] was used. In this representation, the Lipschitz–Hankel functions of the zero order are determined by the Neumann series [27]. Considering the first term only, and substituting the obtained approximate functions $e(t, x, \beta)$ and $f(t, x, \beta)$ into (3.22), the approximate expression for the field $E_2(t, x)$ at large times ($t >> x/v_1$) is found in the following form:

$$E_2(t,x)\approx e^{-\bar{\sigma}_1 t}[B_1^- e^{-i(\Omega t+kx)}+B_4 e^{-i(\Omega t-kx)}] + A_4 e^{-\bar{\sigma}_1 x/v_1} e^{i\omega(t-x/v_1)}+e^{-2\bar{\sigma}_1 t}Q_1(t,x), \tag{3.26}$$

where $A_4 = A_2\frac{\bar{\sigma}_1+i(1-a)\omega}{\bar{\sigma}_1+i\omega}$.

The expressions obtained allow one to analyze the process in the region $0 < x << v_1 t$ where the following waves exist: the direct transmitted wave A_4 induced by the initial wave 1, the inverse wave B_1^- passing from the region $x > v_1 t$ and the direct wave B_4 which is the result of the wave B_1^- reflection from the boundary between the media (Fig. 3.4). The wave A_4 is attenuated with the distance from the boundary as $\sim\exp(-\bar{\sigma}_1 x/v_1)$ in contrast to the waves B_1^+ and B_4 which attenuate with time as $\sim\exp(-\bar{\sigma}_1 t)$.

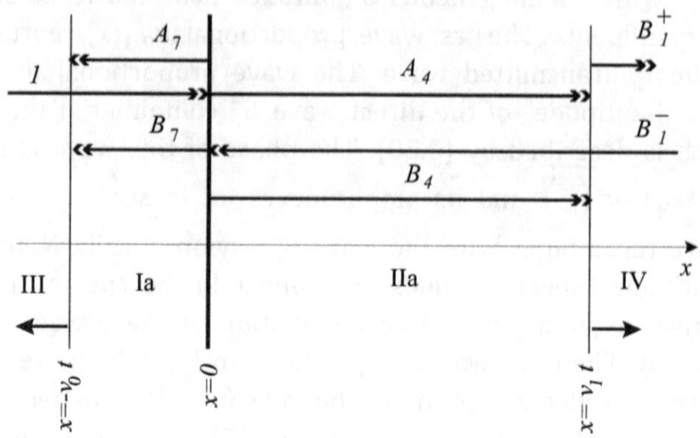

Figure 3.4 The distribution of the secondary waves during the excitation of the medium.

The absence of the wave B_1^+ in (3.26), together with its presence in (3.20), allows one to conclude that it gradually attenuates with the distance from the plane $x = v_1 t$ up to its complete disappearance. The term Q_1 in (3.26) represents the continuous spectrum of waves caused by the medium conductivity jump

$$Q_1(t,x) = -\frac{a^2}{(1-a^2)\omega - 2i\bar{\sigma}_1}\left\{\left(\frac{2i\omega\bar{\sigma}_1}{(\bar{\sigma}_1+i\omega)^2}\frac{a\omega t + (\bar{\sigma}_1+i\omega)x/v_1}{\sqrt{t^2-(x/v_1)^2}}\right.\right.$$
$$\left. - \frac{(1+a^2)a\omega^3 t - \Omega(x/v_1)((1+a^2)\omega\bar{\sigma}_1 - 2i\Omega^2)}{i\Omega\{a^2\omega^2t^2 - \Omega^2(x/v_1)^2\}}\bar{\sigma}_1\sqrt{t^2-(x/v_1)^2}\right)$$
$$I_1(\bar{\sigma}_1\sqrt{t^2-(x/v_1)^2})$$
$$+\left(2\omega - \frac{2(1+a^2)a\omega^3\Omega(t^2-(x/v_1)^2) + i\bar{\sigma}_1{}^2 t(x/v_1)((1+a^2)\omega\bar{\sigma}_1 + 2i\Omega^2)}{\Omega(a^2\omega^2t^2 - \Omega^2(x/v_1)^2)}\right)$$
$$\left. I_0(\bar{\sigma}_1\sqrt{t^2-(x/v_1)^2})\right\}.$$

For long times the continuous spectrum disappears, $\lim_{t\to\infty} Q_1(t,x) = 0$.

Information about the field in the excited medium in the disturbance time interval allows the reaction of the external field on the excitation of the medium behind the boundary to be considered.

The expression describing the electric field in the external region $x \le 0$ at the time interval $t \in [0, \tau]$ is obtained after substituting (3.21) into (3.13):

$$E_3(t,x) = E_0(t,x) + \theta\left(t+\frac{x}{v_0}\right)\left\{\bar{\sigma}_1 + \frac{1}{2}(1-a^2)\frac{\partial}{\partial t}\right\}\left[A_5 e^{i\omega\left(t+\frac{x}{v_0}\right)}\right.$$
$$-\left(\frac{a^2}{2i\Omega} + B_5^+\right)e^{i\Omega\left(t+\frac{x}{v_0}\right)} + \left(\frac{a^2}{2i\Omega} - B_5^-\right)e^{-i\Omega\left(t+\frac{x}{v_0}\right)}$$
$$+ C_5 e^{-\frac{1-a^2}{1+a^2}\bar{\sigma}_1\left(t+\frac{x}{v_0}\right)} + e^{-\bar{\sigma}_1\left(t+\frac{x}{v_0}\right)} A_6 Z\left(t+\frac{x}{v_0}, -\bar{\sigma}_1 - i\omega\right)$$
$$\left. + B_6^+ Z\left(t+\frac{x}{v_0}, i\Omega\right) + B_6^- Z\left(t+\frac{x}{v_0}, -i\Omega\right) + C_6 Z\left(t+\frac{x}{v_0}, \frac{1-a^2}{1+a^2}\bar{\sigma}_1\right)\right]\Bigg\}.$$

(3.27)

The coefficients are equal to

$$A_5 = -8a^2 \frac{2\bar{\sigma}_1 + i(1+a^2)\omega}{[2\bar{\sigma}_1 + i(1-a^2)\omega]^2}, \quad A_6 = 4a^3 \frac{(\omega - 2i\bar{\sigma}_1)\omega}{[2\bar{\sigma}_1 + i(1-a^2)\omega]^2},$$

$$B_5^{\pm} = \pm \frac{a^2(1+a^2)i(\bar{\sigma}_1 + i\Omega)}{2\Omega[(1+a^2)\bar{\sigma}_1 \pm (1-a^2)i\Omega]}, \quad B_6^{\pm} = \pm \frac{a^5 i\omega}{(1+a^2)\bar{\sigma}_1 \pm (1-a^2)i\Omega},$$

$$C_5 = \frac{16a^2\bar{\sigma}_1}{[2\bar{\sigma}_1 + i(1-a^2)\omega]^2}, \quad C_6 = -\bar{\sigma}_1 C_5,$$

and the function $Z(t + x/v_0, \beta)$ is determined by the integral

$$Z\left(t + \frac{x}{v_0}, \beta\right) = -\int_{\alpha - i\infty}^{\alpha + i\infty} \frac{dp}{2\pi i} \frac{1}{(p+\beta)\sqrt{p^2 - \bar{\sigma}_1^2}} e^{p\left(t + \frac{x}{v_0}\right)}. \tag{3.28}$$

An asymptotic approximation ($t >> -x/v_0$) for the function $Z(t + x/v_0, \beta)$ can be derived from the same approximation for $f(t, x, \beta)$ by replacing ($t \to t + x/v_0$) and $x/v_0 \to 0$. Substitution of the approximate expression obtained for the function $Z(t + \frac{x}{v_0}, \beta)$ into (3.27) gives the late-time approximation for the electric field outside the non-stationary half-space:

$$E_3(t,x) \approx E_0(t,x) + \theta(t + x/v_0)\left\{ A_7 e^{i(\omega t + kx)} + B_7 e^{-i\Omega(t + x/v_0)} + \left[\bar{\sigma}_1 + \frac{1}{2}(1-a^2)\frac{\partial}{\partial t}\right] Q_2(t,x) \right\}, \tag{3.29}$$

where

$$A_7 = -4a^2 \frac{\bar{\sigma}_1 + i(1+a^2)\omega - a\omega(\omega - 2i\bar{\sigma}_1)(2\bar{\sigma}_1 + 2i\omega)^{-1}}{2\bar{\sigma}_1 + i(1-a^2)\omega}, \quad B_7 = -i\frac{a^2}{a^2\omega^2 - \bar{\sigma}_1^2}$$

It follows from (3.29) that at zero moment of time the plane $x = -v_0 t$ separates from the boundary $x = 0$ created between the media and moves along the x axis in the negative direction with velocity v_0. This plane conveys the exterior of the field to the left. This field is created by the reflection of the initial wave 1 (as well as the wave A_7) and by the wave B_1^-, which passes through the boundary and gives rise to the wave B_7 (see Fig. 3.5). Both these waves have phase velocities which coincide with that of the primary wave. The frequency and the wavenumber of the wave A_7 coincide with the corresponding characteristics of the initial wave. The frequency and the wavenumber of the wave B_7 are equal to Ω and Ω/v_0, respectively. The function $Q_2(t,x)$ in (3.29) is equal to

$$Q_2(t,x) = -i\frac{a}{2\bar{\sigma}_1 + i(1-a^2)\omega}\frac{t}{t+x/v_0}\left\{\bar{\sigma}_1\left[\frac{4a^2}{2\bar{\sigma}_1 + i(1-a^2)\omega} - \frac{2a^2(1-a^2)\Omega}{2\bar{\sigma}_1\omega - i(1-a^2)\omega^2}\right]\right.$$

$$\times I_1(\bar{\sigma}_1(t+x/v_0)) + \left[\frac{2ia^2(1+a^2)\Omega\bar{\sigma}_1}{2\bar{\sigma}_1\omega - i(1-a^2)\omega^2} - \frac{4a^2(\omega - i\bar{\sigma}_1)}{2\bar{\sigma}_1 + i(1-a^2)\omega}\right.$$

$$\left.\left. - \frac{4|1-a^4|\bar{\sigma}_1}{a(2\bar{\sigma}_1 + i(1-a^2)\omega)} + 2\bar{\sigma}_1 + i(1-a^2)\omega\right] I_0(\bar{\sigma}_1(t+x/v_0))\right\};$$

It describes the continuous spectrum of waves in the region $x < 0$. This continuous spectrum disappears when $t > \infty$.

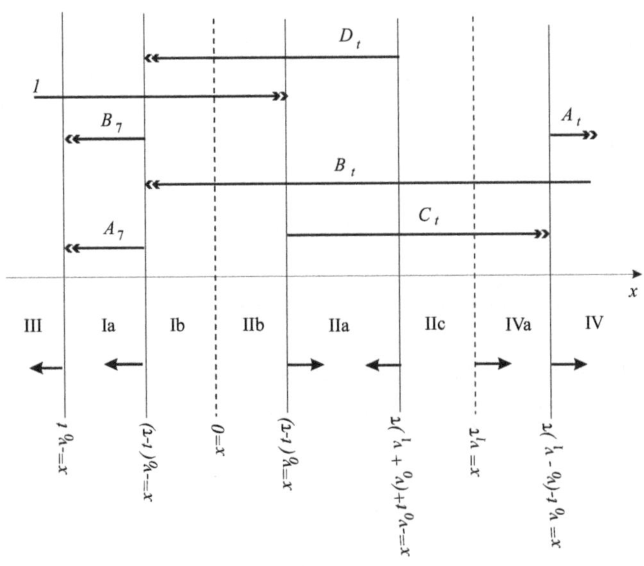

Figure 3.5 The distribution of the secondary waves after the medium excitation decay.

Applying the convolution theorem to (3.28)

$$Z(t+x/v_0,\beta) = -\int_0^{t+x/v_0} d\xi\, e^{-\beta(t+x/v_0-\xi)} I_0(\bar{\sigma}_1\xi) \tag{3.30}$$

and using in (3.27) the first-order (with respect to the magnitude of the product $\omega(t + x/v_0)$) an approximate expression for the field $E_3(t, x)$ near the front of the medium boundary influence $x = - v_0 t$ can be found in the form

$$E_3(t,x) \approx E_0(t,x) + \theta(t+x/v_0)\left\{A_7\, e^{i(\omega t+kx)} + B_7\, e^{-i\Omega(t+x/v_0)}\right.$$

$$\left. -2a\bar{\sigma}_1 \frac{(2a^3-2a-1)\omega^2 - (\bar{\sigma}_1 + i\omega)\bar{\sigma}_1}{(2\bar{\sigma}_1 + i(1-a^2)\omega)^2}\right\}. \tag{3.31}$$

The last term in (3.31) is constant; this differs substantially in character from the corresponding term in (3.29) where the continuous spectrum forms a sophisticated function in the spatial-time coordinates.

The finite duration of the medium disturbance gives rise to a question about residual phenomena after switching off the medium excitation. The electric field in the whole space, left after returning the medium in the region $x \geq 0$ into its initial state, is expressed by the formula

$$E(t,x) = E_0(t,x) + \frac{1}{a^2}\left\{\bar{\sigma}_1 + \frac{1}{2}(1-a^2)\frac{\partial}{\partial t}\right\}$$

$$\int_0^{\tau} dt' \int_0^{\infty} dx'\, \delta(v_0(t-t') - |x-x'|) E_2(t',x') \tag{3.32}$$

Substituting (3.21) into (3.32) shows that the field consists of three groups of components, which occupy different regions in space (see Fig. 3.5),

$$E(t,x) = E_0(t,x) + E_4(t,x) + E_5(t,x) + E_6(t,x). \tag{3.33}$$

The first group consists of the waves located in the regions IIa, IIc and IV. They propagate in the direction of the initial wave and are described by the following expression:

$$E_4(t,x) = -\theta(-t + x/v_0)\theta(x + v_0(\tau - t))E_0(t,x) + \theta(x + v_0(\tau - t))$$

$$\left[\bar{\sigma}_1 + (1/2)(1-a^2)\frac{\partial}{\partial t}\right] \times \left[\left\{\theta(t - x/v_0)A_8 - \frac{\theta(-t + x/v_0)}{k^2 v_1^2}\right.\right.$$

$$\left.(B_3^+ e^{i\Omega\tau} + B_3^- e^{-i\Omega\tau})\right\} E_0(t,x) - a^2 \frac{a^2}{\Omega} e^{-(\bar{\sigma}_1 + i\omega)\tau - i(\omega t + kx)} \sin(\omega\tau) + e^{i\Omega(t - x/v_0)}$$

$$\left\{\frac{\theta(-t + x/v_0)}{k^2 v_1^2} B_3^+ + \theta(t - x/v_0)B_8^+\right\} + e^{-i\Omega(t - x/v_0)}$$

$$\left\{\frac{\theta(-t + x/v_0)}{k^2 v_1^2} B_3^- + \theta(t - x/v_0)B_8^-\right\} + A_8 e(\tau, -(1/a)(t - x/v_0 - \tau),$$

$$-\bar{\sigma}_1 - i\omega) + 2a(2a\bar{\sigma}_1 + (1-a^2)i\omega) \times \left\{B_8^+ e\left(\tau, \frac{-t + x/v_0 + \tau}{a}, i\Omega\right)\right.$$

$$\left.\left.+ B_8^- e\left(\tau, \frac{-t + x/v_0 + \tau}{a}, -i\Omega\right)\right\}\right]\theta(x). \tag{3.34}$$

The second group consists of the waves propagating in the opposite direction and occupying the region Ia,

$$E_5(t,x) = \left[\bar{\sigma}_1 + \frac{1-a^2}{2}\frac{\partial}{\partial t}\right]\left\{A_5 E_0(t,x) + B_5^+ e^{i\Omega(t + x/v_0)} + B_5^- e^{-i\Omega(t + x/v_0)} +\right.$$

$$+ C_5 e^{2\bar{\sigma}_1 (t + x/v_0)/|1-a^2|} - \frac{1}{\Omega} e^{-\bar{\sigma}_1 (t + x/v_0)} - \sin(\Omega(t + x/v_0)) + e^{-\bar{\sigma}_1 (t + x/v_0)}$$

$$A_6 Z(t + x/v_0, -\bar{\sigma}_1 - i\omega) + B_6^+ Z(t + x/v_0, i\Omega) + B_6^- Z(t + x/v_0, -i\Omega)$$

$$\left.+ C_6 Z\left(t + x/v_0, \frac{1+a^2}{|1-a^2|}\bar{\sigma}_1\right)\right\}\theta(-x) \tag{3.35}$$

and the regions Ib, IIa, and IIb,

$$E_6(t,x) = \left[\bar{\sigma}_1 + \frac{1}{2}(1-a^2)\frac{\partial}{\partial t}\right]\left\{\theta(-x) e^{-(\bar{\sigma}_1 - i\omega)\tau - i(\omega t + kx)} \sin(\Omega\tau)\right.$$

$$+ A_5 e\left(\tau, \frac{t + x/v_0 - \tau}{a}, -\bar{\sigma}_1 - i\omega\right) + B_5^+ e\left(\tau, \frac{t + x/v_0 - \tau}{a}, i\Omega\right)$$

$$+B_5^- e\left(\tau, \frac{t+x/v_0-\tau}{a}, -i\Omega\right) + C_5 e\left(\tau, \frac{t+x/v_0-\tau}{a}, \frac{1+a^2}{\left|1-a^2\right|}v\right)$$

$$+A_6 f\left(\tau, \frac{t+x/v_0-\tau}{a}, -\bar{\sigma}_1 - i\omega\right) + B_6^+ f\left(\tau, \frac{t+x/v_0-\tau}{a}, i\Omega\right)$$

$$+B_6^- f\left(\tau, \frac{t+x/v_0-\tau}{a}, -i\Omega\right) + C_6 f\left(\tau, \frac{t+x/v_0-\tau}{a}, \frac{1+a^2}{\left|1-a^2\right|}\bar{\sigma}_1\right)\Bigg\}\Bigg]. \tag{3.36}$$

where $A_8 = -\frac{1}{a}\frac{(1-a^2)\omega - 2ai\bar{\sigma}}{(1-a^2)\omega - 2i\bar{\sigma}_1}$ and $B_8^{\pm} = \frac{i}{2\Omega}\frac{(1+a^2)\omega \pm 2i\Omega}{(1-a^2)\omega - 2i\bar{\sigma}_1}$.

Consequently, the two planes $x = \pm v_0(t - \tau)$ are separated from the boundary $x = 0$ at the moment $t = \tau$, propagating in opposite directions with the velocity v_0, as the result of restoring the medium parameters in the half-space $x \geq 0$ to their initial magnitudes. By the same reason, the plane $x = v_1 t$ moving in the region $x \geq 0$ splits into two scattered planes: $x = \pm v_0 t + (v_1 \mp v_1)\tau$ that separate the regions with different field structures.

Using the approximate expressions for the integrals (3.23), the late-time ($t >> x/v_1$) approximation for the field is obtained as

$$\begin{aligned} E(t,x) \approx\ & A_t(t,x)E_0(t,x) + B_t(t,x)\mathrm{e}^{-i(\omega t+kx)} + C_t(t,x)\mathrm{e}^{i(\omega t-kx)/a} \\ & + D_t(t,x)\mathrm{e}^{-i(\omega t+kx)/a} + \theta(v_0(\tau-t)-x)E_3(t,x) \\ & + \sum_{n=3}^{4}\theta((v_1+(-1)^n v_0)\tau + (-1)^{n-1}v_0 t - x) \\ & \times \theta(x+(-1)^n v_0(\tau-t))\mathrm{e}^{-\bar{\sigma}_1\tau}Q_n(t,x,\tau). \end{aligned} \tag{3.37}$$

which describes the field components originated by splitting the waves B_1^-, B_4 and B_1^+ into the direct and inverse waves, respectively. The waves formed, having the same phases, interfere with each other, composing the direct wave A_t,

$$A_t(t,x) = \mathrm{e}^{-(\bar{\sigma}_1+i\omega)\tau}\left(\bar{\sigma}_1 - \frac{i\omega(1-a^2)}{2}\right)(B_8^+\mathrm{e}^{i\Omega\tau} + B_8^-\mathrm{e}^{-i\Omega\tau}) \tag{3.38}$$

as well as the inverse wave B_t,

$$B_t(t,x) = -a^2\frac{2\bar{\sigma}_1 - i(1-a^2)\omega}{2\Omega}\theta(x + v_0(t-\tau))$$
$$\times e^{-(\bar{\sigma}_1 - i\omega)\tau}\{\cos(\Omega\tau) + i\theta((v_1+v_0)\tau - v_0t - x)\sin(\Omega\tau)\}. \quad (3.39)$$

The phase characteristics of these waves coincide with the phase characteristics of the primary wave.

The direct wave A_t exists in the region IV only. The inverse wave B_t has the following amplitudes in the regions IIa, Ib and IIb

$$-a^2\frac{2\bar{\sigma}_1 - i(1-a^2)\omega}{2\Omega}e^{-(\bar{\sigma}_1 - i\omega)\tau}\sin(\Omega\tau) \quad (3.40)$$

and in the regions IV, IIc, and IVa

$$-a^2\frac{2\bar{\sigma}_1 - i(1-a^2)\omega}{2\Omega}e^{-(\bar{\sigma}_1 - i\omega)\tau}. \quad (3.41)$$

The splitting of the wave A_4 in (3.26) gives rise to the direct wave C_t, which exists in the regions IIa, IIc and IVa,

$$C_t(t,x) = \theta(x + v_0(\tau - t))\theta((v_1 - v_0)\tau + v_0t - x)e^{-(\bar{\sigma}_1 + i(1+1/a)\omega)\tau}A_8 \quad (3.42)$$

and the inverse wave D_t which exists in the regions Ia, IIa and IIb

$$D_t(t,x) = \theta(x + v_0(t-\tau))\theta((v_1+v_0)\tau - v_0t - x)e^{-(\bar{\sigma}_1 - i(1+1/a)\omega)\tau}A_7. \quad (3.43)$$

The frequencies and wavenumbers of these waves are equal to ω/a and k/a, respectively.

The functions Q_n in (3.37) are given in Appendix: Lipshitz–Hankel Functions (in the end of this chapter) and describe the spatial distribution of the continuous wave spectrum which arises at the moment $t = \tau$ and disappears when $t \to \infty$.

The term $E_3(t, x)$ in the formula (3.37) describes the field in the region Ia (Fig. 3.5). It is equal to the corresponding expression for the field in this region (Fig. 3.4) before the medium return to its initial state. It implies a conservation of the field structure in the regions Ia and III.

Hence, two pairs of the direct and inverse waves, namely, A_t, B_t and C_t, D_t (Fig. 3.5), and the wave continuous spectrum remain in the space after the medium returns to its initial state. The frequencies and wavenumbers of the waves A_t and B_t coincide with the frequency and wavenumber of the initial wave. The frequencies and wavenumbers of the waves C_t and D_t differ from the corresponding characteristics of the initial wave by the coefficient $\sqrt{\varepsilon_1/\varepsilon_0}$. The initial wave and the wave B_t remain in the whole space in the steady-state regime when all the interfaces between the regions go to infinity. The asymptotic expression for the electric field gets the following form:

$$E(t,x) = E_0(t,x) - a^2 \frac{2\overline{\sigma}_1 - i(1-a^2)\omega}{2\Omega} e^{-(\overline{\sigma}_1 - i(\omega-\Omega))\tau - i(\omega t + kx)} \tag{3.44}$$

The module and phase of the inverse wave $B_t = |B_t|\exp(i\Phi)$ are equal to

$$|B_t| = a\sqrt{\frac{4b^2a^2 + (1-a^2)^2}{1-b^2a^2}} e^{-ba^2T} \tag{3.45}$$

$$\Phi = (1 - a\sqrt{1-b^2a^2})T + artg\frac{a^2-1}{2ba^2}, \tag{3.46}$$

where the normalized values $b = 2\pi\overline{\sigma}_1/(\varepsilon\omega)$, $T = \omega\tau$ are introduced.

The dependence of the inverse wave amplitude module on the medium permittivity jump a is shown in Fig. 3.6 for various magnitudes of the dissipation coefficient b. The solid line stands for $b = 0$, the dotted line corresponds to $b = 0.1$, and the dot-dashed line stands for $b = -0.1$. The negative magnitude of b corresponds to a model of an active medium. It follows from Fig. 3.6 that the amplification of the inverse wave occurs for any sign of the dissipation coefficient. This effect is caused by the energy transfer from the source, which induces the medium parameter variations, to the electromagnetic field.

Hence, the impulse excitation of the restricted medium parameters during a finite time interval creates a complex structure of the field composed of plane harmonic waves. This structure is a system of space regions that are divided by the planes moving with corresponding phase velocities. Each region is characterized by a discrete set of waves as well as by a continuous set. Asymptotically, i.e., a long time after the non-stationary behavior is switched off, this set reduced to the inverse wave, which remains in the whole space. The phase characteristic of this wave is the same as that of the primary wave, but its amplitude is different. The magnitude of this amplitude can exceed the amplitude of the initial wave.

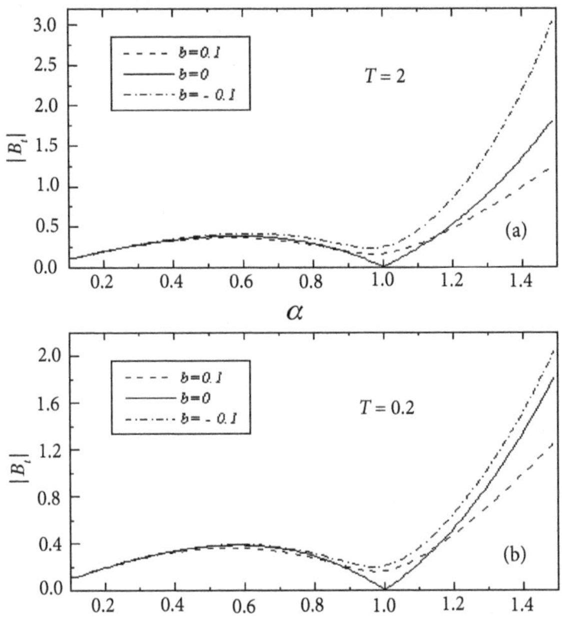

Figure 3.6 The dependence of the module of the inverse wave B_t on the normalized variation of the permittivity a.

3.2.2 Splitting of a Pulse in a Half-Space with Time-Varying Conductivity

The influence of a non-stationary conductive medium boundary on a pulse with a rectangular envelope

$$E_1(t,x) = \theta(vt + vt_0 - x)\theta(x - vt - vt_0 + vT), \tag{3.47}$$

where t_0 is the start of the pulse and T its duration is considered in this section. A jump change of conductivity to the value σ occurs in the region $x \geq 0$ at zero moment and the pulse is assumed to be in this region at this moment; this means $t_0 + T < 0$. In this case, the field E_2 of the second stage (when there is no boundary influence) and the field E_3 of the third stage (when there is boundary influence) are given by the expressions:

$$E_2(t,x) = \theta(vt + vt_0 - x)\theta(x - vt - vt_0 + vT)E_0 + \sigma E_0 \int_{\eta-i\infty}^{\eta+i\infty} \frac{dp}{2\pi i} \sum_{m=0}^{1} \left\{ \theta(x + vt_0 - mvT) \left[\frac{e^{(p-\sigma)t-(t_0-mT+x/v)\sqrt{p^2-\sigma^2}}}{(p-\sigma)(\sqrt{p^2-\sigma^2}-q+\sigma)} - \frac{e^{(p-\sigma)(t-t_0+mT-x/v)}}{\sigma(p-\sigma)} \right] + \theta(-x - vt_0 + mvT) \frac{e^{(p-\sigma)t+(t_0-mT+x/v)\sqrt{p^2-\sigma^2}}}{(p-\sigma)(\sqrt{p^2-\sigma^2}+p-\sigma)} \right\} \tag{3.48}$$

$$E_3(t,x) = E_2(t,x) - \sigma E_0 \int_{\eta-i\infty}^{\eta+i\infty} \frac{dp}{2\pi i} \frac{p-\sigma-\sqrt{p^2-\sigma^2}}{(p+\sigma)(\sqrt{p^2-\sigma^2}+p-\sigma)} \sum_{m=0}^{1} e^{(p-\sigma)t+(t_0-mT-x/v)\sqrt{p^2-\sigma^2}} \tag{3.49}$$

Integration with respect to p gives

$$E_2(t,x) = E_0\theta(vt_0 - x)\theta(x - vt_0 + vT)e^{-2\sigma t} + \sum_{m=0}^{1}(-1)^m \theta(t - |x/v| - t_0 + mT)E'(t,x,t_0 - mT) \tag{3.50}$$

$$E_3(t,x) = E_2(t,x) - \sum_{m=0}^{1}(-1)^m \theta(t - x/v - t_0 + mT)E''(t,x,t_0 - mT), \tag{3.51}$$

where

$$E'(t,x,\tau) = \frac{1}{2}E_0 e^{-2\sigma t}\left[I_0(\sigma\sqrt{t^2 - (x/v - \tau)^2}) + \sigma(x/v - \tau)\int_{x/v-\tau}^{t} (x/v - \tau)^2]^{-1/2} I_1(\sigma\sqrt{u^2 - (x/v - \tau)^2})e^{\sigma(u-t)}du \right] + \frac{1}{2}E_0 e^{-2\sigma t}\begin{cases} e^{\sigma(x/v-\tau)}, & \text{if } \tau \le x/v \le t + \tau \\ e^{-\sigma(x/v-\tau)}, & \text{if } t + \tau \le x/v \le \tau \end{cases} \tag{3.52}$$

and

$$E''(t,x,\tau)=\frac{1}{2}E_0e^{-2\sigma t}\left\{\left(\frac{vt-v\tau-x}{vt+v\tau+x}\right)^{1/2}I_1(\sigma\sqrt{t^2-(x/v+\tau)^2})\right.$$
$$+\int_{-x/v-\tau}^{t}\left[\sigma(x/v+\tau)\frac{I_0(\sigma\sqrt{u^2-(x/v+\tau)^2})}{u-x/v-\tau}\right.$$
$$\left.\left.-\frac{I_1(\sigma\sqrt{u^2-(x/v+\tau)^2})}{\sqrt{u^2-(x/v+\tau)^2}}\right]e^{\sigma(u-t)}du\right\} \tag{3.53}$$

where I_0, I_1 are modified Bessel functions.

These expressions describe the splitting of a rectangular electromagnetic pulse following a time jump of conductivity in the half-space $x \geq 0$. The picture of the transformed field gives the time-spatial diagram in Fig. 3.7.

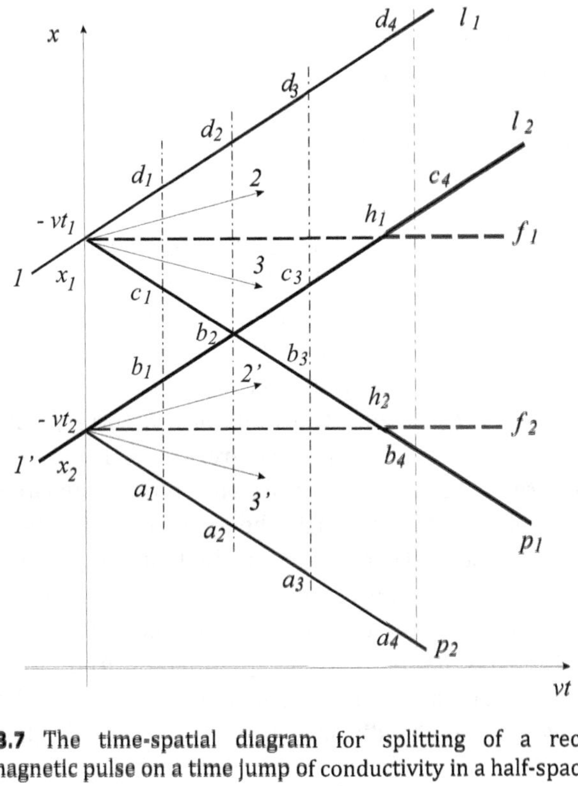

Figure 3.7 The time-spatial diagram for splitting of a rectangular electromagnetic pulse on a time jump of conductivity in a half-space $x \geq 0$.

The beams 1 and 1′ determine the world lines of the fronts of the initial pulse until the appearance of conductivity, i.e., for $t < 0$. The temporal conductivity jump leads to the appearance of the beams 2 and 2′ passing through the conductivity time discontinuity and the beams 3 and 3′ reflected from it. The sources of these beams are located at the points $x = -vt_0$ and $x = -v(t_0 + T)$. The package of passed time-spatial beams 2 and 2′ occupies the region $s_1b_1l_1$, $-vt_0 < x < v(t - t_0)$, and $s_2b_2l_2$, $-v(t_0 + T) < x < v(t - t_0 + T)$, respectively. The reflected beams 3 and 3′ occupy the regions $p_1b_1s_1$, $-v(t + t_0) < x < -vt_0$, and $p_2b_1s_2$, $-vt - v(t_0 - T) < x < -v(t_0 + T)$. Dashed lines in Fig. 3.7 show the boundaries between the regions with different expressions for the field. The field is continuous on these boundaries. Solid lines designate the boundaries where the field has a discontinuity. The field is continuous also on the boundaries between the regions of passed and reflected beams $x = -vt_0$ and $x = -v(t_0 + T)$.

Transformation of the rectangular pulse can be interpreted as its splitting by the conductivity jump into time-transmitted and time-reflected pulses with inhomogeneous structures. The reflection angle from the quasi-plane $t = 0$ in the space-time coordinates is equal to the incidence angle. The lines l_1l_2 and p_1p_2 are the world lines of the fronts of the direct and reversed pulses.

The reversed pulse moving to the medium boundary reflects from it. This reflected pulse is described by the second term in (3.51). The fronts of the direct, reversed and reflected pulses induced by the medium boundary are the sources of the secondary waves whose phase velocities are given by $v_{ph} = vt/(-t_0 + mT + x/v)^{-1}$. The value $m = 0$ corresponds to the leading front of the initial pulse whereas $m = 1$ corresponds to the rear front. Inhomogeneity of the phase velocities leads to leakage of the field through the rear pulse fronts and the formation of traces. Thereby, the passed and the reflected by the temporal jump pulses move as a whole in opposite directions with the same velocity, whereas each point inside these pulses has its phase velocity different from the others. These pulses separate at the moment $t = T/2$ when their rear fronts have passed the distance that is equal to the half pulse duration. Until this moment the region of their intersection is filled by the homogeneous field $E_0 \exp(-2\sigma t)$. After these pulses split the region $p_1 a l_2$ between their rear fronts is filled by the inhomogeneous field which represents the traces of these pulses. These traces excite the electromagnetic oscillations of the continuous spectrum in that region where they have passed. Redistribution of the energy in the direct and reversed pulses occurs in such a way that their amplitudes become equal each other and decrease with time.

The field outside the conductive medium is given by

$$E_r(t,x) = E_0 e^{-2\sigma t} \sum_{m=0}^{1} \theta(v(t - t_0 + mT) + x)$$
$$\times \left(\frac{vt + x - vt_0 + mvT}{vt + x + vt_0 + mvT} \right)^{1/2} I_1(\sigma\sqrt{(t + x/v)^2 - (t_0 - mT)^2}). \quad (3.54)$$

It follows from this expression that the temporal asymptotic of the field in the pulse leaving the conductive medium is given by exponential law

$$E_r(t,x) \approx \frac{1}{2} E_0 e^{-\sigma t} \theta(t + t_0 - T + x/v), \quad (3.55)$$

whereas the field near the medium boundary decreases by the power law:

$$E_r(t,x) \approx E_0[1 + \sigma(t_0 - T/2)] e^{\sigma x/v} [2\pi\sigma(t + x/v)]^{-3/2} \quad (3.56)$$

Therefore, the reversed pulse appearing as a result of the initial pulse splitting reflects partly from the medium boundary inside the conductor and goes out partly representing the return of the initial pulse. This returned pulse leaves the trace where the field decreases by the power law whereas the field in the pulse itself decreases by the exponential law. The fact that the field in the trace decreases more slowly than that in the pulse itself leads to attachment of the field to the medium boundary. It means that eventually the field in the stationary dielectric half-space will be observed near the boundary only.

3.3 Jump Changes of Plasma Density in a Plasma Half-Space with a Plane Boundary

3.3.1 Plasma Density's Jump Change in a Half-Space

The resolvent built in Section 2.4 describes the electrodynamic processes in non-stationary but unbounded homogeneous plasma. As applied to the initial-boundary problem this resolvent corresponds to the second (intermediate) stage of the interaction evolution. However, in the presence of boundaries the main stage is the third, final one. The transient process after one jump of plasma density in a half-space is investigated in the paper [21] where the problem is solved by means of the Riemann function's method in a space-time representation. In the more complicated cases, examination is confined by certain approximations. For example, diffraction of electromagnetic waves on a plasma ellipsoid with a variable plasma density was investigated in [31] in a quasi-static approach.

In this section, the initial-boundary-value problem is investigated for the case of a dispersive medium, namely the problem with a plasma half-space restricted by a flat boundary in which plasma density experiences jump changes in time such that a plasma frequency changes from ω_{e1} to ω_{e2}. Unlike the cases studied in the literature, the half-space, from which the initial wave $\mathbf{E}_0$ impinges on the boundary, is filled with stationary plasma. To describe such a situation one has to use an equation corresponding to the plasma background examined in Chapter 1. In this case, the propagation operator in impulse representation is defined by the expression:

$$\left\langle \mathbf{p}\middle|\hat{\Gamma}_e\middle|\mathbf{p}'\right\rangle_{ij} = -4\pi\frac{c^2k_ik_j + (p^2+\omega_e^2)\delta_{ij}}{p^2+c^2k^2+\omega_e^2}\frac{p^2}{p^2+\omega_e^2}\Delta(\mathbf{p}-\mathbf{p}'), \qquad (3.57)$$

where ω_e is a background plasma density in the stationary plasma half-space and $\Delta(\mathbf{p}-\mathbf{p}') = \frac{2\pi\delta(\mathbf{k}-\mathbf{k}')}{p-p'}$. A polarization vector of plasma is defined by the operator

$$\frac{1}{\varepsilon\varepsilon_0}(\mathbf{P}-\mathbf{P}_{\text{ex}}) = (\omega_{e2}^2-\omega_{e1}^2)\int_0^t dt'(t-t')\mathbf{E}(t',\mathbf{r}). \qquad (3.58)$$

The matrix elements for the equation kernel for the intermediate stage of the evolution are given by

$$\left\langle \mathbf{p}\left|\hat{\mathrm{K}}_2\right|\mathbf{p}'\right\rangle_{ij} = -\frac{c^2k_ik_j + (p^2+\omega_e^2)\delta_{ij}}{p^2+c^2k^2+\omega_e^2}\frac{\omega_{e2}^2-\omega_{e1}^2}{p^2+\omega_{e2}^2}\Delta(\mathbf{p}-\mathbf{p}'), \tag{3.59}$$

where Re$(p - p') > 0$. The solution to the equation for the resolvent (3.7) with this kernel results in

$$\left\langle \mathbf{p}\left|\hat{R}_2\right|\mathbf{p}'\right\rangle_{ij} = -\frac{c^2k_ik_j + (p^2+\omega_e^2)\delta_{ij}}{p^2+c^2k^2+\omega_e^2}\frac{\omega_{e2}^2-\omega_{e1}^2}{p^2+\omega_{e2}^2}\Delta(\mathbf{p}-\mathbf{p}'), \tag{3.60}$$

where Re$(p - p') > 0$.

To construct the third stage resolvent, we confine the investigation to the one-dimensional scalar case. This means that in the formulae (3.59), (3.60) we discard the tensor part k_ik_j and carry out descending with respect to two transverse spatial coordinates. It gives the kernel of the second stage, in both the impulse and coordinate representations

$$\left\langle \mathbf{p}\left|\hat{\mathrm{K}}_2\right|\mathbf{p}'\right\rangle_{ij} = -\frac{\omega_{e2}^2-\omega_{e1}^2}{p^2+c^2k^2+\omega_e^2}\Delta(\mathbf{p}-\mathbf{p}'), \tag{3.61}$$

$$\left\langle \mathbf{x}\left|\hat{\mathrm{K}}_2\right|\mathbf{x}'\right\rangle_{ij} = -\frac{\omega_{e2}^2-\omega_{e1}^2}{2c}J_0\left(\omega_e\sqrt{(t-t')^2-\frac{(x-x')^2}{c^2}}\right)\theta\left(t-t'-\frac{|x-x'|}{c}\right), \tag{3.62}$$

and the second stage resolvent

$$\left\langle \mathbf{p}\left|\hat{R}_2\right|\mathbf{p}'\right\rangle_{ij} = -\frac{\omega_{e2}^2-\omega_{e1}^2}{p^2+c^2k^2+\omega_{e2}^2}\frac{2\pi\delta(k-k')}{p-p'}, \tag{3.63}$$

$$\left\langle \mathbf{x}\left|\hat{R}_2\right|\mathbf{x}'\right\rangle = -\frac{\omega_{e2}^2-\omega_{e1}^2}{2c}J_0\left(\omega_{e2}\sqrt{(t-t')^2-\frac{(x-x')^2}{c^2}}\right)\theta\left(t-t'-\frac{|x-x'|}{c}\right), \tag{3.64}$$

It is worth noting that the expressions for the resolvent obtained as a solution to Eq. (3.7) are exact ones. The possibility to obtain such exact expressions is determined by the specific nature of the medium parameter change, namely, by the step-wise character of this change. If the medium parameters change arbitrarily, then obtaining the resolvent becomes more complex and can only be approximated. In this case, we can use the direct definition of the resolvent as the series

$$\hat{R} = \sum_{n=1}^{\infty} \hat{K}^n, \tag{3.65}$$

where $\hat{K}$ is a kernel of the integral equation for the field.

As a confirmation of the equivalence between these two approaches to constructing the resolvent let us find the resolvent by means of summation of the series Eq. (3.65) for the case considered above for the plasma density jump change to ω_{e2} = const. To make calculations easier we confine ourselves to the one-dimensional problem with the background as vacuum ($\varepsilon = 1$). We shall construct the resolvent in a coordinate representation where the kernel is defined by the integral

$$\left\langle \mathbf{x} \middle| \hat{\mathrm{K}}_2 \middle| \mathbf{x}' \right\rangle = \int d\mathbf{x}' \left\langle \mathbf{x} \middle| \hat{\Gamma} \middle| \mathbf{x}'' \right\rangle \left\langle \mathbf{x}'' \middle| \hat{V}_e \middle| \mathbf{x}' \right\rangle, \tag{3.66}$$

with $\left\langle \mathbf{x} \,|\, \hat{\Gamma} \,|\, \mathbf{x}' \right\rangle = -\dfrac{1}{2c}\dfrac{\partial^2}{\partial t^2}\theta\left(t - t' - \frac{|x-x'|}{c}\right)$. The matrix elements of the medium operator are

$$\left\langle \mathbf{x} \middle| \hat{V}_e \middle| \mathbf{x}' \right\rangle = \omega_{e2}^2 \delta(x - x')(t - t')\theta(t - t'). \tag{3.67}$$

Integrating in (3.66) gives

$$\left\langle \mathbf{x} \,|\, \hat{K}_2 \,|\, \mathbf{x}' \right\rangle = -\frac{\omega_{e2}^2}{2c}\theta\left(t - t' - \frac{|x - x'|}{c}\right). \tag{3.68}$$

The matrix elements of the operator $\hat{K}_2^n$ are obtained after $(n - 1)$-multiple integrations

$$\left\langle \mathbf{x} \,|\, \hat{K}_2^n \,|\, \mathbf{x}' \right\rangle = \int d\mathbf{x}_{n-1} \left\langle \mathbf{x} \,|\, \hat{K}_2 \,|\, \mathbf{x}_{n-1} \right\rangle$$
$$\ldots \int d\mathbf{x}_2 \left\langle \mathbf{x}_3 \,|\, \hat{K}_2 \,|\, \mathbf{x}_2 \right\rangle \int d\mathbf{x}_1 \left\langle \mathbf{x}_2 \,|\, \hat{K}_2 \,|\, \mathbf{x}_1 \right\rangle \left\langle \mathbf{x}_1 \,|\, \hat{K}_2 \,|\, \mathbf{x}' \right\rangle \tag{3.69}$$

as

$$\int d\mathbf{x}_1 \theta\left(t_2 - t_1 - \frac{|x_2 - x_1|}{c}\right)\theta\left(t - t' - \frac{|x_1 - x'|}{c}\right) = \theta\left(t_2 - t' - \frac{|x_2 - x'|}{c}\right)\int_{D_2} d\mathbf{x}_1$$

(3.70)

As illustrated in Fig. 3.8a, the expression (3.69) can be rewritten in the form

$$\langle \mathbf{x}_3 | \hat{K}_2^n | \mathbf{x}' \rangle = \left(-\frac{\omega_{e2}^2}{2c}\right)^2 \theta\left(t - t' - \frac{|x - x'|}{c}\right)\int_D d\mathbf{x}_{n-1} \int_{D_{n-1}} d\mathbf{x}_{n-2} \cdots \int_{D_2} d\mathbf{x}_1,$$

(3.71)

where the region D_k is defined by the intersection of the cones $t_\kappa - t_{\kappa-1} - \frac{|x_\kappa - x_{\kappa-1}|}{c} > 0$ and $t_{\kappa-1} - t' - \frac{1}{c}|x_{\kappa-1} - x'| > 0$ and represents a parallelogram. To calculate the integrals in Eq. (3.71), it is convenient to use the new variables $u_i = \frac{1}{\sqrt{2}}(ct_i + x_i)$ and $w_i = \frac{1}{\sqrt{2}}(ct_i - x_i)$, in which the regions of integration are transforming into rectangles D_i (Fig. 3.8b).

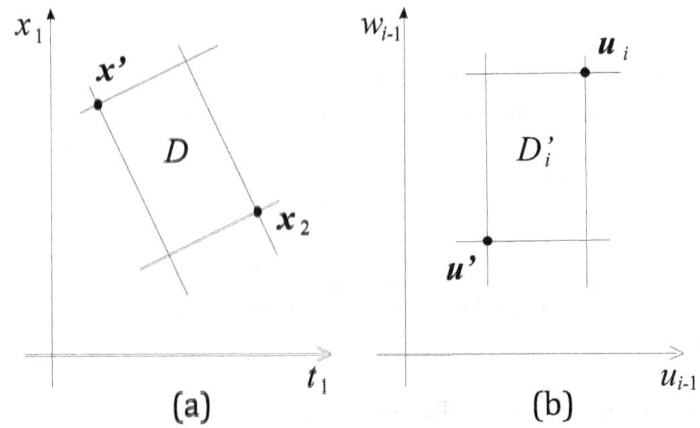

Figure 3.8 An example of regions for integration in (3.71).

After calculating the integrals in Eq. (3.71) and substituting them into the series (Eq. (3.65)), we have

$$\langle \mathbf{x}|\hat{R}_2|\mathbf{x}'\rangle = \sum_{n=1}^{\infty}\left(-\frac{\omega_{e2}^2}{2c}\right)^n \theta\left(t-t'-\frac{|x-x'|}{c}\right)\frac{[(u-u')(w-w')]^{n-1}}{c^{n-1}((n-1)!)^2}$$
$$= -\frac{\omega_{e2}^2}{2c}J_0\left(2\sqrt{\frac{\zeta_2}{c}}\sqrt{(u-u')(w-w')}\right)\theta\left(t-t'-\frac{|x-x'|}{c}\right)$$
$$= -\frac{\omega_{e2}^2}{2c}J_0\left(\sqrt{2\zeta_2 c}\sqrt{(t-t')^2-\frac{(x-x')^2}{c^2}}\right)\theta\left(t-t'-\frac{|x-x'|}{c}\right) \tag{3.72}$$

that is, an expression matching the one given in Eq. (3.64) with $\omega_{e1} = 0$ for the resolvent obtained as a solution to the operator equation.

Let us follow constructing the resolvent for a boundary value problem which is described by the kernel $\hat{K}_3 = \chi\hat{K}_2\chi$ (where $\chi = \theta(x)$) of the equation for the third stage and has the form

$$\langle \mathbf{p}|\hat{K}_3|\mathbf{p}'\rangle = \frac{\omega_{e2}^2-\omega_{e1}^2}{p^2+c^2k^2+\omega_e^2}\frac{1}{p-p'}\frac{i}{k-k'-io}$$
$$+\frac{\omega_{e2}^2-\omega_{e1}^2}{2}\frac{c}{\varphi(p)}\frac{1}{\varphi(p)-ick'}\frac{1}{p-p'}, \tag{3.73}$$

where $\varphi(p)=\sqrt{p^2+\omega_e^2}$.

Using the same technique as above, we represent the resolvent $\hat{R}_3$ in the form taking into account a reflected component

$$\langle \mathbf{p}|\hat{R}_3|\mathbf{p}'\rangle = \frac{\omega_{e2}^2-\omega_{e1}^2}{p^2+c^2k^2+\omega_{e2}^2}\frac{1}{p-p'}\frac{i}{k-k'-io}$$
$$+c\frac{\omega_{e2}^2-\omega_{e1}^2}{2\varphi_2(p)}\left[\frac{1}{\varphi_2(p)-ick}+\frac{R(p)}{\varphi_2(p)+ick}\right]\frac{1}{\varphi_2(p)-ick'}\frac{1}{p-p'}, \tag{3.74}$$

where $\varphi_2(p)=\sqrt{p^2+\omega_{e2}^2}$, $\mathrm{Re}(p-p')>0$.

Substituting this expression into Eq. (3.7) with the kernel (3.73) we find a coefficient

$$R(p)=\frac{\varphi(p)-\varphi_2(p)}{\varphi(p)+\varphi_2(p)}. \tag{3.75}$$

As in the case of dielectric, the expression (3.75) coincides with the reflectance for a monochromatic wave reflected from a flat boundary of stationary plasma

$$R(\omega)=\frac{\sqrt{\varepsilon_2(\omega)}-\sqrt{\varepsilon(\omega)}}{\sqrt{\varepsilon_2(\omega)}+\sqrt{\varepsilon(\omega)}}=\frac{\sqrt{\omega^2-\omega_{e2}^2}-\sqrt{\omega^2-\omega_e^2}}{\sqrt{\omega^2-\omega_{e2}^2}+\sqrt{\omega^2-\omega_e^2}}, \tag{3.76}$$

if one proceeds from a Fourier transform variable to a Laplace transform variable $p \to i\omega$ and takes into account the fact that reflection occurs from the plasma's side; therefore, it is necessary to use $R(\omega)$ with an opposite sign.

The resolvent in the coordinate representation has the form

$$\langle \mathbf{x}|\hat{R}_3|\mathbf{x}'\rangle=\theta(x)\langle \mathbf{x}|\hat{R}_2|\mathbf{x}'\rangle\theta(x')+\frac{\omega_{e2}^2-\omega_{e1}^2}{2c}\theta(x)\int_{-i\infty}^{i\infty}\frac{dp}{2\pi i}R(p)J(p)e^{p(t-t')}\theta(x'), \tag{3.77}$$

where $R(p)$ is defined by the formula (3.75), and

$$J(p)=\frac{1}{\varphi_2(p)}e^{-\frac{x+x'}{c}\varphi_2(p)}, \tag{3.78}$$

The inverse Laplace transform of this function is equal to

$$\int_{-i\infty}^{i\infty}\frac{dp}{2\pi i}J(p)e^{p\tau}=J_0\left(\omega_{e2}\sqrt{\tau^2-\frac{(x-x')^2}{c^2}}\right)\theta\left(\tau-\frac{x+x'}{c}\right). \tag{3.79}$$

Thus, the second term in Eq. (3.77) is a contribution of a reflected characteristic with the weight $J(p)$ determined by the medium dispersion.

In the case of vacuum background ($\omega_e = 0$), the resolvent structure remains the same; however, the calculations become simpler. In particular, an integral in (3.77) reduces to the Bessel function and the resolvent takes the following form:

$$\langle \mathbf{x}|\hat{R}_3|\mathbf{x}'\rangle=\theta(x)\langle \mathbf{x}|\hat{R}_2|\mathbf{x}'\rangle\theta(x')+\frac{\omega_{e2}^2}{2c}\theta(x)\cdot\frac{t-t'-\frac{x-x'}{c}}{t-t'+\frac{x+x'}{c}}J_2$$

$$\left(\omega_{e2}\sqrt{(t-t)^2-\frac{(x+x')^2}{c^2}}\right)\theta\left(t-t'-\frac{x+x'}{c}\right)\theta(x'). \tag{3.80}$$

3.3.2 The Resolvent of the Second Stage of Evolution with Two-Step Change of Plasma Density

The resolvent method developed allows solving not only a problem with a single jump of a plasma density but also a problem with series of arbitrary jumps at arbitrary instants. One such problem is the key problem, in which the change of a plasma density has an impulse nature. This situation is described by the function for the plasma frequency

$$\omega_e^2(t) = \omega_{e2}^2 \cdot \theta(\mathrm{T}-t) + \omega_{e3}^2 \theta(t-\mathrm{T})$$

that means that after time interval T the plasma density turns to the value ω_{e3}. The kernel of the operator equation for the second stage resolvent in the case of vacuum background is defined by the integral

$$\langle \mathbf{x} | \hat{\mathrm{K}}_2 | \mathbf{x}' \rangle = \int d\mathbf{x}_1 \langle \mathbf{x} | \hat{\Gamma}_e | \mathbf{x}_1 \rangle \langle \mathbf{x}_1 | \hat{V}_\mathrm{T} | \mathbf{x}' \rangle, \tag{3.81}$$

where

$$\hat{V}_\mathrm{T} = -\int_0^t [\omega_{e2}^2 \theta(\mathrm{T}-t') + \omega_{e3}^2 \theta(t'-\mathrm{T})](t-t')dt'. \tag{3.82}$$

In the impulse representation

$$\langle \mathbf{p} | \hat{\mathrm{K}}_2 | \mathbf{p}' \rangle = \frac{-1}{p^2 + c^2 k^2} [\omega_{e2} + (\omega_{e3} - \omega_{e2}) e^{-(p-p')T}] \Delta(\mathbf{p}-\mathbf{p}'), \tag{3.83}$$

where $\mathrm{Re}(p-p') > 0$, $\Delta(\mathbf{p}-\mathbf{p}') = \dfrac{2\pi\delta(k-k')}{(p-p')}$.

Substituting this expression into the equation for the resolvent, and assuming that all the singularities of the function $\langle \mathbf{p} | \hat{R}_2 | \mathbf{p}' \rangle$ with respect to the variable p are to the left from the path of integration, we can integrate partly by a contour closed to the right and obtain

$$(p^2 + c^2k^2 + \omega_{e2}^2) \langle \mathbf{p} | \hat{R}_2 | \mathbf{p}' \rangle + (\omega_{e3}^2 - \omega_{e2}^2)$$
$$\int_{-i\infty}^{i\infty} \frac{dp_1}{2\pi i} e^{-(p-p_1)T} \cdot \frac{\langle p_1, \mathbf{k} | \hat{R}_2 | \mathbf{p}' \rangle}{p - p_1} = -\{\omega_{e2}^2 + (\omega_{e3}^2 - \omega_{e2}^2) e^{-(p-p')T}\} \Delta(\mathbf{p}-\mathbf{p}'). \tag{3.84}$$

Based on the structure of this equation its solution should be found in the form

$$\langle \mathbf{p}|\hat{R}_2|\mathbf{p}'\rangle = \left[\frac{-\omega_{e2}^2}{p^2+c^2k^2+\omega_{e2}^2} + \Psi(\mathbf{p},\mathbf{p}')e^{-(p-p')T}\right]\Delta(\mathbf{p}-\mathbf{p}'). \quad (3.85)$$

where

$$\Psi(\mathbf{p},\mathbf{p}') = \frac{\omega_{e2}^2-\omega_{e3}^2}{p^2+c^2k^2+\omega_{e3}^2} - \frac{\omega_{e2}^2(\omega_{e2}^2-\omega_{e3}^2)(p-p')}{p^2+c^2k^2+\omega_{e3}^2} \cdot \int_{-i\infty}^{i\infty} \frac{dp_1}{2\pi i} \frac{e^{(p'-p_1)T}}{(p^2+c^2k^2+\omega_{e2}^2)(p-p_1)(p_1-p')}. \quad (3.86)$$

Finally, the coordinate representation of the second stage resolvent has the form

$$\langle \mathbf{x}|\hat{R}_2|\mathbf{x}'\rangle = \theta(\mathrm{T}-t')\theta(\mathrm{T}-t)\langle \mathbf{x}|\hat{R}^{(2)}|\mathbf{x}'\rangle + \theta(t'-\mathrm{T})\langle \mathbf{x}|\hat{R}^{(3)}|\mathbf{x}'\rangle + \theta(\mathrm{T}-t')\theta(t-\mathrm{T})\Psi)(\mathbf{x},\mathbf{x}'), \quad (3.87)$$

where the first two items are analogous to the resolvent $\hat{R}_2$ from (3.64)

$$\langle \mathbf{x}|\hat{R}^{(2,3)}|\mathbf{x}'\rangle = -\frac{\omega_{e2,3}^2}{2c} J_0\left(\omega_{e2,3}\sqrt{(t-t')^2-\frac{(x-x')^2}{c^2}}\right)\theta\left(t-t'-\frac{|x-x'|}{c}\right)$$

and the last item is defined by the coordinate representation of the function (3.86)

$$\Psi(\mathbf{x},\mathbf{x}') = -\frac{\omega_{e2}^2}{2c}\int_{-\infty}^{\infty}\frac{dk}{2\pi}e^{ik(x-x')}\left\{\frac{\cos\Omega_3(t-\mathrm{T})\sin\Omega_2(\mathrm{T}-t')}{\Omega_2} + \frac{\sin\Omega_3(t-\mathrm{T})\cos\Omega_2(\mathrm{T}-t')}{\Omega_3}\right\}, \quad (3.88)$$

where $\Omega_{3,2} = \sqrt{c^2k^2+\omega_{e3,2}^2}$. The time interval corresponding to the second stage of the evolution is defined by the inequality $t < x/c$.

3.3.3 The Field after One Jump of Plasma Density

Let us examine the field in cases considered above when, in the half-space $x \geq 0$ at zero moment, plasma is transformed under external forces such that the new plasma frequency is ω_{e2}. The initial field in the form of a plane wave $E_0 = \exp[i(\omega t - kx)]$ existing in the half-space $x < 0$ penetrates into the plasma before zero moment where it induces the plane wave

$$E_1 = Ae^{i(\omega t - kx)},\ A = \frac{2\omega(\omega - ck)}{\omega_{e1}^2},\ k = \frac{1}{c}\sqrt{\omega^2 - \omega_{e1}^2}. \tag{3.89}$$

The equation for the field has the following form:

$$E(\mathbf{x}) = F(\mathbf{x}) - \frac{2\pi}{c}\int d\mathbf{x}'\left\langle \mathbf{x}\middle|\hat{\Gamma}_e\middle|\mathbf{x}'\right\rangle \theta(x')P(\mathbf{x}'), \tag{3.90}$$

where $\left\langle \mathbf{x}\,|\,\hat{\Gamma}_e\,|\,\mathbf{x}'\right\rangle = \frac{\partial^2}{\partial t^2}\theta\left(t - t' - \frac{|x - x'|}{c}\right)$ and $P(\mathbf{x})$ is the plasma polarization. The free term of Eq. (3.90) is defined by the following expression:

$$F = E_0 - \frac{2\pi}{c}\int_{-\infty}^{0} dt' \int_{0}^{\infty} dx < \mathbf{x}\middle|\hat{\Gamma}_e\middle|\mathbf{x}' > P_1(\mathbf{x}') - \frac{2\pi}{c}\int d\mathbf{x}' < \mathbf{x}\middle|\hat{\Gamma}_e\middle|\mathbf{x}' > \chi(\mathbf{x}')P_0(\mathbf{x}'), \tag{3.91}$$

where $P_1(\mathbf{x})$ is the plasma polarization before zero moment, $P_0(\mathbf{x})$ is the relict plasma polarization and E_0 is the initial wave impinging on the plasma half-space. If plasma density changes from ω_{e1} to ω_{e2} at zero moment then a solution to Eq. (3.90) is made by means of the resolvent (3.80).

The wave (3.89) defines a particular form for the free terms $F_{2,3}$ that have a meaning of "initial waves" for the second and third stages of the interaction evolution. F_2 is expressed by

$$F_2 = -\frac{1}{2\omega\omega_1}\left\{\frac{\omega_{e1}^2\omega_1 + \omega_{e2}^2(\omega - \omega_1)}{\omega_1 - \omega}e^{i\omega_1 t} - \frac{\omega_{p1}^2\omega_1 + \omega_{e2}^2(\omega + \omega_1)}{\omega_1 + \omega}e^{-i\omega_1 t}\right\}e^{-ikx}, \tag{3.92}$$

where $\omega_1 = \sqrt{\omega^2 - \omega_{e1}^2}$.

By the means of the resolvent (3.64), we find the field in the time interval $0 \le t < \frac{x}{c}$ in which an influence of the boundary does not take place:

$$E_2 = A_1 e^{i(\omega_2 t - kx)} + A_2 e^{i(\omega_2 t + kx)}, \tag{3.93}$$

where $\omega_2 = \sqrt{c^2k^2 + \omega_{e2}^2} = \sqrt{\omega^2 - \omega_{e1}^2 + \omega_{e2}^2}$, $A_{1,2} = \dfrac{\omega \pm \omega_2}{2\omega}$.

It follows from this expression that after plasma appearance, the field E_1 splits into two waves propagating in opposite directions. These waves have the same wavenumber as the wave E_1 but a different frequency. The frequency as well as the amplitude of waves in (3.93) depends on the incident wave frequency and on the plasma density. With $\omega_{e2} = \omega_{e1}$, the second wave in (3.93) disappears and the first one turns into E_1.

For simplification of computations further on we are confined here by the case of $\omega_{e1} = 0$. This would not result in a loss of generality because with a plasma density stepwise change we can always assume that up to some moment of time plasma density is equal to zero and after that it takes any given value by discrete steps. Now let us define the field penetrating into the plasma in the interval $t \ge x/c$, in which the boundary influence takes place. To calculate the field it is convenient to represent the resolvent $\hat{R}_3$ (3.80) in the form of the integral:

$$\langle \mathbf{x} | \hat{R}_3 | \mathbf{x}' \rangle = -\frac{\omega_{e2}^2}{2c} \int_{-i\infty}^{i\infty} \frac{dp}{2\pi i} \frac{e^{-\varphi_2(p)|x-x'|/c}}{\varphi_2(p)} e^{p(t-t')}\theta(x')$$
$$+ \frac{\omega_{e2}^2}{2c} \int_{-i\infty}^{i\infty} \frac{dp}{2\pi i} \frac{p - \varphi_2(p)}{p + \varphi_2(p)} \frac{1}{\varphi_2(p)} e^{p(t-t') - \varphi_2(p)(x+x')/c}\theta(x'), \tag{3.94}$$

where $\varphi_2(p) = \sqrt{p^2 + \omega_{e2}^2}$. Application of this operator to the free term (3.91)

$$F_3 = E_0 + \frac{\omega_{e2}^2}{2\omega^2}\left(1 - e^{-i\frac{\omega}{c}(x+ct)}\right), \tag{3.95}$$

gives

$$E_3 = E_2(t,x) - \frac{\omega_{e2}^2}{\omega} \int_{-i\infty}^{i\infty} \frac{dp}{2\pi i} \frac{ip(p+i\omega)e^{pt - \frac{x}{c}\varphi_2(p)}}{(p - i\omega)(\varphi_2 + p)(p^2 + \omega_2^2)}. \tag{3.96}$$

To calculate the integral in the interval $t \geq x/c$ we make branch cuts C which provide a choice of the root branch satisfying the condition $\mathrm{Re}\varphi_2(p) > 0$ in the plane of the complex variable $p = \xi + i\eta$ (Fig. 3.9).

It is convenient to make these cuts along the hyperbola $\xi\eta = -\delta\omega_{e2}$ from the points $\pm i(\omega_p + i\delta)$ where δ is an infinitesimal addition that takes into account damping in plasma. After the calculation of residues at the points of the singularities placed on the upper sheet of the Riemann's surface, an expression for the field is reduced to the form

$$E_3 = A_3 e^{i\omega t - ik_2 x} + A_2 e^{-i\omega_2 t + i\frac{\omega}{c}x} + A_2 B_1 e^{-i\omega_2 t + i\frac{\omega}{c}x}$$
$$-\frac{\omega_{e2}^2}{\omega}\cdot\int_c \frac{dp}{2\pi i}\frac{ip(p+i\omega)e^{pt-\frac{x}{c}\varphi_2(p)}}{(p-i\omega)(\varphi_2+p)(p^2+\omega_2^2)}, \tag{3.97}$$

where $A_3 = \frac{2\omega}{\omega+ck_2}$, $ck_2 = \sqrt{\omega^2 - \omega_{e2}^2}$, $B_1 = (\omega-\omega_2)/(\omega+\omega_2)$. Here it is taken into account that $j_2(\pm i\omega_2) = \pm i\omega$.

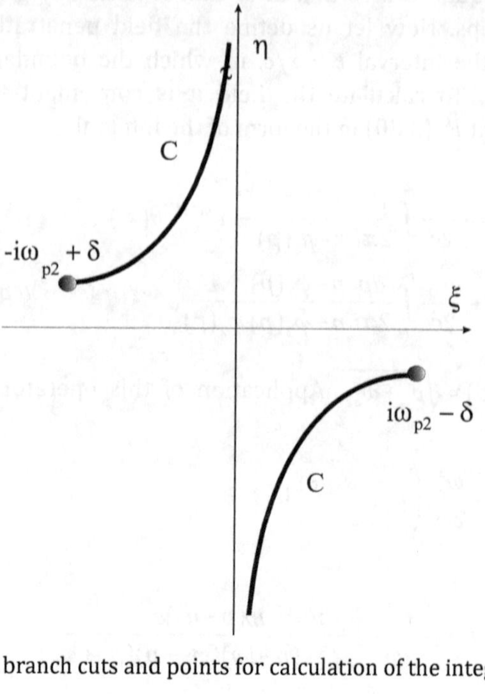

Figure 3.9 The branch cuts and points for calculation of the integral (3.97).

As we can see from this expression, the field E_3 for $t \geq x/c$ contains a discrete spectrum as well as a continuous spectrum, which is defined by the integral over the branch cut. The discrete wave spectrum is analogous to the corresponding spectrum in the case of the dielectric half-space and it consists of three components. The first component in (3.97) is the wave initiated in plasma by the wave E_0 falling on its boundary after zero moment, and coincides with a wave penetrating into the stationary plasma. The second wave is a reverse wave that appears as a result of splitting of the wave E_1 at the moment of the plasma density jump. A reflection of this wave from within the plasma boundary gives the third wave of the discrete spectrum. The whole field E_3 on the world line $x = ct$ is equal to the field E_2.

3.3.4 Transformation of an Electromagnetic Wave by Two-Step Change of the Plasma Density

Construction of the resolvent on the third stage of the evolution (on the interval $t \geq x/c$) in the case of a two-step change of plasma density represents some considerable difficulties. However, it is possible to avoid these difficulties by means of a method that consists of transferring an initial point of time. This gives an opportunity to solve not only problems with a plasma density's double jump, but also the case of arbitrary number of jumps at arbitrary instants. For example, we can illustrate this on a scalar one-dimensional problem that was defined at the beginning of this section. We consider also the case of a vacuum background.

Now let plasma density change stepwise to the value of ω_{e3} at some moment of time T. Then the plasma frequency is given by the function $\omega_e^2(t) = \omega_{e2}^2\theta(T-t) + \omega_{e3}^2\theta(t-T)$. Equation (3.90) takes the form

$$E = F - \frac{\omega_{e2}^2}{2c}\int_0^T dt' \int_0^\infty dx' \theta\left(t - t' - \frac{|x-x'|}{c}\right) E(t',x')$$

$$- \frac{\omega_{e3}^2}{2c}\int_T^\infty dt' \int_0^\infty dx' \theta\left(t - t' - \frac{|x-x'|}{c}\right) E(t',x') \qquad (3.98)$$

if the differentiation operator in (3.90) is transferred from the first multiplier to the second one.

The field E in the first integral in (3.98) is defined on the interval $[0, T]$ and is known. Therefore, this integral can be attributed to the free term

$$\tilde{F} = F - \frac{\omega_{e2}^2}{2c}\int_0^T dt' \int_{-\infty}^{\infty} dx' \theta\left(t - t' - \frac{|x - x'|}{c}\right)\chi(\mathbf{x}')E(t', x'). \tag{3.99}$$

Transformation of the initial point to the moment T according to the formulae $t = \tau + T$, $t' = \tau' + T$ brings (3.98) to the form

$$E(\tau, x) = \tilde{F}(\tau, x) - \frac{\omega_{e3}^2}{2c}\int_0^{\infty} d\tau \int_0^{\infty} dx' \theta\left(\tau - \tau' - \frac{|x - x'|}{c}\right)E(\tau', x'). \tag{3.100}$$

As the equation obtained has the same kernel as Eq. (3.90), its resolvent can be defined by the same formula (3.80) if we exchange $\omega_{e2}^2 \to \omega_{e3}^2$. It is obvious that with a number of jumps this procedure can be repeated for a required number of times.

Now let's examine the two-step plasma density change in time when the plasma frequency is defined by the function

$$\omega_{e(t)}^2 = \omega_{e2}^2\theta(t_1 - t) + \omega_{e3}^2\theta(t - t_1). \tag{3.101}$$

We assume that before zero moment plasma did not exist, i.e., that $\omega_{e1} = 0$. In this case, in the interval $0 \le t < t_1$ the field penetration into the plasma is defined by the expressions (3.93) and (3.97). By virtue of this field we find the new free term (3.99)

$$\begin{aligned}
<\mathbf{p}|F_2> = {} & \frac{2\pi\delta\left(k + \frac{\omega}{c}\right)}{p - i\omega}e^{i\omega t_1} + \frac{\omega_{e3}^2 - \omega_{e2}^2}{\omega}\frac{1}{p^2 + c^2k^2}\frac{1}{k + \frac{\omega}{c} - io} \\
& + \frac{\omega_{e2}^2}{\omega}\int_{-i\infty}^{i\infty}\frac{dp'}{2\pi i}\left\{\frac{p^2 + c^2k^2}{p - p'} + \frac{(p + p')(i\omega p' - \omega_2^2)}{p'^2 + \omega_2^2}\right. \\
& + ic\omega_{e2}^2\frac{\left(k + \frac{\omega}{c} - ic\right)p'(p' + i\omega)(p + p')}{(p'^2 + \omega_2^2)(p' - i\omega)(p' + \varphi_2')(\varphi_2' + ick)} \\
& \left. + \frac{\omega_{e3}^2 - \omega_{e2}^2}{\omega_{e2}^2}\frac{(i\omega p' - \omega_2^2)(p'^2 + c^2k^2)}{p'(p'^2 + \omega_2^2)}\right\} \\
& \frac{e^{p't_1}}{(p^2 + c^2k^2)(p'^2 + c^2k^2)\left(k + \frac{\omega}{c} - io\right)}
\end{aligned} \tag{3.102}$$

By virtue of this free term and the resolvent $\hat{R}_3$ (3.80), we define the field E which has different forms in different space-time zones shown in Fig. 3.10.

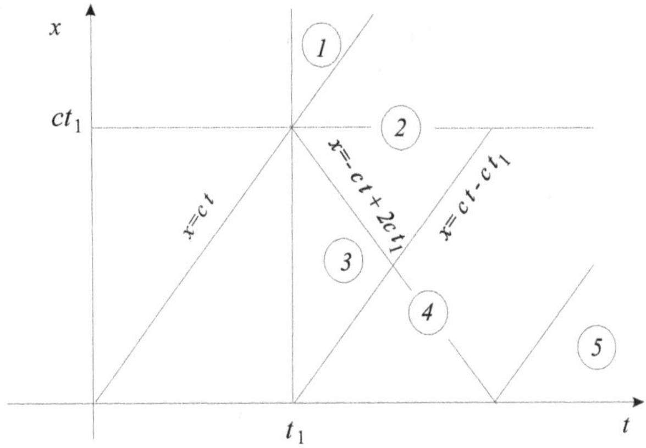

Figure 3.10 Space-time zones of different forms for the field.

In the first zone, $t_1 \le t < x/c$, which corresponds to an unbounded problem, the field E_2 consists of four waves

$$\tilde{E}_2^{(1)} = A_1(C_1 e^{i\omega_3(t-t_1)} + C_2 e^{-i\omega_3(t-t_1)})e^{i\omega_2 t_1 - i\frac{\omega}{c}x}$$
$$+ A_2(C_2 e^{i\omega_3(t-t_1)} + C_1 e^{-i\omega_3(t-t_1)})e^{-i\omega_2 t_1 - i\frac{\omega}{c}x}, \qquad (3.103)$$

where $C_{1,2} = (\omega_2 \pm \omega_3)/(2\omega_2)$. These waves are formed as a result of splitting the waves A_1 and A_2 at the moment of the second jump of the plasma density and they represent two pairs of plane waves propagating in opposite directions. They keep the wavenumber of the waves $A_{1,2}$ but have a new frequency $\omega_3 = \sqrt{\omega^2 + \omega_{e3}^2}$.

Assuming in (3.103) that $\omega_{e3} = 0$, we get the case of a plasma impulse corresponding to the plasma existence only during the interval t_1:

$$\tilde{E}_2^{(1)} = [(\omega_2 + \omega)^2 e^{i(\omega_2 - \omega)t_1} - (\omega_2 - \omega)^2 e^{i(\omega - \omega_2)t_1}]\frac{e^{i\omega\left(t - \frac{x}{c}\right)}}{4\omega\omega_2}$$
$$+ [e^{i(\omega-\omega_2)t_1} - e^{i(\omega+\omega_2)t_1}]\frac{\omega_{p2}^2}{4\omega\omega_2}e^{-i\omega\left(t + \frac{x}{c}\right)} \qquad (3.104)$$

The influence of the boundary is absent in the zones 2 and 3 after the second jump of the plasma density; however the field in these zones has a continuous spectrum, determined by the influence of the plasma boundary in the time interval $[0, t_1]$. A discrete spectrum in the third zone is formed as a result of splitting of the waves A_2 and A_3 on the line $t = t_1$. The field in this zone takes the form

$$\begin{aligned}\tilde{E}_2^{(3)} &= A_2(C_2 e^{i\omega_3(t-t_1)} + C_1 e^{-i\omega_3(t-t_1)})e^{i\omega_2 t_1 - i\frac{\omega}{c}x} \\ &+ A_2 B_1(C_1 e^{-i\omega_3(t-t_1)} + C_2 e^{i\omega_3(t-t_1)})e^{-i\omega_2 t_1 - i\frac{\omega}{c}x} \\ &+ A_3(D_1 e^{i\omega_{23}(t-t_1)} + D_2 e^{-i\omega_{23}(t-t_1)})e^{i\omega t_1 - ik_2 x} + I_1 + I_2, \end{aligned} \tag{3.105}$$

where $\omega_{23} = \sqrt{\omega^2 - \omega_{e2}^2 + \omega_{e3}^2}$, $D_{1,2} = (\omega \pm \omega_{23})/(2\omega)$

$$I_{1,2} = -\frac{\omega_{e2}^2}{2\omega}\int_C \frac{dp}{2\pi i}\frac{i(p+i\omega)(p \pm \sqrt{p^2 + \omega_{e2}^2 - \omega_{e3}^2})}{(p^2+\omega_2^2)(p-i\omega)(p+\varphi_2)} e^{pt - \frac{x}{c}\varphi_2(p) \pm (t-t_1)\sqrt{p^2+\omega_{e2}^2-\omega_{e3}^2}}. \tag{3.106}$$

In the zone 2, $t - t_1 \le \frac{x}{c} \le 2t_1 - t$, the field consists of the waves passing into this zone from the zones 1 and 3

$$\begin{aligned}\tilde{E}_2^{(2)} &= (A_1 C_2 e^{-i\omega_3(t-t_1)} + A_2 C_1 e^{-i\omega_3(t-t_1) - 2i\omega_2 t_1})e^{i\omega_2 t_1 - i\frac{\omega}{c}x} \\ &+ (A_2 C_2 e^{i\omega_3(t-t_1) - 2i\frac{\omega}{c}x} + A_2 B_1 C_1 e^{-i\omega_3(t-t_1)})e^{-i\omega_2 t_1 + i\frac{\omega}{c}x} \\ &+ A_3 D_1 e^{i\omega_{23}(t-t_1)\, i\omega t_1 - ik_2 x} + I_2. \end{aligned} \tag{3.107}$$

The influence of the plasma boundary after the moment t_1 becomes apparent in the fourth zone $(0 \le x < c(t - t_1))$. In this zone, below the line $x = -ct + 2ct_1$, the field is formed by the discrete and continuous spectrum waves, passing from the third zone and also by the wave initiated in the plasma by the wave E_0 falling on its boundary after the moment t_1. The frequency of this wave does not change but the wavenumber corresponds to the new plasma frequency $ck_3 = \sqrt{\omega^2 - \omega_{e3}^2}$:

$$\begin{aligned}\tilde{E}_3^{(4')} &= A_4 e^{i\omega t_1 - ik_2 x} + (A_2 C_1 e^{-i\omega_3(t-t_1) - i\frac{\omega}{c}x} + A_2 B_1 C_2 e^{-i\omega_3(t-t_1) + i\frac{\omega}{c}x}) e^{-i\omega_2 t_1} \\ &+ A_3 D_2 e^{-i\omega_{23}(t-t_1) + i\omega t_1 - ik_2 x} + A_2 C_1 F_1 e^{-i\omega_3(t-t_1) + i\omega_2 t_1 - i\frac{\omega}{c}x} \\ &+ A_2 B_1 C_2 F_1 e^{i\omega_3(t-t_1) - i\omega_2 t_1 - i\frac{\omega}{c}x} e^{-i\omega_2 t_1 - i\frac{\omega}{c}x} + A_3 D_2 F_2 e^{-i\omega_{23}(t-t_1) + i\omega t_1 + ik_2 x} + I_1 + I_3.\end{aligned} \tag{3.108}$$

Here,

$$A_4 = \frac{2\omega}{\omega + ck_3}; \; F_1 = \frac{\omega - \omega_3}{\omega + \omega_3}; \; F_2 = \frac{ck_2 - \omega_{23}}{ck_2 + \omega_{23}};$$

$$\begin{aligned}I_3 &= \frac{(\omega_{e2}^2 - \omega_{e3}^2)\omega_{e2}^2}{\omega} \int_{-i\infty}^{i\infty} \frac{dp'}{2\pi i} \frac{i(p' + i\omega)e^{p't_1}}{(p'^2 + \omega_2^2)(p' - i\omega)(p' + \varphi_2')} \cdot \\ &\cdot \int_C \frac{dp}{2\pi i} \frac{p(p' + \varphi_2')}{(p + \psi_3)(p - p')(p^2 - p'^2 + \omega_{e3}^2 - \omega_{e2}^2)} e^{p(t-t_1) - \frac{x}{c}\psi_3(p)} \\ &+ \int_C \frac{dp}{2\pi i} \left\{ \frac{\omega_{e3}^2 - \omega_{e2}^2}{2\omega\omega_2} \frac{ip(p + i\omega)}{(p + \psi_3)(p^2 + \omega_3^2)} \left[\frac{\omega - \omega_2}{p + i\omega_2} e^{i\omega_2 t_1} - \frac{\omega - \omega_2}{p - i\omega_2} e^{-i\omega_2 t_1} \right] e^{-pt_1} \right. \\ &\left. - \frac{\omega_{e2}^2}{\omega} \frac{ip(p + i\omega)}{(p + \psi_3)(p^2 + \omega_3^2)(p - i\omega)} \right\} e^{pt - \frac{x}{c}\psi_3(p)},\end{aligned}$$

where $\psi_3 = \sqrt{p^2 + \omega_{e3}^2}$ and $\psi_3(i\omega) = i\sqrt{\omega^2 - \omega_{e3}^2}$.

In the discrete spectrum, the waves with the amplitude $F_{1,2}$ are formed as a result of waves reflected from the plasma boundary and passing from the third zone. The same reflected waves, stationary wave with the wavenumber k_3, and the waves passing from the second zone form the field in the fourth zone above the line $x = -ct + 2ct_1$ in the stripe $-2ct_1 + ct \le x < -ct_1 + ct$:

$$\begin{aligned}\tilde{E}_3^{(4'')} &= A_4 e^{i\omega t_1 - ik_2 x} + A_2 C_1 F_1 e^{-i\omega_3(t-t_1) - i\omega_2 t_1 + i\frac{\omega}{c}x} \\ &+ A_2 B_1 C_2 F_1 e^{i\omega_3(t-t_1) - i\omega_2 t_1 - i\frac{\omega}{c}x} e^{-i\omega_2 t_1} + A_3 D_2 F_2 e^{-i\omega_{23}(t-t_1) + i\omega t_1 + ik_2 x} \\ &+ A_1 C_2 e^{-i\omega_3(t-t_1) + i\omega_2 t_1 - i\frac{\omega}{c}x} + A_2 C_1 e^{-i\omega_3(t-t_1) - i\omega_2 t_1 - i\frac{\omega}{c}x} + I_3,\end{aligned} \tag{3.109}$$

where $\psi_3(-i\omega) = -i\sqrt{\omega^2 - \omega_{e3}^2}$.

After the moment $2t_1$ the steady-state field is formed by the stationary wave with a wavenumber k_3 and by the waves passing from the second zone and reflected from the plasma boundary. This field occupies the region $0 \le x \le ct - 2ct_1$ and is defined by the expression

$$\tilde{E}_3^{(5)} = A_4 e^{i\omega t - ik_2 x} + (A_1 C_2 e^{i\omega_2 t_1} + A_2 C_1 e^{-i\omega_2 t_1}) e^{-i\omega_3 (t-t_1) - i\frac{\omega}{X}x}$$
$$+ (A_2 C_1 F_1 + A_1 C_2 F_1) e^{-i\omega_3 (t-t_1) + i\omega_2 t_1 + i\frac{\omega}{X}x} + I_3 \qquad (3.110)$$

It is worth noting that the fields $\tilde{E}_{2,3}$ change continuously on all lines of the assigned zones.

Thus, the second jump of the plasma density essentially complicates the field structure. A transient process in the field penetrating into the plasma lasts until the instant $t = 2t_1 - x/c$, and near the plasma boundary, its duration is equal to the time interval between the first and the second plasma density jumps.

3.4 The Evolution of an Electromagnetic Field in the Dielectric Layer after Its Creation

In the previous sections, initial-boundary-value 1D problems for electromagnetic waves were considered for a half-closed region. A closed region will be studied in the following section on an example of a plane layer with a time-varying medium (Fig. 3.11). We consider the plane electromagnetic wave $E_0(t, x)$ that is normally incident on a plane homogeneous dielectric layer of thickness a in which the permittivity changes abruptly under some external forces to the value ε_2 at the zero moment of time. The characteristic function of this region is equal to $\chi = \theta(x)\theta(a - x)$.

The equation describing the electromagnetic field inside and outside the layer at an arbitrary moment of time has the form

$$E = E_0 - \frac{1}{2\varepsilon v}\frac{\partial^2}{\partial t^2}\int_{-\infty}^{\infty} dt' \int_{-\infty}^{\infty} dx' \left[\varepsilon_1 - \varepsilon + (\varepsilon_2 - \varepsilon_1)\theta(t')\right]\chi(x')\theta\left(t - t' - \frac{|x - x'|}{v}\right)E, \qquad (3.111)$$

where ε is the permittivity of the medium outside the layer; ε_1 and ε_2 are the medium permittivity inside the layer before and after the zero time moment, respectively, $v = c/\sqrt{\varepsilon}$ is the wave phase velocity, $\chi(t, x)$ is the characteristic function that is equal to unity inside the layer and zero outside the layer and $\theta(t)$ is the unit Heaviside step-function.

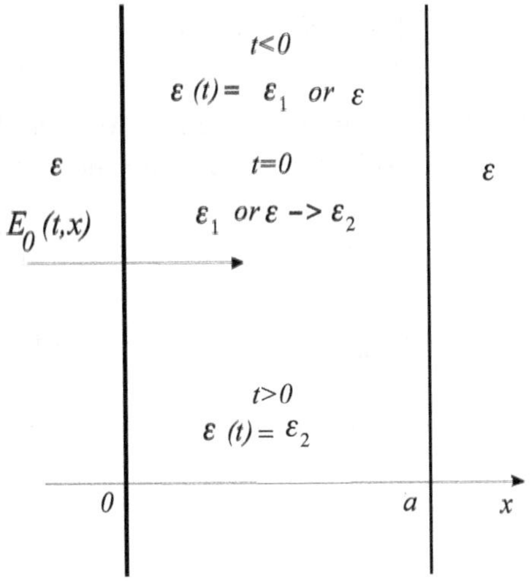

Figure 3.11 The geometry of the evolutionary problem.

Before zero moment of time, the field can be treated as the stationary problem,

$$E_1 = E_0 - \frac{v^2 - v_1^2}{2vv_1^2}\frac{\partial^2}{\partial t^2}\int_{-\infty}^{\infty} dt' \int_{-\infty}^{\infty} dx' \chi(x')\theta\left(t - t' - \frac{|x - x'|}{v}\right)E_1, \quad (3.112)$$

where $v_1 = c/\sqrt{\varepsilon_1}$; $\chi(x) = \theta(x)\theta(a - x)$, which has a well-known solution in the case of the harmonic wave incidence $E_0 = E_0 e^{i\omega(t-x/v)}$:

$$E_1 = A_1 e^{i\omega(t-x/v_1)} + A_2 e^{i\omega(t+x/v_1)}, \quad (3.113)$$

where $A^{(\pm)} = \dfrac{\pm 2(v \pm v_1)v_1 E_0}{(v \pm v_1)^2 - (v \mp v_1)^2 \exp(\mp i\omega 2a/v_1)}$.

If the permittivity has undergone the abrupt changing from the value ε_1 to the value ε_2 at zero moment, the field equation takes the following form:

$$E = F - \frac{v^2 - v_2^2}{2vv_2^2}\frac{\partial^2}{\partial t^2}\int_{0}^{\infty} dt' \int_{-\infty}^{\infty} dx' \chi(x')\theta\left(t - t' - \frac{|x - x'|}{v}\right)E, \quad (3.114)$$

where $v_2 = c/\sqrt{\varepsilon_2}$ and the free term is determined by the field history in the layer before zero moment of time,

$$F = E_0 - \frac{v^2 - v_1^2}{2vv_1^2}\frac{\partial^2}{\partial t^2}\int_{-\infty}^{0} dt' \int_{-\infty}^{\infty} dx' \chi(x')\theta\left(t - t' - \frac{|x - x'|}{v}\right)E_1. \qquad (3.115)$$

Substituting the expression (3.113) for the field E_1 into (3.115), we obtain the explicit expression for the free term

$$F = \theta(a/2 - |x - a/2| - vt)[(v - v_1)e^{-i\omega_1(t + x/v)} + (v + v_1)e^{i\omega_1(t - x/v)}]\frac{v}{2v_1^2}A^{(+)}$$

$$+ \theta(vt - x)\theta(a - x - vt)\theta(x)\frac{v}{2v_1^2}[(v - v_1)A^{(+)}e^{-i\omega_1(t + x/v)} + (v + v_1)A^{(-)}e^{i\omega_1(t + x/v)}]$$

$$+ \theta(x - vt)\theta(vt - a + x)\theta(a - x)\frac{v}{2v_1^2}[(v + v_1)A^{(+)}e^{i\omega_1(t - x/v)} + (v - v_1)A^{(-)}e^{-i\omega_1(t - x/v)}]$$

$$+ \theta(vt - x)[\theta(vt - a + x)\theta(x)\theta(a - x) + \theta(a - x - vt)]e^{i\omega(t - x/v)},$$

(3.116)

where $\omega_1 = \omega v/v_1$. It is seen that the free term is determined by different expressions in four regions instead of two regions in the case of one plane boundary. The slope of the region boundaries are determined in (3.116) by the value of the phase velocity v but it can give rise to the false impression that the field after the permittivity change will be determined by various expressions in such regions. In truth, the position of zones with different real fields, as given in (3.116), is spurious. The real slope of these boundaries is determined by the *new* phase velocity v_2. This value determines the position of zones with various expressions for the fields.

A zone that is independent of the layer boundary influence is determined by the inequality $\frac{a}{2} - \left|x - \frac{a}{2}\right| - v_2 t > 0$. The resolvent in this zone is the same as in the case of an unbounded medium and is given by the expression (3.4). Outside this zone the resolvent is constructed by taking into account re-reflections of the resolvent characteristics from the world lines of the layer boundaries. This is shown in Fig. 3.12. The weight contribution of every reflection of the characteristic is equal to $R = \frac{v - v_2}{v + v_2}$ as shown in Section 3.1.

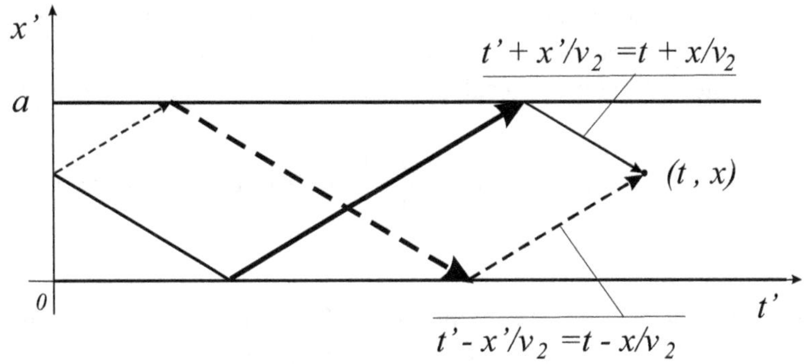

Figure 3.12 Re-reflections of the resolvent characteristics.

The result of the re-reflections of the resolvent cone beams is a division of the whole stripe determined by the layer characteristic function $\chi(x)$ on zones labeled with two indexes m, n in Fig. 3.13. The first index of the zone number indicates how many times the lower beam of the resolvent cone has been reflected from the world line of the layer (the dashed beam in Fig. 3.12) and the second one is related with the upper beam (the solid one in Fig. 3.12).

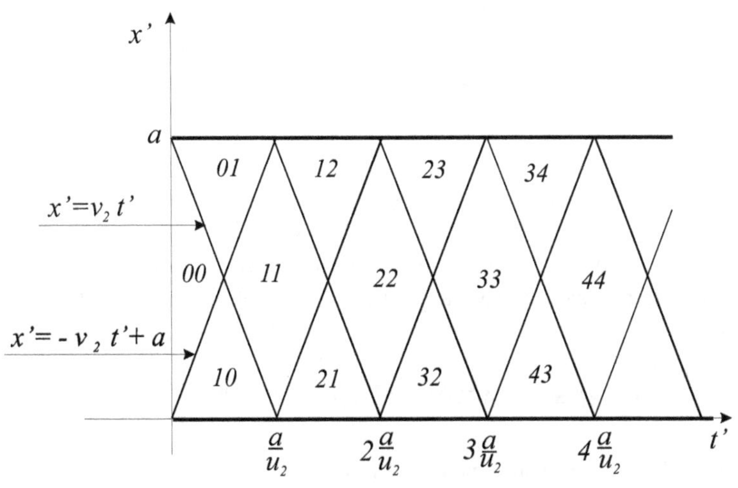

Figure 3.13 Spatial-temporal zones corresponding to a certain number of reflections of the resolvent cone beams from each boundary of the layer after the permittivity jump.

The resolvent out the region where the field is not influenced by the layer boundaries is constructed proceeding from the expression (3.9) for the resolvent taking into account one boundary

$$\langle \mathbf{x}|\hat{R}|\mathbf{x}'\rangle = \theta(x)\left\{\langle \mathbf{x}|\hat{R}^{(1)}|\mathbf{x}'\rangle - \alpha\frac{v_1}{v}\frac{\partial}{\partial t}R\frac{\partial}{\partial t}\theta\left(t-t'+\frac{x+x'}{v_1}\right)\right\}\theta(x'), \tag{3.117}$$

where $\alpha = (\varepsilon_2 - \varepsilon)/2\varepsilon v$ and $R=\frac{v-v_2}{v+v_2}$. Arguments of the unit functions in (3.117) define the equations for the zones lines

$$x' = \pm v_2 t' \pm na, \quad n = 0, \pm 1, \pm 2, \ldots. \tag{3.118}$$

The resolvent determining the field in the zone mn has the following form

$$\hat{R} = (\hat{R}_1 + \hat{R}_{mn})\theta(x')\theta(a-x'), \tag{3.119}$$

where

$$\langle \mathbf{x}|\hat{R}_{mn}|\mathbf{x}'\rangle = -\alpha\frac{v_2}{v}\frac{\partial^2}{\partial t^2}\left\{\sum_{k=1}^{m}R^k\theta\left[v_2t - v_2t' - x + (-1)^{k-1}\left(\frac{a}{2}-x'\right) - \frac{2k-1}{2}a\right]\right.$$
$$\left. + \sum_{k=1}^{n}R^k\theta\left[v_2t - v_2t' + x + (-1)^k\left(\frac{a}{2}-x'\right) - \frac{2k+1}{2}a\right]\right\} \tag{3.120}$$

If the jump of the permittivity does not occur at zero moment of time ($\varepsilon_2 = \varepsilon_1$), the field after that moment can be found by means of the free term in (3.116) using the resolvent (3.120). It leads to the well-known expression for the field in the stationary layer:

$$E_{mn} = F + \overline{R}F = E_1 = A^{(+)}e^{i\omega(t-x/v_2)} + A^{(-)}e^{i\omega(t+x/v_2)}, \tag{3.121}$$

where $A^{(\pm)}$ are the same as in (3.113).

Let us now investigate the field in the layer which is formed abruptly at zero moment of time in the region $0 \le x \le a$. The new permittivity inside the layer is ε_2 and it is formed in the background with permittivity ε (i.e., we assume that $\varepsilon_1 = \varepsilon$).

The free term in (3.114) coincides with the expression for the incident wave in this case.

The zone 00 in Fig. 3.12 is the region in spatial-temporal coordinates where the field is not influenced by the layer boundaries. The field in this zone is determined by the same expression as the field in the infinite medium, after substituting $v \to v$ in it

$$E = C_1 e^{i\omega_2(t-x/v_2)} + C_2 e^{-i\omega_2(t+x/v_2)}, \quad C_{1,2} = \frac{v_2}{v}\frac{v_2 \pm v}{2v}E_0, \quad \omega_2 = \omega\frac{v_2}{v} \tag{3.122}$$

In the zone mm, with m = 1, 2, 3, ..., where the field is resulting from an equal number of reflections from the layer's boundaries, the resolvents (3.120) gives

$$E_{mm} = C_0 e^{i\omega(t-x/v_2)} + \sum_{k=1}^{m-1} R^k C_0 e^{i\omega(t-(-1)^k x/v_2)} \Phi^{\frac{k-2+i_k}{2}}$$
$$+ R^m C_1 e^{i\omega_2(t-(-1)^m x/v_2)} \Phi^{\frac{m+i_m}{2}\frac{v_2}{v}} + R^m C_2 e^{-i\omega_2(t+(-1)^m x/v_2)} \Phi^{-\frac{m-i_m}{2}\frac{v_2}{v}}, \tag{3.123}$$

where $C_{1,2}$ and ω_2 are as in (3.122) and R is the same as that determined in (3.117),

$$C_0 = \frac{2v_2}{v+v_2}E_0, \quad \Phi = e^{-i\omega\frac{2a}{v_2}} \text{ and } i_m = \frac{1}{2}(1-(-1)^m). \tag{3.124}$$

It can be concluded from (3.123) that the structure of the internal field in each zone consists of three groups of waves. The first one contains the waves, which are proportional to C_0, being excited by the field which is incident on the new layer after the zero moment. The second and third groups contain the waves C_1 and C_2 correspondingly which are the result of the incident wave splitting due to the permittivity jump. The field asymptotic for $t \to \infty$ can be easily derived from Eq. (3.123) by finding its limit for $m \to \infty$. To do this, we group the waves of the same propagation direction and use the sum of the geometric progression

$$E_{mm} = C_0\tilde{R}(e^{i\omega(t-x/v_2)} + Re^{i\omega(t+x/v_2)-i\omega 2a/v_2}) + i_m C_0 R^{m-1} e^{i\omega(t-x/v_2)-i(m-1)\omega a/v_2}$$
$$+ R^m C_1 e^{i\omega_2(t-(-1)^m x/v_2)} \Phi^{\frac{m+i_m}{2}\frac{v_2}{v}} + R^m C_2 e^{-i\omega_2(t+(-1)^m x/v_2)} \Phi^{-\frac{m-i_m}{2}\frac{v_2}{v}}, \tag{3.125}$$

where $\tilde{R} = \sum_{k=0}^{(m-2-i_m)/2} (R^2 e^{-i\omega 2a/v_2})^k = \dfrac{1-(R^2 e^{-i\omega 2a/v_2})^{\frac{m-i_m}{2}}}{1-R^2 e^{(-i\omega 2a/v_2)}}$.

The limit of (3.125) exists always because $|R| < 1$. As a result, this field asymptotic has the following form:

$$E = \frac{C_0}{1-R^2\Phi}\{e^{i\omega(t-x/v_2)} + R\Phi e^{i\omega(t+x/v_2)}\} = A^{(+)} e^{i\omega(t-x/v_2)} + A^{(-)} e^{i\omega(t+x/v_2)} \tag{3.126}$$

The last formula describes the ordinary stationary field in a dielectric layer and coincides with the expression (3.121). It means that the waves with the frequency ω_2 "fall out" because of the multiple reflections and so the influence of the non-stationarity which occurred at zero time moment tends to zero when $t \to \infty$.

3.5 Electromagnetic Field in a Layer with Non-Linear and Time-Varying Medium

Interactions between optical pulses and active semiconductor waveguides, including pulse generation, amplification, compression, reshaping, and filtering, as well as four-wave mixing, are of significant importance in quantum electronics and optical communications technologies. All of these studies concern ultrafast electromagnetic transients in complex media, and their solution in the time domain is the subject of much research.

As a rule, an exact consideration of electromagnetic processes in media with arbitrary time-varying parameters is possible only in special cases. Numerical methods, on the other hand, can be used to investigate such problems in the general case. The FDTD method has received wide recognition [9] for solving the differential problem. FDTD methods are extremely popular because of their relative simplicity and clarity, and because they lend themselves to elegant numerical algorithms. Indeed, it is difficult to identify areas in electromagnetic theory where FDTD-based methods have not been applied. However, this method is a purely numerical one and, as such, it does not allow an analytical investigation of the problem. Furthermore, the classical Yee FDTD meshing scheme is limited by the accumulation of phase errors for electrically large domains and late-time analysis [10].

An alternative approach is to solve the problem using integral equations, in particular the integral Volterra equations, which are equivalent to Maxwell's equations in the time domain and initial and boundary conditions [12, 22]. The most attractive feature of this approach is that the Volterra integral equation of the second kind allows the construction of a solution by a convergent sequence of approximations. It permits analytical as well as numerical investigations, and in some cases of the latter, this method has been found to give better accuracy and stability than FDTD [13]. To investigate truly non-stationary phenomena, we must consider initial value problems. A Volterra integral equation allows one to take into account the initial conditions in a natural way and the numerical investigation of initial problems has been implemented for an unbounded medium.

In this section, the development of integral-equation based software for the computer modeling of electromagnetic wave propagation in a spatially bounded time-varying medium is described. This software allows investigations of phenomena for different initial fields and arbitrary changes in material properties. To show its usefulness and applicability to optical problems, we consider the behavior of electromagnetic signals in a 1D layer that is created at some moment of time within a background medium. We consider in particular detail the cases where the medium comprising the layer is a non-linear dielectric or a linear one with time-varying parameters.

3.5.1 Integral Equation Representation for Numerical Calculation

Consider an initial electromagnetic field $E_0(t, x)$ in a background medium having a relative permittivity ε. At moment $t = 0$ a 1D layer is created in this background medium in the region $0 \leq x \leq a$.

The medium inside the layer after its creation is described by the polarization $P(t, x)$. In the case of a signal field $E_0(t, x)$ propagating along the normal to the layer, all electromagnetic values depend only on the spatial coordinate x and the time t and the problem becomes a one-dimensional one. The geometry of the problem is shown in Fig. 3.11.

After the layer is created the electromagnetic field inside and outside of the layer is described by the equation that follows from (3.2)

$$E(t,x) = E_0(t,x) - \frac{1}{2v}\frac{\partial}{\partial t}\int_0^{\infty} dt' \int_0^{a} dx' \delta\left(t - t' - \frac{|x - x'|}{v}\right)\left[\frac{1}{\varepsilon\varepsilon_0}P(t',x') - \frac{\varepsilon - 1}{\varepsilon}E(t',x')\right], \tag{3.127}$$

where $v = c/\sqrt{\varepsilon}$, c is velocity of the light in vacuum, $\delta(t)$ is the Dirac delta-function, a is the width of the layer and ε_0 is the permittivity of free-space. An equation similar to (3.127) but obtained by another way for the scalar case has been used in [34] for a similar problem.

For numerical calculation it is convenient to switch to dimensionless variables: $\tau = vt/a$, $\xi = x/a$ and also to normalize the electromagnetic field quantities $\tilde{E}(\tau,\xi) = E(a\tau/v, a\xi)/\max|E_0|$ and the polarization function $\tilde{P}(\tau,\xi) = P(a\tau/v, a\xi)/\max|E_0|$. Further, we will omit the tilde for clarity. Then Eq. (3.127) takes the following form in dimensionless variables:

$$E(\tau,\xi) = E_0(\tau,\xi) - \frac{1}{2}\frac{\partial}{\partial \tau}\int_0^{\infty} d\tau' \int_0^{1} d\xi' \delta(\tau - \tau' - |\xi - \xi'|)\left[\frac{1}{\varepsilon\varepsilon_0}P(\tau,\xi) - \frac{\varepsilon - 1}{\varepsilon}E(\tau',\xi')\right] \tag{3.128}$$

Taking into account that $\delta(\tau - \tau' - |\xi - \xi'|) = \delta(\tau - \tau' - \xi + \xi') + d(\tau - \tau' + \xi - \xi')$ one can calculate the inner integral analytically. The form of the result obtained is different for various zones identified by the two indexes in the system of coordinates (τ, χ), shown in Fig. 3.14.

In each of these zones the transitional process has a qualitatively different character, i.e., the behavior of an electromagnetic field and its description will be governed by the location of the point of observation.

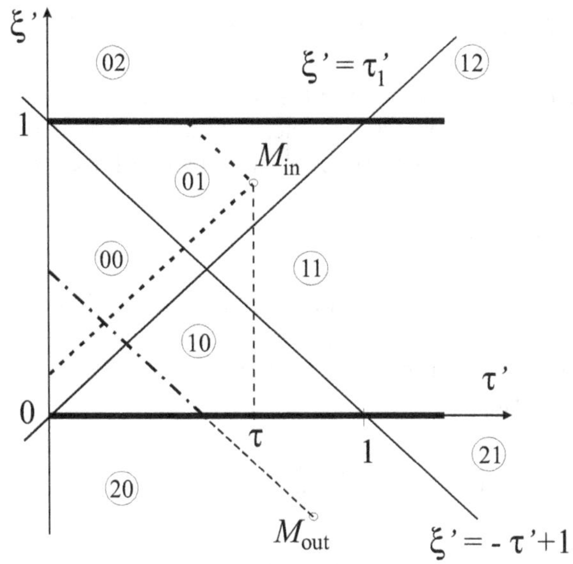

Figure 3.14 Scheme of the integration problem.

If the point of observation belongs to the layer $0 \leq x \leq 1$, as the point M_{in} in Fig. 3.14 then the field has to be found as solutions to the equations that have the following forms in different zones. In zone {00}:

$$E(\tau,\xi)=E_0(\tau,\xi)-\frac{1}{2}\frac{\partial}{\partial\tau}(F_{00}^{(+)}+F_{00}^{(-)}), \tag{3.129}$$

where $F_{00}^{(\pm)}=\int\limits_0^{\tau}Q(\tau',\xi\mp(\tau-\tau'))d\tau'$. (3.130)

Here, $Q(\tau,\xi)=\frac{1}{\varepsilon\varepsilon_0}P(\tau,\xi)-\frac{\varepsilon-1}{\varepsilon}E(\tau,\xi)$, (3.131)

where $P(\tau,\ \xi)\ =\ P_L(\tau,\ \xi)\ +\ P_{NL}(\tau,\ \xi)$. The linear part of the normalized polarization allows a medium with spatial-time-varying permittivity to be described,

$$P_L(\tau,\xi)\ =\ \varepsilon_0[\varepsilon(\tau,\xi)-1]E(a\tau/v,a\xi)/\max|E_0|, \tag{3.132}$$

as well as the non-linear normalized part which describes material non-linearity

$$\tilde{P}_{NL}(\tau,\xi) = P_{NL}/\max|E_0| = \chi_{NL}^{(3)} E^3(a\tau/v, a\xi)/\max|E_0| = \gamma \tilde{E}^3(\tau,\xi). \quad (3.133)$$

Further we will omit the tilde over normalized variables. The normalized dimensionless coefficient of non-linearity γ relates to a susceptibility $\chi_{NL}^{(3)}$ of the third order material non-linearity as

$$\gamma = \chi_{NL}^{(3)} \max|E_0|^2. \quad (3.134)$$

The integration paths in (3.130) intersect the line $\tau' = 0$ only. This means that in this zone the electromagnetic field depends only on initial conditions and does not feel the influence of the layer boundaries.

In zone {10} $$E(\tau,\xi) = E_0(\tau,\xi) - \frac{1}{2}\frac{\partial}{\partial\tau}(F_{10} + F_{00}^{(-)}) \quad (3.135)$$

the integral $$F_{10} = \int_{\tau-\xi}^{\tau} Q(\tau',\xi - \tau + \tau')d\tau' \quad (3.136)$$

takes into account the influence of the lower boundary of the layer.

In zone {01} $$E(\tau,\xi) = E_0(\tau,\xi) - \frac{1}{2}\frac{\partial}{\partial\tau}(F_{00}^{(+)} + F_{01}) \quad (3.137)$$

the integral $$F_{01} = \int_{\tau+\xi-1}^{\tau} Q(\tau',\xi + \tau - \tau')d\tau' \quad (3.138)$$

takes into account the influence of the upper boundary of the layer.

Finally, in zone {11} the qualitative changes of the transient finish and the field is described by the equation

$$E(\tau,\xi) = E_0(\tau,\xi) - \frac{1}{2}\frac{\partial}{\partial\tau}(F_{10} + F_{01}) \quad (3.139)$$

In this zone, the field feels the influence of the initial conditions indirectly.

The field outside the layer, for example at the point M_{out} in Fig. 3.14, is given simply by the integral formula that has different meaning in the different zones. In zones {20} and {02} the formulae

$$\{20\}: \quad E(\tau,\xi) = E_0(\tau,\xi) - \frac{1}{2}\frac{\partial}{\partial\tau}\int_0^{\tau+\xi} Q(\tau',\xi + \tau - \tau')d\tau' \quad (3.140)$$

$$\{02\}: \quad E(\tau,\xi) = E_0(\tau,\xi) - \frac{1}{2}\frac{\partial}{\partial \tau}\int_{0}^{\tau-\xi+1} Q(\tau',\xi-\tau+\tau')d\tau' \tag{3.141}$$

take into account the influence of the lower (upper) boundary of the layer only while in zones {21} and {12} the formulae

$$\{21\}: \quad E(\tau,\xi) = E_0(\tau,\xi) - \frac{1}{2}\frac{\partial}{\partial \tau}\int_{\tau+\xi-1}^{\tau+\xi} Q(\tau',\xi+\tau-\tau')d\tau' \tag{3.142}$$

$$\{12\}: \quad E(\tau,\xi) = E_0(\tau,\xi) - \frac{1}{2}\frac{\partial}{\partial \tau}\int_{\tau-\xi}^{\tau-\xi-1} Q(\tau',\xi-\tau+\tau')d\tau' \tag{3.143}$$

take into account the influence of both layer boundaries.

Note, that calculation of the field value at points outside the layer requires integration only inside the layer. This allows computation of far-field radiation with the same algorithm; special transformations of near- to far-field zones, such as that presented in [35], are not required. One advantage of the integral equation approach is that there is no need to compute the entire region between the medium layer and the far-field point.

3.5.2 Numerical Algorithm for Solving Integral Equations

(i) Time marching scheme

For numerical calculation of the field it is convenient to rewrite the problem Eqs. (3.129)–(3.143) in the form of two expressions to omit redundant super- and subscripts in later investigations. Thus we write:

$$E(\tau,\xi) = E_0(\tau,\xi) - \frac{1}{2}\frac{\partial}{\partial \tau}(H(\tau,\xi) + L(\tau,\xi)) \tag{3.144}$$

where the functions F are replaced by lower L and upper H integrals correspondingly (see Fig. 3.14) with integration limits. Although these limits appear complicated they are nevertheless easily implemented on a PC. The lower integral is calculated along the path between the lower boundary of the layer and the observation point

$$\tilde{L}(\tau,\xi) = \int\limits_{\substack{0<\tau'<\tau \\ 0<\xi-\tau+\tau'<1}} Q(\tau',\xi-\tau+\tau'), \tag{3.145}$$

the upper integral is calculated along the path between the observation point and the upper boundary of the layer

$$\tilde{H}(\tau,\xi) = \int\limits_{\substack{0<\tau'<\tau \\ 0<\xi+\tau-\tau'<1}} Q(\tau',\xi+\tau-\tau'). \tag{3.146}$$

Expressions (3.144)–(3.146) are the starter formulae for the implementation of the numerical calculation schemes. Introducing the further notation

$$\begin{gathered}\tau_n = hn,\ \xi_i = hi,\ E_i^n = E(\tau_n,\xi_i) \\ \tilde{L}_i^n = \tilde{L}(\tau_n,\xi_i),\ \tilde{H}_i^n = \tilde{H}(\tau_n,\xi_i),\end{gathered} \tag{3.147}$$

allows us to discretize Eq. (3.144) in the form

$$E_i^n = E_{0i}^{\ n} - \frac{1}{2}\frac{\partial}{\partial\tau}[\tilde{L}_i^n + \tilde{H}_i^n] \tag{3.148}$$

for a regular quadratic mesh, as presented in Fig. 3.15. Here h is a time/spatial step, n gives the time at the n-th step and i-th gives the position of the i-th spatial point. The backward differentiation formula

$$\frac{\partial}{\partial\tau}\tilde{L}_i^n = \frac{\tilde{L}_i^n - \tilde{L}_i^{n-1}}{h} \tag{3.149}$$

is used for simplicity, however a better one can replace it if necessary with minor modification.

The main iterative procedure, obtained directly from Eq. (3.148), is presented in the following formula:

$$E_i^n = E_{0i}^n - \frac{1}{2}\left[L_i^n - L_i^{n-1} + H_i^n - H_i^{n-1}\right] \tag{3.150}$$

$$\text{where } L_i^n = \sum_{m=l}^{n-1} k_m Q_{i-n+m}^m,\ H_i^n = \sum_{m=l}^{n-1} k_m Q_{i+n-m}^m \tag{3.151}$$

Here l is the lower time index that is determined by the integration limits in (3.145), (3.146), k_m are the integration weight coefficients and the function

$$Q_i^n = \left(\frac{\varepsilon_{1i}^n}{\varepsilon} - 1\right)E_i^n + \gamma(E_i^n)^3 \tag{3.152}$$

follows from (3.131) and takes into account the polarization of the layer medium.

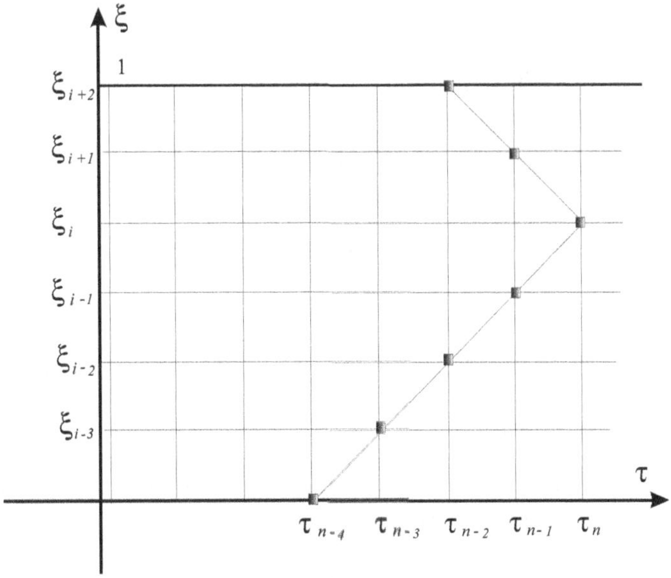

Figure 3.15 Square time-spatial calculation mesh. The integration paths (two diagonal lines) are limited with the time boundary $\tau' \in [0, \tau]$ and with layer boundaries $\mathbf{x}' \in [0, 1]$.

The differentiation denominator h in (3.147) and the integration step h inside the integrals in (3.146) cancel each other in Eq. (3.148).

Using a well-known semi-open trapezoidal formula for integration of (3.145), (3.146)

$$\int_{x_1}^{x_n} f(x) = h\left(\frac{1}{2}f_1 + f_2 + \cdots + f_{n-2} + \frac{3}{2}f_{n-1}\right) + O\left(\frac{1}{N^2}\right)$$

we can re-express (3.151) as

$$L_i^n = \frac{1}{2}Q_{i-n+l}^l + \sum_{m=l+1}^{n-2} Q_{i-n+m}^m + \frac{3}{2}Q_{i-l}^{n-1} = LS_{i-2}^{n-2} + \frac{3}{2}Q_{i-l}^{n-1}, \tag{3.153}$$

where the variable $LS_{i-2}^{n-2} = \frac{1}{2}Q_{i-n+l}^l + \sum_{m=l+1}^{n-2} Q_{i-n+m}^m$ is an accumulation one. The accumulation variable LS_i^n is used to prevent superfluous operations when calculating the L integral. The iterative relationships for calculation of LS_i^n are

$$LS_i^0 = \frac{1}{2}Q_i^0 \text{ and } LS_i^n = LS_{i-1}^{n-1} + Q_i^n \text{ for } n > 0. \tag{3.154}$$

Similar relationships for H are

$$H_i^n = \frac{1}{2}Q_{i+n-l}^l + \sum_{m=l+1}^{n-2} Q_{i+n-m}^m + \frac{3}{2}Q_{i+l}^{n-1} = HS_{i+2}^{n-2} + \frac{3}{2}Q_{i+l}^{n-1} \tag{3.155}$$

$$\text{and } HS_i^0 = \frac{1}{2}Q_i^0 \text{ and } HS_i^n = HS_{i-1}^{n-1} + Q_i^n \text{ for } n > 0. \tag{3.156}$$

(ii) Initial conditions

In this scheme E^0 and E^1 should be computed by "special" procedures, because there are not enough points to use higher accuracy formulae:

Step $\tau = \tau_0$: $L_i^0 = 0,\ H_i^0 = 0,\ E_i^0 = E_{0i}^0$, and calculation of the polarization Q^0 by using (3.152);

Step $\tau = \tau_1$: calculation of the L, H integrals by using only one point: $L_i^1 = Q_{i-1}^0$, $H_i^1 = Q_{i+1}^0$, then calculation of field E by $E_i^1 = E_{0i}^1 - \frac{1}{2}(Q_{i-1}^0 + Q_{i+1}^0)$ and calculation of the polarization Q^1 by using (3.152). Initialization of integration sums LS_i^0 and HS_i^0 by (3.154) and (3.156) accordingly.

(iii) Main iteration operations

The general iteration on time step τ_n for the calculation of E_i^n will perform the following operations:

Calculation of the integration sums L^n, H^n using (3.153) and (3.155);

Calculation of the E^n according to (3.150);

Updating of LS^{n-2} to LS^{n-1} by using (3.154), and similarly calculation of the HS^{n-1} through (3.156);

Storing of L^n as L^{n-1} for future steps;

Calculation of the polarization Q^n (by using (3.152) and the newly computed E^n) and its retention for future use.

(iv) Calculation of near-boundary points

Two boundary points cannot be calculated directly by the main iteration procedure, because there are not enough points to calculate one of the integrals (L or H). In this case the integral is computed using formulae involving fewer points:

The lower-boundary point: $L_0^n = 0$, and the field E is calculated by using the procedure as above;

The next, first near-boundary point: $L_1^n = Q_0^{n-1}$ and the field E is then calculated as usual.

The H integral is calculated in a similar way for upper boundary points.

(v) Memory requirements

The problem size is NT * NX points where NT is the number of nodes on the t axis, and NX is the number of nodes on the x axis. The memory resources required for calculation are given in Table 3.1. The coefficient 8 is the memory size in bytes of a double precision float number. It must be noted that for every iteration step the algorithm requires only information from a few previous time layers (2 or 3, depending on the integration formula used). This is because integration sums for previous time layers are accumulated in the LS and HS variables. The computer memory required by the algorithm is determined by the calculation of Q_i^n in (3.153), (3.155) and is independent of the simulation period in time. This is because the accumulation variables LS_i^n and HS_i^n that are used do not require any additional memory for calculation of the integrals (3.145), (3.146) and this decreases the calculation complexity.

Table 3.1 Memory requirements for the algorithm

Name	Size (bytes)
Temporary data, required for the algorithm execution (7 arrays)	NX*8*7
Resulting values of E don't need to be stored in an operative memory for the algorithm execution and can be stored in an external memory.	NX*NT*8

(vi) Integration on a rectangular mesh

The time and spatial steps (in normalized quantities) in the algorithm above are equal each other (i.e., a quadratic mesh is used). Such a discretization provides fulfillment of the Courant stability condition [9].

Discretization $\Delta\tau \neq \xi$ cannot be implemented in a straightforward manner. This is because a shift of integration paths from mesh points would make it very hard to build an iterative procedure for computation of the integrals in (3.148). However, there are several possible ways to solve this problem. One of them is to modify the space-time mesh in the manner shown in Fig. 3.16. Here, the time step is a factor of two smaller than the spatial step and the spatial mesh on the odd time steps is shifted to a half step along the spatial axis.

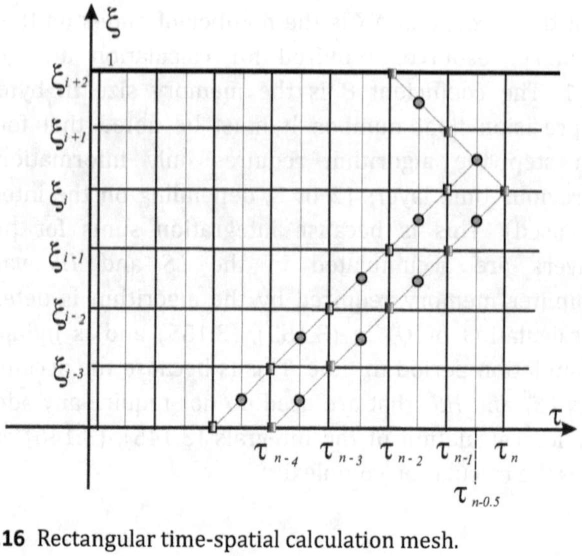

Figure 3.16 Rectangular time-spatial calculation mesh.

The calculation algorithm is similar to the integration algorithm for the quadratic mesh described in the previous section. Because the odd time step points are spatially shifted onto the half-cell from the points on the even time steps, the calculation routines for even and odd time iterations differ slightly.

(vii) Validation of the algorithm

Let us consider the case when the layer permittivity changes abruptly at $\tau = 0$ from ε to ε_1 and the initial field is a plane wave $E_0 = \cos(2\pi\eta(\tau - \xi))$. The normalized frequency $\eta = \omega a/(3\pi v) = a/\lambda$ shows the number of full wavelengths λ per layer width a where λ is the wavelength in the background medium. The exact solution for this case is constructed in the previous Section 3.4 and for the central spatial point of the layer it is

$$E = C_0 e^{i2\pi\eta(\tau-\xi\alpha)} + \sum_{k=1}^{m-1} R^k C_0 e^{i2\pi\eta(\tau-(-1)^k\xi\alpha)} \Phi^{\frac{k-2+i_k}{2}}$$

$$+ R^m C_1 e^{i2\pi\eta(\tau-(-1)^m\xi\alpha)/\alpha} \Phi^{\frac{m+i_m}{2\alpha}} + R^m C_2 e^{-i2\pi\eta(\tau+(-1)^m\xi\alpha)/\alpha} \Phi^{-\frac{m-i_m}{2\alpha}}, \quad (3.157)$$

where $C_{1,2} = \dfrac{1 \pm \alpha}{2\alpha^2}$, $R = \dfrac{\alpha - 1}{\alpha + 1}$, (3.158)

$$C_0 = \frac{2}{\alpha+1}, \quad \Phi = e^{-i4\alpha\pi\eta}, \quad \alpha = \sqrt{\varepsilon_1/\varepsilon}, \quad i_m = \frac{1}{2}(1-(-1)^m)$$

and $m = \text{Int}(\tau)$ is the integer part of τ.

The following parameters are used to perform an accuracy check: $\varepsilon = 9$, $\varepsilon_1 = 11$, normalized frequency of the initial wave $\eta = 1$. This normalized frequency corresponds to $f = 100$ THz if the layer width is equal to 1 μm.

The exact solution for this case and the numerical solutions near the jump region obtained using the different integration steps h are presented in Fig. 3.17. The propagated electromagnetic field contains a jump in its amplitude at the junction point of zones shown in Fig. 3.14, which is caused by the layer creation at time $t = 0$. A good description of the overall behavior is shown even for the coarsest mesh size. Near the field discontinuity itself, the numerical solution shows good convergence to the exact one with decreasing mesh size.

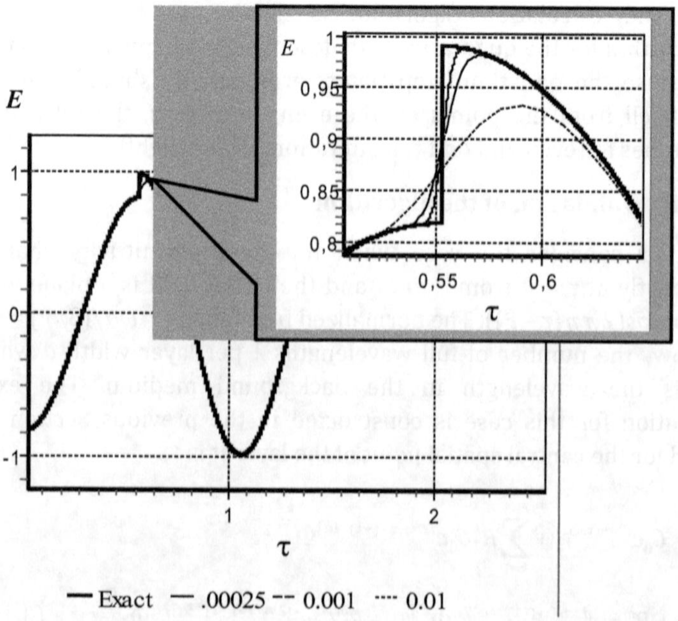

Figure 3.17 Comparison of the numerical solutions, obtained for different mesh sizes, with the exact one.

3.5.3 Numerical Results

(i) Plane wave propagation in a linear or non-linear dielectric layer

Now we make a numerical investigation of the propagation of a plane wave with normalized amplitude in a layer of linear or non-linear dielectric. The initial wave and the linear medium in the dielectric layer have the same parameters as above. Non-linear media with $\gamma = 0.2$, $\gamma = 1$, $\gamma = 5$ in the notation of expression (3.134) are studied. The linear material corresponds to $\gamma = 0$. To estimate the normalized non-linearity coefficient γ we take into account that the susceptibility $\chi_{NL}^{(3)}$ can have the resonance magnitudes of $\sim 10^{-9} \sim 10^{-16}$ [m^2/V^2] and $\approx 10^{-19}$ [m^2/V^2] for non-resonance magnitudes [4]. Laser light field intensities I of 10^{10} [W/m^2] and higher can now be generated routinely with a table-top high-power chirp pulse amplification laser system [9]. As $\max|\bar{E}_0|^2 \sim I10^2[V^2/m^2]$ then we have from (3.134) that γ can take the magnitudes of $\approx 10^3 \approx 10^{-4}$.

The investigated duration τ_{max} of the process is 100.0 in dimensionless quantities (corresponding to 1 *ps*) and the integration step is h = 0.01. For NX = 100 and NT = 10 000 the minimum required memory is equal to 6 Kb, and to completely store the results -8 Mb. The run-time is equal to ~2.8 s on a PC with a Celeron 700 MHz processor.

A comparison of the cases of linear and non-linear dielectric layers, and the change of the wave shape and its deceleration in the non-linear dielectric layers is shown in Fig. 3.18a. As could be anticipated, increasing the material non-linearity leads to more shape distortion and signal retardation. The corresponding changes in the frequency spectrum caused by the non-linearity are obtained by FFT. Additional multiple frequencies appear in the propagating signal and their amplitudes grow with increasing γ. The spectrum obtained for the case γ = 1 is presented in Fig. 3.18b.

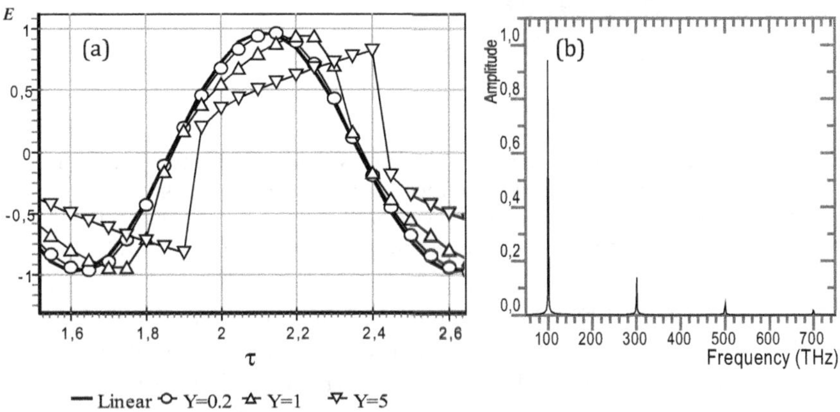

Figure 3.18 (a) Transformation of a time behavior of the electromagnetic signal at the point ξ = 1 in a layer of a linear and non-linear dielectrics. (b) Spectrum of the distorted wave at the point ξ = 1 inside the dielectric layer with non-linearity coefficient γ = 1. Additional multiple harmonics appear in the signal spectrum due to influence of the non-linearity.

(ii) Dielectric layer irradiated by a Gaussian pulse and a soliton-like signals

The proposed algorithm does not place any restrictions on the form of the investigated signal. In this section we present a numerical investigation of the electromagnetic field in the layer irradiated by a Gaussian pulse and then compare results with propagation of a soliton-like signal. The linear and non-linear layers and background medium have the same parameters as above. Let us start with the case when the layer is irradiated by a Gaussian pulse, defined as $e^{-\frac{(\tau-\tau_0-\xi)^2}{2\sigma^2}}$ with parameters $\sigma = 1$ and $\tau = 1$. The integration step is $h = 0.005$ and the duration $\tau_{\max}$ of investigated interval in dimensionless quantities is 5.

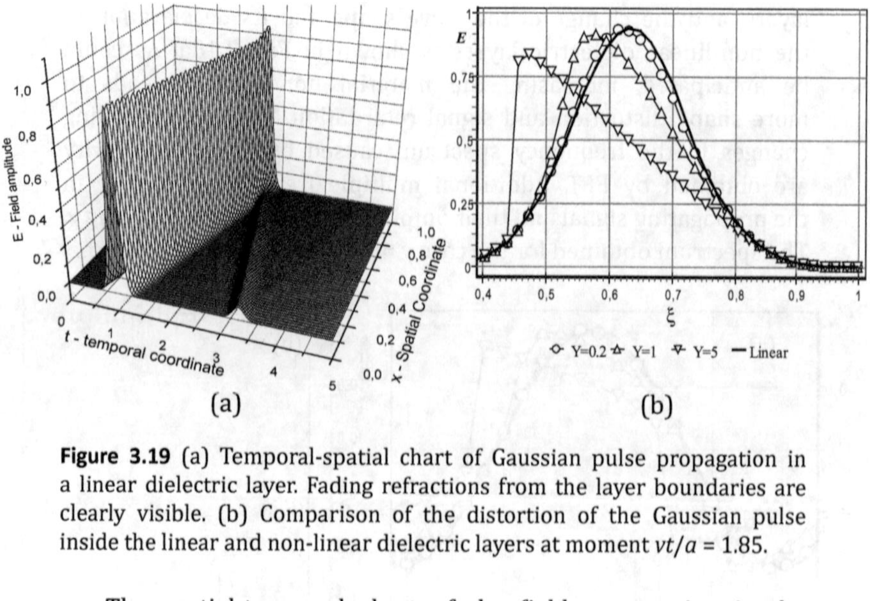

(a) (b)

Figure 3.19 (a) Temporal-spatial chart of Gaussian pulse propagation in a linear dielectric layer. Fading refractions from the layer boundaries are clearly visible. (b) Comparison of the distortion of the Gaussian pulse inside the linear and non-linear dielectric layers at moment $vt/a = 1.85$.

The spatial-temporal chart of the field propagation in the linear dielectric layer is presented in Fig. 3.19a. Fading refractions from the layer boundaries are clearly visible and the refraction coefficient is in a good agreement with expected analytical value $(k = (\varepsilon_1 - \varepsilon)/(\varepsilon_1 + \varepsilon) = 2/13 \approx 0.15)$.

The comparison of the Gaussian pulse propagation in the linear and non-linear dielectric layers is presented in Fig. 3.19b. Again, signal retardation and shape distortion are increasing with increased non-linearity coefficient.

Consider now a propagation of a soliton-like signal defined by $E = \text{sech}(\frac{\tau-\xi}{T})\cos(\frac{\xi}{\Lambda})$ where T is the normalized signal width and Λ is reference wavelength. Also let us compare the influence of the medium non-linearity on this signal and on the Gaussian pulse. The results for parameters $T = 0.1$, $\Lambda = 0.01$ are presented in Fig. 3.20. As one can see the Gaussian pulse signals are significantly transformed due to influence of the non-linearity, while soliton-like signal keeps its shape unchanged.

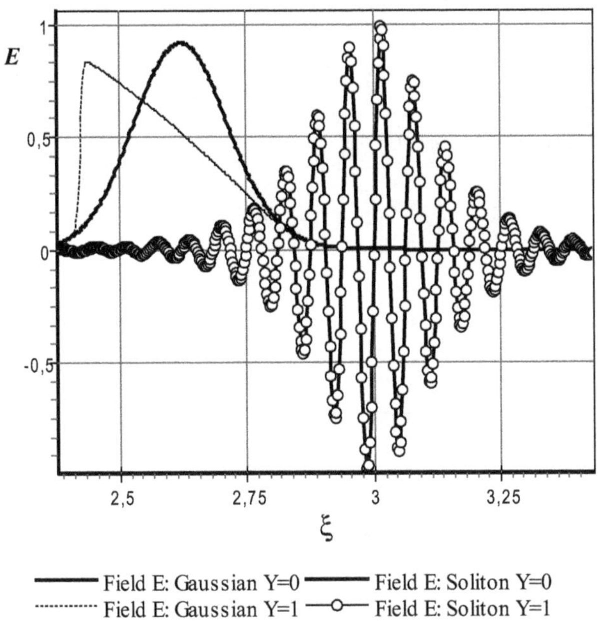

Figure 3.20 Comparison of the soliton-like signal propagation in the linear and non-linear media with the Gaussian pulse at time moment τ = 3. The Gaussian pulse signal is significantly transformed due to influence of non-linearity while soliton signal keeps its shape unchanged.

(iii) Plane wave propagation in a linear dielectric layer with time-varying parameters

The transformation of the electromagnetic signal can be implemented in a parametric way when the medium parameters are changed by extrinsic sources, for example by another non-linear wave. To illustrate this we consider the transformation of the same initial plane wave and the layer when the permittivity of the medium is modulated in time by the harmonic law:

$\varepsilon_1/\varepsilon = 11/9 + m\sin(2\pi\Theta(\tau - \xi))$. Here the parameter $\Theta = a/\Lambda$ is the normalized modulation frequency where Λ is the wavelength of the initial field and m is the modulation depth.

The transformation of the signal spectrum for this case is presented in Fig. 3.21. Additional multiple harmonics appear in the signal. The amplitudes of the extra harmonics grow with the modulation depth and their frequencies are strongly dependent on the modulation frequency.

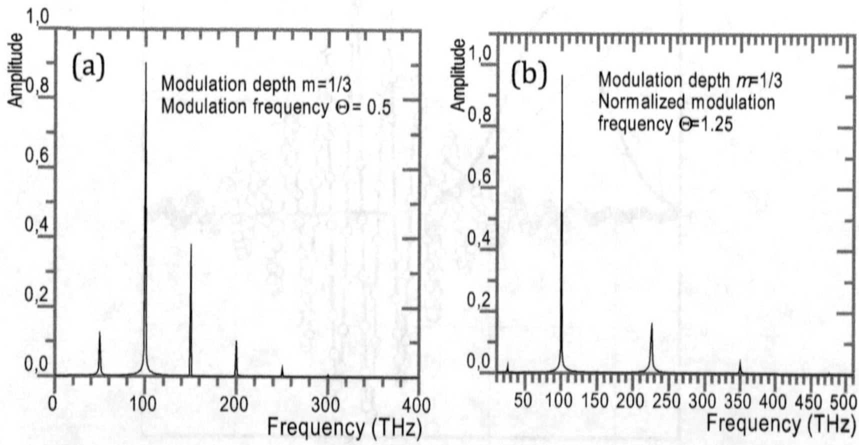

Figure 3.21 Spectrum of the transformed signal at the point ξ = 1 inside the layers with different temporal medium modulation frequencies: vt/a = 0.5 (a), vt/a = 1.25 (b).

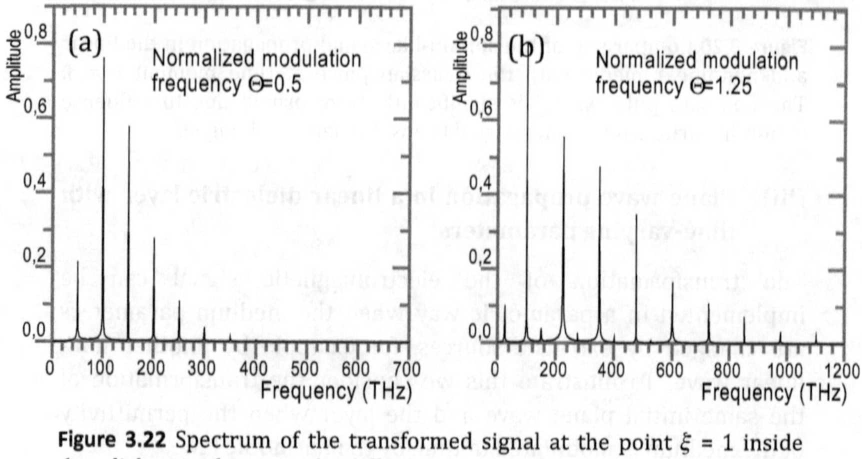

Figure 3.22 Spectrum of the transformed signal at the point ξ = 1 inside the dielectric layer with different time-spatial medium modulation frequencies: vt/a = 0.5 (a), vt/a = 1.25 (b). The modulation depth for all cases is m = 1/3.

Now consider a similar situation when the permittivity of the medium inside the layer is modulated in space as well as in time by the harmonic law $\varepsilon_1/\varepsilon = 11/9 + m\sin(2\pi\Theta(\tau - \xi))$. The influence of the medium modulation on the signal spectrum in this case is presented in Fig. 3.22. In comparison with the previous case discussed in this section, where only temporal modulation of the medium parameters was present, here the number of additional harmonics and their amplitudes are greater. This can be explained by the additional phase modulation at each point of the layer.

3.5.4 Comparison of the FDTD and Volterra Integral Equations in Time Domain Approaches

This section contains comparison of the accuracy and stability finite difference in time domain (FDTD) method and the Volterra integral equations method for calculation of electromagnetic signal propagation in an active media. This is achieved by studying test problems using both methods and comparing obtained results with the exact analytical solution. We consider test problems with known exact solutions [30] for validation purposes.

The simplified Volterra integral equation derived from (3.128) is

$$E(\tau,\xi) = a^2\left\{E_0(\tau,\xi) - \frac{1}{2}\int_0^{\tau} d\tau' \frac{1-a^2}{a^2}\frac{\partial}{\partial\xi}\left[E(\tau',\xi+\tau-\tau) - E(\tau',\xi-\tau+\tau')\right]\right\}. \tag{3.159}$$

where $a^2(\tau) = \varepsilon/\varepsilon_1$, $\tau = \kappa vt$, $\xi = \kappa x$, k is a scaling factor.

A differential equation with initial conditions for the problem is obtained from the Maxwell equations:

$$\frac{\partial^2}{\partial\tau^2}\left(\frac{1}{a^2}E(\tau,\xi)\right) - \frac{\partial^2}{\partial\xi^2}E(\tau,\xi) = 0,\ E_0(0,\xi) = a(0)E(0,\xi), \tag{3.160}$$

$$\left.\frac{\partial E_0(\tau,\xi)}{\partial\tau}\right|_{\tau=0} = \left.\frac{\partial}{\partial\tau}\left(a(\tau)E(\tau,\xi)\right)\right|_{\tau=0} \tag{3.161}$$

Consider a plane wave as an initial field. For all calculations the grid spacing is $\Delta t = 0.05$. Different variations of the medium parameters are investigated:

(1) Step changing of the medium parameters

Consider the case when medium permittivity changes abruptly at $\tau = 0$ from ε to ε_1. An exact solution for this case is known as [30]:

$$E(\tau,\xi) = \frac{a^2}{2}[(a+1)\cos\omega(a\tau - \xi) + (a-1)\cos\omega(a\tau + \xi)] \tag{3.162}$$

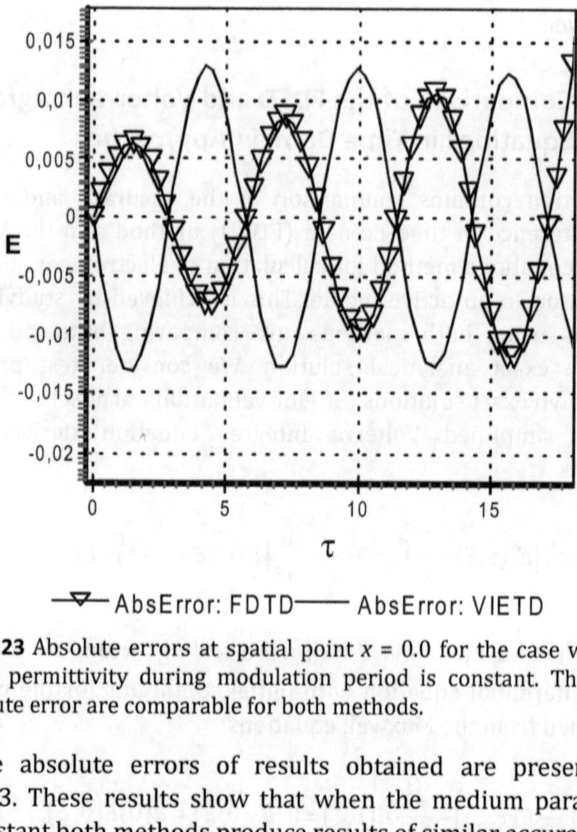

Figure 3.23 Absolute errors at spatial point $x = 0.0$ for the case when the medium permittivity during modulation period is constant. The values of absolute error are comparable for both methods.

The absolute errors of results obtained are presented in Fig. 3.23. These results show that when the medium parameters are constant both methods produce results of similar accuracy.

(2) Modulation of the medium by a power law

Consider a medium with $\varepsilon_1/\varepsilon = (1 + g\tau)^4$. An exact solution for this case is known [6]:

$$E(\tau,\xi) = \frac{e^{-i\xi}}{(1+g\tau)^3}\left(\cos\left(\frac{\tau}{1+g\tau}\right) + (i-g)\sin\left(\frac{\tau}{1+g\tau}\right)\right) \tag{3.163}$$

The relative error of the computed fields for the case $g = 0.2$ is presented in Fig. 3.24. The results show that integral equation algorithm has better accuracy than FDTD, which is more sensitive to a steep changing of the medium parameters.

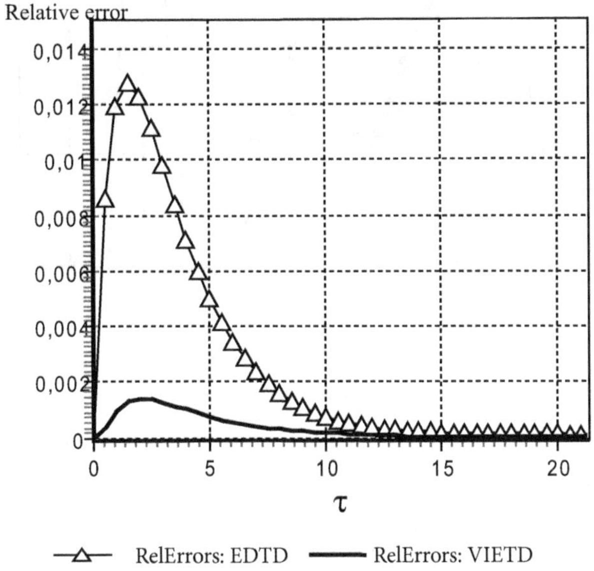

Figure 3.24 Relative errors, $x = -1.0$ for the case when medium permittivity change by power law. The relative numerical error is calculated as the ratio of the absolute error to the greatest value of field.

(3) Modulation of the medium by periodic impulses

Finally consider a temporal variation of medium parameters as a finite sequence of periodic rectangular pulses. We assume that the medium parameter $a(\tau)$ acquires the value A_1 respectively on the disturbance intervals $(n - 1)T < \tau < T_1 + (n - 1)T$, $n = 1, \ldots, N$ and the value A_0 on the quiescence intervals $T_1 + (n - 1)T < \tau < nT_1$, $n = 1, \ldots, N$:

$$a(\tau) = A_0 + (A_1 - A_0)\sum_{k=1}^{N}\{\theta(\tau - (k-1)T) - \theta(\tau - T_1 - (k-1)T)\} \quad (3.164)$$

The exact solution to this problem is given in [30]. The resulting fields for the case $A_0 = 1$, $A_1 = 1.2$, $T_1 = 2$, $T_2 = 0.5$, $N = 20$ are presented in Fig. 3.25.

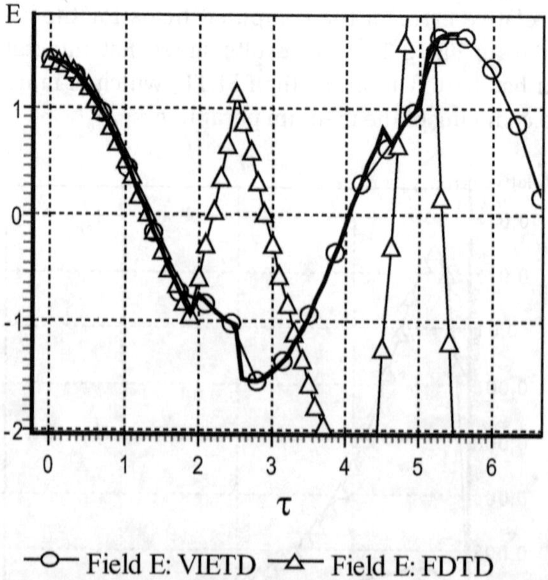

Figure 3.25 Electric field E at the spatial point $x = 0.0$ for the medium which is modulated by a sequence of periodic rectangular pulses (Bold line represents exact analytical solution). The FDTD solution is not stable when the medium parameter changes abruptly.

The calculated results indicate that the FDTD solution is not stable when the medium parameters change abruptly, whereas the integral equation algorithm successfully overcomes points of discontinuity. The numerical results obtained in this section indicate that the FDTD algorithm is more sensitive to significant changes in the medium parameters, whereas the integral equation algorithm provides better accuracy and is more stable when medium parameters change abruptly. So, a Volterra integral equations time-domain approach used for computing of an electromagnetic field inside as well as outside of a non-stationary flat dielectric layer has the following attractive features: it allows investigation of the propagation of different types of incident fields with arbitrary signal shape, amplitude and duration; it can handle non-linear dielectric layers with arbitrary time-spatial dependence of the medium parameters; it is capable of computing the far-field propagation of the scattering signal. With minor modifications, this model can handle media with other types of non-linearity, dissipative media with losses, plasma media, Lorentz media, etc. To implement these modifications in the model only the function *P*, describing the medium, should be redefined. No modifications to the main algorithm are required. This approach can be extended for solving 2D and 3D electrodynamics problems, but it is obvious that the computational complexity in these cases will be much greater.

3.6 Triple Asymmetry in Time-Spatial Structure of an Airy Pulse in Non-Stationary Environment

In the previous sections, transformation of an electromagnetic wave was considered in conditions of double asymmetry availability: a temporal asymmetry of medium parameters behavior and a spatial asymmetry of medium parameters distribution. An initial electromagnetic wave was assumed symmetric in the form of a harmonic wave.

Here, a triply asymmetric electrodynamic problem will be considered: the transformation of an initially asymmetric Airy pulse on a boundary between two dielectric media in one of which the refractive index experiences a jump change with time. There has been active interest in the paraxial propagation of optical Airy pulses [38] for almost a decade due to their unusual properties. At the same time interest in ordinary Airy pulses has not been lost due to their asymmetry form, especially if there is another asymmetry in the phenomenon considered. Such asymmetry may be caused by the appearance of a medium boundary or by the non-stationarity of its features [39–45]. Parametric phenomena in active media have attracted a great deal of attention for a long time in connection with the generation and amplification of electromagnetic waves due to the time modulation of medium parameters [46–55]. Investigation of transient electromagnetic phenomena is important for the control of an electromagnetic signal by the temporal adjustment of medium parameters in connection with the development of optoelectronic systems. The fundamental question of what is the spatiotemporal limit of a macroscopic model that describes the optoelectronic interaction at the interface between different media has become relevant for time-dependent photoemission from solid surfaces [56]. To construct the desired behavior of the field one has to use a complex modulation of medium parameters. However, the main features of the wave transformation can be revealed by considering the simple case of an abrupt change of medium parameters with time. It is known, since the first publication by [56], that a single abrupt change of medium parameters leads to a changing of the wave frequency and amplitude or the pulse shape. Analysis of the results obtained with known changes in wavelength, wave energy and pulse compression, allowed the authors of [57–59] to claim the identification of a correct set of boundary conditions that seemingly resolved a discrepancy existing in the literature. But, as noted by [60] "the boundary conditions discarded in [59] as incorrect have been used in the literature for rapidly growing plasma" and the material model of [59] misinterprets results from the literature by opposing two sets of boundary conditions that are related to different material models of the temporal boundary.

The time-domain concept enables the reshaping of an ultrashort optical pulse into a desired complex (amplitude and phase) and arbitrary temporal pulse waveform using a single temporal amplitude modulator between two media [61]. The propagation and transformation of electromagnetic waves through spatially homogeneous, yet smoothly time-dependent media within the framework of classical electrodynamics is explored in [62]. By modeling the smooth transition as a phenomenologically realistic change of the dielectric permittivity, an analytically exact solution to Maxwell's equations is derived that shows the possibility of the amplification and attenuation of waves, and associates this with the decrease and increase of the time-dependent permittivity. Such an energy exchange between waves and non-stationary media paves the way towards controllable light-matter interaction in time-varying structures. Wave control is usually performed by spatially engineering the properties of a medium. Because time and space play similar roles in wave propagation, manipulating time boundaries provides a complementary approach [62]. The relevance of this concept was experimentally demonstrated by introducing instantaneous time mirrors which explain the effect of any time disruption on wave propagation. It was shown that sudden changes of the medium properties generate instant wave sources that emerge instantaneously from the entire space at the time disruption. The time-reversed waves originate from these 'Cauchy sources', which are the counterpart of Huygens virtual sources on a time boundary. Recently time-domain techniques have been actively discussed in the literature for solving such electromagnetic problems, mainly owing to their superiority in solving wide-band problems in comparison to frequency-domain methods. Numerical demonstration of the wavelength conversion of light via the simple dynamic refractive index tuning of an optical cavity in a photonic crystal has been described by [63].

In the previous sections, transformation of an electromagnetic wave was considered in conditions of double asymmetry availability: a temporal asymmetry of medium parameters behavior and a spatial asymmetry of medium parameters distribution. An initial electromagnetic wave was assumed symmetric in the form of a harmonic wave.

3.6.1 The Problem Equations

The problem of the electromagnetic field transformation in a medium whose refractive index changes in time is solved by the Volterra integral equation method which takes into account the initial and boundary conditions for the electromagnetic field. It is important that the spot shape considered is described by an Airy function, and not by a Gaussian function, since there are in fact no free propagating Gaussian beams. The reason for this is that a Gaussian profile has a Gaussian Fourier spectrum which is never zero, and only asymptotically approaches zero as the wavenumber tends to infinity. This pulse is symmetric with respect to its argument, that is, it behaves identically when its argument tends to plus or minus infinity. Thus, for a Gaussian profile one needs to include evanescent components, even if their contribution is small. The oscillations in the Airy profile arise from the hard cut-off at high frequencies [64].

We consider the transformation of an electromagnetic pulse $E_0(t, x)$ caused by a plane boundary of a dielectric half-space $(x > 0)$ which appears at zero moment of time in a previously homogeneous medium characterized by a refractive index n. After zero moment the medium becomes inhomogeneous due to a temporal jump-like change of the refractive index from n

to n_1 in the half-space $x \geq 0$ (Fig. 3.26a). The initial pulse $E_0(t, x)$ propagates normally to the boundary of the non-stationary dielectric. Such a problem statement gives asymmetry in time and space: i.e., the situation is different in the future ($t \to \infty$) and in the past ($t \to -\infty$) as well as in the direction $x \geq 0$ and the opposite direction, $x \leq 0$. A third asymmetry is raised by the initial pulse as the Airy function of the first kind

$$E_0(t, x) = Ai(-t/T + x/vt - p) \tag{3.165}$$

as discussed below. Here, $v = c/n$ is the wave velocity in the primary medium and c is the light velocity in vacuum, T is a time scale and $p = (x_0/v - t_0)/T$ is a start parameter conditioned by the moment of the pulse initiation t_0 and the source point location x_0.

The behavior of the Airy function in the opposite directions

$$\text{Ai}(t) \approx \begin{cases} 2^{-1}\pi^{-1/2}t^{-1/4}\exp(-2t^{3/2}/3), \; t \to +\infty \\ \pi^{-1/2}|t|^{-1/4}\sin(2|t|^{3/2}/3 + \pi/4), \; t \to -\infty \end{cases} \tag{3.166}$$

gives the third asymmetry in the problem statement (Fig. 3.26b). So, we describe a triply asymmetric electromagnetic phenomenon: an asymmetric initial electromagnetic signal, an asymmetric object of diffraction and the asymmetric behavior of this object in time.

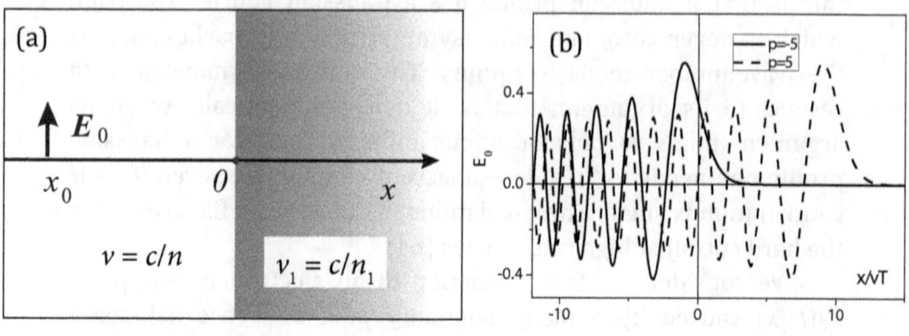

Figure 3.26 The geometry of the problem. The medium with a refractive index n_1 appears at zero moment (a). The spatial form of the initial pulse at the moment $t/T = 5$ for two values of the start parameter: $p = -5$ (solid line) and $p = 5$ (dash line) (b).

The sign of the start parameter p has a great influence on the pulse transformation process. A location of the initial pulse front in (3.165) can be determined by the inequality $-t + x/v - pT > 0$. It defines a region in time-spatial coordinates with the near zero field that is marked by the bold 'double point and dashed' line in Fig. 3.27. If $p > 0$ then the initial field is deep inside the region where a dielectric has changed at zero moment of time (Fig. 3.27a). On the other hand, for $p < 0$, the initial front does not reach at $t = 0$ the medium boundary and the pulse front does not see the medium appearing at this moment at all (Fig. 3.27b).

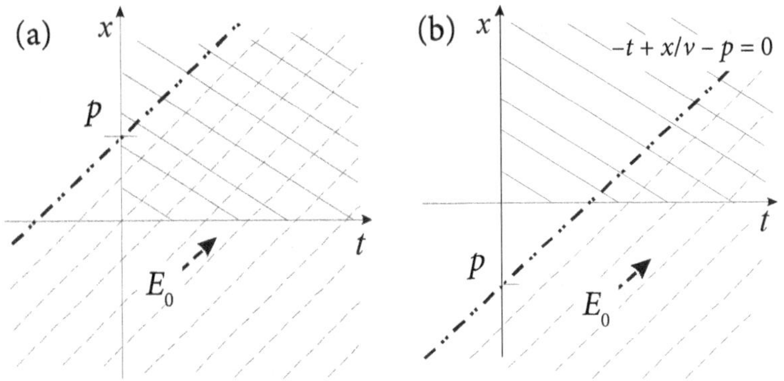

Figure 3.27 The initial pulse position with respect to a medium with changing refractive index at the zero moment of time for the start parameter of different signs: (a) $p > 0$, (b) $p < 0$. Hatching by solid lines marks a time-spatial region corresponding to a new state of the medium where the permittivity changes at zero moment of time. Hatching by dash lines marks a time-spatial region occupied by the initial pulse.

To solve the whole problem it is convenient to use the Volterra integral equation method which takes into account the initial and boundary conditions for the electromagnetic field and is universal with respect to the form of the initial pulse. In the case when the medium refractive index changes abruptly at a zero moment of time inside a region $x > 0$, that is the medium becomes inhomogeneous, the field is described by the equation

$$E(t,x) = E_0(t,x) + \frac{v_1^2 - v^2}{2vv_1^2}\frac{\partial^2}{\partial t^2}\int_0^\infty dt'\int_0^\infty dx'\theta\left(t - t' - \frac{|x - x'|}{v}\right)E(t',x'), \tag{3.167}$$

where $\theta(t)$ is the Heaviside unit function. Inside the non-stationary medium, $x \geq 0$, where the refractive index experiences a temporal jump, the expression (3.167) is the integral equation for the field inside this part of the medium. The solution to Eq. (3.167) when the observation point (t, x) belongs to the non-stationary medium $(t > 0, x > 0)$, gives a transmitted field. If the observation point (t, x) is outside the non-stationary medium, $(t > 0, x < 0)$, then (3.167) is the formula for calculating the reflected field using the transmitted field found. For $t < 0$ the kernel in (3.167) is equal to zero and, so, $E(t, x) = E_0(t, x)$.

3.6.2 Transformed Pulses

To find the transmitted field in the newly created medium we use the resolvent method

$$E(t,x) = E_0(t,x) + \int_0^{\infty} dt' \int_0^{\infty} dx' R(t,t',x,x') E_0(t',x'), \tag{3.168}$$

with the resolvent operator kernel

$$\hat{R} = \frac{v_1^2 - v^2}{2v^2 v_1}\theta(x)\frac{\partial^2}{\partial t^2}\left[\theta\left(t - t' - \frac{|x - x'|}{v_1}\right) + \frac{v - v_1}{v + v_1}\theta\left(t - t' - \frac{x + x'}{v_1}\right)\right]\theta(x') \tag{3.169}$$

Integration of (3.168) gives the field in the transient medium

$$\int_0^{\infty} dt' \int_0^{\infty} dx' \hat{R}(t,t',x,x') E_0(t',x') = -E_0(t,x) + E_T(t,x), \quad t > 0, \ x > 0, \tag{3.170}$$

where the first term $-E_0(t, x)$ is a virtual pulse which extinguishes the initial pulse $E_0(t, x)$ in agreement with the Ewald–Oseen extinction theorem [18]). The second term is, in fact, the transmitted field which consists of four terms.

The first term

$$E_{T1} = \frac{2v_1}{v_1 + v}\mathrm{Ai}\left(-t/T + \frac{x}{v_1 T} - p\right) \tag{3.171}$$

represents transmission of the initial pulse through the medium boundary that appears (Fig. 3.28). It propagates in the same direction as the initial pulse but with a new velocity corresponding to the new value of the refractive index. The amplitude of this wave is equal to the Fresnel wave one.

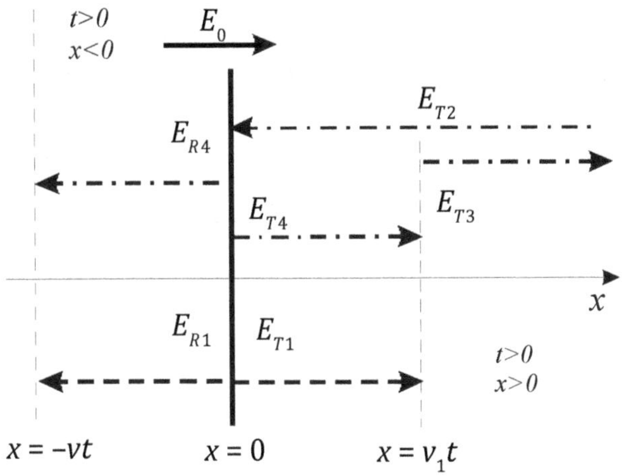

Figure 3.28 An arrangement of diffracted pulses. Ordinary transmitted and reflected pulses are marked by dash arrows, pulses caused by a temporal change of a medium are marked by dash-dot arrows.

The terms

$$E_{T2} = \frac{v_1(v_1 - v)}{2v^2}\mathrm{Ai}\left(\frac{v_1}{vT}\left(t + \frac{x}{v_1}\right) - p\right)$$

$$E_{T3} = \frac{v_1(v_1 + v)}{2v^2}\mathrm{Ai}\left(-\frac{v_1}{vT}\left(t - \frac{x}{v_1}\right) - p\right), \tag{3.172}$$

are the results of the initial pulse $E_0(t, x)$ splitting into two pulses propagating into opposite directions as a consequence of a refractive index time jump at the zero moment of time [21, 47, 48, 56]. They propagate in opposite directions but the wave E_{T3} exists in the region $x > v_1t$ only as it has been created in the changed medium. This pulse keeps the same propagation direction as the initial pulse, contrary to the pulse E_{T2}, which has the opposite direction, towards the boundary that has appeared.

This pulse reflects from the boundary that appeared ($x = 0$) giving rise to the pulse E_{T4} reflected towards the newly created medium in the region $0 < x < v_1 t$

$$E_{T4} = -\frac{v_1(v-v_1)^2}{2v^2(v+v_1)}\mathrm{Ai}\left(\frac{v_1}{vT}\left(t-\frac{x}{v_1}\right)-p\right) \tag{3.173}$$

So, the whole transmitted field is represented by the formula

$$E_T = \theta(x)[\theta(t-x/v_1)(E_{T1}+E_{T4})+E_{T2}+\theta(x/v_1-t)E_{T3}] \tag{3.174}$$

Calculation of the integral in (3.167), by using the formula (3.174) for the inner field inside the transient half-space, gives the field of reflected pulses as

$$E_R = \theta(t)\theta(-x)\theta(t+x/v)\{E_{R1}+E_{R2}\} \tag{3.175}$$

Here,

$$E_{R1} = \frac{v_1-v}{v+v_1}\mathrm{Ai}\left[-\left(t+\frac{x}{v}\right)/T-p\right],\ E_{R2} = \frac{v_1}{v}\frac{v_1-v}{v+v_1}\mathrm{Ai}\left[\frac{v_1}{v}\left(t+\frac{x}{v}\right)/T-p\right] \tag{3.176}$$

The first pulse E_{R1} has the ordinary argument $(-t + x/v)/T - p$ and the other pulse E_{R2} has a (new) argument that is changed by the time jump of the refractive index: $v_1(t + x/v)vT - p$.

The schematic arrangement of all diffracted pulses is shown in Fig. 3.28. The temporal jump of the refractive index leads to a split of the initial pulse into two new ones E_{T2} and E_{T3} propagating in opposite directions. The pulse E_{T2}, existing in the region $x > v_1 t$, moves away from the boundary. The pulse E_{T2} moving towards the boundary passes partly through this boundary and gives rise to the pulse E_{R2} in the region of the reflected field $x < 0$. The pulse E_{T2} reflects also partly from the boundary inside the non-stationary medium and gives rise to the pulse E_{T4}. This pulse reflects from the medium appearing boundary and exists in the region $0 < x < v_1 t$. The initial pulse E_0 induces the pulse E_{T1} in the region $0 < x < v_1 t$ and the reflected pulse E_{R1} in the region $-vt < x < 0$.

It is easy to make sure from (3.174) and (3.175) that the transformed electric field is continuous on the spatial boundary between the two media, $(E_0 + E_R)|_{x=-0} = E_T|_{x=+0}$, as well as at the moment of the temporal jump of the refractive index, $(v_1/v)^2 E|_{t=-0} = E|_{t=+0}$.

The character of the transmitted and reflected pulses depends very strongly on the sign of the starting parameter p. This can be seen easily for the reflected pulses existing in the time-spatial zone $-vt < x < 0$ which is provoked by the boundary appearance and shown in Fig. 3.29.

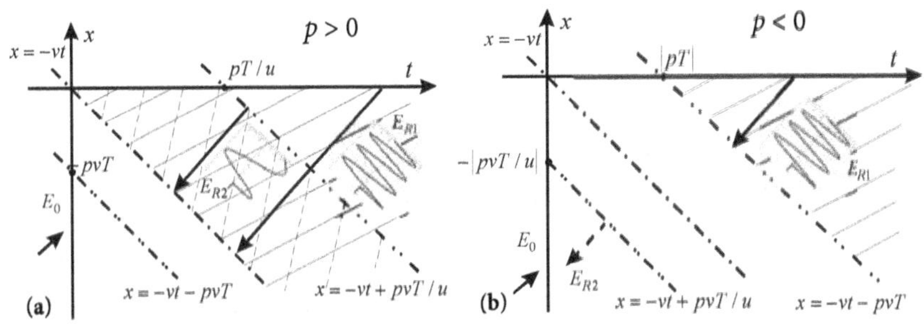

Figure 3.29 The time-spatial zones in the reflected field region for the different signs of the start parameter. In the case of the positive start parameter (Fig. 3.4a), $p > 0$, the double hatching marks the region of two ways existence, E_{R1} and E_{R2} (the dash-dot lines are this region boundary). In the opposite case, $p < 0$, such a zone is free from the field. In both cases the parameter $u = n/n_1$ is equal to 0.8.

For a positive value of the starting parameter, $p > 0$, both components of the reflected pulse, E_{R1} and E_{R2}, are present, but the reflected pulse with the changed phase $E_{R2} = \frac{v_1}{v}\frac{v_1 - v}{v + v_1}\text{Ai}\left(\frac{v_1}{vT}\left(t + \frac{x}{v}\right) - p\right)$ occupies only the area restricted by inequalities $x/v + t - pTv/v_1 < 0$ and $x/v + t > 0$ (the double hatching in Fig. 3.29a). In the case of a negative parameter, $p < 0$, the transitional area is a stripe in the time-spatial coordinates $(x/v + t - pT < 0,$ $x/v + t > 0)$ which is free from the field (Fig. 3.29b).

The field distribution of all diffracted pulses near the boundary of the transient medium $E = E_T + E_R$ is shown in Fig. 3.30: (a) for $p = 5$ and (b) for $p = -5$.

The character of the whole field distribution is complex, and very different for two cases, $p > 0$ and $p < 0$. The complicated character of the field distribution is explained by the different arguments in the Airy functions: $\pm(v_1t\pm x)/vT-p$ and $\pm(t+x/v_1)/T-p$ in the transmitted field and $-(t + x/v)/T - p$ and $v_1(t + x/v)/T - p$ in the reflected field.

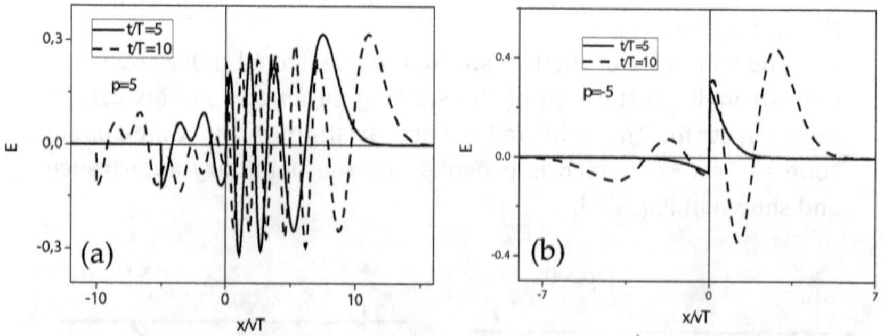

Figure 3.30 The transmitted and reflected pulses in the transitional zone near the boundary at two moments, $t/T = 5$ (solid) and $t/T = 10$ (dash); (a) for $p = 5$, (b) for $p = -5$. The parameter $u = n/n_1$ is equal to 0.8.

The contribution of the pulses determined by the temporally changing medium manifests itself clearly if one omits the pulses with ordinary arguments $-(t + x/v_1)/T - p$ in the transmitted field and $-(t + x/v)/T - p$ in the reflected field and considers only the pulses with the new arguments $\pm(v_1t \pm x)/vT - p$ and $v_1(t + x/v)/vT - p$. It means

$$E_{Tnew} = \theta(x)[\theta(t - x/v_1)E_{T4}(v_1(t - x/v_1)/vT - p) + E_{T2}(v_1(t + x/v_1)/vT - p) + \theta(x/v_1 - t)E_{T3}(-v_1(t - x/v_1)/vT - p)] \tag{3.177}$$

for the transmitted field and

$$E_{Rnew} = \theta(t)\theta(-x)\theta(t + x/v)E_{R2}(v_1(t + x/v)/vT - p) \tag{3.178}$$

for the reflected field.

The field spatial distribution at the moments $t/T = 5$ and $t/T = 10$ is shown in Fig. 3.31: for start parameters $p = 5$ (a) and for start parameter $p = -5$ (b). It is seen that in the stripe $(x/v + t - pT < 0, x/v + t > 0)$, the reflected field is almost zeroth. Figure 3.32 shows that the complicated field character, seen in the reflected and transmitted pulses in this figure, is explained by the addition of the pulses with ordinary arguments,

$$E_{\text{ord}} = \theta(x)\theta\left(t - \frac{x}{v_1}\right)\frac{2v_1}{v_1 + v}\text{Ai}\left[-\left(t - \frac{x}{v_1}\right)/T + p\right] +$$
$$\theta(-x)\theta\left(t + \frac{x}{v}\right)\frac{v_1 - v}{v + v_1}\text{Ai}\left[-\left(t + \frac{x}{v}\right)/T + p\right] \tag{3.179}$$

to the pulses with new arguments (3.177) and (3.178).

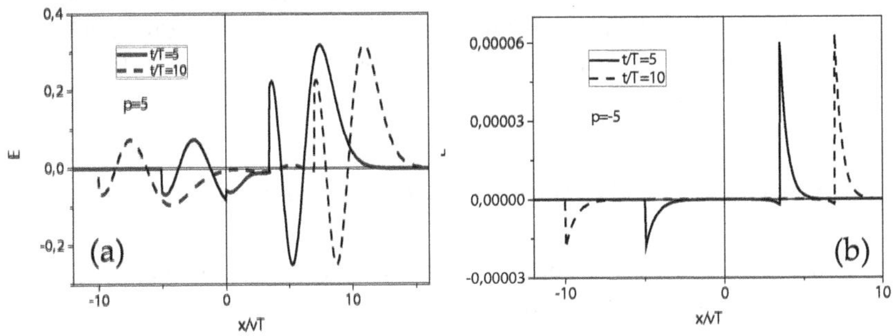

Figure 3.31 Transmitted and reflected pulses with the new arguments in the transitional zone at two moments, $t/T = 5$ (solid) and $t/T = 10$ (dash). The start parameter equals to $p = 5$ (a) and $p = -5$ (b). The parameter $u = n/n_1$ is equal to 0.8.

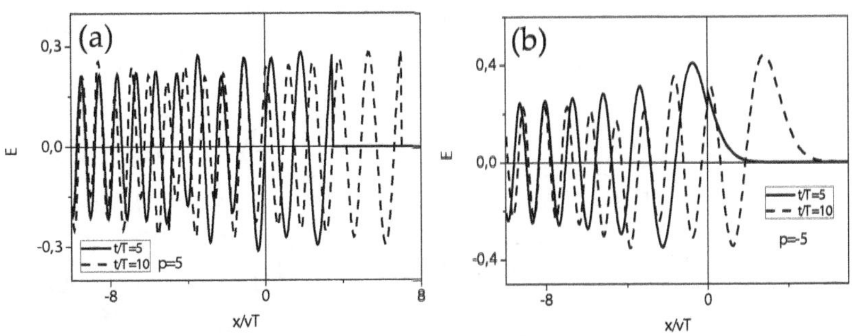

Figure 3.32 Transmitted and reflected pulses in the transitional zone with ordinary arguments at two moments, $t/T = 5$ (solid) and $t/T = 10$ (dash) for the start parameter: $p = 5$ (a) and $p = -5$ (b). The parameter $u = n/n_1$ equals to 0.8.

So, the considered electrodynamic problem with three types of asymmetry shows their individuality. One type is determined by a form of an initial signal whose behavior to the past and to the future is qualitatively different. The second and third types of asymmetry are dictated by the state of a medium which turns from a homogeneous medium into an inhomogeneous one at the zero moment of time. The consideration uses the Volterra integral equation method which takes into account the initial and boundary conditions for the electromagnetic field and is universal with respect to the form of the initial pulse. It is shown that the whole diffraction processes is very strongly controlled by the start parameter introduced. The peculiarities of this start parameter influence on the diffraction process show that all diffracted pulses consist of two kinds of pulses: pulses conditioned by the medium inhomogeneity and pulses conditioned by the temporally changing features in the medium.

Appendix: Lipshitz–Hankel Functions

It is shown in the work [65] that the functions $e(t, x, \beta)$ and $f(t, x, \beta)$ can be represented in the form

$$
\begin{aligned}
e(t,x,\beta) = \theta(t - x/v_1)e^{-\beta t}\Big\{ & ch((x/v_1)\sqrt{\beta^2 - \bar{\sigma}_1^2}) \\
& + (2i\bar{\sigma}_1\sqrt{t^2 - (x/v_1)^2})^{-1}\Big[(\beta x/v_1 - t\sqrt{\beta^2 - \bar{\sigma}_1^2})\exp(a_+\xi)\Im_{e0}(a_+,\xi) \\
& + (\beta x/v_1 + t\sqrt{\beta^2 - \bar{\sigma}_1^2})\exp(a_-\xi)\Im_{e0}(a_-,\xi)\Big]\Big\}
\end{aligned} \tag{A.1}
$$

$$
\begin{aligned}
f(t,x,\beta) = \theta(t - x/v_1)(\beta^2 - \bar{\sigma}_1^2)^{-1/2}\Big\{ & \exp(-\beta t)sh((x/v_1)\sqrt{\beta^2 - \bar{\sigma}_1^2}) \\
& + (2i\bar{\sigma}_1\sqrt{t^2 - (x/v_1)^2})^{-1}\Big[(\beta x/v_1 - t\sqrt{\beta^2 - \bar{\sigma}_1^2})\exp(a_+\xi)\Im_{e0}(a_+,\xi) \\
& - (\beta x/v_1 + t\sqrt{\beta^2 - \bar{\sigma}_1^2})\exp(a_-\xi)\Im_{e0}(a_-,\xi)\Big]\Big\}
\end{aligned} \tag{A.2}
$$

where $a_\pm = (-\beta t \pm (x/v_1)\sqrt{\beta^2 - \bar{\sigma}_1^2})(i\bar{\sigma}_1\sqrt{t^2 - (x/v_1)^2})^{-1}, \xi = i\bar{\sigma}_1\sqrt{t^2 - (x/v_1)^2}$,

$\Im_{en}(a_\pm, z) = \int_0^z ds e^{-a_\pm s} s^n J_n(s)$ is the Lipshitz–Hankel function of the order n, $J_n(z)$ is the Bessel function of the order n. It is assumed here that $\mathrm{Re}(\sqrt{a_\pm^2 + 1}) > 0$, if $\mathrm{Re}(a_\pm) \geq 0$, and $\mathrm{Re}(\sqrt{a_\pm^2 + 1}) < 0$, if $\mathrm{Re}(a_\pm) < 0$.

The Lipshitz–Hankel function of the zero order $\Im_{e0}(a_\pm, z)$ is determined by the Neumann series [66, 67]

$$\Im_{e0}(a_\pm, z) = (a_\pm^2+1)^{-1/2} + \frac{\exp(-a_\pm z)}{\Gamma(1/2)(a_\pm^2+1)}\sum_{k=0}^{\infty}(2(a_\pm^2+1)^{-1}z^{-1})^k\Gamma(k+1/2)\left[J_{k+1}(z)-a_\pm J_k(z)\right], \tag{A.3}$$

The late-time approximation for the functions $Z(t + x/v_0, \beta)$ has the following form:

$$Z(t+x/v_0,\beta)\approx -\theta(t+x/v_0)(\beta^2-\bar{\sigma}_1^2)^{-1/2}\left\{\exp(-\bar{\sigma}_1(t+x/v_0)) - \frac{i\sigma_\xi}{\beta^2-\bar{\sigma}_1{}^2}\frac{t\sqrt{\beta^2-\bar{\sigma}_1{}^2}}{t+x/v_0}\left[I_1(\bar{\sigma}_1(t+x/v_0))\right.\right.\cdot \tag{A.4}$$

The functions describing the wave continuous spectra in different regions of space

$$Q_3(t,x,\tau) = -\frac{ia^2(t-x/v_0-\tau)}{(a^2-1)\omega-2i\bar{\sigma}_1}\left(\left(\frac{2i\bar{\sigma}_1}{\bar{\sigma}_1+i\omega}\frac{t-x/v_0-\tau}{a\sqrt{a^2\tau^2-(t-x/v_0-\tau)^2}}\right.\right.$$
$$\left.-(\bar{\sigma}_1/2a)\sqrt{\tau^2-(1/a^2)(t-x/v_0-\tau)^2}\frac{(a^2+1)\omega+2i\bar{\sigma}_1}{a^4\omega^2\tau^2-\Omega^2(t-x/v_0-\tau)^2}\right)$$
$$\times I_1(\bar{\sigma}_1\sqrt{\tau^2-(1/a^2)(t-x/v_0-\tau)^2})-I_0(\bar{\sigma}_1\sqrt{\tau^2-(1/a^2)(t-x/v_0-\tau)^2})$$
$$\left.\left.\times(1/a)\left(\frac{2}{t-x/v_0-\tau}-\frac{\bar{\sigma}_1^2\Omega\tau[(a^2+1)\omega+2ia\bar{\sigma}_1]}{a^4\omega^2\tau^2-\Omega^2(t-x/v_0-\tau)^2}\right)\right)\right). \tag{A.5}$$

$$Q_4(t,x,\tau) = \frac{a^3}{2\bar{\sigma}_1+i(1-a^2)\omega}$$
$$\times\left(\left(\frac{4i\bar{\sigma}_1}{(\bar{\sigma}_1+i\omega)^2}\frac{a\omega(\omega-2i\bar{\sigma}_1)\tau-2(\bar{\sigma}_1+i\omega)(2\bar{\sigma}_1+i(1+a^2)\omega)(t+x/v_0-\tau)}{(2\bar{\sigma}_1+i(1-a^2)\omega)\sqrt{a^2\tau^2-(t+x/v_0-\tau)^2}}\right.\right.$$
$$+ia(1+a^2)\frac{2a^4\bar{\sigma}_1\omega\tau+(2\bar{\sigma}_1{}^2+(1-a^2)\omega^2)(t+x/v_0-\tau)}{(2\bar{\sigma}_1-i(1-a^2)\omega)(a^4\omega^2\tau^2-\Omega^2(t+x/v_0-\tau)^2)}$$
$$\times\bar{\sigma}_1\sqrt{a^2\tau^2-(t+x/v_0-\tau)^2}-16a\,|1-a^2|\sqrt{a^2\tau^2-(t+x/v_0-\tau)^2}$$
$$\left.\times\bar{\sigma}_1\frac{a\tau-i(1+a^2)(t+x/v_0-\tau)}{(2\bar{\sigma}_1+i(1-a^2)\omega)\left((1+a^2)^2(t+x/v_0-\tau)^2-4a^4\tau^2\right)}\right)$$
$$\times I_1(\bar{\sigma}_1\sqrt{\tau^2-(1/a^2)(t+x/v_0-\tau)^2})-\left(8\frac{2\bar{\sigma}_1+i(1+a^2)\omega}{2\bar{\sigma}_1-i(1+a^2)\omega}-a(1+a^2)\bar{\sigma}_1\right.$$

$$-\frac{2\omega\Omega(a^2\tau^2-(t+x/v_0-\tau)^2)-i\bar{\sigma}_1\tau(t+x/v_0-\tau)\Omega^{-1}(2\bar{\sigma}^2+(1+a^2)\omega^2)}{(2\bar{\sigma}_1-i(1+a^2)\omega)(a^4\omega^2\tau^2-\Omega^2(t+x/v_0-\tau)^2)}$$

$$\left.\times I_0\left(\bar{\sigma}_1\sqrt{\tau^2-(1/a^2)(t+x/v_0-\tau)^2}\right)\right). \tag{A.6}$$

References

1. Kuo, S., and Ren, A. (1993) Experimental study of wave propagation through a rapidly created plasma, *IEEE Trans. Plasma Sci.*, **21**(1), pp. 53–56.
2. Hawkins, R. J., and Kallman, J. S. (1994) Lasing in tilted-waveguide semiconductor laser amplifiers, *Opt. Quantum Electronics*, **26**, pp. 207–217.
3. Hagness, S. C., Joseph, R. M., and Taflove, A. (1996) Subpicosecond electrodynamics of distributed Bragg reflector microlasers: Results from finite difference time domain simulations, *Radio Sci.*, **31**(4), 931–941.
4. Kalluri, D. K. (1998) *Electromagnetics of Complex Media (Frequency Shifting by a Transient Magnetoplasma Medium)* CRC Press: Boca Raton, Boston, New York.
5. Bakunov, M. I., and Maslov, A. V. (1999) Frequency upshifting of electromagnetic radiation via oblique incidence on an ionization front, *IEEE Trans. Plasma Sci.*, **27**(3), 655–663.
6. Chi, J. W. D., Chao, C. L., and Rao, M. K. (2001) Time-domain large-signal investigation on non-linear interactions between an optical pulse and semiconductor waveguide, *IEEE J. Quantum Electronics*, **37**(10), 1329–1336.
7. Jeong, Y., and Lee, B. (2001) Characteristics of second-harmonic generation including third-order non-linear interactions, *IEEE J. Quantum Electron.*, **37**(10), 1292–1300.
8. Shifman, Y., and Leviatan, Y. (2001) On the use of spatio-temporal multiresolution analysis in method of moments solutions of transient electromagnetic scattering, *IEEE Trans. Antennas Propag.*, **49**(8), 1123–1129.
9. Taflove, A. (1995) *Computational Electrodynamics: The Finite-Difference Time-Domain Method*, Boston, Artech House.
10. Young, J. L., Gaitoude, D., and Shang, J. S. (1997) Toward the construction of a fourth order difference scheme for transient

EM wave simulation: Staggered grid approach, *IEEE Trans. Antennas Propag.*, **45**(11), 1573–1580.

11. Nerukh, A. G., and Khizhnyak, N. A. (1991) *Modern Problems in Nonstationary Macroscopic Electrodynamics*, Kharkov, "Test-Radio" Publishing House.
12. Nerukh, A. G., Scherbatko, I. V., and Marciniak, M. (2001) *Electromagnetics of Modulated Media with Applications to Photonics*, National Institute Telecommunications Publishing House, Warsaw, 268 p.
13. Fedotov, F. V., Nerukh, A. G., Scherbatko, I. V. (2001) Comparison of the FDTD and direct-integrating methods for electrodynamic problem in time-varying medium, *5th International Conference on Applied Electromagnetics* 8–10 October, Niš, Yugoslavia.
14. Nerukh, A. G., Scherbatko, I. V., and Nerukh, D. A. (1997) Using evolutionary recursion to solve an electromagnetic problem with time-varying parameters, *Microwave Opt. Technol. Lett.*, **14**(1), 31–36.
15. Yemelyanov, K. M., Fedotov, F. V., and Nerukh, A. G. (2001) Numerical simulations of electromagnetic transients in nonstationary plasma layers, *Proceedings of 3rd International Conference on Transparent Optical Networks, June 18–21, 2001,* Cracow, Poland, 226–229.
16. Tamm, I. E. (1966) *Fundamentals of Electricity Theory*, Moscow, "Nauka", 624 p.
17. Vladimirov, V. S. (1967) *Equations of Mathematical Physics* Moscow: "Nauka", 437 p. (in Russian).
18. Born, M., and Wolf, E. (1964) *Principles of Optics*, Oxford-London-Edinburgh-New York-Paris-Frankfurt: Pergamon Press, 856 p.
19. Nerukh, A. G., Rybin, O. N., and Scherbatko, I. V. (1999) Action of pulse excitation of bounded medium on plane electromagnetic wave, *J. Tehnicheskoi Fiziki*, **69**(8), 84–92 (in Russian).
20. Averkov, S. I., and Boldin, V. P. (1980) Waves in nondispersive nonstationary inhomogeneous media, *Radiophys. Quantum Electron.* (English transl.), **23**(9), 1060–1066 (in Russian pagination).
21. Borisov, V. (1987) *Transient Electromagnetic Waves*, Leningrad State University Press, Leningrad, 240 p. (in Russian).
22. Nerukh, A. G., and Khizhnjak, N. A. (1991) *Modern Problems of Transient Macroscopic Electrodynamics*, Test-Radio Publ., Kharkov (in Russian).

23. Barsukov, K. A., and Grigorieva, N. Y. (1996) To the theory of electromagnetic field pressing, *J. Tehnicheskoi Fiziki*, **66**(7), 134–140. (in Russian).

24. Borisov, V. V. (1989) Electromagnetic signal transformation on sharp change of medium properties in time, *Geomagnetizm Aeronomy*, **29**(5), 730–737 (in Russian).

25. Nerukh, A. G., and Shavorikina, I. Y. (1992) Splitting of electromagnetic impulse under a conductivity jump of restricted medium, *J. Tehnicheskoi Fisiki*, **62**(5), 108–118 (in Russian).

26. Dvorak, S. L., and Dudley, D. G. (1995) Propagation of ultra-wide-band electromagnetic pulses through dispersive media, *IEEE Trans. Electromagnetic Compatibility*, **37**(2), 192–200.

27. Dvorak, S. L., and Kuester, E. F. (1990) Numerical computation of the incomplete Lipschitz-Hankel integral Je_0 (a,z), *J. Comput. Phys.*, **87**(2), 301–327.

28. Pao, H.-Y., Dvorak, S. L., and Dudley, D. G. (1996) An accurate and efficient analysis for transient plane waves obliquely incident on a conductive half space (TE case), *IEEE Trans. Antennas Propag.*, **44**(7), 925–932.

29. Lavrentiev, M. A., and Shabat, B. V (1987) *Methods of Theory of Complex Variables*, Moscow: "Nauka", 688 p. (in Russian).

30. Borisov, V. V. (1987) *Nonsteady Electromagnetic Waves*, Leningrad, Leningrad University Publishing House, 240 p. (in Russian).

31. Tyzhnenko, A. G., and Khyzhnyak, N. A. (1973) Scattering of radiowaves by a plasma ellipsoid with variable plasma density, *Radiotekhnika i Electronika*, **18**(No 1), 19–25 (in Russian).

32. Batemen, H., and Erdely, A. (1954) *Tables of Integral Transforms*, New York, Toronto, London: McGraw-Hill Book Comp., Inc., 344 p.

33. Bolomey, J. C., Durix, C., and Lesselier, D. (1978) Time domain integral equation approach for inhomogeneous and dispersive slab problems, *IEEE Trans. Antennas Propag.*, **AP-26**(5), 658–667.

34. Borisov, V. V. (1987) *Nonsteady Electromagnetic Waves.* (In Russian), Leningrad, Leningrad University Publishing House, 240 p.

35. Felsen, L. B., and Whitman, G. M. (1970) Wave propagation in time-varying media, *IEEE Trans. Antennas Propag.*, **AP-18**, 242–253.

36. Morgenthaler, F. R. (1958) Velocity modulation of electromagnetic waves, *IRE Trans. Microwave Theory Tech.*, **MTT-6**, 167–172.

37. Bera, A., Roussey, M., Kuittinen, M., and Honkanen, S. (2016) Slow-light enhanced electro-optic modulation with an on-chip silicon-hybrid Fano system, *Opt. Lett.*, **41**, 2233–2236.

38. Siviloglou, G. A., and Christodoulides, D. N. (2007) Accelerating finite energy Airy beams, *Opt. Lett.*, **32**, 979–981.

39. Barwick, S. (2011) Reduced side-lobe Airy beams, *Opt. Lett.*, **36**, 2827–2829.

40. Kaminer, I., Lumer, Y., Segev, M., and Christodoulides, D. N. (2011) Causality effects on accelerating light pulses, *Opt. Exp.*, **19**, 23132–23139.

41. Kaminer, I., Bekenstein, R., Nemirovsky, J., and Segev, M. (2012) Nondiffracting accelerating wave packets of Maxwell's equations, *Phys. Rev. Lett.*, **108**, 163901.

42. Aleahmad, P., Cessa, H. M., Kaminer, I., Segev, M., Demetrios, N., and Christodoulides, D. N. (2016) Dynamics of accelerating Bessel solutions of Maxwell's equations, *J. Opt. Soc. Am. A*, **33**, 2047–2052.

43. Alonso, M., and Bandres, M (2014) Generation of nonparaxial accelerating fields through mirrors. I: Two dimensions, *Opt. Exp.*, **22**, 7124–7132.

44. Liang, Y., Hu, Y., Ye, Z., Song, D., Lou, C., Zhang, X., Xu, J., Morandotti, R., and Chen, Z. (2014) Dynamical deformed Airy beams with arbitrary angles between two wings. *J. Opt. Soc. Am. A*, **31**, 1468–1472.

45. Li, P., Liu, S., Peng, T., Xie, G., Gan, X., and Zhao, J. (2014) Spiral autofocusing Airy beams carrying power-exponent-phase vortices, *Opt. Exp.*, **22**, 7598–7606.

46. Censor, D. (1987) Electgromagnetic propagation and scattering in time-dependence moving medium, *Phys. Rev. A. Gen. Phys.*, **35**, 2869–2877.

47. Felsen, L. B., and Whitman, G. M. (1970) Wave propagation in time-varying media. *IEEE Trans. Antennas Propag.*, **AP-18**, 242–253.

48. Fante, R. L. (1971) Transmission of electromagnetic waves into time-varying media. *IEEE Trans. Antennas Propag.*, **AP-19**, 417–424.

49. Ostrovsky, L. A., and Stepanov, N. S. (1971) Non-resonance parametric phenomena in distributive systems (in Russian), *Radiophys. Quantum Electronics* (English transl.), **14**, 489–529.

50. Averkov, S. I., and Boldin, V. P. (1980) Waves in nondispersive non-stationary inhomogeneous media, *Radiophys. Quantum Electronics.* (English transl.), **23**, 1060–1066.

51. Stolyarov, S. N. (1983) Resonance transform of waves in periodic non-stationary medium, *Radiophys. Quantum Electronics.* (English transl.), **26**, 514–516.

52. Borisov, V. V. (1987) *Nonsteady Electromagnetic Waves* (In Russian), Leningrad, Leningrad University Publishing House, 240 p.

53. Nerukh, A. G. (1995) Evolutionary approach in transient electrodynamics problems, *Radio Sci.*, **30**, 481–491.

54. Kalluri, D. K. (1999) *Electromagnetics of Complex Media: Frequency Shifting by a Transient Magnetoplasma Medium,* CRC Press, Boca Raton, 292 p.

55. Nerukh, A., Remayeva, T., Sakhnenko, N. (2009) Frequency change of partial spherical waves induced by time change of the medium permittivity, *Opt. Quantum Electronics.* **41**, 327–335.

56. Zhao, X. C., Ouyang, X. P., Xu, Y. D., Han, H. T., Zhang, Z. C., Wang, T., Zha, G. Q., and Ouyang, X. (2012) Time response of $Cd_{0.9}Zn_{0.1}Te$ crystals under transient and pulsed irradiation, *AIP Adv.,* **2**, 012162-1-6.

57. Xiao, Y., Agrawal, G. P., and Maywar, D. N. (2011) Spectral and temporal changes of optical pulses propagating through time-varying linear media, *Opt. Lett.*, **36**, 505–507.

58. Xiao, Y., Agrawal, G. P., and Maywar, D. N. (2012): Nonlinear pulse propagation: A time-transformation approach, *Opt. Lett.*, **37**, 1271–1273.

59. Xiao, Y., Maywar, D. N., and Agrawal, G. P. (2014) Reflection and transmission of electromagnetic waves at a temporal boundary, *Opt. Lett.*, **39**, 574–576.

60. Bakunov, M. I., and Maslov, A. V. (2014) Reflection and transmission of electromagnetic waves at a temporal boundary: Comment, *Opt. Lett.*, **39**, 6029.

61. Fernández-Ruiz, M. R., Huh, J., and Azaña, J. (2016) Time-domain Vander–Lugt filters for in-fiber complex (amplitude and phase) optical pulse shaping, *Opt. Lett.*, **41**, 2121–2124.

62. Hayrapetyan, A. G., Götte, J. B., Grigoryan, K. K., Fritzsche, S., and Petrosyan, R. G. (2016) Electromagnetic wave propagation in spatially homogeneous yet smoothly time-varying dielectric media, *J. Quant. Spectrosc. Radiative Transfer,* **178**, 158–166.

63. Lucchini, M., Castiglioni, L., Kasmi, L., Kliuiev, P., Ludwig, A., Greif, M., Osterwalder, J., Hengsberger, M., Gallmann, L., and Keller, U. (2015)

Light-matter interaction at surfaces in the spatiotemporal limit of macroscopic models, *Phys. Rev. Lett.*, **115**, 137401.

64. Novotny, L., and Hecht, B. (2006) *Principles of Nano-Optics*, Cambridge University Press, 539 p.
65. Pao, H.-Y., Dvorak, S. L., and Dudley, D. G. (1996) An accurate and efficient analysis for transient plane waves obliquely incident on a conductive half space (TM case), *IEEE Trans. Antennas Propagation*, **44**(7), 925–932.
66. Dvorak, S. L., and Dudley, D. G. (1995) Propagation of ultra-wide-band electromagnetic pulses through dispersive media, *IEEE Trans. Electromagn. Compatibility*, **37**(2), 192–200.
67. Dvorak, S. L., and Kuester, E. F. (1990) Numerical computation of the incomplete Lipschitz-Hankel integral $Je_0(a,z)$, *J. Computat. Phys.*, **87**(2), 301–327.

Chapter 4

3D+T Problems with Electromagnetic Transients

4.1 The 3D Resolvent for a Problem with a Plane Boundary of a Dielectric Half-Space

The resolvent construction used in the previous sections for the 1D case is developed here for the 3D one. The resolvent must satisfy Eq. (3.7)

$$\langle \mathbf{p} | \hat{R} | \mathbf{p}' \rangle = \langle \mathbf{p} | \hat{K} | \mathbf{p}' \rangle + \int d\mathbf{p}_1 \langle \mathbf{p} | \hat{K} | \mathbf{p}_1 \rangle \langle \mathbf{p}_1 | \hat{R} | \mathbf{p}' \rangle \tag{4.1}$$

The initial problem in an unbounded medium is described by the resolvent

$$\left\langle \mathbf{x} \middle| \hat{R}_e \middle| \mathbf{x}' \right\rangle_{ij} = \frac{\varepsilon_1 - \varepsilon}{\varepsilon_1} \left(\frac{\partial^2}{\partial x_i \partial x_j} - \frac{1}{v_1^{\,2}} \delta_{ij} \frac{\partial^2}{\partial t^2} \right) \frac{\delta(t - t' - |\mathbf{r} - \mathbf{r}'|/v_1)}{|\mathbf{r} - \mathbf{r}'|}, \tag{4.2}$$

where the support domain of the δ-function is the light cone Γ. Let us consider now the situation when there is a boundary that divides space into two half-spaces; the left half-space is occupied by the background dielectric with the permittivity ε, while in the right half-space the permittivity changes its magnitude from ε

Non-Stationary Electromagnetics: An Integral Equations Approach (Second Edition)
Alexander Nerukh and Trevor Benson

ISBN 978-981-4774-95-6 (Hardcover), 978-0-429-02773-4 (eBook)
www.panstanford.com

into ε_1 at zero moment of time. In this case the integration in (4.1) is determined over the intersection of the past light cone and the world domain of the region $V(t)$, i.e., the intersection of the cone Γ and the region $\chi(\mathbf{x}') = \theta(x') = 1$, as shown in Fig. 4.1 for a projection on the (t, x)-plane. The part of this cone that is not included into the region $\theta(x) = 1$ gives rise to a reflection of this part from the boundary (the double shaded region Γ' in Fig. 4.1.) This reflected part has to contribute to the resolvent of the boundary-value problem. As it was shown above in the 1D problems, this reflection has got a weighting coefficient. Suppose this reasoning is also true for the 3D problem. Let us try to find this weight coefficient when the boundary divides the space in two half-spaces.

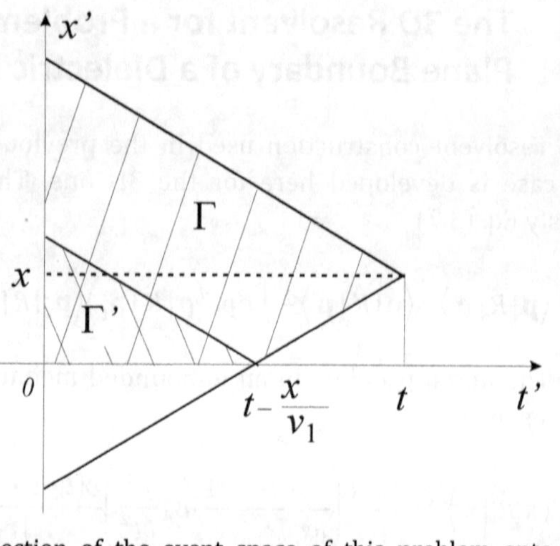

Figure 4.1 The projection of the event space of this problem onto the plane *x*, *t*.

The reflected region can be defined by the term, which is symmetric to the outgoing region with respect to the boundary,

$$\theta(x)\hat{S}\frac{\delta\left(t-t'-\frac{1}{v_1}\sqrt{(x+x')^2+(y-y')^2+(z-z')^2}\right)}{\sqrt{(x+x')^2+(y-y')^2+(z-z')^2}}\theta(x'), \tag{4.3}$$

where the reflection from the boundary is taken into account by substituting x' for $-x'$, and S is an arbitrary weight operator that has to be obtained.

We find the impulse representation for the expression (4.3) assuming now that $\hat{S} = \hat{D}_1 = \nabla\nabla - \frac{1}{v_1^2}\frac{\partial^2}{\partial t^2}$ analogously to (4.2). Here $\nabla = (\frac{\partial}{\partial x}, \frac{\partial}{\partial y}, \frac{\partial}{\partial z})$ is the differential operator nabla. The impulse representations of the multipliers in (4.3) have the following forms:

$$\left\langle \mathbf{p} \left| \frac{\delta\left(t - t' - \frac{1}{v_1}\sqrt{(x+x')^2 + (y-y')^2 + (z-z')^2}\right)}{\sqrt{(x+x')^2 + (y-y')^2 + (z-z')^2}} \right| \mathbf{p}' \right\rangle$$

$$= \frac{8\pi^2 v_1^{\,2}}{p^2 + v_1^{\,2} k^2}\delta(k_1 + k_1')\Delta(\mathbf{p}_\perp - \mathbf{p}_\perp') \tag{4.4}$$

$$\langle \mathbf{p}|\hat{D}_1|\mathbf{p}'\rangle_{ij} = -\left(k_i k_j + \frac{p'^2\delta_{ij}}{v_1^{\,2}}\right)2\pi\delta(k_1 - k_1')\Delta(\mathbf{p}_\perp - \mathbf{p}_\perp') \tag{4.5}$$

$$\langle \mathbf{p}|\theta(x)|\mathbf{p}'\rangle = \frac{-i}{k_1 - k_1' - i_0}\Delta(\mathbf{p}_\perp - \mathbf{p}_\perp'), \tag{4.6}$$

where $\Delta(\mathbf{p}_\perp - \mathbf{p}_\perp') = (2\pi)^2\frac{\delta(\mathbf{k}_\perp - \mathbf{k}_\perp')}{p - p'}$ and $\mathbf{k}_\perp = (k_2, k_3)$. Infinitesimal addition in the denominator of $\langle \mathbf{p}|\theta(x)\mathbf{p}'|\rangle$ takes into account the possible complex-valued character of the variable k_1. Its sign is determined by the radiation condition so that $\exp[i(k_1 + i_0)x] \to 0$ when $x \to \infty$.

An operator product of the multipliers gives for (4.3)

$$\left\langle \mathbf{p} \left| \theta(x)\hat{D}_1 \frac{\delta\left(t - t' - \frac{1}{v_1}\sqrt{(x+x')^2 + (y-y')^2 + (z-z')^2}\right)}{\sqrt{(x+x')^2 + (y-y')^2 + (z-z')^2}} \theta(x') \right| \mathbf{p}' \right\rangle_{ij}$$

$$= \frac{-2\pi v_1}{\varphi_1}\frac{v_1^{\,2} s_i s_j + p^2\delta_{ij}}{(\varphi_1 + iv_1k_1)(\varphi_1 - iv_1k_1')}\Delta(\mathbf{p}_\perp - \mathbf{p}_\perp'), \tag{4.7}$$

where $s_1 = i\frac{\varphi_1}{v_1}$, $s_2 = k_2$, $s_3 = k_3$, $\varphi_1 = \sqrt{p^2 + v_1^2 k_\perp^2}$ and $\mathrm{Re}\,\varphi_1 > 0$.

4.1.1 The Resolvent for the Inner Problem

The impulse representation for the electric part of the kernel (1.68) is given by the formula

$$\left\langle \mathbf{p}\left|\hat{K}_e\right|\mathbf{p}'\right\rangle_{ij} = \frac{\varepsilon_1 - \varepsilon}{\varepsilon}\frac{v^2 k_i k_j + p^2\delta_{ij}}{p^2 + v^2k^2}\Delta(\mathbf{p}-\mathbf{p}'), \tag{4.8}$$

where $\Delta(\mathbf{p}-\mathbf{p}') = (2\pi)^3\frac{\delta(\mathbf{k}-\mathbf{k}')}{p-p'}$ and it is assumed that

$$\frac{1}{\varepsilon_0\varepsilon}(\hat{P}_1 - \hat{P}_{\text{ex}}) = \frac{\varepsilon_1 - \varepsilon}{\varepsilon}.$$

Using the formula (4.5), we find the impulse representation for the kernel for half-restricted space

$$\left\langle \mathbf{p}\,|\,\theta(x)\hat{K}_e\theta(x')\,|\,\mathbf{p}'\right\rangle_{ij} = \frac{\varepsilon_1 - \varepsilon}{\varepsilon}\left[\frac{v^2k_ik_j + p^2\delta_{ij}}{p^2 + v^2k^2}\frac{i}{k_1 - k_1' - i_0}\right.$$
$$\left. + \frac{v}{2\varphi}\frac{v^2\xi_i\xi_j + p^2\delta_{ij}}{(\varphi - ivk_1)(\varphi - ivk_1')}\right]\Delta(\mathbf{p}_\perp - \mathbf{p}_\perp'), \tag{4.9}$$

where $\xi_1 = -i\frac{\varphi}{v}$, $\xi_2 = k_2$, $\xi_3 = k_3$, $\varphi = \sqrt{p^2 + v^2k_\perp^2}$ and $\mathrm{Re}\,\varphi > 0$.

Comparison of (4.9) with the equivalent 1D problem (Sections 3.1.9 and 3.1.10) shows that the assumed additional term determined by the influence of the medium boundary is defined by the expression with the denominator conjugate to that of the second term in (4.9).

It is convenient to use matrix notation, making Eq. (4.9) take the form

$$\left\langle \mathbf{p}\left|\theta(x)\hat{K}_e\theta(x')\right|\mathbf{p}'\right\rangle = \frac{\varepsilon_1 - \varepsilon}{\varepsilon}\left[\frac{v^2Q + p^2I}{p^2 + v^2k^2}\frac{i}{k_1 - k_1' - i0}\right.$$
$$\left. + \frac{v}{2\varphi}\frac{v^2Y + p^2I}{(\varphi - ivk_1)(\varphi - ivk_1')}\right]\Delta(\mathbf{p}_\perp - \mathbf{p}_\perp'). \tag{4.10}$$

Here, the block matrices are introduced as follows:

$$Q = \begin{pmatrix} k_1^{\ 2} & k_1\mathbf{k}_\perp \\ k_1\tilde{\mathbf{k}}_\perp & \hat{k}_\perp \end{pmatrix}, \quad I = \begin{pmatrix} 1 & 0 \\ 0 & \overline{I} \end{pmatrix}, \quad \overline{I} = \begin{pmatrix} 1 & 0 \\ 0 & 1 \end{pmatrix},$$

$$\hat{k}_\perp = \begin{pmatrix} k_2^2 & k_2k_3 \\ k_3k_2 & k_3^{\ 2} \end{pmatrix}, \quad \mathbf{k}_\perp = (k_2, k_3), \quad \tilde{\mathbf{k}}_\perp = \begin{pmatrix} k_2 \\ k_3 \end{pmatrix}, \tag{4.11}$$

$$Y_a = \begin{pmatrix} -\dfrac{\varphi_a^{\ 2}}{v_a^{\ 2}} & -i\dfrac{\varphi_a}{v_a}\mathbf{k}_\perp \\ -i\dfrac{\varphi_a}{v_a}\tilde{\mathbf{k}}_\perp & \hat{k}_\perp \end{pmatrix}, \tag{4.12}$$

where $\varphi_a = \sqrt{p^2 + v_a^2 k_\perp^2}$.

Then the assumed expression for the resolvent with the additional term takes the form

$$\langle \mathbf{p} | \theta(x)\hat{R}_e\theta(x') | \mathbf{p}' \rangle = \langle \mathbf{p} | \hat{R}_e^{(1)} | \mathbf{p}' \rangle + \langle \mathbf{p} | \hat{R}_e^{(2)} | \mathbf{p}' \rangle \tag{4.13}$$

$$\langle \mathbf{p} | \hat{R}_e^{(1)} | \mathbf{p}' \rangle = \frac{\varepsilon_1 - \varepsilon}{\varepsilon_1} \frac{v_1^2 Q + p^2 I}{p^2 + v_1^{\ 2}k^2} \frac{i}{k_1 - k_1' - i0} \Delta(\mathbf{p}_\perp - \mathbf{p}_\perp'), \tag{4.14}$$

$$\langle \mathbf{p} | \hat{R}_e^{(2)} | \mathbf{p}' \rangle = \frac{\varepsilon_1 - \varepsilon}{\varepsilon_1} \frac{v_1}{2\varphi_1} \left\{ \frac{v_1^{\ 2}Y_1 + p^2 I}{\varphi_1 - iv_1k_1} + \frac{X}{\varphi_1 + iv_1k_1} \right\} \frac{1}{\varphi_1 - iv_1k_1'} \Delta(\mathbf{p}_\perp - \mathbf{p}_\perp'), \tag{4.15}$$

where X is an unknown matrix. Suppose that all singularities of this matrix with respect to p are located in the left half-plane of this complex variable and that the X matrix does not depend on k_1. Then the product $\hat{K}\hat{R}$ in Eq. (4.1) can be calculated explicitly.

Calculation of this product shows that the first part of the resolvent (4.14) takes into account the influence of the boundary implicitly. It means that the magnitude of the integral in the product $\hat{K}\hat{R}$ is determined by the singular points of the first multiplier. However, only this part of the resolvent $\hat{R}_e^{(1)}$ cancels the free term of the equation, i.e., the cancellation theorem of Ewald–Oseen [1] is satisfied:

$$\left\langle \mathbf{p} | \hat{R}_e^{(1)} | \mathbf{p}' \right\rangle + \int d\mathbf{p}'' \left\langle \mathbf{p} | \theta(x) \hat{K}_e \theta(x') | \mathbf{p}'' \right\rangle \left\langle \mathbf{p}'' | \hat{R}_e^{(1)} | \mathbf{p}' \right\rangle \Big|_{\text{in the } \hat{K} \text{ pole}}$$
$$= \left\langle \mathbf{p} | \theta(x) \hat{K}_e \theta(x') | \mathbf{p}' \right\rangle \tag{4.16}$$

Note that coupling of the multipliers in the integrals of the product $\hat{K}\hat{R}$, caused by the singularities with respect to the variable k_1, takes place only if they are located on different sides of the real axis. Every multiplier in the product $\hat{K}\hat{R}^{(1)}$ has one singularity but only the singularity of the first multiplier contributes to the cancellation theorem. It determines the passivity of the first part of the resolvent $\hat{R}_e^{(1)}$. The explicit influence of the boundary is determined by the second part of the resolvent $\hat{R}_e^{(2)}$.

The part of Eq. (4.1) that remains after the relation (4.16) is satisfied allows one to obtain the X matrix. Assuming its structure as $X = v_1^2 Z + p^2 w_e I$, the equation transforms to the matrix form:

$$v_1^2\left(\frac{p^2}{v\varphi_1 - v_1\varphi} I + \frac{1}{\varphi_1 + iv_1k_1} U\right) Z + p^4\left[1 + \frac{v\varphi_1 + v_1\varphi}{v\varphi_1 - v_1\varphi} w_e\right] I + \frac{p^2}{\varphi_1 + iv_1k_1}$$
$$w_e U = \begin{pmatrix} p^2\varphi_1^2 + v_1\varphi(v\varphi_1 + v_1\varphi)k_\perp^2 & -iv\varphi_1(\varphi\varphi_1 + vv_1k_\perp^2)\mathbf{k}_\perp \\ iv_1\varphi(\varphi\varphi_1 + vv_1k_\perp^2)\tilde{\mathbf{k}}_\perp & -[p^2 v_1{}^2 + v\varphi_1(v\varphi_1 + v_1\varphi)]\hat{\mathbf{k}}_\perp \end{pmatrix}, \tag{4.17}$$

where the matrix U is equal to

$$U = \begin{pmatrix} \varphi\varphi_1(\varphi + ivk_1) + i\varphi(v\varphi_1 + v_1\varphi)k_1 & iv\varphi(\varphi_1 - iv_1k_1)\mathbf{k}_\perp \\ iv\varphi(\varphi_1 - iv_1k_1)\tilde{\mathbf{k}}_\perp & v[v_1(\varphi + ivk_1) + v\varphi_1 + v_1\varphi]\hat{\mathbf{k}}_\perp \end{pmatrix}.$$

If the Z matrix is assumed to have the same structure as U

$$Z = \begin{pmatrix} g & b\mathbf{k}_\perp \\ a\tilde{\mathbf{k}}_\perp & f\hat{\mathbf{k}}_\perp \end{pmatrix}. \tag{4.18}$$

then the expression for the unknown coefficient follows from the condition that the term with the unit matrix in (4.17) is equal to zero,

$$w_e = \frac{v_1\varphi - v\varphi_1}{v_1\varphi + v\varphi_1}. \tag{4.19}$$

This coincides with the expression for the reflectance of the perpendicular polarized plane wave, if substituting p for $i\omega$ (ω is the wave frequency). Indeed, if ϑ and ϑ' are the incident and refracted angles, respectively, then $\cos\vartheta \to \frac{1}{k}\sqrt{\frac{\omega^2}{v^2} - k_\perp^2} \to \varphi$ and $\cos\vartheta' \to \varphi_1$. Then, from Fresnel's formulae, it follows for the case of the perpendicular polarization that

$$\frac{v_1 \cos\vartheta - v\cos\vartheta'}{v_1 \cos\vartheta + v\cos\vartheta'} = \frac{v_1\varphi - v\varphi_1}{v_1\varphi + v\varphi_1} = w_e.$$

The remaining part of Eq. (4.17) yields the expression for the Z matrix elements,

$$g = \frac{v^2 - v_1^2}{v_1^2}\varphi_1 \frac{\varphi p^2 - v_1 k_\perp^2 \psi_1}{\psi_1\psi_2},$$

$$b = -i\frac{v^2 - v_1^2}{v_1^2}\varphi_1 \frac{\varphi_1\psi_1 - vp^2}{\psi_1\psi_2} = -a,$$

$$f = -\frac{v^2 - v_1^2}{v_1^2}\frac{\varphi_1\psi_1 + vp^2}{\psi_1\psi_2},$$

where $\psi_1 = v\varphi_1 + v_1\varphi$, $\psi_2 = v\varphi + v_1\varphi_1$.

Substituting these elements into the matrix (4.18) gives the matrix

$$Z = \frac{w_m}{v_1^2}\begin{pmatrix} -v_1^2 k_\perp^2 & -i\varphi_1 v_1 \mathbf{k}_\perp \\ i\varphi_1 v_1 \mathbf{k}_\perp^* & -v_1^2 \hat{\mathbf{k}}_\perp \end{pmatrix} + \frac{v^2 - v_1^2}{v_1^2}\frac{p^2}{\psi_1\psi_2}\begin{pmatrix} \varphi\varphi_1 - vv_1 k_\perp^2 & 0 \\ 0 & -2vv_1\hat{\mathbf{k}}_\perp \end{pmatrix}. \tag{4.20}$$

Here, the coefficient

$$w_m = \frac{v\varphi - v_1\varphi_1}{v\varphi + v_1\varphi_1} \tag{4.21}$$

coincides with the reflectance of the parallel polarized plane wave if one substitutes p for $i\omega$ as above (ω is the wave frequency). Indeed, the same expression for the coefficient can be obtained from Fresnel's formulae for the case of parallel polarization

$$\frac{v\cos\vartheta - v_1\cos\vartheta'}{v\cos\vartheta + v_1\cos\vartheta'} = \frac{v\varphi - v_1\varphi_1}{v\varphi + v_1\varphi_1} = w_m.$$

If the unit matrix is represented as a sum

$$I = \begin{pmatrix} 1 & 0 \\ 0 & 0 \end{pmatrix} + I_{\perp}, \text{ where } I_{\perp} = \begin{pmatrix} 0 & 0 & 0 \\ 0 & 1 & 0 \\ 0 & 0 & 1 \end{pmatrix}, \tag{4.22}$$

then final expression for the matrix X takes the following form:

$$X = -v_1^2 w_m \begin{pmatrix} k_{\perp}^2 & i\dfrac{\varphi_1}{v_1}\mathbf{k}_{\perp} \\ -i\dfrac{\varphi_1}{v_1}\tilde{\mathbf{k}}_{\perp} & \hat{k}_{\perp} \end{pmatrix} + 2vv_1 w_{em} \begin{pmatrix} 0 & 0 \\ 0 & \hat{k}_{\perp} \end{pmatrix} + p^2 w_e I_{\perp}, \tag{4.23}$$

where the coefficient

$$w_{em} = \frac{v_1\varphi - v\varphi_1}{v\varphi + v_1\varphi_1} \tag{4.24}$$

is an analogue of some mixed reflectance.

Thus, the expressions (4.13), (4.15) and (4.23) completely determine the 3D resolvent for the problem with a dielectric. This resolvent takes into account not only standard directions of the field polarization, but also the intermediate polarization by virtue of the middle term in Eq. (4.23). The action of this term on the field results in the following:

$$\begin{pmatrix} 0 & 0 \\ 0 & \hat{k}_{\perp} \end{pmatrix} \begin{pmatrix} E_1 \\ \mathbf{E}_{\perp} \end{pmatrix} = \tilde{\mathbf{k}}_{\perp}(\mathbf{k}_{\perp}\mathbf{E}_{\perp}) \tag{4.25}$$

This action is non-zero, because the vector $\mathbf{k}_{\perp}$ is the variable of the Fourier transformation but not a wavenumber vector.

The resolvent for the problem with a plasma half-space characterized by a plasma frequency ω_e and restricted by a plane boundary can be easily obtained by substituting $1 + \omega_e^2/p^2$ for ε in all formulae above.

The resolvent for the problem with a dielectric dissipation can be obtained immediately from the resolvent for the dielectric constructed above, using an effective permittivity, i.e., by substituting $\varepsilon_1 \to \varepsilon_1\left(1 + \frac{\bar{\sigma}_1}{p}\right)$ where $\bar{\sigma}_1 = \frac{\sigma_1}{\varepsilon_0\varepsilon_1}$. The result is

$$\left\langle \mathbf{p} \middle| \theta(x)\hat{R}_{ej}\theta(x') \middle| \mathbf{p}' \right\rangle = \left\langle \mathbf{p} \middle| \hat{R}_{ej}^{(1)} \middle| \mathbf{p}' \right\rangle + \left\langle \mathbf{p} \middle| \hat{R}_{ej}^{(2)} \middle| \mathbf{p}' \right\rangle, \tag{4.26}$$

where

$$\left\langle \mathbf{p} \middle| \hat{R}_{ej}^{(1)} \middle| \mathbf{p}' \right\rangle = \frac{p(1-\varepsilon/\varepsilon_1)+\bar{\sigma}_1}{p+\bar{\sigma}_1} \times \frac{v_1^2 Q + p(p+\bar{\sigma}_1)I}{p(p+\bar{\sigma}_1)+v_1^2 k^2} \frac{i}{k_1 - k_1' - i0} \Delta(\mathbf{p}_\perp - \mathbf{p}_\perp'), \quad (4.27)$$

$$\left\langle \mathbf{p} \middle| \hat{R}_{ej}^{(2)} \middle| \mathbf{p}' \right\rangle = \frac{p(1-\varepsilon/\varepsilon_1)+\bar{\sigma}_1}{p+\bar{\sigma}_1} \times \frac{v_1}{2\xi_1} \left\{ \frac{v_1^2 L_1 + p(p+\bar{\sigma}_1)I}{\xi_1 - iv_1 k_1} + \frac{X'}{\xi_1 + iv_1 k_1} \right\} \frac{1}{\xi_1 - iv_1 k_1'} \Delta(\mathbf{p}_\perp - \mathbf{p}_\perp'). \quad (4.28)$$

Here

$$X' = -v_1^2 w_{mj} \begin{pmatrix} k_\perp^2 & i\frac{\xi_1}{v_1}\mathbf{k}_\perp \\ -i\frac{\xi_1}{v_1}\tilde{\mathbf{k}}_\perp & \hat{k}_\perp \end{pmatrix} + 2vv_1 w_{emj} \begin{pmatrix} 0 & 0 \\ 0 & \hat{k}_\perp \end{pmatrix} + p(p+\bar{\sigma}_1) w_{ej} I_\perp, \quad (4.29)$$

$$L_1 = \begin{pmatrix} -\frac{\xi_1^2}{v_1^2} & -i\frac{\xi_1}{v_1}\mathbf{k}_\perp \\ -i\frac{\xi_1}{v_1}\tilde{\mathbf{k}}_\perp & \hat{k}_\perp \end{pmatrix} \quad \text{and} \quad \xi_1 = \sqrt{p(p+\bar{\sigma}_1)+v_1^2 k_\perp^2}. \quad (4.30)$$

Analogously, the reflectances have the forms

$$w_{mj} = \frac{v\varphi(p+\bar{\sigma}_1) - v_1 p\xi_1}{v\varphi(p+\bar{\sigma}_1) + v_1 p\xi_1}, \quad (4.31)$$

$$w_{emj} = (p+\bar{\sigma}_1)\frac{v_1\varphi - v\xi_1}{v(p+\bar{\sigma}_1)\varphi + v_1 p\xi_1}, \quad (4.32)$$

$$w_{ej} = \frac{v_1\varphi - v\xi_1}{v_1\varphi + v\xi_1}. \quad (4.33)$$

We now find the coordinate representation of the resolvent by virtue of the formula

$$\langle \mathbf{x}|\hat{R}|\mathbf{x}'\rangle = \int d\mathbf{p} \int d\mathbf{p}' \langle \mathbf{x}\,|\,\mathbf{p}\rangle \langle \mathbf{p}|\hat{R}|\mathbf{p}'\rangle \langle \mathbf{p}'\,|\,\mathbf{x}\rangle \tag{4.34}$$

Integration with respect to variables $\mathbf{p}'$ and k_1 is performed completely. In the first part of the resolvent (4.14), the integral with respect to these variables results in the same expression as for the infinite medium problem. The boundary is taken into account in this expression stating formally the multiplier equal to the characteristic function $\theta(x')$:

$$\int \frac{dk_1}{2\pi}\frac{dp'}{2\pi i}\frac{d\mathbf{k}'}{(2\pi)^3} e^{ik_1x - p't' - i\mathbf{k}'\mathbf{r}'} \frac{1}{p^2 + v_1^2k^2} \frac{i}{k_1 - k_1' - i0} \Delta(\mathbf{p}-\mathbf{p}')$$

$$= -\frac{1}{2v_1\varphi_1} e^{-pt' - i\mathbf{k}_\perp \mathbf{r}'_\perp - \frac{\varphi_1}{v_1}|x-x'|} \theta(x') \tag{4.35}$$

The integral of the first term of the second part in Eq. (4.15), $\frac{1}{2\varphi_1}\frac{v_1^2 Y_1 + p^2 I}{\varphi_1 - iv_1k_1}$, is of a zero contribution to the resolvent, because the real part of the root φ_1 is fixed, $\mathrm{Re}\,\varphi_1 > 0$:

$$\int \frac{dk_1}{2\pi} e^{ik_1x} \frac{1}{\varphi_1 - iv_1k_1} = 0, \tag{4.36}$$

The second term of this part takes into account the boundary explicitly with the reflection coefficients. Hence, the integral has the form of a wave, propagating away from the boundary,

$$\int \frac{dk_1}{2\pi}\frac{dp'}{2\pi i}\frac{d\mathbf{k}'}{(2\pi)^3} e^{ik_1x - p't' - i\mathbf{k}'\mathbf{r}'} \frac{1}{\varphi_1 + iv_1k_1} \frac{1}{\varphi_1 - iv_1k_1'} \Delta(\mathbf{p}-\mathbf{p}')$$

$$= \theta(x)\frac{1}{v_1^2} e^{-pt' - i\mathbf{k}_\perp \mathbf{r}'_\perp - \frac{\varphi_1}{v_1}(x+x')} \theta(x') \tag{4.37}$$

Using the formulae

$$\int \frac{dp}{2\pi i} e^{p(t-t') - \varphi_1\frac{|x-x'|}{v_1}} \frac{1}{\varphi_1} = J_0(k_\perp \sqrt{v_1^2(t-t')^2 - |x-x'|^2})\,\theta\!\left(t - t' - \frac{|x-x'|}{v_1}\right),$$

and $\int \frac{d\alpha}{2\pi} e^{ik_\perp|\mathbf{r}_\perp - \mathbf{r}'_\perp|\cos\alpha} = J_0(k_\perp |\mathbf{r}_\perp - \mathbf{r}'_\perp|)$,

the first part of the resolvent (4.14) takes the form

$$\left\langle \mathbf{x} \middle| \hat{R}_e^{(2)} \middle| \mathbf{x}' \right\rangle = \frac{1}{2v_1}\left(v_1^2 \hat{Q} - \frac{\partial^2}{\partial t^2}\hat{I} \right) W_0, \tag{4.38}$$

where $\hat{Q} = \begin{pmatrix} \frac{\partial^2}{\partial x^2} & \frac{\partial}{\partial x}\nabla_\perp \\ \frac{\partial}{\partial x}\tilde{\nabla}_\perp & \tilde{\nabla}_\perp \nabla_\perp \end{pmatrix}$, $\nabla_\perp = \left(\frac{\partial}{\partial y}, \frac{\partial}{\partial z} \right)$, $\tilde{\nabla}_\perp = \begin{pmatrix} \frac{\partial}{\partial y} \\ \frac{\partial}{\partial z} \end{pmatrix}$.

and the function W_0 can take one of the following forms:

$$W_0 = \frac{v_1^2 - v^2}{v^2} \int d\mathbf{p}_\perp e^{p(t-t') + i\mathbf{k}_\perp(\mathbf{r}_\perp - \mathbf{r}') - \varphi_1 \frac{|x-x'|}{v_1}} \frac{1}{\varphi_1}\theta(x'), \tag{4.39}$$

$$W_0 = \frac{v_1^2 - v^2}{v^2} \int \frac{d\mathbf{k}_\perp}{(2\pi)^2} e^{i\mathbf{k}_\perp(\mathbf{r}_\perp - \mathbf{r}'_\perp)} J_0(k_\perp \sqrt{v_1^2(t-t')^2 - |x-x'|^2})$$
$$\times \theta\left(t - t' - \frac{|x-x'|}{v_1} \right) \theta(x') \tag{4.40}$$

$$W_0 = \frac{v_1^2 - v^2}{v^2} \int_0^\infty \frac{dk_\perp}{2\pi} k_\perp J_0(k_\perp |\mathbf{r}_\perp - \mathbf{r}'_\perp|)$$
$$\times J_0(k_\perp \sqrt{v_1^2(t-t')^2 - |x-x'|^2})\theta\left(t - t' - \frac{|x-x'|}{v_1} \right)\theta(x'), \tag{4.41}$$

$$W_0 = \frac{v_1^2 - v^2}{v^2} \frac{\delta(|\mathbf{r} - \mathbf{r}'| - v_1(t-t'))}{2\pi|\mathbf{r} - \mathbf{r}'|}. \tag{4.42}$$

The transition from one form to the other can be performed easily. For example, the integral of the product of the Bessel's functions (4.41), which is equal to the following:

$$W_0 = \frac{v_1^2 - v^2}{v^2} \frac{\delta(|\mathbf{r}_\perp - \mathbf{r}'_\perp| - \sqrt{v_1^2(t-t')^2 - (x-x')^2})}{2\pi\sqrt{v_1^2(t-t')^2 - (x-x')^2}} \theta\left(t - t' - \frac{|x-x'|}{v_1} \right),$$

proves the identity

$$\frac{\delta(|\mathbf{r}_\perp-\mathbf{r}_\perp'|-\sqrt{v_1^2(t-t')^2-(x-x')^2})}{\sqrt{v_1^2(t-t')^2-(x-x')^2}}\theta\left(t-t'-\frac{|x-x'|}{v_1}\right)=\frac{\delta(|\mathbf{r}-\mathbf{r}'|-v_1(t-t'))}{4\pi v_1|\mathbf{r}-\mathbf{r}'|},$$

which can be also obtained applying the delta-function properties.

The coordinate representation of the second part of the resolvent can be finally represented as an expansion with respect to the polarization,

$$\langle\mathbf{x}|\hat{R}_e^{(2)}|\mathbf{x}'\rangle=\frac{1}{2v_1}\left\{v_1^2\begin{pmatrix}\nabla_\perp^2 & \frac{\partial}{\partial x}\nabla_\perp\\ -\frac{\partial}{\partial x}\tilde{\nabla}_\perp & \tilde{\nabla}_\perp\nabla_\perp\end{pmatrix}\hat{W}_m-2vv_1\begin{pmatrix}0 & 0\\ 0 & \tilde{\nabla}_\perp\nabla_\perp\end{pmatrix}\hat{W}_{em}+\frac{\partial^2}{\partial t^2}\hat{I}_\perp\hat{W}_e\right\},\tag{4.43}$$

where the reflectance operators are equal to

$$\hat{W}_a=\frac{v_1^2-v^2}{v^2}\int d\mathbf{p}_\perp e^{p(t-t')+i\mathbf{k}_\perp(\mathbf{r}_\perp-\mathbf{r}')-\varphi_1\frac{x+x'}{v_1}}\frac{w_a}{\varphi_1}\theta(x')$$

$$=\frac{v_1^2-v^2}{v^2}\int_0^\infty\frac{dp}{2\pi i}\frac{dk_\perp}{2\pi}k_\perp\frac{w_a}{\varphi_1}J_0(k_\perp|\mathbf{r}_\perp-\mathbf{r}_\perp'|)e^{p(t-t')-\varphi_1\frac{x+x'}{v_1}}\theta(x').\tag{4.44}$$

Evaluations of this expression are given in more detail below:

$$\hat{W}_a=\frac{v_1^2-v^2}{v^2}\int_{-i\infty}^{i\infty}\frac{dp}{2\pi i}\int_0^\infty\frac{dk}{2\pi}w_a\frac{k}{\varphi_1}J_0(kR_\perp)e^{pT-\varphi_1\frac{x+x'}{v_1}}\theta(x')$$

$$=\frac{v_1^2-v^2}{v^2}\int_0^\infty\frac{dk}{2\pi}kJ_0(kR_\perp)\int_0^T dT'\tilde{w}_a(T',k)\int_{-i\infty}^{i\infty}\frac{dp}{2\pi i}\frac{1}{\varphi_1}e^{p(T-T')-\varphi_1\frac{x+x'}{v_1}}\theta(x')$$

$$=\frac{v_1^2-v^2}{v^2}\int_0^\infty\frac{dk}{2\pi}kJ_0(kR_\perp)\int_0^T dT'\tilde{w}_a(T',k)J_0\left(k\sqrt{v_1^2(T-T')^2-(x+x')^2}\right)$$

$$=\theta\left(T-T'-\frac{x+x'}{v_1}\right)\theta(x'),$$

where $\tilde{w}_a(T',k)=\int_{-i\infty}^{i\infty}\frac{dp}{2\pi i}w_a(p,k)e^{pT'}$.

The coordinate representation of the resolvent for a dissipative dielectric has the form

$$\left\langle \mathbf{x} \middle| \hat{R}_{ej}^{(1)} \middle| \mathbf{x}' \right\rangle = -\int d\mathbf{p}_\perp \frac{p(1-\varepsilon/\varepsilon_1)+\bar{\sigma}_1}{p+\bar{\sigma}_1} \frac{v_1^2 \hat{Q}' + p(p+\bar{\sigma}_1)I}{2v_1\xi_1}$$

$$= e^{p(t-t') + i\mathbf{k}_\perp(\mathbf{r}_\perp - \mathbf{r}'_\perp) - \xi_1 \frac{|x-x'|}{v_1}} \theta(x') \tag{4.45}$$

$$\left\langle \mathbf{x} \middle| \hat{R}_{ej}^{(2)} \middle| \mathbf{x}' \right\rangle = \int d\mathbf{p}_\perp \frac{p(1-\varepsilon/\varepsilon_1)+\bar{\sigma}_1}{p+\bar{\sigma}_1} \frac{e^{p(t-t') + i\mathbf{k}_\perp(\mathbf{r}_\perp - \mathbf{r}'_\perp) - \xi_1 \frac{x+x'}{v_1}}}{2v_1\xi_1}$$

$$\times \left\{ -v_1^2 w_{mj} \begin{pmatrix} k_\perp^2 & i\frac{\xi_1}{v_1}\mathbf{k}_\perp \\ -i\frac{\xi_1}{v_1}\tilde{\mathbf{k}}_\perp & \hat{k}_\perp \end{pmatrix} + 2vv_1 w_{emj} \begin{pmatrix} 0 & 0 \\ 0 & \hat{k}_\perp \end{pmatrix} + p(p+\bar{\sigma}_1) w_{ej} I_\perp \right\} \theta(x')$$

$$\tag{4.46}$$

where $\hat{Q}' = \begin{pmatrix} -\frac{\partial^2}{\partial x^2} & -i\frac{\partial}{\partial x}\mathbf{k}_\perp \\ -i\frac{\partial}{\partial x}\tilde{\mathbf{k}}_\perp & \hat{k}_\perp \end{pmatrix}$.

4.1.2 The Resolvent for the External Problem

According to the general relation stated in Section 1.3.5, the field beyond the region $V(t)$ is equal to

$$\mathbf{E}_{\text{ex}} = \mathbf{F}_{\text{ex}} + \hat{N}\mathbf{F}_{\text{in}}, \tag{4.47}$$

where the reflection operator (the external resolvent) has the form

$$\hat{N} = \hat{K}_{\text{ex}}(\hat{I} + \hat{R}), \tag{4.48}$$

and $\mathbf{F}_{\text{ex,in}}$ is the free term of the integral equation outside and inside V, respectively.

To obtain an explicit form of this operator, we use the coordinate representation

$$\left\langle \mathbf{x} \middle| \hat{K}_{\text{ex}} \middle| \mathbf{x}' \right\rangle = \frac{1}{4\pi}\theta(-x)\frac{\varepsilon_1 - \varepsilon}{\varepsilon}\left(\nabla\nabla - \frac{1}{v^2}\frac{\partial^2}{\partial t^2}\hat{I}\right)\frac{\delta\left(t - t' - \frac{|\mathbf{r}-\mathbf{r}'|}{v}\right)}{|\mathbf{r}-\mathbf{r}'|}\theta(x'). \tag{4.49}$$

and the mixed representation of the resolvent following from Eqs. (4.14), (4.15), (4.38), and (4.43),

$$\left\langle \mathbf{x}\left|\hat{R}_e\right|\mathbf{x}'\right\rangle = \int d\mathbf{p} e^{pt+i\mathbf{kr}}\left\langle \mathbf{p}\left|\hat{R}_e\right|\mathbf{x}'\right\rangle. \tag{4.50}$$

where

$$\left\langle \mathbf{p}\left|\hat{R}_e\right|\mathbf{x}'\right\rangle = \frac{\varepsilon_1-\varepsilon}{\varepsilon_1}\left[-\frac{v_1^2 Q+p^2 I}{p^2+v_1^2k^2}e^{-ik_1x'}+\frac{1}{2\varphi_1}\left\{\frac{v_1^2Y_1+p^2I}{\varphi_1-iv_1k_1}+\frac{X}{\varphi_1+iv_1k_1}\right\}e^{-\frac{\varphi_1}{v_1}x'}\right]e^{-pt'-i\mathbf{k}_\perp\mathbf{r}'_\perp} \tag{4.51}$$

Then the matrix elements of the product $\left\langle \mathbf{x}\,|\,\hat{K}_{ex}\hat{R}_e\,|\,\mathbf{x}'\right\rangle$ are determined by the formula

$$\left\langle \mathbf{x}\,|\,\hat{K}_{ex}\hat{R}_e\,|\,\mathbf{x}'\right\rangle = \theta(-x)\frac{\varepsilon_1-\varepsilon}{4\pi\varepsilon}\hat{D}\int d\mathbf{x}'' f(\mathbf{x}-\mathbf{x}'')\theta(x'')\int d\mathbf{p}e^{pt+i\mathbf{kr}}\left\langle \mathbf{p}\left|\hat{R}_e\right|\mathbf{x}'\right\rangle, \tag{4.52}$$

where $\hat{D}=\left(\nabla\nabla-\frac{1}{v^2}\frac{\partial^2}{\partial t^2}\hat{I}\right)$, $f(\mathbf{x}-\mathbf{x}'')=\dfrac{\delta\left(t-t''-\frac{|\mathbf{r}-\mathbf{r}''|}{v}\right)}{|\mathbf{r}-\mathbf{r}''|}$.

After substituting (4.50) into (4.52) and executing the integrals with respect to the variables t'', $\mathbf{r}''$ and k_1, the product takes the following form:

$$\left\langle \mathbf{x}\,|\,\hat{K}_{ex}\hat{R}_e\,|\,\mathbf{x}'\right\rangle = \theta(-x)\frac{(v_1^2-v^2)^2}{v_1^2}\hat{D}\int d\mathbf{p}_\perp\frac{1}{2\varphi}\left[v\frac{v_1^2Y+p^2I}{(v_1^{\,2}-v^2)p^2}e^{\frac{\varphi}{v}(x-x')}\right.$$
$$\left.+\frac{1}{2\varphi_1}\left\{\frac{v_1^2Y_1+p^2I}{v\varphi_1-v_1\varphi}+\frac{X}{v\varphi_1+v_1\varphi}\right\}e^{\frac{\varphi}{v}x-\frac{\varphi_1}{v_1}x'}\right]e^{p(t-t')+i\mathbf{k}_\perp(\mathbf{r}_\perp-\mathbf{r}'_\perp)}\theta(x') \tag{4.53}$$

The action of the operator $\hat{D}$ on the integrand in (4.53) gives the transformation of this operator into the matrix $\hat{D}\to-(Y+\frac{p^2}{v^2}I)$. After integrating the first item in the square brackets in (4.53) and taking into account that $(Y+\frac{p^2}{v^2}I)Y=0$, we obtain

$$\frac{(v_1^2-v^2)^2}{v_1^2}\hat{D}\int d\mathbf{p}_\perp\frac{v}{2\varphi}\frac{v_1^2Y+p^2I}{(v_1^2-v^2)p^2}e^{\frac{\varphi}{v}(x-x')p(t-t')+i\mathbf{k}_\perp(\mathbf{r}_\perp-\mathbf{r}'_\perp)}\theta(x') = -\left\langle \mathbf{x}\left|\hat{K}_{ex}\right|\mathbf{x}'\right\rangle$$

Executing the integral of the term in the braces in (4.53) yields

$$\left\langle \mathbf{x}\left|\hat{K}_{\text{ex}}\hat{R}_{e}\right|\mathbf{x}'\right\rangle = -\left\langle \mathbf{x}\left|\hat{K}_{\text{ex}}\right|\mathbf{x}'\right\rangle + \theta(-x)\frac{v_1^2 - v^2}{vv_1}\int d\mathbf{p}_{\perp}$$

$$\left\{\frac{vv_1}{v\varphi + v_1\varphi_1}\begin{pmatrix} -k_{\perp}^2 & -i\dfrac{\varphi_1}{v_1}\mathbf{k}_{\perp} \\ -i\dfrac{\varphi}{v}\tilde{\mathbf{k}}_{\perp} & \hat{k}_{\perp} \end{pmatrix} + \frac{p^2}{v\varphi_1 + v_1\varphi}I_{\perp}\right\}$$

$$e^{p(t-t') + \frac{\varphi}{v}x - \frac{\varphi_1}{v_1}x' + i\mathbf{k}_{\perp}(\mathbf{r}_{\perp} - \mathbf{r}'_{\perp})}\theta(x') \tag{4.54}$$

Substituting (4.54) into (4.48) gives the reflectance operator

$$\left\langle \mathbf{x}\left|\hat{N}\right|\mathbf{x}'\right\rangle = \theta(-x)\frac{v_1^2 - v^2}{v^2 v_1}\int d\mathbf{p}_{\perp}\frac{1}{2\varphi_1}\{vv_1 u_m P + p^2 u_e I_{\perp}\}e^{p(t-t') + \frac{\varphi}{v}x - \frac{\varphi_1}{v_1}x' + i\mathbf{k}_{\perp}(\mathbf{r}_{\perp} - \mathbf{r}'_{\perp})}\theta(x') \tag{4.55}$$

The matrix in this expression has the form

$$P = \begin{pmatrix} -k_{\perp}^2 & -i\dfrac{\varphi_1}{v_1}\mathbf{k}_{\perp} \\ -i\dfrac{\varphi}{v}\tilde{\mathbf{k}}_{\perp} & \hat{k}_{\perp} \end{pmatrix}. \tag{4.56}$$

The coefficients

$$u_m = \frac{2v_1\varphi}{v\varphi + v_1\varphi_1}, \quad u_e = \frac{2v_1\varphi}{v\varphi_1 + v_1\varphi} \tag{4.57}$$

are the analogues for a plane wave transmittance through a plane boundary (Fresnel's coefficients) in the case of the parallel and perpendicular polarization, respectively. The two terms in the braces in Eq. (4.55) yield the expansion of the external field onto purely lateral and mixed components. Indeed, if the field is expressed as $\boldsymbol{F}_{\text{in}} = (F_1, \mathbf{F}_{\perp})$ we have

$$\{vv_1 u_m P + p^2 u_e I_{\perp}\}F_{\text{in}}$$

$$= vv_1 u_m\left[\begin{pmatrix} -k_{\perp}^2 F_1 - i\dfrac{\varphi_1}{v_1}(\mathbf{k}_{\perp}\mathbf{F}_{\perp}) \\ 0 \\ 0 \end{pmatrix} + \mathbf{k}_{\perp}\left(-i\frac{\varphi_1}{v_1}F_1 + -(\mathbf{k}_{\perp}\mathbf{F}_{\perp})\right)\right] + p^2 u_e \mathbf{F}_{\perp}. \tag{4.58}$$

It follows from this, that the coefficient u_e characterizes the scattering of the lateral component only while u_m characterizes the formation of the longitudinal as well as lateral components and their intermixing.

Changing the matrix p for the operator form

$$\hat{P} = \begin{pmatrix} -k_\perp^2 & -\frac{\varphi_1}{v_1}\nabla_\perp \\ -\frac{\varphi}{v}\tilde{\nabla}_\perp & -\tilde{\nabla}_\perp\nabla_\perp \end{pmatrix}. \tag{4.59}$$

in Eq. (4.55) allows one to execute the integrals with respect to the angle variables. As a result, we finally obtain this operator in the form of a double integral,

$$\langle \mathbf{x}|\hat{N}|\mathbf{x}'\rangle = \theta(-x)\frac{v_1^2 - v^2}{v^2 v_1}\int_{-i\infty}^{i\infty}\frac{dp}{2\pi i}\int_0^\infty \frac{dk_\perp}{2\pi}\frac{1}{2\varphi_1}\{vv_1 u_m \hat{P} + p^2 u_e I_\perp\}$$
$$\times J_0(k_\perp|\mathbf{r}_\perp - \mathbf{r}'_\perp|)e^{p(t-t')+\frac{\varphi}{v}x-\frac{\varphi_1}{v_1}x'}\theta(x'). \tag{4.60}$$

The integrands in Eqs. (4.55) and (4.60) on the p plane (this particular plane is important for the field time behavior) have singularities as the branch points $\pm i\frac{k_\perp}{v_1}$ and $\pm i\frac{k_\perp}{v}$ which are located only on the imaginary axis. From this, it follows that the integral is non-zero only in the region $t - t' + \frac{x}{v} - \frac{x'}{v_1} > 0$. It is conditioned that the reflection operator satisfies the causality principle.

The resolvents and reflection operator constructed allow investigation of the transformation of an arbitrary initial field in the case when the permittivity changes in time continuously in the half-space bounded by the plane boundary. The investigation can be carried out using the procedure described in the next section, where once again the continuous dependence is approximated by a sequence of step functions.

4.2 Fresnel's Formulae in Time-Domain for a Plane Interface between Two Dielectrics

A key problem in electromagnetic waves scattering is the reflection/transmission of plane harmonic waves by/through a plane boundary between two media. The solution to this problem can be used for the analysis of more complicated problems. One such problem, significant for applications and for understanding transient phenomena, is an electromagnetic wave transformation caused by the permittivity and conductivity time changes in one of the two half-spaces. In this section, the operators will be obtained for the description of such phenomena using the resolvent obtained above for the case of a plane boundary between two media.

It is well known that the interaction of a harmonic plane wave with a plane boundary between two media is described by Fresnel's formulae in the frequency domain. For complex time-spatial electromagnetic signals, their transmission and reflection might be very complicated and its study would be convenient to do by Fresnel's formulae in a time-spatial domain. Such formulae, and corresponding integral operators, are derived from the Volterra integral equation method in this section.

Let us consider a typical problem when an initial electromagnetic field is generated by an arbitrary source in a space divided by a plane boundary that coincides with the plane $x = 0$ of the coordinates system. The medium with the permittivity ε' is placed to the right of this interface, $x > 0$. To the left of the interface there is a medium with the permittivity ε, as shown in Fig. 4.2. $\mathbf{E}_0'$ denotes the field in the right half-space and $\mathbf{E}_0$ the field in the left one.

We consider the key problem when the medium properties in the right half-space undergo a jump at zero moment of time. The new medium parameters are the relative permittivity ε_1 and the conductivity σ_1. $\mathbf{E}$ denotes the field in the right non-stationary half-space after the change of the medium parameters (a transformed field) and $\mathbf{E}_r$ is the field in the left half-space in which the medium properties are constant (a reflected field) (see Fig. 4.2). If the medium is homogeneous in the entire space before the zero moment of time, that is $\varepsilon' = \varepsilon$, then $\mathbf{E}_0' = \mathbf{E}_0$ and we have a problem about a field transformation caused by the newly formed boundary between two media. We shall proceed further from this statement of the problem because it is possible to come to the problem with an arbitrary change time of medium properties merely by translation of the time counting point.

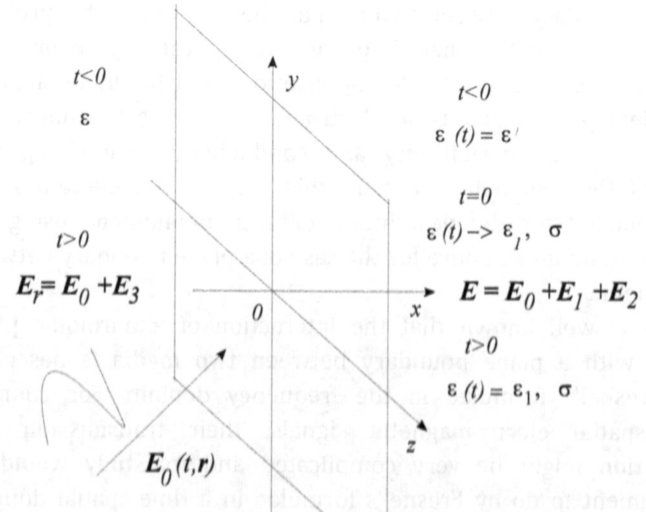

Figure 4.2 The geometry of the problem.

The transmitted and the reflected signals are determined by the resolvent operators and can be represented as follows:

$$\mathbf{E} = \mathbf{E}_0 + \hat{R}\mathbf{E}_0 = \mathbf{E}_0 + \mathbf{E}_1 + \mathbf{E}_2, \quad \mathbf{E}_r = \mathbf{E}_0 + \hat{N}\mathbf{E}_0 = \mathbf{E}_0 + \mathbf{E}_3 \tag{4.61}$$

The resolvent operator which determines the field in the non-stationary half-space $x > 0$ is given by the integrals

$$\hat{R}_{ej} = \theta(x)\int_0^\infty dt' \int_{x'>0} d\mathbf{r}'\left\langle \mathbf{x}\middle|\hat{R}_{ej}^{(1)}\middle|\mathbf{x}'\right\rangle + \theta(x)\int_0^\infty dt' \int_{x'>0} d\mathbf{r}'\left\langle \mathbf{x}\middle|\hat{R}_{ej}^{(2)}\middle|\mathbf{x}'\right\rangle, \tag{4.62}$$

where $x = (t, \mathbf{r})$ and the kernels are defined by the expressions (4.38) and (4.43).

The first part of the operator determines the field via the formula

$$\mathbf{E}_1 = \int \left\langle \mathbf{x}\middle|\hat{R}_{ej}^{(1)}\middle|\mathbf{x}'\right\rangle \mathbf{E}_0(\mathbf{x}')d\mathbf{x}'$$
$$= \frac{1}{2v_1}\left[v_1^2\hat{Q} - \frac{\partial}{\partial t}\left(\frac{\partial}{\partial t} + 2\sigma\right)\hat{\text{I}}\right]\int_0^\infty dt' \int_0^\infty dx' \int_\infty d\mathbf{r}' W_0(t-t', \mathbf{r}_\perp - \mathbf{r}'_\perp, |x-x'|)\mathbf{E}_0(t', \mathbf{r}'), \tag{4.63}$$

where

$$W_0(\tau,R_\perp,x-x')=\int_0^\infty\frac{dk}{2\pi}kJ_0(kR_\perp)\int_{-i\infty}^{i\infty}\frac{dq}{2\pi i}\frac{q(1-n^2)+(1+n^2)\sigma}{(q+\sigma)\xi}e^{(q-\sigma)\tau-\xi\frac{x-x'}{v_1}},$$

(4.64)

$\tau = t - t'$, $R_\perp = |\mathbf{r}_\perp - \mathbf{r}_\perp'|$, $\mathbf{r}_\perp = (y, z)$. This part of the field is similar to the one corresponding to the infinite medium.

The second part determines the field

$$\mathbf{E}_2=\int\left\langle\mathbf{x}\left|\hat{R}_{ej}^{(2)}\right|\mathbf{x}'\right\rangle\mathbf{E}_0(\mathbf{x}')d\mathbf{x}'$$
$$=\frac{1}{2v_1}\left[v_1^2\,\hat{Q}_\perp\mathbf{W}_m-2vv_1\,\hat{T}\mathbf{W}_{em}+\frac{\partial}{\partial t}\left(\frac{\partial}{\partial t}+2\sigma\right)\hat{I}_\perp\mathbf{W}_e\right], \quad (4.65)$$

which is completely determined by the presence of the boundary. Its contribution is equal to zero when the medium properties are changed in time in the whole space, not only in the half-space.

The differential matrices in (4.64) and (4.65) are equal to

$$\hat{Q}=\begin{pmatrix}\dfrac{\partial^2}{\partial x^2} & \dfrac{\partial}{\partial x}\nabla_\perp\\ \dfrac{\partial}{\partial x}\tilde{\nabla}_\perp & \tilde{\nabla}_\perp\nabla_\perp\end{pmatrix},$$

$$\hat{Q}_\perp=\begin{pmatrix}\nabla_\perp^2 & \dfrac{\partial}{\partial x}\nabla_\perp\\ -\dfrac{\partial}{\partial x}\tilde{\nabla}_\perp & \tilde{\nabla}_\perp\nabla_\perp\end{pmatrix},\quad \hat{T}=\begin{pmatrix}0 & 0\\ 0 & \tilde{\nabla}_\perp\nabla_\perp\end{pmatrix},\quad \hat{I}_\perp=\begin{pmatrix}0 & 0\\ 0 & \hat{I}\end{pmatrix}, \quad (4.66)$$

where $\nabla_\perp=\left(\frac{\partial}{\partial y},\frac{\partial}{\partial z}\right)$, $\tilde{\nabla}_\perp=\begin{pmatrix}\frac{\partial}{\partial y}\\ \frac{\partial}{\partial z}\end{pmatrix}$, and $\hat{I}$ is the unit matrix.

The rest of the notations correspond to: $v_1 = c/\sqrt{\varepsilon_1}$, $v = c/\sqrt{\varepsilon}$, $\sigma = \sigma_1/(2\varepsilon_0\varepsilon_1)$, ε_0 is the vacuum permittivity, and c is the light velocity in vacuum.

The vector functions

$$\mathbf{W}_a=\int_0^\infty dt'\int_0^\infty dx'\int_\infty d\mathbf{r}_\perp' W_a(t-t',\mathbf{r}_\perp-\mathbf{r}_\perp',|x-x'|)\mathbf{E}_0(t',x',\mathbf{r}_\perp') \quad (4.67)$$

are defined by the functions (4.43) which for a dissipative dielectric can be written in such a form

$$W_a(\tau,R_\perp,X)=\int_0^\infty \frac{dk}{2\pi}kJ_0(kR_\perp)\int_{-i\infty}^{i\infty}\frac{dq}{2\pi i}w_a(q,k)\frac{q(1-n^2)+(1+n^2)\sigma}{(q+\sigma)\xi}e^{(q-\sigma)\tau-\xi\frac{X}{v_1}}, \tag{4.68}$$

where the index a takes the values m, em, e. $j_0(.)$ is the Bessel function, $\xi=(q^2-\sigma^2+v_1^2k^2)^{1/2}$, $n^2=\varepsilon/\varepsilon_1=v_1^2/v^2$ and $X = x + x'$. The weight factor in the integral (4.68) is one of the analogues to Fresnel's formulae for the reflection coefficients (4.69)–(4.71), in which the substitution $p = q - \sigma$ and $\overline{\sigma}_1 = 2\sigma$ must be done:

$$w_m(q,k)=\frac{v(q+\sigma)\varphi-v_1(q-\sigma)\xi}{v(q+\sigma)\varphi+v_1(q-\sigma)\xi}, \tag{4.69}$$

$$w_{em}(q,k)=\frac{(q+\sigma)(v_1\varphi-v\xi)}{v(q+\sigma)\varphi+v_1(q-\sigma)\xi}, \tag{4.70}$$

$$w_e(p,k)=\frac{v_1\varphi-v\xi}{v_1\varphi+v\xi}. \tag{4.71}$$

Here, $\varphi = ((\theta-\sigma)^2 + v^2k^2)^{1/2}$.

The field beyond the non-stationary half-space, $x < 0$, is determined by the operator that is obtained from Eq. (4.72) by the same procedure as the operators for an internal field,

$$\mathbf{E}_3=\int_0^\infty dt'\int_{x'>0} d\mathbf{r}'\langle \mathbf{x}|\hat{N}|\mathbf{x}'\rangle \mathbf{E}_0(\mathbf{x}')=\frac{1}{2v_1}\left[-vv_1\hat{\mathrm{S}}_\perp\mathbf{U}_m+\frac{\partial}{\partial t}\left(\frac{\partial}{\partial t}+2\sigma\right)\hat{\mathrm{I}}_\perp\mathbf{U}_e\right], \tag{4.72}$$

where the differential matrix has the form

$$\hat{\mathrm{S}}=\begin{pmatrix} -\nabla_\perp^2 & -\frac{\partial}{\partial x'}\nabla_\perp \\ \frac{\partial}{\partial x}\tilde{\nabla}_\perp & \tilde{\nabla}_\perp\nabla_\perp \end{pmatrix}, \tag{4.73}$$

and the vector functions $\mathbf{U}_a$ are defined by the formula

$$\mathbf{U}_a = \int_0^\infty dt' \int_0^\infty dx' \int_\infty d\mathbf{r}'_\perp U_a(t-t', \mathbf{r}_\perp - \mathbf{r}'_\perp, |x-x'|)\mathbf{E}_0(t', x', \mathbf{r}'_\perp), \tag{4.74}$$

where

$$U_a(\tau, R_\perp, x, x') = \int_0^\infty \frac{dk}{2\pi} kJ_0(kR_\perp) \int_{-i\infty}^{i\infty} \frac{dq}{2\pi i} u_a(q,k) \frac{n^2(q-\sigma)-(q+\sigma)}{n(q+\sigma)\varphi} e^{(q-\sigma)\tau+\varphi\frac{x}{v}-\xi\frac{x'}{v_1}}. \tag{4.75}$$

The weight factors in the integrals (4.75) are defined by the formulae (4.56), in which the substitution $p = q - \sigma$ and $\bar{\sigma}_1 = 2\sigma$ must be done:

$$u_m(q,k) = \frac{2n(q+\sigma)\varphi}{(q+\sigma)\varphi + n(q-\sigma)\xi}, \tag{4.76}$$

$$u_e(q,k) = \frac{2n\varphi}{n\varphi + \xi}. \tag{4.77}$$

These formulae are also analogous to Fresnel's formulae for refractive coefficients in the cases of parallel and perpendicular polarization, respectively.

4.3 Inclined Incidence of a Plane Wave on a Plane Boundary of the Time-Varying Medium

In Sections 3.1–3.6, the influence of the non-stationary medium boundary on the electromagnetic wave was considered in the 1D case. The presence of the non-stationary medium boundary in the 3D case produces more essential variations in the space distribution of the field transformed by the time-varying medium. It will be shown here that a primary harmonic plane wave produces a continuous wave spectrum and the formation of a secondary wave with the polarization features that have a different qualitative nature than that for the stationary wave ones. If we consider transformation of point source radiation in transient a medium with a plane boundary, the focusing point of the collapsing secondary spherical waves is located at the mirror point of the source, symmetrical to the plane boundary of the non-stationary medium. Unlike the case of an unbounded medium, the field at this focusing point has a peak but a finite value.

If a plane monochromatic wave is normally incident on the newly created plane boundary of the non-dissipative medium whose permittivity varies abruptly in time, the transformed field consists of monochromatic waves only. Inclined incidence changes significantly the picture of the phenomena, because field transformation results in the appearance of a transient wave. In this section, we treat this process in more detail.

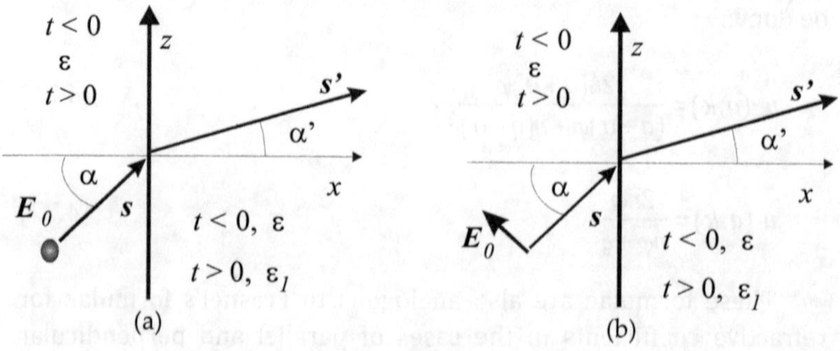

Figure 4.3 The geometry of the problem: (a) perpendicular polarization of the incident wave; (b) parallel polarization of the incident wave.

Consider a plane monochromatic wave $\mathbf{E}_0(\mathbf{x}) = \mathbf{E}_0 \exp[i(\omega t - \mathbf{sr})]$ where $\mathbf{s} = (s_1, \mathbf{s}_\perp)$ is the wave vector with magnitude $s = \omega/v$ that is incident on a plane medium boundary with angle of incidence α. The problem geometry and accepted notations for the various polarizations of the incident wave are commonly used and shown in Fig. 4.3. Now we consider a more general problem, a transient problem, when the time-varying medium occupies the half-space $x \geq 0$ and its permittivity varies abruptly from the background value ε to the value ε_1 at the zero moment of time, $t = 0$. The field in this medium after this zero moment is determined by the integral operators (4.63) and (4.65).

4.3.1 The Field Caused by the Permittivity Time Jump

This field is given by the action of the first part of the resolvent operator (4.38) on the incident field

$$\mathbf{E}_1 = \hat{R}_e^{(1)}\mathbf{E}_0(\mathbf{x}) = \int d\mathbf{x}'\left\langle \mathbf{x}\left|\hat{R}_e^{(1)}\right|\mathbf{x}'\right\rangle \mathbf{E}_0(\mathbf{x}')$$

$$= \frac{v_1^2 - v^2}{v^2}\left(-v_1^2\nabla\nabla + \frac{\partial^2}{\partial t^2}\hat{I}\right)\int_{-i\infty}^{i\infty}\frac{dp}{2\pi i}\frac{e^{pt-is_\perp r_\perp}}{2\phi_1(p-i\omega)}\left[\frac{2\phi_1 e^{-is_1 x}}{p^2+v_1^2 s^2} - \frac{e^{-\frac{\phi_1}{v_1}x}}{\phi_1 - iv_1 s_1}\right]\mathbf{E}_0 \tag{4.78}$$

where the vector $\mathbf{r}_\perp = (y, z)$, is located in the plane $x = 0$, i.e., $v_1 = c/\sqrt{\varepsilon_1}$ and $\phi_1 = (p^2 + v_1^2 s_\perp^2)^{1/2}$ with $\mathrm{Re}\phi_1 > 0$.

Integration of the first term in square brackets in Eq. (4.78), representing the resolvent of the unbounded medium, provides two field components

$$\mathbf{E}_1^{(1)}(t,\mathbf{r}) = -\mathbf{E}_0 e^{i\omega t - is\mathbf{r}} + \frac{v_1}{2v}\left(\frac{v_1 + v}{v}e^{i\omega_1 t} + \frac{v_1 - v}{v}e^{-i\omega_1 t}\right)\mathbf{E}_0 e^{-is\mathbf{r}}. \tag{4.79}$$

The first component in Eq. (4.79) is an antipode to the incident wave that annihilates the initial field according to the extinction theorem [1]. The second component in (4.79) has the same form as the transformed field in the unbounded medium resulting in a splitting the incident wave into two waves (direct and inverse ones) with new frequency $\omega_1 = \omega v_1/v$.

The boundary influence creates the second term in square brackets in (4.78). Its integration can be represented as a sum of residues in the integrand poles and the integrals along the cuts drawn from the branching points $p = +iv_1 s_\perp$ of the function ϕ_1 in the complex plane p. We consider the choice of the cuts similar to [2]. The correct procedure of the choice can be followed for a dissipative dielectric. For carrying out this procedure, the effective permittivity $\varepsilon_1 \to \varepsilon_1 + i\sigma/(\varepsilon_0\omega)$ is introduced with an infinitesimal conductivity σ. Thus, we have $v_1 s\perp = a + ib$, where $a > 0$ and $b > 0$ in the square root for ϕ_1 in (4.78). If $p = \xi +$ in then $\phi_1^2 = \xi^2 + a^2 - \eta^2 - b^2 + 2i(\xi\eta + ab)$. The regions for which both the real and imaginary parts of ϕ_1^2 maintain their signs are shown in Fig. 4.4.

As $\phi_1^2 = |\phi_1^2| \exp(iArg\phi_1^2)$ the condition $\mathrm{Re}\phi_1 > 0$ is fulfilled for $|Arg\phi_1^2| < \pi$. Now draw the cut on the plane of ϕ_1^2 in such a way that the module of the argument of this variable is less than π (see Fig. 4.5).

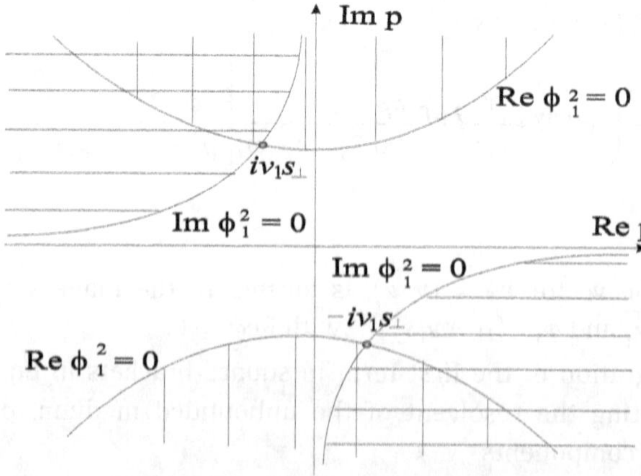

Figure 4.4 The regions for which the real and imaginary parts of ϕ_1^2 maintain their signs: Reϕ_1^2 < 0 in the vertical shading region; Imϕ_1^2 < 0 in the horizontal shading region.

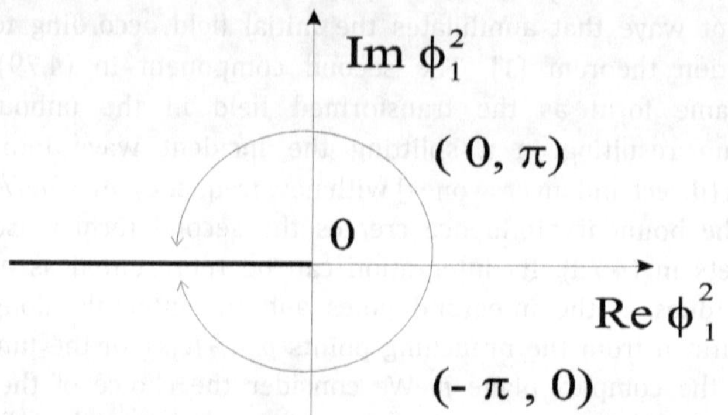

Figure 4.5 The cut in the complex plane ϕ_1^2 providing the condition $|Arg\phi_1^2| < \pi$.

Along the cut sides Reϕ_1^2 < 0, Imϕ_1^2 = 0. Let us choose in the complex plane p the lines which satisfy the similar conditions Re p < 0 and Im p = 0. The cuts along these lines define the Riemann surface with Reϕ_1 > 0 on its upper side. The imaginary part of ϕ_1 possesses various signs in the regions shown in Figs. 4.4 and 4.5. So, the cuts drawn as in Fig. 4.6 allow one to execute the straightforward integration in (4.78).

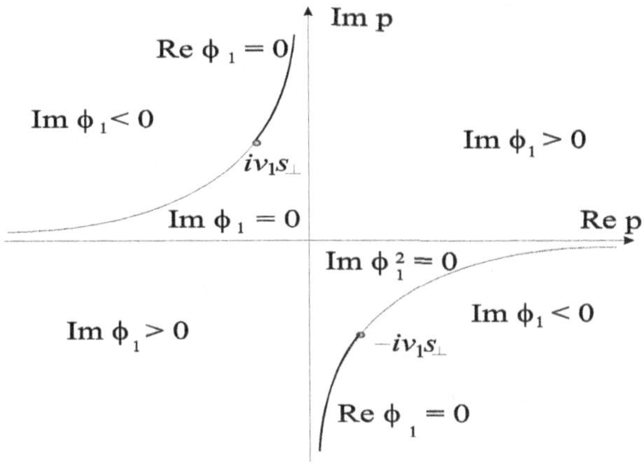

Figure 4.6 The cuts in the complex plane *p*, satisfying the condition $\text{Re}\phi_1 > 0$.

Integrating the second term in the square brackets in (4.78), we obtain after extracting explicitly the residue in the pole $p = i\omega$:

$$\int_{-i\infty}^{i\infty} \frac{dp}{2\pi i} \frac{e^{pt}}{2\phi_1(p-i\omega)} \frac{1}{\phi_1 - iv_1 s_1} e^{-\frac{\phi_1}{v_1}x} = \frac{\theta(t-x/v_1)}{2{v_1}^2 s_1'(s_1 - s_1')} e^{i(\omega t - s_1' x)} + \Im_1. \quad (4.80)$$

It has been taken into account here that

$$\phi_1(i\omega) = (-\omega^2 + v_1^2 s_\perp^2)^{1/2} = v_1(-s'^2 + s_\perp^2)^{1/2} = iv_1 s_1',$$

where s_1' is a component of the wave vector $\mathbf{s}'$ in the new medium with an absolute value equal to $s' = \omega/v_1$. This component is normal to the medium boundary. $\Im_1$ in Eq. (4.80) represents the field component which is determined by other poles of the integrand in Eq. (4.80) as well as by its branching points.

To choose the sign of the square root in ϕ_1, make a similar consideration to that above. It follows that the condition $\text{Re}\phi_1 > 0$ provides $\phi_1(\pm i\omega) = \pm iv_1 s_1'$. Indeed, in this case there is the expression $p^2 + \frac{\sigma}{\varepsilon_0 \varepsilon_1} p + v_1^2 k^2$ in the denominator of the resolvent (4.79), which has the roots $k = \pm i\frac{\phi_1}{v_1}$, where $\phi_1 = \sqrt{p^2 + \frac{\sigma}{\varepsilon_0 \varepsilon_1} p}$. If $p = iw$ then the root is approximately equal to

$$\phi_1(i\omega) = \sqrt{-\omega^2 + i\frac{\sigma}{\varepsilon_0\varepsilon_1}\omega} \cong \pm i\omega\left(1 - i\frac{\sigma}{2\varepsilon_0\varepsilon_1\omega}\right), \tag{4.81}$$

The requirement Re$\phi_1 > 0$ provides the following choice of the positive sign for ϕ_1: $\phi_1(i\omega) = +i\omega(1 - i\frac{\sigma}{2\varepsilon_0\varepsilon_1\omega})$. If $p = -iw$ then we have

$$\phi_1(-i\omega) = \sqrt{-\omega^2 - i\frac{\sigma}{\varepsilon_0\varepsilon_1}\omega} \cong \pm i\omega\left(1 + i\frac{\sigma}{2\varepsilon_0\varepsilon_1\omega}\right) \tag{4.82}$$

and the same requirement for the real part Re$\phi_1 > 0$ now yields

$$\phi_1(-i\omega) = -i\omega\left(1 + i\frac{\sigma}{2\varepsilon_0\varepsilon_1\omega}\right).$$

Applying the differential operator in Eq. (4.78) to the first term determined in Eq. (4.80), we obtain a component of the secondary field which has the same phase as the stationary refracted wave:

$$\mathbf{E}_1^{(2)}(\mathbf{x}) = \theta(t - x/v_1)\frac{v^2 - v_1^2}{v^2}\frac{\mathbf{s}'(\mathbf{s}'\mathbf{E}_0) - s'^2\mathbf{E}_0}{2s_1'(s_1 - s_1')}e^{i(\omega t - \mathbf{s}'\mathbf{r})} \tag{4.83}$$

Turn now to considering the transient field. It is determined by another singularity in the integral and is described by the following expression:

$$\Im_1 = \frac{v^2 - v_1^2}{v^2}\left(-v_1^2\nabla\nabla + \frac{\partial^2}{\partial t^2}\hat{I}\right)\int_{-i\infty}^{i\infty}\frac{dp}{2\pi i}\frac{e^{pt - is_\perp \mathbf{r}_\perp}}{2\phi_1(p - i\omega)}\frac{1}{\phi_1 - iv_1 s_1}e^{-\frac{\phi_1}{v_1}x}\mathbf{E}_0\Bigg|_{p\neq i\omega}. \tag{4.84}$$

Besides the branching points there are the poles defined by the equation $\phi_1 - iv_1s_1 = 0$. Formally this equation has two roots, but these roots are located on different sheets of the Riemann surface of the function $\phi_1(p)$. According to the formula (4.81) it follows that

$$\phi_1(i\omega_1) = \sqrt{-\omega_1^2 + v_1^2 s_\perp^2} = i\sqrt{\omega_1^2 - v_1^2 s_\perp^2} = iv_1 s_1,$$

and therefore the equation $\phi_1 - iv_1s_1 = 0$ has only one solution $p = iw_1$ (if taking into account that $v_1 s = \omega_1 = \frac{v_1}{v}\omega$) on the upper sheet of the Riemann surface where the condition $\mathrm{Re}\phi_1 > 0$ is satisfied.

The residue in the pole $p = iw$ in the integral (4.84) determines the following wave:

$$\Im_1'(\mathbf{x}) = -\theta(t - x/v_1)\frac{v_1(v+v)}{2v^2}\mathbf{E}_0 e^{i(\omega_1 t - \mathbf{sr})}. \tag{4.85}$$

This wave annihilates the corresponding wave in $\mathbf{E}_1^{(1)}$ (4.79) in the region $0 < x < v_1t$. Consequently, from two splitting waves in this region only the wave propagating toward the boundary survives. The rest of the integral (4.84), determined by the integration along the cut lines on the complex plane p, provides a transient field in the region $0 < x < v_1t$.

Thus, the first part of the secondary field, resulting from applying the first part of the resolvent operator to the incident field, consists of the following components:

- The component $-\mathbf{E}_0(\mathbf{x})$ that cancels the incident wave in the entire half-space $x \geq 0$.
- Two splitting waves in the region $x > v_1t$, whose sum is described by the function $\mathbf{E}_1^{(1)}(\mathbf{x})$, with new frequency ω_1. They are exactly the same ones as in the unbounded transient medium.
- The inverse wave in the region $0 < x < v_1t$, with the new frequency ω_1.
- The part of the ordinary refracted wave $\mathbf{E}_2^{(1)}(\mathbf{x})$ with the original frequency of the incident wave ω in the region $0 < x < v_1t$.
- The transient field in the region $0 < x < v_1t$.

4.3.2 The Field Caused by the Boundary Presence Only

The action of the second part of the resolvent operator on an incident field is determined by the integral (4.65), which can be written in the form

$$\mathbf{E}_2 = \hat{R}^{(2)}\mathbf{E}_0(\mathbf{x}) = \int d\mathbf{x}'\left\langle \mathbf{x}\middle|\hat{R}^{(2)}\middle|\mathbf{x}'\right\rangle \mathbf{E}_0(\mathbf{x}')$$
$$= -\frac{v_1^2 - v^2}{v^2}\int_{-\mathrm{i}\infty}^{\mathrm{i}\infty}\frac{dp}{2\pi\mathrm{i}}\frac{e^{pt-\mathrm{i}\mathbf{s}_\perp\mathbf{r}_\perp}}{2\phi_1(p-i\omega)}\frac{\hat{\mathrm{X}}_1 + p^2 w_e \hat{\mathrm{I}}_\perp}{\phi_1 + iv_1 s_1}e^{-\phi_1 x/v_1}\,\mathbf{E}_0 = \mathbf{E}_2^{(1)} + \Im_2, \tag{4.86}$$

where the representation $\hat{X}_1 + p^2 w_e I_\perp$ can be deduced from X in (4.65) after substituting ϕ_1 for φ_1. In this representation, we have

$$\hat{\mathrm{X}}_1 = -v_1^2 w_m \begin{pmatrix} s_\perp^2 & i\dfrac{\phi_1}{v_1}\mathbf{s}_\perp \\ -i\dfrac{\phi_1}{v_1}\tilde{\mathbf{s}}_\perp & \hat{\mathrm{s}}_\perp \end{pmatrix} + 2vv_1 w_{em}\begin{pmatrix} 0 & 0 \\ 0 & \hat{\mathrm{s}}_\perp \end{pmatrix}, \tag{4.87}$$

where $\hat{s}_\perp = \begin{pmatrix} s_2^2 & s_2 s_3 \\ s_2 s_3 & s_3^2 \end{pmatrix}$.

The first term in (4.86) is the residue in the pole $p = i\omega$,

$$\mathbf{E}_2^{(1)}(\mathbf{x}) = \theta(t - x/v_1)\frac{v_1^2 - v^2}{v^2}\frac{\hat{\mathrm{X}}'}{2v_1^2 s_1'(s_1 + s_1')}\mathbf{E}_0 e^{i(\omega t - \mathbf{s}'\mathbf{r})}. \tag{4.88}$$

Here, $\hat{\mathrm{X}}'$ is found from $\hat{\mathrm{X}}_1 + p^2 w_e \hat{\mathrm{I}}_\perp$ after substituting $i\omega$ for p. The term $\Im_2$ occurs owing to the other poles and branching points in (4.86).

Consider the transient field, which is determined by other singularities than the pole $p = i\omega$ in the integrand of the integral:

$$\Im_2 = \frac{v^2 - v_1^2}{v^2}\int_{-\mathrm{i}\infty}^{\mathrm{i}\infty}\frac{\mathrm{dp}}{2\pi\mathrm{i}}\frac{\mathrm{e}^{\mathrm{pt}-\mathrm{i}\mathbf{s}_\perp\mathbf{r}_\perp}}{2\phi_1(p-i\omega)}\frac{\overline{X}}{\phi_1 + iv_1 s_1}e^{-\frac{\phi_1}{v_1}x}\mathbf{E}_0\Bigg|_{p\neq i\omega}. \tag{4.89}$$

In addition to the branching points in Eq. (4.89), there are the poles defined by the equation $\phi_1 + iv_1 s_1 = 0$. It can be deduced from the previous subsection that this equation has only one solution $p_2 = -i\omega_1$ in the upper sheet of the Riemann surface because $\phi_1(-i\omega_1) = -iv_1 s_1 \phi_1$.

The residue in the pole $p_2 = -i\omega_1$ in the integral (4.89) determines the wave, which represents the reflection of the inverse wave proportional to $\sim\exp[-i(\omega_1 t + \mathbf{sr})]$ in Eq. (4.79), from the medium boundary. In the case of perpendicular polarization, this wave is described by the formula

$$\mathfrak{I}'_{2\perp} = \mathbf{E}^{(-)}_{1r} = -\theta(t - x/v_1)\frac{(m-\cos\alpha)^2}{2(n+1)} n\mathbf{E}_{0\perp} e^{-i(\omega_1 t + s_1 x - \mathbf{s}_\perp \mathbf{r}_\perp)}, \tag{4.90}$$

where $m(n^2 - \sin^2\alpha)^{1/2}$ and $n = v_1/v$.

In the case of parallel polarization, this wave has different amplitudes for different components of the field vector

$$\mathfrak{I}'_{2||} = \mathbf{E}^{(-)}_{1r} = \theta(t - x/v_1)$$

$$= \frac{n(1-n)(m-\cos\alpha)(m\cos\alpha - \sin^2\alpha)}{2(m+n^2\cos\alpha)}\begin{pmatrix}\sin\alpha\\ 0\\ \cos\alpha\end{pmatrix} E_{0||} e^{-i(\omega_1 t + s_1 x - \mathbf{s}_\perp \mathbf{r}_\perp)} \tag{4.91}$$

The above expressions indicate that the inverse wave is reflected from the medium boundary according to the conventional reflection law. Particularly, the amplitude $\mathfrak{I}'_{2||}$ is equal to zero if $tga = n$, which corresponds to the Brewster angle for the internal waves.

So, the second part of the resolvent determines the part of the conventionally refracted wave $\mathbf{E}^{(1)}_2(\mathbf{x})$, which corresponds to the incident wave interaction with the new stationary medium and has the frequency of the incident wave and, in addition, the transient field, which contains the inverse wave reflection from the boundary as well.

4.3.3 The Evolution of the Refracted Field

Both parts of the ordinary refracted wave fuse into the stationary refracted wave, $\mathbf{E}^{(2)}_1(\mathbf{x}) + \mathbf{E}^{(1)}_2(\mathbf{x}) = \mathbf{E}_{\text{refract}}(\mathbf{x})$, which has the frequency of the incident wave, and its description coincides with the Fresnel formula. This wave exists in the region bounded by two planes: the boundary between the media and the plane $x = v_1 t$ (see Fig. 4.7).

The expressions (4.83) and (4.88) are different for different orientations of the incident wave field vector with respect to the created medium boundary. In the case of the perpendicularly polarized incident field $\mathbf{E}_0 = (0, E_0, 0)$, the two components given by (4.83) and (4.88),

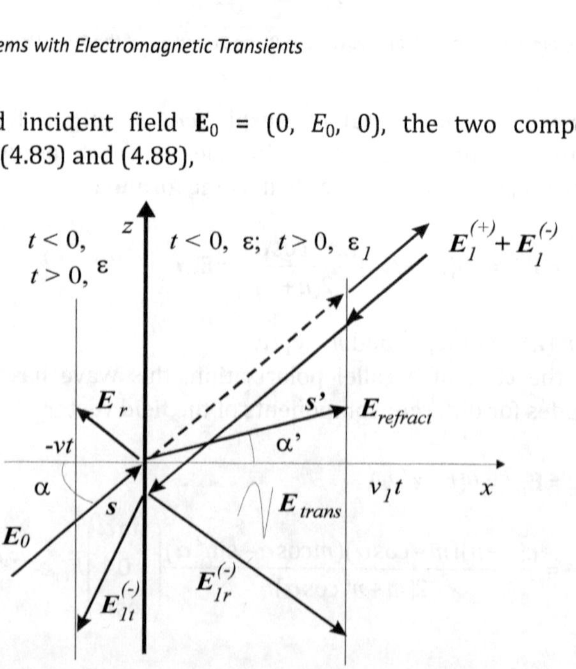

Figure 4.7 The space distribution of the secondary waves.

$$\mathbf{E}_1^{(2)}(\mathbf{x}) = \theta(t - x/v_1)\frac{v_1^2 - v^2}{v^2}\frac{s'^2}{2s_1'(s_1 - s_1')}\mathbf{E}_0 e^{i(\omega t - \mathbf{s'r})} \qquad (4.92)$$

$$\mathbf{E}_2^{(1)}(\mathbf{x}) = \theta(t - x/v_1)\frac{v^2 - v_1^2}{v^2}\frac{s'^2(s_1 - s_1')}{2s_1'(s_1 + s_1')^2}\mathbf{E}_0 e^{i(\omega t - \mathbf{s'r})} \qquad (4.93)$$

compose the ordinary refracted wave whose amplitude corresponds to the well-known Fresnel formulae:

$$E_1^{(2)} + E_2^{(1)} = \frac{2s_1}{s_1 + s_1'}E_0 = \frac{2v_1\cos\alpha}{v_1\cos\alpha + v\cos\alpha'}E_0 = \frac{2\cos\alpha\sin\alpha'}{\sin(\alpha + \alpha')}E_0 \qquad (4.94)$$

In the case of parallel polarization of the incident field, if its vector is situated in the incidence plane, $\mathbf{E}_0 = (E_{01}, 0, E_{03})$, where $E_{01} = E_0 s_\perp/s$, $E_{03} = -E_0 s_1/s$ (see Fig. 4.7), the field vector is equal to

$$\mathbf{E}_{||} = \mathbf{E}_1^{(2)} + \mathbf{E}_2^{(1)} = \frac{v^2 - v_1^2}{2s_1' v^2}\left(\frac{\mathbf{s}'(\mathbf{s}'\mathbf{E}_0)}{s_1 - s_1'} - \frac{\hat{X}_1\mathbf{E}_0}{v_1^2(s_1 + s_1')}\right) + \frac{2s_1\mathbf{E}_0}{s_1 - s_1'}. \qquad (4.95)$$

In order to find the relations between the vector modulus for the case of parallel polarization of the incidence, we pre-multiply both sides of Eq. (4.95) by the vector $\mathbf{s}'$ as a vector product, taking into account that $\mathbf{s}' \times \mathbf{E}_{||} = \mathbf{e}_2 s' E_{||}$ and $\mathbf{s}' \times \mathbf{E}_{0||} = \mathbf{e}_2 \frac{s_1 s_1' + s_3^2}{s} E_{0||}$, where $\mathbf{e}_2$ is the unit vector in the y-direction. It follows that

$$E_{||} = \frac{2s_1 v v_1}{v^2 s_1 + v_1^2 s'_1} E_{0||} = \frac{2\cos\alpha \sin\alpha'}{\sin\alpha\cos\alpha + \sin\alpha'\cos\alpha'} E_{0||}, \tag{4.96}$$

which coincides with the Fresnel formulae as well.

The integrals in Eqs. (4.84) and (4.89) along the cut lines drawn from the branching points of the function ϕ_1 describe the transient field caused by the permittivity jump in time. These integrals tend to zero if $t \to \infty$. The steady-state monochromatic waves, which are the same as in the case of the normal incidence, are separated from this transient spectrum in their asymptotic behavior when $t \to \infty$. The exact expression for the steady-state waves in the non-stationary half space, $x \geq 0$, for an arbitrarily polarized incident wave is yielded by virtue of the residues in the poles of the integrand (4.78) with (4.84) and (4.86) with (4.89). Hence, the asymptotic expression for the field being determined by the first part of the resolvent arises from the integral in (4.78) having the following form:

$$\hat{\mathrm{R}}_e^{(1)} \mathbf{E}_0(\mathbf{x}) = -\mathbf{E}_0(\mathbf{x}) + \frac{v_1(v_1 - v)}{2v^2} \mathbf{E}_0 e^{-i(\omega_1 t + \mathbf{sr})} + \frac{s_1' + s_1}{2s_1'} \mathbf{E}_0 e^{i(\omega t - \mathbf{s'r})}. \tag{4.97}$$

It follows from Eq. (4.91) that an asymptotic complete quenching of the direct wave with the frequency equal to ω_1 also takes place in the case of a primary wave incident on the medium boundary at an arbitrary angle of incidence. It can be seen that the complete extinction of the direct wave, proportional to $\sim e^{in\omega t}$, in the region $0 \leq x < v_1 t$ takes place immediately after the permittivity jump only for the normal incidence, when $\cos a = 1$.

The spatial and temporal field distributions after the permittivity time jump will be investigated in a more simple way, apart from its polarization properties. To do this we shall examine a perpendicular polarization of the incident wave, when the vector nature of the field (i.e., polarization properties) does not influence the result. In this case, only the last term in Eq. (4.86) remains, and the integrals in (4.84) and (4.89) can be transformed using their convolution representation,

$$\int_{-i\infty}^{i\infty} \frac{dp}{2\pi i} \frac{e^{-\frac{\phi_1}{v_1}x}}{2\phi_1(p-i\omega)} \frac{1}{\phi_1 - iv_1 s_1} e^{pt} = \frac{h(t-x/v_2)\varepsilon_1}{2\omega^2(\varepsilon_1-\varepsilon)} \left\{ e^{-i\omega(t-x/v_2)} + \frac{\varepsilon}{\varepsilon_1} A(t-x/v_1) \right.$$
$$\left. + \int_{x/v_1}^{t} \left[-\frac{v_1 s_\perp x}{\xi} J_1(s_\perp \xi) \pm iv_1 s_1 J_0(s_\perp \xi) \right] \left[e^{-i\omega(t-\tau)} + \frac{\varepsilon}{\varepsilon_1} A(t-\tau) \right] d\tau \right\}, \tag{4.98}$$

where $A(t) = \dfrac{v_1}{2v}\left(\dfrac{v_1+v}{v} e^{i\omega_1 t} + \dfrac{v_1 - v}{v} e^{-i\omega_1 t} \right) e^{i\omega t}$,

$J_{0,1}$ are the Bessel functions and $\xi = \sqrt{(v_1\tau)^2 - x^2}$. The integrals in (4.98) differ from zero in the region $0 \le x \le v_1 t$ only. It determines that in the whole region located to the right of the moving plane $x = v_1 t$ the transformed field has a form identical to that in the same time-varying but infinite medium. It consists of the direct and reversed waves with frequency ω_1 only, as in (4.79).

A vector nature of the field also does not influence the transformation of the normal plane wave incidence ($s_\perp = 0$) with any orientation of vector $\mathbf{E}_0$. In this case, the coefficient w_e becomes a constant, $w_e = (v_2 - v)/(v_2 + v)$, and the integrals (4.98) can be exactly calculated. As a result, in the region $0 \le x \le v_1 t$ we obtain an ordinary refracted wave with the frequency ω, and the inverse wave with the frequency ω_1, as well as its reflection from the formed medium boundary. It is significant that the transient field is absent for normal incidence, and the monochromatic wave regime in the region $0 \le x \le v_1 t$ is settled just immediately after the permittivity suddenly changes. It should be pointed out that the transformed field in the case considered takes a form that is identical to that for the scalar case.

In the case of inclined incidence, a steady-state field does not appear immediately and the transient field with a complex form appears. This transient is absent in the case of normal incidence. The monochromatic wave regime reaches the steady state in the region $0 \le x \le v_1 t$ immediately after the permittivity jump in this case. In the case of an inclined incidence a new region of transient field genesis, $t^2 - x^2/v_1^2 \leq (v_1 s_\perp)^{-2}$, is formed near the medium boundary; it follows from analyzing the integrals (4.78) with (4.84) and (4.86) with (4.89). This process is described exactly by the spatial-temporal Fresnel formulae (the extension of the well-known stationary Fresnel formulae for the case of time-varying media). These formulae can be used to calculate the transient fields numerically or analyze it qualitatively. As an example, the creation of the dual-wave structure in the case of an initial plane wave can be shown by the function $\bar{\mathbf{W}}_0$, which determines the component $\mathbf{E}_1$ of the transmitted field. It follows from (4.38) and (4.43) that

$$\bar{\mathbf{W}}_0 = \mathbf{E}_0 e^{i\omega t - is_1 x}\left\{\int_0^t dt' \int_{-v_1 t'}^{v_1 t'} J_0(s_\perp \sqrt{v_1^2 t'^2 - x'^2})e^{-i\omega t' + is_1 x'} dx'\right.$$

$$\left. -\theta(t' - x/v_1)\int_{x/v_1}^{t} dt' \int_{x}^{v_1 t'} J_0(s_\perp \sqrt{v_1^2 t'^2 - x'^2})e^{-i\omega t' + is_1 x'} dx'\right\}. \quad (4.99)$$

The first integral in (4.99) can be easily calculated [3], providing components with the same character as those in the expression (4.79)

$$\frac{2v_1}{\omega^2(n^2-1)} - \frac{v}{\omega^2(n-1)} e^{i\omega(n-1)t} - \frac{v}{\omega^2(n+1)} e^{-i\omega(n+1)t}. \quad (4.100)$$

The calculation of the second integral in (4.99) can be made in the approximation $(x/v_1 t) << 1$. We obtain under this condition that

$$\begin{aligned}\bar{\mathbf{W}}_0 = &\frac{4\pi v_1}{\omega^2(1-n^2)}\mathbf{E}_0 e^{-is\mathbf{r}}\theta\left(\frac{x}{v_1} - t\right)\left(-e^{i\omega t} + \frac{1+n}{2n}e^{i\omega_1 t} + \frac{1-n}{2n}e^{-i\omega_1 t}\right)\\ &+ \frac{4\pi v_1}{\omega^2(1-n^2)}\mathbf{E}_0 e^{-is\mathbf{r}}\theta\left(t - \frac{x}{v_1}\right)\left(\frac{1-\cos\alpha}{2}\frac{1+n}{2n}e^{i\omega_1 t} - \frac{1+\cos\alpha}{2}\frac{1-n}{2n}e^{-i\omega_1 t}\right)\\ &- \frac{4\pi v_1}{\omega^2(1-n^2)}\mathbf{E}_0 e^{-is_\perp \mathbf{r}}\theta\left(t - \frac{x}{v_1}\right)\left(\frac{1+n}{2n}e^{i\omega t - is(\cos\alpha - 1 + 1/n)x} - \frac{1-n}{2n}e^{i\omega t + is(\cos\alpha + 1 + 1/n)x}\right),\end{aligned} \quad (4.101)$$

where $\omega_1 = n\omega$.

The time behavior of the wave being described in (4.99) is shown in Figs. 4.8–4.11. Figure 4.8 illustrates the temporal behavior of the total first part of the transmitted field at the point x where $\omega x/v$ = 15 if n = 1.7, i.e., the point located at a distance of 15 wavelengths from the medium boundary. The boundary influence front arrives at this point at the moment $t_{\text{front}} = x/v_1$ which determines the value $\omega t_{\text{front}} = 8.8$ in the case considered. The transient part of this field is illustrated in Fig. 4.9. As one can see from Figs. 4.8 and 4.9, the temporal behavior of the field has a regular character until the moment t_{front}, because the transient field is absent until this moment, and the total field consists of two stationary waves $\mathbf{E}_1^{(\pm)}$ propagating in opposite directions.

After the time t_{front} the transient field is "switched on" in the inclined incidence case ($\alpha = \pi/3$), and the temporal behavior of the field becomes more complicated. In the case of the normal incidence, the field temporal behavior has another character but a regular one. Figures 4.10 and 4.11 show the same behavior for the case of $n = v_1/v = 0.7$, when the medium gets a larger optical density after the permittivity jump. It holds for $\omega t_{\text{front}} = 21.4$ in this case.

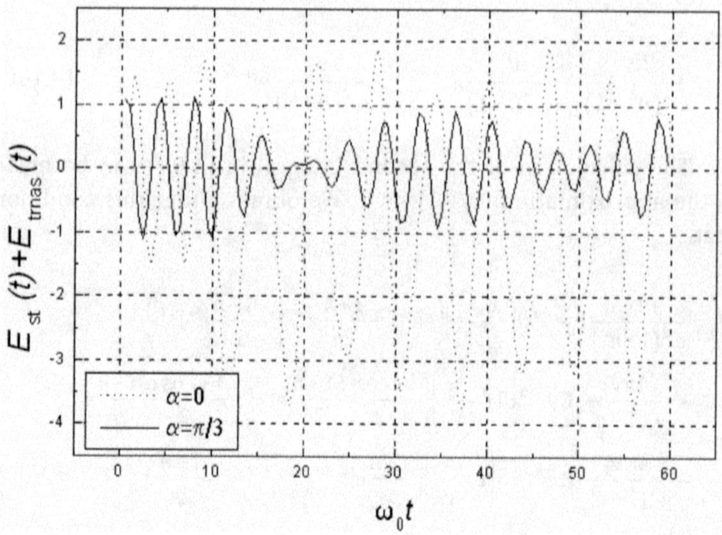

Figure 4.8 The temporal behavior of the total first part of the transmitted field for the case $n = 1.7$, $\omega_0 x/v = 15$ and $\omega_0\, t_{\text{front}} = 8.8$.

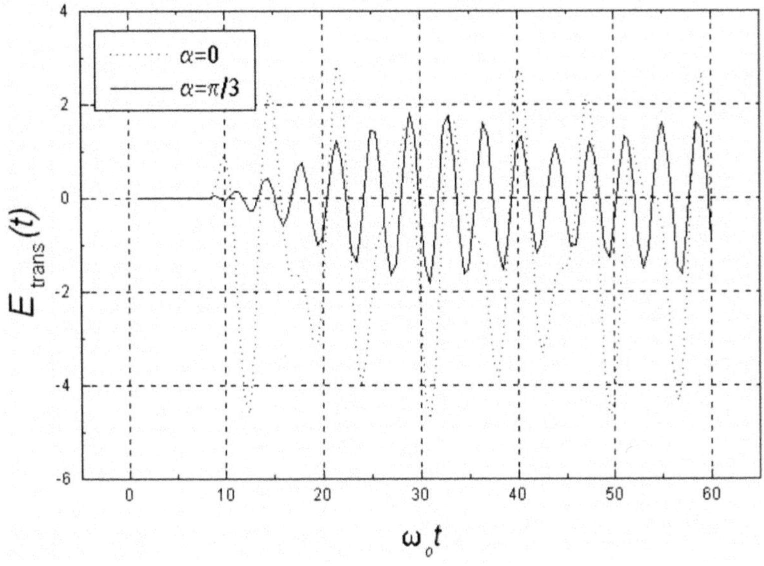

Figure 4.9 The transient field in the first part of the transmitted field for the case $n = 1.7$, $\omega_0 x/v = 15$ and $\omega_0\, t_{front} = 8.8$.

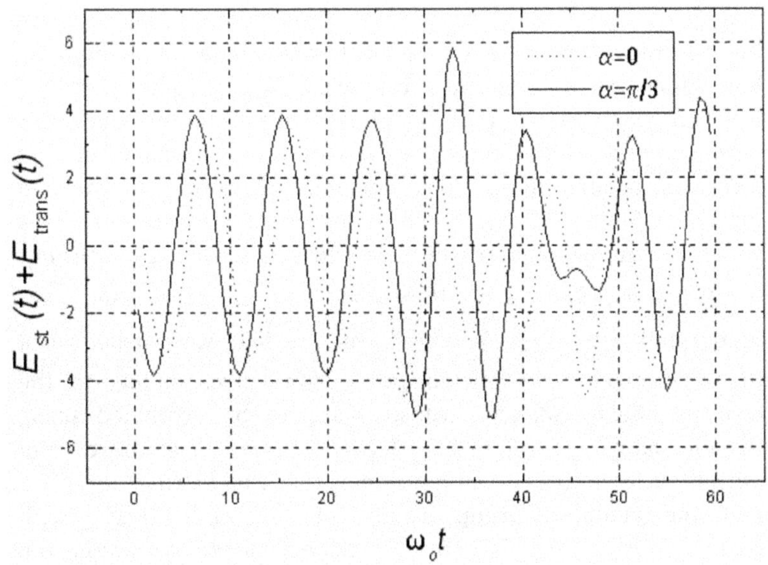

Figure 4.10 The temporal behavior of the total first part of the transmitted field for the case $n = 0.7$, $\omega_0 x/v = 15$ and $\omega_0\, t_{front} = 21.4$.

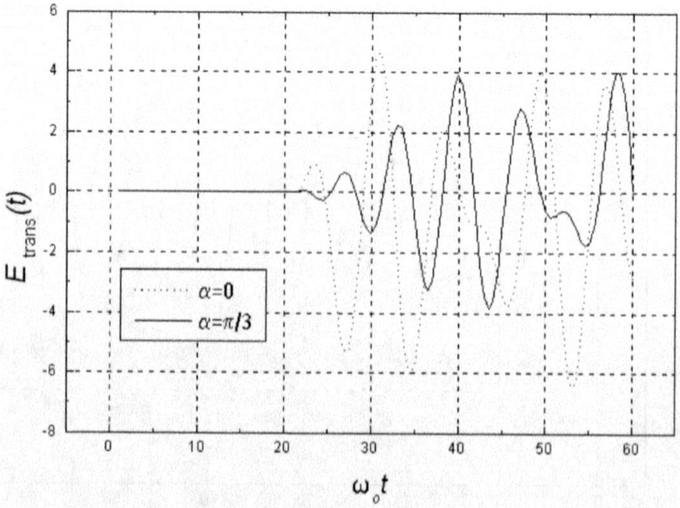

Figure 4.11 The transient field in the first part of the transmitted field for the case $n = 0.7$, $\omega_0 x/v = 15$ and $\omega_0 t_{front} = 21.4$.

It follows from (4.101) that in addition to the direct wave proportional to $-\exp[i(\omega_1 t - \mathbf{sr})]$ and existing in both regions $x \geq v_1 t$ and $0 \leq x < v_1 t$, there is another direct monochromatic wave in the region $0 \leq x < v_1 t$ with the same frequency ω_1, but possessing a different wave vector, $\mathbf{s}_\perp$, being proportional to $\sim\exp[i(\omega_1 t - sx - \mathbf{s}_\perp \mathbf{r}_\perp)]$. In the case of normal incidence, this wave suppresses the direct wave with the frequency ω_1, but complete compensation does not take place for an inclined incidence. Similar to these dual waves, there are also two waves which are complementary ones to those described by $\exp[i(\omega_1 t - \mathbf{sr})]$ and $\exp[i(\omega t - \mathbf{s}'\mathbf{r})]$. Here $\mathbf{s}' = (s_1', \mathbf{s}_\perp)$ is the refracted wave vector and $s_1' = s\sqrt{(v/v_1)^2 - \sin^2\theta}$. Such a dual-wave structure of the field takes place within the region corresponding to the transient field "infancy" whose size can be estimated using the inequality $s_\perp x \leq 1$ or $t^2 - (x/v_1)^2 < (v_1 s_\perp)^{-2}$. In the case of normal incidence, this region occupies the whole strip $0 \leq x < v_1 t$, and the complete compensation of the direct wave with frequency ω_1 takes place in this region immediately after the permittivity jump. The amplitudes of "the other wave doubles" are equal to zero in this case. In the case of an inclined incidence, the region $t^2 - (x/v_1)^2 < (v_1 s_\perp)^{-2}$ occupies only a part of the strip, and the continuous wave spectrum is formed outside this region. The settled monochromatic waves, identical to those for a normal incidence, are separated asymptotically from this spectrum. Indeed, the asymptotic expression for the field is determined by the resolvent first term and can be deduced from integrals (4.98) replacing the upper limits by ∞. In this case, we obtain an expression similar to (4.97).

4.3.4 The Field Outside the Non-Stationary Medium

The electromagnetic field in the stationary half-space $x < 0$ can be described using the formulae (4.72) and (4.74). The wave with the frequency ω gives rise to the ordinary reflected wave, and all the waves with the frequency ω yield a wave with the same frequency that comes out from the non-stationary dielectric and is defined by the expression

$$\mathbf{E}_{1t}^{(1-)}(\mathbf{x}) = \mathbf{E}_{1t} e^{-i\left[\omega_1 t - (\omega/v) Hx - \mathbf{s}_\perp \mathbf{r}_\perp\right]}. \tag{4.102}$$

If the vector $\mathbf{E}_0$ is normal to the plane of incidence then the wave amplitude is $\mathbf{E}_{1t\perp} = \frac{n(n-1)\cos\alpha}{H+\cos\alpha} \mathbf{E}_{0\perp}$. For the parallel polarization, it follows that

$$\mathbf{E}_{1t||} = \frac{n(n-1)\cos\alpha}{m+\cos\alpha} \frac{1}{m\cos\alpha + \sin^2\alpha} \begin{pmatrix} -\sin\alpha \\ 0 \\ m \end{pmatrix} E_{0||}. \tag{4.103}$$

Similar to the internal field, the external field also contains a damped continuous wave spectrum existing outside the non-stationary dielectric in addition to the waves described by (4.103). This continuous spectrum of waves is produced by the transition process in the internal field and is described by the proper integrals in (4.84) and (4.89) over the complex plane.

4.4 Refocusing of Point Source Radiation by the Plane Boundary of a Time-Varying Dielectric

The focusing properties of the transient medium, due to the abrupt change of its permittivity in time in an unbounded medium, represent new features in the presence of the medium boundaries. Consider the radiation generated by the point harmonic source $\mathbf{j}_{\text{extr}} = \mathbf{J}(t)\delta(\mathbf{r} - \mathbf{r}_0)$ located at the point $\mathbf{r}_0 = (-x_0, 0, 0)$. Substituting $\mathbf{J}(t) = \mathbf{J}_0 \exp(i\omega t)$ into Eq. (2.32) (Fig. 2.4), we obtain the well-known expression for the radiation field of this source in the homogeneous medium. Let the permittivity change abruptly in time from the background value ε to the value ε_1 in the half-space $x \geq 0$. For the sake of simplicity, and without loss of generality, the transformation of one part of the source field, $\mathbf{E}_0 = (1/R_0)\,\mathbf{I}\exp(-i\kappa R_0 + i\omega t)$ with $\kappa = \omega/v$ and $R_0 = |\mathbf{r} - \mathbf{r}_0|$, is examined. Substituting this field into the integral operators, we obtain

$$\hat{R}_1\mathbf{E}_0 = 4(n^2-1)\left(-v_1^2\nabla\nabla + \frac{\partial^2}{\partial t^2}\right)\mathbf{I}\int_{-i\infty}^{i\infty}\frac{dp}{2\pi i}e^{pt}\int_0^{\infty}\frac{dk_\perp}{2p}k_\perp \mathbf{J}_0(k_\perp\rho)$$
$$\times\left[\frac{e^{-i\psi x}}{(p-i\omega)(p^2+v_1^2k_\perp^2)} - \frac{v_1 e^{-\frac{\phi_1}{v_1}x}}{2\phi_1(p-i\omega)(\phi_1 - iv_1\psi)}\right]\frac{e^{-i\psi x_0}}{\mathrm{i}\psi}, \tag{4.104}$$

$$\hat{R}_2\mathbf{E}_0 = 4(n^2-1)\int_{-i\infty}^{i\infty}\frac{dp}{2\pi i}e^{pt}\int_0^{\infty}\frac{dk_\perp}{2\pi}k_\perp\,\hat{\mathrm{M}}\mathbf{I}\,J_0(k_\perp\rho)\frac{v_1 e^{-\frac{\phi_1}{v_1}x}}{2\phi_1(p-i\omega)(\phi_1 - iv_1\psi)}\frac{e^{-i\psi x_0}}{i\psi}, \tag{4.105}$$

where $\hat{\mathrm{M}} = v_1^2\begin{pmatrix}\nabla_\perp^2 & \frac{\partial}{\partial x}\nabla_\perp \\ -\frac{\partial}{\partial x}\nabla_\perp^* & \nabla_\perp^*\nabla_\perp\end{pmatrix}w_m - 2vv_1\begin{pmatrix}0 & 0 \\ 0 & \nabla_\perp^*\nabla_\perp\end{pmatrix}w_{em} + \frac{\partial^2}{\partial t^2}\begin{pmatrix}0 & 0 \\ 0 & \hat{\mathrm{I}}_\perp\end{pmatrix}w_e$,

$\psi = \sqrt{\kappa^2 - k_\perp^2}$ for $k_\perp < \kappa$ and $\psi = i\sqrt{k_\perp^2 - \kappa^2}$ for $k_\perp > \kappa$, $n = \sqrt{\varepsilon/\varepsilon_1}$, and $\rho = \sqrt{y^2 + z^2}$.

The formulae (4.104) and (4.105) yield the total field in the non-stationary half-space. The integrand pole residues yield monochromatic waves with the old frequency ω and the new one ω_1. The integrals along the cut lines drawn from the branch points yield the continuous spectrum of waves that describes the transition process. This continuous spectrum of waves is attenuated as $t \to \infty$.

It follows from (2.51) (Fig. 2.4) that the source $\mathbf{J}(t) = \mathbf{J}_0\exp(i\omega t)$ located in the infinite medium gives rise to a wave that collapses to the source point, having the new frequency ω_1. Let us examine the spatial distribution of this field in a steady-state regime. To do this, we find the integral components in Eqs. (4.104) and (4.105) which are caused by the residues in the poles $p^{(\pm)} = \pm i\omega_1$.

Calculation of the residue in the pole $p^{(\pm)}$ in Eq. (4.104) provides

$$\hat{R}_1\mathbf{E}_0\Big|_{p=i\omega_1} = 4(n^2-1)\left(-v_1^2\nabla\nabla + \frac{\partial^2}{\partial t^2}\right)\mathbf{I}\left[-\frac{e^{-i(\omega_1 t+\kappa R_0)}}{2\omega^2 n(n+1)R_0} + \theta(x-v_1 t)\right.$$
$$\left.\frac{e^{i(\omega_1 t-\kappa R_0)}}{2\omega^2 n(n+1)R_0} + \theta(v_1 t - x)\frac{e^{-i\omega_1 t}}{2\omega^2 n(n+1)}\int_0^{\kappa} J_0(k_\perp\rho)\frac{e^{-i(x+x_0)\psi}}{i\psi}k_\perp dk_\perp\right]$$

As it holds that [3],

$$\int_0^{\kappa} J_0(k_\perp\rho)\frac{e^{-ia\psi}}{\psi}k_\perp dk_\perp = -(1+i)\frac{\sin(\kappa\sqrt{\rho^2+a^2})}{\sqrt{\rho^2+a^2}} - \frac{1-\cos(\kappa a)-\sin(\kappa a)}{a},$$

we can obtain the following expression:

$$\hat{R}_1\mathbf{E}_0\Big|_{p=i\omega_1} = 4(n^2-1)\left(v_1^2\nabla\nabla - \frac{\partial^2}{\partial t^2}\right)\mathbf{I}\left\{\left[e^{-i\omega_1 t} - \theta(x-v_2 t)\right]e^{i\omega_1 t}\right]\frac{e^{-i\kappa R_0}}{2\omega^2 n(n+1)R_0}$$
$$+\,\theta(v_1 t - x)\frac{e^{-i\omega_1 t}}{2\omega^2 n(n+1)}\left[(1+i)\frac{\sin(\kappa\sqrt{\rho^2+(x+x_0)^2})}{\sqrt{\rho^2+(x+x_0)^2}}\right.$$
$$\left.\left.+\frac{1-\cos(\kappa(x+x_0))-\sin(\kappa(x+x_0))}{(x+x_0)}\right]\right\} \quad (4.106)$$

This expression contains a spherical standing wave being centered at the source point $x = -x_0$. Since the source is located outside the non-stationary half-space, this center is imaginary because the expression (4.106) is defined within the half-space $x > 0$.

Calculating the residue in the pole $p^{(-)}$ in Eq. (4.105) provides the following expression:

$$\hat{R}_2\mathbf{E}_0\Big|_{p=i\omega_1} = \left[\langle\hat{M}\mathbf{I}\rangle_1 + \langle\hat{M}\mathbf{I}\rangle_2\right]\frac{4(1-n^2)i\kappa}{2\omega_1(\omega+\omega_1)}e^{-i\omega_1 t} - \langle\hat{M}\mathbf{I}\rangle_1$$
$$\left[\frac{2\sin\left(\kappa\sqrt{\rho^2+(x-x_0)^2}\right)}{(1-i)\sqrt{\rho^2+(x-x_0)^2}} + \frac{1-\cos(\kappa(x-x_0))+\sin(\kappa(x-x_0))}{x_0-x}\right]$$
$$\frac{4(n^2-1)e^{-i\omega_1 t}}{2\omega_1(\omega+\omega_1)} \tag{4.107}$$

where $\langle\hat{M}\mathbf{I}\rangle_n$ are the values of vector $\hat{M}\mathbf{I}$ at certain points of the interval $0 < k_\perp < \kappa$. When calculating the expression (4.107), it is taken into account that

$$\int_0^\kappa \hat{M}\mathbf{I}\, J_0(k_\perp\rho)\frac{e^{ia\psi}}{i\psi}k_\perp dk_\perp = \int_0^\kappa \hat{M}\mathbf{I}\, J_0(k_\perp\rho)\frac{k_\perp dk_\perp}{i\psi}$$
$$+\int_0^\kappa \hat{M}\mathbf{I}\left[J_0(k_\perp\rho)\frac{e^{ia\psi}}{i\psi} - \frac{1}{i\psi}\right]k_\perp dk_\perp$$
$$= \langle\hat{M}\mathbf{I}\rangle_1\int_0^\kappa J_0(k_\perp\rho)\frac{e^{ia\psi}}{i\psi}k_\perp dk_\perp - i\kappa\langle\hat{M}\mathbf{I}\rangle_2.$$

The last expression also involves a spherical standing wave but unlike that in Eq. (4.106) its center is located at the point $x = x_0$, which is symmetric to the source point with respect to the boundary being located in the region where this wave exists (see Fig. 4.12).

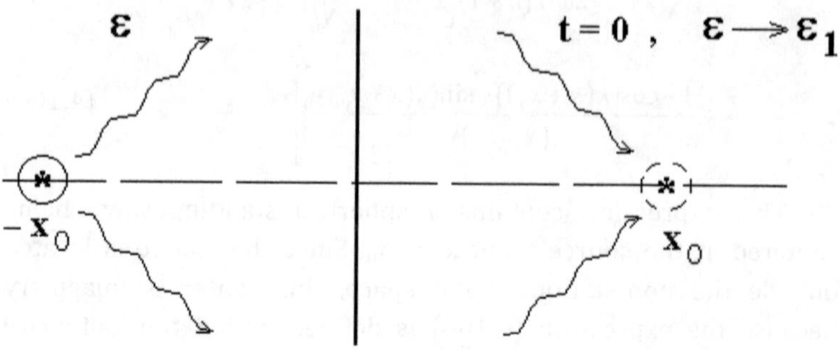

Figure 4.12 Relocation of the focus point of the secondary waves.

So, one can conclude that the plane boundary of a non-stationary dielectric relocates the focus point of secondary waves into its mirror image with respect to the boundary. It is essential that the field at the focus point has a peak, but this peak has a finite value, unlike for the case of an infinite medium. The space distribution of the field of this secondary wave is shown in Figs. 4.13 and 4.14. In these figures, the value of E is defined by the expression in the first square brackets in Eq. (4.107). Figure 4.13 indicates the case when $kx_0 = 1$ and Fig. 4.14 indicates the case when $kx_0 = 10$.

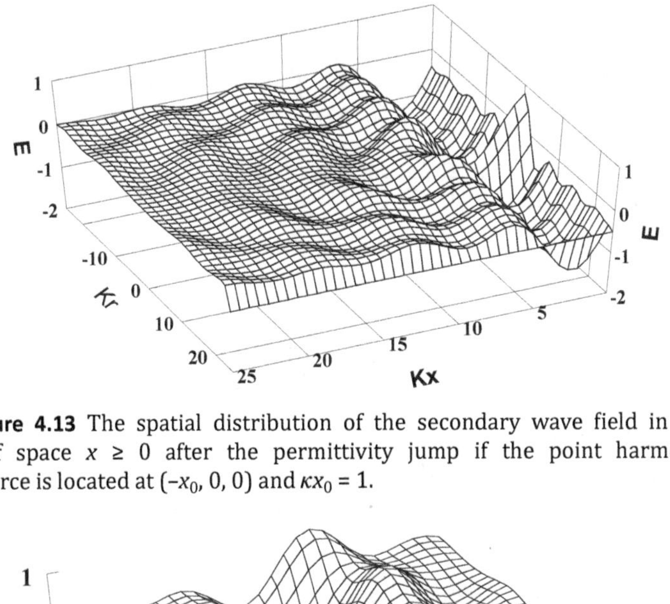

Figure 4.13 The spatial distribution of the secondary wave field in the half space $x \geq 0$ after the permittivity jump if the point harmonic source is located at $(-x_0, 0, 0)$ and $\kappa x_0 = 1$.

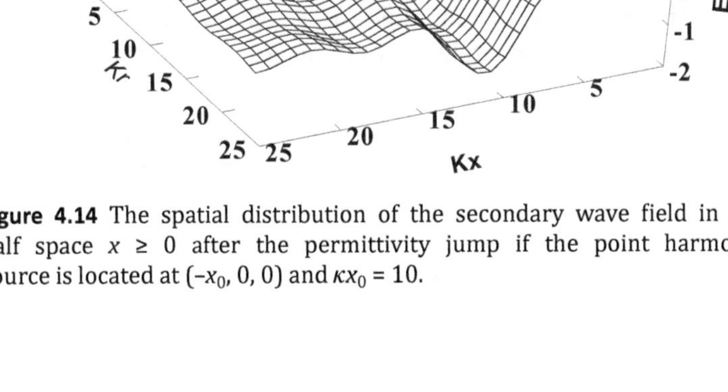

Figure 4.14 The spatial distribution of the secondary wave field in the half space $x \geq 0$ after the permittivity jump if the point harmonic source is located at $(-x_0, 0, 0)$ and $\kappa x_0 = 10$.

Comparing this result to the incident plane wave case, we can see that the reflection of the inverse wave from the boundary just represents an analogue to the refocused wave in the point source case.

This refracting wave passes through the medium boundary and forms a field component with the transformed frequency ω_1 outside the non-stationary half-space in both cases. Substituting the internal field $\mathbf{E}_{\text{in}}$, determined by (4.104) and (4.105), in the formula (4.47) we obtain the expression for the reflected field $\mathbf{E}_r = E_0 + \hat{K}_{\text{ex}}\mathbf{E}_{\text{in}}$ where $\hat{K}_{\text{ex}}$ is determined by (4.49):

$$\mathbf{E}_r = \mathbf{E}_0(\mathbf{x}) + 2\left(\frac{v_1^2 - v^2}{v_1}\right)^2 \left(\nabla\nabla - \frac{1}{v^2}\frac{\partial^2}{\partial t^2}\right) \int_{-i\infty}^{i\infty} \frac{dp}{2\pi i} e^{pt} \int_0^{\infty} \frac{e^{x\varphi/v - ix_0\psi}}{(p - i\omega)\varphi i\psi}$$
$$\left[\frac{v_1^2 Q_1 + p^2 I}{(p^2 + v_1^2\kappa^2)(\varphi + iv\psi)} + \frac{v_1}{2\varphi_1}\left[\frac{v_1^2 Q_2 + p^2 I}{\varphi_1 - iv_1\psi} + \frac{\hat{\mathrm{M}}}{\varphi_1 + iv_1\psi}\right]\frac{1}{v_1\varphi + v\varphi_1}\right]$$
$$\mathbf{I} J_0(k_\perp \rho) k_\perp dk_\perp, \tag{4.108}$$

where $Q_1 = \begin{pmatrix} \psi^2 & \psi\nabla_\perp \\ \psi\nabla_\perp^* & \nabla_\perp^*\nabla_\perp \end{pmatrix}$, $Q_2 = \frac{1}{v_1^2}\begin{pmatrix} -\varphi_\perp^2 & -iv_1\varphi_1\nabla_\perp \\ -iv_1\varphi_1\nabla_\perp^* & v_\perp^2\nabla_\perp^*\nabla_\perp \end{pmatrix}$,

$\varphi_a = \sqrt{p^2 + (v_a k_\perp)^2}$.

Calculating the residues in the poles in (4.108), we find that the spatial distribution of the field is determined by the integral

$$G(x,\rho) = \int_0^{\kappa} J_0(k_\perp \rho)\frac{e^{-ix_0\psi - ix\overline{\psi}}}{i\psi\overline{\psi}} k_\perp dk_\perp, \tag{4.109}$$

where $\overline{\psi} = \sqrt{(n\kappa)^2 - k_\perp^2}$ if $k_\perp < n\kappa$, and $\overline{\psi} = i\sqrt{k_\perp^2 - (n\kappa)^2}$ if $k_\perp > n\kappa$.

If the permittivity jump value is small then the integral can be calculated approximately, e.g. example if $n \geq 1$ it is equal to

$$G(x,\rho) \approx \left(1 + \frac{1}{2n^2} + \frac{1}{2n^2\kappa^2}\frac{d^2}{dx^2}\right)\left[(1-i)\frac{\sin(\kappa\sqrt{\rho^2 + (x+x_0)^2})}{\sqrt{\rho^2 + (x+x_0)^2}}\right.$$
$$\left. - \frac{1 - \cos(\kappa(x+x_0)) - \sin(\kappa(x+x_0))}{x + x_0}\right]. \tag{4.110}$$

It follows that the wave doing out from the non-stationary medium is focusing at the source point. However, the wave amplitude is an infinitesimally small quantity, being of the order of the factor $(n-1)$, so $G(x, \mathbf{r}) \to \kappa$ when $\rho = 0$ and $(x + x_0) \to 0$.

If considering a high permittivity jump ($n >> 1$), the function $G(x,\rho)$ in (4.109) takes the form

$$G(x,\rho) \approx -\frac{e^{-in\kappa x}}{n\kappa}\left[(1+i)\frac{\sin(\kappa\sqrt{\rho^2+x_0^2})}{\sqrt{\rho^2+x_0^2}} + \frac{1-\cos(\kappa x_0)-\sin(\kappa x_0)}{x_0}\right], \tag{4.111}$$

meaning that the wave going out of the non-stationary medium is focused along the normal to the boundary drawn from the source point.

4.5 Formation of Point Source Image by Time Change of Plasma

Time variations in plasma are typical and must be taken into account in many applications. They include the following: (i) Varying plasma in time is important for tuning the resonant characteristics of microdisk resonators by free carrier plasma injections; this has a wide range of potential applications including active switches or tunable filters [5–9]. (ii) The use of resonators composed of negative permittivity materials, such as plasma, can form the basis of effective small antenna elements. (iii) Plasma is used in a light-modulated photo-induced method for designing non-mechanical millimeter wave scanning devices. (iv) Transient phenomena in plasma lead to frequency up-shifting and wave generation. (v) The time variations are very important for remote sensing of plasma. Therefore, accurate time domain modeling for the investigation of the electromagnetic field with time-varying plasma is of great importance.

In this section, the transformation of a point source radiation by the bounded plasma created at some moment in time is investigated.

Let us consider the radiation of a point source located at the point $(-a, 0, 0)$. This source radiation is taken in a simplified form

$\mathbf{E}_0 = \mathbf{d}e^{i(\omega t - \kappa R)}/R$ where ω is the radiation frequency, $\kappa = \omega/v$, $v = c/\sqrt{\varepsilon}$, ε is the permittivity of the environment, c is the light velocity in vacuum, $R = \sqrt{(x+a)^2 + \rho^2}$, and $\rho^2 = y^2 + z^2$. To simplify the calculation, without the loss of the essential features, the vector **d** is considered constant, thus, the directional features of the radiator are not taken into account. We consider the field transformation by the plasma generated in the half-space $x \geq 0$ at zero moment of time. This plasma is characterized by the plasma frequency w_e and the plane boundary of the newly created plasma is placed at $x = 0$, normal to the x axis (Fig. 4.15). The plasma permittivity is given by $\varepsilon(\omega_e, \omega) = \varepsilon_2 - \omega_e^2/\omega^2$, where ε_2 describes the dispersion-free part of the new medium after the zero moment in the half-space $x > 0$.

Using the resolvent found in Section 3.6 the problem solution in the half-space $x > 0$ can be presented by two terms: $\mathbf{E} = \mathbf{E}_1 + \mathbf{E}_2$. The first is defined by the plasma change itself and the second is determined by the presence of the plasma boundary.

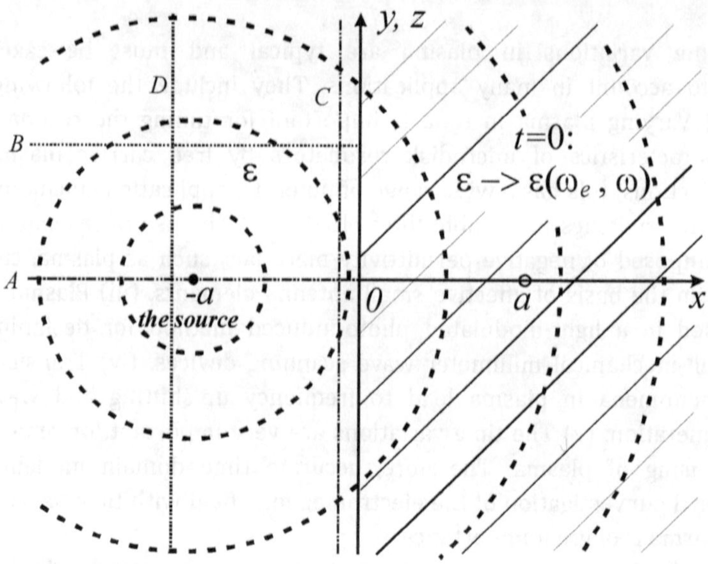

Figure 4.15 The system at zero moment.

The exact solution for the first term is given by the integral

$$\mathbf{E}_1 = \int_{\sigma-i\infty}^{\sigma+i\infty} \frac{dp}{2\pi i} \int_0^{\infty} dk \left(\varepsilon - \varepsilon_2 - \frac{\omega_e^2}{p^2} \right) \left[\frac{p^2 c^2}{\varepsilon_2 p^2 + \omega_e^2} \hat{Q} + \frac{\partial^2}{\partial t^2} \hat{I} \right] \mathbf{d}$$
$$\times \frac{k J_0(k\rho) e^{pt - i a \varphi(k)}}{(p - i\omega) i \varphi(k)} \left\{ \frac{e^{-ix\varphi(k)}}{\varepsilon_2 p^2 + \omega_e^2 + c^2 \kappa^2} - \frac{1}{2\psi(p,k)} \frac{e^{-\psi(p,k)x/c}}{\psi(p,k) - ic\varphi(k)} \right\}. \tag{4.112}$$

Here, $\hat{Q} = -\begin{pmatrix} \partial^2/\partial x^2 & \partial^2/\partial x \partial y & \partial^2/\partial x \partial z \\ \partial^2/\partial y \partial x & \partial^2/\partial y^2 & \partial^2/\partial y \partial z \\ \partial^2/\partial z \partial x & \partial^2/\partial z \partial y & \partial^2/\partial z^2 \end{pmatrix}$ is the matrix operator, $\hat{I} = \begin{pmatrix} 1 & 0 & 0 \\ 0 & 1 & 0 \\ 0 & 0 & 1 \end{pmatrix}$ is the unit matrix, $J_0(k\rho)$ is the Bessel function of the first kind, $\psi(p,k) = \sqrt{\varepsilon_2 p^2 + \omega_e^2 + c^2 k^2}$ and $\varphi(k) = \sqrt{\kappa^2 - k^2}$ if $k < \kappa$ and $\varphi(k) = -i\sqrt{k^2 - \kappa^2}$ for $k > \kappa$.

The exact solution for the second term has a more complex form

$$\mathbf{E}_2 = -\int_{\sigma-i\infty}^{\sigma+i\infty} \frac{dp}{2\pi i} \int_0^{\infty} dk \left(\varepsilon - \varepsilon_2 - \frac{\omega_e^2}{p^2} \right) \left[\frac{p^2 c^2}{\varepsilon_2 p^2 + \omega_e^2} \hat{Q}_{\|} V_{\|} + \frac{2c^2 p}{\sqrt{\varepsilon\left(\varepsilon_2 p^2 + \omega_e^2\right)}} \hat{Q}_{\times} V_{\times}\right.$$
$$\left. + \frac{\partial^2}{\partial t^2} \hat{I}_{\perp} V_{\perp} \right] \mathbf{d} \times \frac{k J_0(k\rho) e^{pt - i a \varphi(k)}}{2\psi(p,k)(p - i\omega) i \varphi(k)} \frac{e^{-\psi(p,k)x/c}}{\psi(p,k) - ic\varphi(k)}, \tag{4.113}$$

where the three matrix operators are

$$\hat{Q}_{\|} = -\begin{pmatrix} \partial^2/\partial y^2 + \partial^2/\partial z^2 & \partial^2/\partial x \partial y & \partial^2/\partial x \partial z \\ -\partial^2/\partial y \partial x & \partial^2/\partial y^2 & \partial^2/\partial y \partial z \\ -\partial^2/\partial z \partial x & \partial^2/\partial z \partial y & \partial^2/\partial z^2 \end{pmatrix},$$

$$\hat{Q}_{\times} = -\begin{pmatrix} 0 & 0 & 0 \\ 0 & \partial^2/\partial y^2 & \partial^2/\partial y \partial z \\ 0 & \partial^2/\partial z \partial y & \partial^2/\partial z^2 \end{pmatrix},$$

$$\hat{I}_{\perp} = \begin{pmatrix} 0 & 0 & 0 \\ 0 & 1 & 0 \\ 0 & 0 & 1 \end{pmatrix}.$$

Each operator is associated with a specific polarization of the radiation; this is emphasized by the functions $V_{\dagger}$ that are the analogous to Fresnel's coefficients: for the parallel polarization

$$V_{||}(k)=\frac{\varepsilon_2\phi(k)-(\varepsilon+\omega_e^2/\omega^2)\varphi(k)}{\varepsilon_2\phi(k)+(\varepsilon+\omega_e^2/\omega^2)\varphi(k)}, \tag{4.114}$$

and for the mixed and perpendicular polarizations

$$V_{\times}(k)=\frac{\sqrt{\varepsilon_2}\sqrt{\varepsilon+\omega_e^2/\omega^2}\left[\phi(k)-\varphi(k)\right]}{\varepsilon_2\phi(k)+(\varepsilon+\omega_e^2/\omega^2)\varphi(k)},\quad V_{\perp}(k)=\frac{\phi(k)-\varphi(k)}{\phi(k)+\varphi(k)}, \tag{4.115}$$

In these formulae $\phi(k)=-i\sqrt{k^2-\kappa^2(\varepsilon+\omega_e^2/\omega^2)}\Big/\varepsilon_2$ if $k^2-\kappa^2$ $(\varepsilon+\omega_e^2/\omega^2)>0$ and $\phi(k)=\sqrt{\kappa^2(\varepsilon+\omega_e^2/\omega^2)-k^2}\Big/\varepsilon_2$ for the opposite case, i.e., $k^2-\kappa^2$ $(\varepsilon+\omega_e^2/\omega^2)>0$.

The integrands in (4.112) and (4.113) have poles that create monochromatic waves with the frequencies ω, $\bar{\omega}_e=\omega_e\Big/\sqrt{\varepsilon_2}$ and $\omega_1=\sqrt{(\omega_e^2+\varepsilon\omega)}\Big/\sqrt{\varepsilon_2}$. The frequency ω wave is the same one as in the stationary case, when the source radiation is transformed by the plane boundary of the stationary plasma. The frequency $\bar{\omega}_e$ corresponds to the excitation of plasma oscillations induced by the appearance of the plasma. However, of special interest are the waves of frequency ω_1. It is worth noting that these waves are excited in the plasma independently of the relation between the plasma frequency and the source frequency.

The waves of this frequency arise in the unbounded plasma also due to the change of the plasma permittivity. In this case, these new waves are the result of the initial wave splitting into two waves propagating in opposite directions. One of these waves propagates as the initial wave going out of the radiation source. The other converges to the source point and its amplitude becomes infinite at this point. One can say that it is the consequence of infinity in all directions of the region the converging wave energy is gathered from. The solutions obtained to the problem considered show that the situation changes essentially when the region of interaction of the radiation and the matter is bounded.

To investigate this further we calculate the integrals in Eqs. (4.112) and (4.113) in that part which corresponds to the frequency ω_1. The calculation of the residues at the points

$p = \pm i\omega_1$ and integration with respect to k in (4.112) gives the waves of the frequency ω_1 determined by the plasma change itself:

$$\mathbf{E}_1^{(1)} = -\frac{(\varepsilon-\varepsilon_2)\omega_1^2+\omega_e^2}{2\varepsilon_2\omega_1^3(\omega_1+\omega)}\left[\frac{\omega_1^2c^2}{\varepsilon\omega}\hat{Q}+\frac{\partial^2}{\partial t^2}\hat{I}\right]\mathbf{d}\left\{\frac{e^{-\omega_1 t}}{\omega_1+\omega}+\frac{e^{\omega_1 t}}{\omega_1-\omega}\right\}F_1(a+x,\rho), \tag{4.116}$$

where

$$F_1(x,\rho)=\int_0^{\kappa} dk\frac{kJ_0(k\rho)e^{-ix\varphi(k)}}{\varphi(k)}=(1-i)\frac{\sin(\kappa\sqrt{x^2+\rho^2})}{\sqrt{x^2+\rho^2}}+i\frac{\cos(\kappa x)+\sin(\kappa x)-1}{x} \tag{4.117}$$

The spatial distribution of the field given by Eq. (4.112) does not have a maximum in the half-space $x \geq 0$, so it does not provide wave convergence to the source point.

The calculation in Eq. (4.113) gives only one wave of the frequency ω_1

$$\mathbf{E}_2^{(1)} = \frac{(\varepsilon-\varepsilon_2)\omega_1^2+\omega_e^2}{2\varepsilon_2\omega_1^3}\frac{e^{-\omega_1 t}}{\omega_1+\omega}\left\{\frac{\omega_1^2c^2}{\varepsilon\omega}\hat{Q}_{||}\mathbf{d}F_{||}(a-x,\rho)\right.$$
$$\left.+\frac{\partial^2}{\partial t^2}\hat{I}_{\perp}\mathbf{d}F_{\perp}(a-x,\rho)-\frac{2c^2\omega_1}{\varepsilon\omega}\hat{Q}_{\times}\mathbf{d}F_{\times}(a-x,\rho)\right\} \tag{4.118}$$

but it has a very noteworthy dependence on the spatial coordinates which is determined by the functions

$$F_{\dagger}(x,\rho)=\int_0^{\kappa} dkV_{\dagger}(k)\frac{kJ_0(k\rho)e^{-ix\varphi(k)}}{\varphi(k)}. \tag{4.119}$$

These functions cannot be calculated exactly because of the sophisticated Fresnel's coefficients (4.115) but they can be analyzed using the mean value theorem. It gives

$$F_{\dagger}^{\text{mean}}(x,\rho)=V_{\dagger}^{\text{mean}}F_1(x,\rho), \tag{4.120}$$

where $V_{\dagger}^{\text{mean}}$ is the mean value of Fresnel's coefficient for the integral in Eq. (4.119). The application of the mean value theorem does not bring large errors

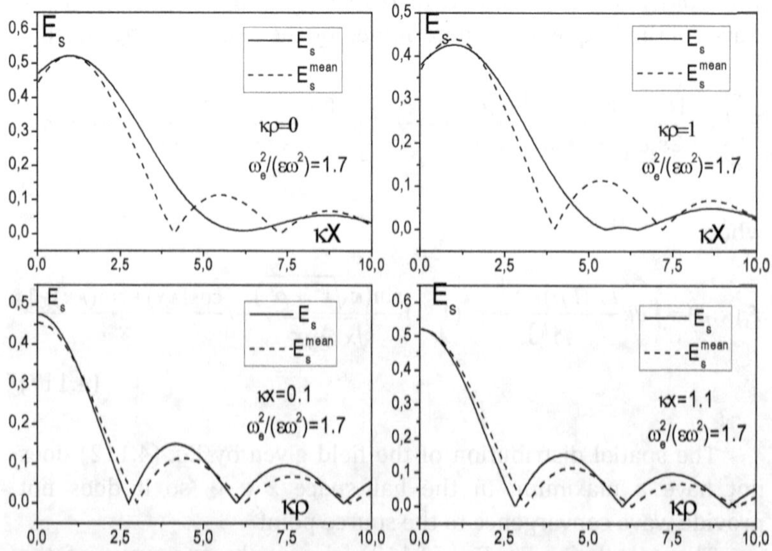

Figure 4.16 The exact and the approximate calculations of the functions (4.119) (solid line) and (4.120) (dashed line).

even when we take a constant value for $V_{\dagger}$ since the value of $V_{\dagger}$ has a small sweep in the interval of change for the integration variable k. This can be seen in Fig. 4.16, where a constant value for $V_{\perp}^{\text{mean}}$ is used. Figure 4.16 shows that the function $F_{\perp}^{\text{mean}}(x,\rho)$ does not differ substantially from the exact function $F_{\perp}(x, \rho)$ but Eq. (4.120) allows to use the explicit expression (4.117) for the qualitative analysis of the spatial field distribution. It follows that the amplitude of the ω_1 wave takes its maximum at the point ($x = a$, $\rho = 0$) which is the mirror symmetrical image of the source point with respect to the plasma plane boundary.

The spatial distribution of the frequency ω_1 field calculated numerically by the functions (4.114) and (4.115) shown in Fig. 4.17 confirms the overall qualitative analysis and emphasizes specific details. The field is concentrated in the vicinity of the source image point but is less pronounced for an over-critical (the plasma frequency is greater than the source frequency) than for an under-critical (the plasma frequency is less than the source frequency) plasma. In addition, the concentration range and the point of the concentration maximum depend on the polarization. For the parallel polarization characterized by $V_{||}$

(the electric field vector is parallel to the incident plane) the region of the field concentration stretches out in the case of the over-critical plasma and confines itself to a spot for the under-critical plasma. In contrast, focusing of the mixed polarization field and the perpendicular polarization field is the same for both plasma densities but the focusing points are different for these polarizations. The mixed polarization field is focusing at the point symmetrical to the source point with respect to the boundary and it coincides with the qualitative analysis, whereas the focus point for the perpendicular polarization field is shifted deep into the plasma.

In the case of both plasma frequencies considered, the time change of the medium property in the half-space forms a spatial image of the source on the other side of the plane boundary of the medium. This effect is similar to the effect of a plane lens made from a double negative material.

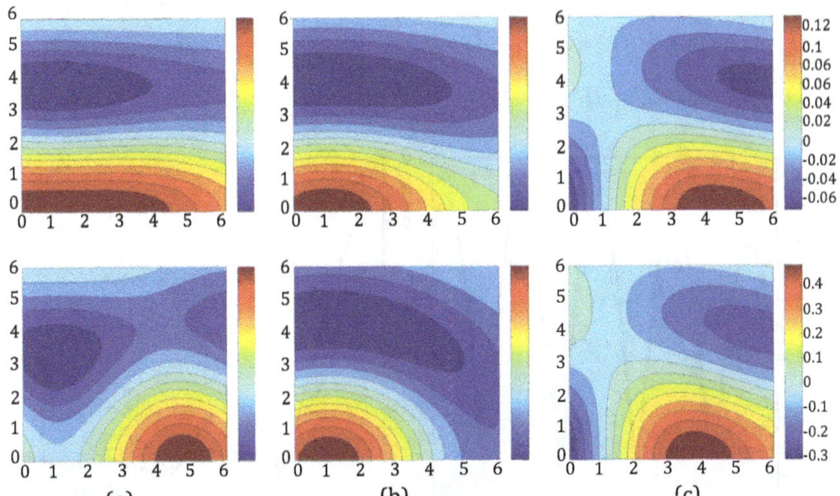

Figure 4.17 The spatial distribution of the field inside the plasma for various Fresnel coefficients: (a) $V_{||}$; (b) $V_{\perp}$; (c) $V_{\times}$ (the x-axis is horizontal, the ρ-axis is vertical). The upper row for over-critical plasma ($\kappa_a = 1$, $\varepsilon = \varepsilon_2 = 1$, $\omega/\omega_e = 1/2$). The lower row for under-critical plasma ($\kappa_a = 1$, $\varepsilon = \varepsilon_2 = 1$, $\omega/\omega_e = 2/1$).

It is worth noting that the presence of the medium boundary leads to a finite field value at the maximum point, this is in contrast to the medium time change in unbounded space.

The reflected field is also finite in the half-space free from plasma. The analysis shows that in this region the spatial distribution of the ω_1 wave is described by the functions

$$G_\dagger(x,\rho) = T_\dagger^{\text{mean}} \int_0^{\kappa} dk \frac{kJ_0(k\rho)e^{-ix\sqrt{\varepsilon\omega_1^2/c^2-k^2}-ia\varphi(k)}}{\varphi(k)}, \tag{4.121}$$

where, similar to $V_\dagger^{\text{mean}}$, the coefficients $T_\dagger^{\text{mean}}$ are the mean values of the Fresnel coefficients

$$T_\parallel = \frac{\varepsilon\varepsilon_2}{\phi_r(k)+(1+\varepsilon\omega_e^2/\omega^2)\varphi(k)},$$

$$T_\perp = \frac{\sqrt{\varepsilon_2}}{\sqrt{\varepsilon+\omega_e^2/\omega^2}\,[\phi_r(k)+\varphi(k)]}. \tag{4.122}$$

Here $\phi_r(k) = -i\sqrt{k^2-\kappa^2(\varepsilon+\omega_e^2/\omega^2)/\varepsilon_2}$ if $k^2 - \kappa^2\,(\varepsilon + \omega_e^2/\omega^2)/\varepsilon_2 > 0$ and $\phi_r(k) = \sqrt{k^2+\kappa^2(\varepsilon+\omega_e^2/\omega^2)/\varepsilon_2}$ for the opposite case, $k^2 - \kappa^2\,(\varepsilon + \omega_e^2/\omega^2)/\varepsilon_2 < 0$.

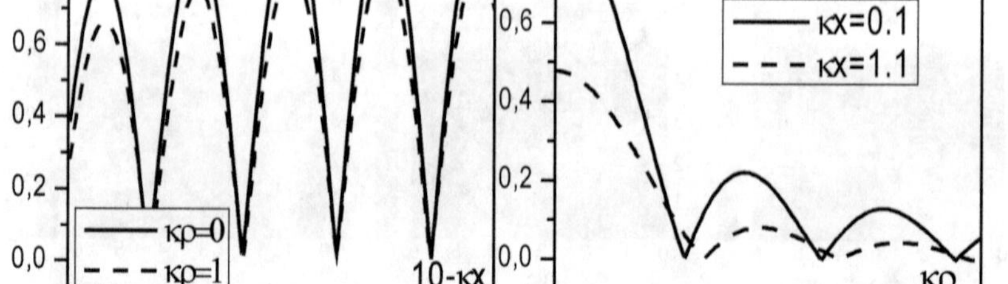

Figure 4.18 The field decreases along the normal to the boundary (right) and concentrates near the normal passing through the source point (left).

The field distribution along the normal to the medium boundary is shown in Fig. 4.18. The field value decreases starting from the plasma boundary (Fig. 4.18a, the solid line corresponds to the line A and the dotted line to the line B in Fig. 4.15). In contrast to the unbounded case, there is no conversion to infinity of the waves converging to the source point though they concentrate near the line passing through this point (Fig. 4.18b; the solid line corresponds to the line C and the dotted line to the line D in Fig. 4.15).

4.6 Frequency Change of Partial Spherical Waves Induced by Time Change of Medium Permittivity

A sphere is another basic element of micro-cavity devices and other structures. A wide range of structures, including nano-spheres and nano-shells as well as plasmonic resonators, optical nano-antennas and deep sub-wavelength size resonators, can be approximated by such a simple structure as a layered spherical region. The interactions between the light and the medium evolve in time in restricted space regions and their detailed investigations require a solution of initial-boundary-value problems. Impressive achievements in the practical solutions of such problems in experiments and the production of working devices presently exceed the theoretical development for such non-stationary problems. The latter are restricted to 1D or 2D layered structures and, more importantly, to the fields that change harmonically in time. Real objects are three-dimensional and ultrafast processes have a wide bandwidth, calling for a solution of three-dimensional (3D) initial-boundary problems. The simplest structure, adequate for many problems, is a layered spherically symmetric region. It should also be noted that the detailed descriptions of the geometrical features and the material properties that are frequently required for understanding and optimization of these micro-structured configurations present a substantial challenge to modeling and simulation tools. Some of the restrictions faced by numerical modelers when modeling 3D resonator structures are discussed in [10]. These include mesh generation, techniques for terminating the computational workspace with sufficient accuracy, and algorithm dispersion. However, the challenges are often dominated by computational resources.

In this section, a solution to the initial electromagnetic problem for the Maxwell's equations in a spherical symmetric medium is investigated by virtue of an effective mathematical tool developed for this purpose. This section is devoted to the investigation of the key processes in the interaction of an electromagnetic field with spherical particles containing time-varying media. The transformation of a plane harmonic wave caused by a time jump of the permittivity inside a dielectric sphere is investigated. Exact expressions for the transformed field are obtained as a solution of an initial and boundary value electromagnetic problem for Maxwell's equations based on longitudinal and transverse spherical vector functions.

4.6.1 Field Representation

An unconditioned ratio of the sphere radius to the spatial scale of the electromagnetic field is assumed. The investigation takes into account the full vector nature of the electromagnetic field by using expansions over the set of orthogonal vector spherical functions. We focus on the excited spectrum of the cavity resonances and the transient radiation in the near- and far-fields.

A 3D initial spherical symmetric problem for Maxwell's equations when the medium permittivity changes in time inside a region $V(t)$ is formulated in the form of the Volterra integral equation in the time domain

$$\mathbf{E}(t,\mathbf{r}) = \mathbf{F}(t,\mathbf{r}) + \frac{1}{4\pi}\left(\nabla\nabla - \frac{1}{v^2}\frac{\partial^2}{\partial t^2}\right)\int_0^t dt' \int_{V(t')} d\mathbf{r}' \hat{G}(t-t',\mathbf{r}-\mathbf{r}')\hat{V}_e(t')\mathbf{E}(t',\mathbf{r}'), \tag{4.123}$$

Here $\mathbf{F}(t, \mathbf{r})$ is the integral equation free term representing the initial electromagnetic field $\mathbf{E}_0(t, \mathbf{r})$ that existed in the background with permittivity ε before zero moment of time. The other values in (4.123) are the dyadic Green's function in the time domain $\hat{G} = \frac{1}{|\mathbf{r}-\mathbf{r}'|}\delta(t-t'-\frac{|\mathbf{r}-\mathbf{r}'|}{v})\hat{I}$ where $\hat{I}$ is the identity tensor and the environment operator $\hat{V}_e$ takes into account the medium properties including their changes in time.

The solution of the problem by virtue of equation (4.123) consists of two stages. At the first stage, the observation point $\mathbf{r}$ is considered as belonging to the region $V(t)$ and the solutions to Eq. (4.123) give the field inside this region. The second stage takes place when the observation point **r** is outside of the region $V(t)$. At this stage, Eq. (4.123) is the formula for the calculation of the exterior field through the interior one.

The main goal of this section is to investigate the principal transformations in the electromagnetic field when the medium property changes in a 3D bounded region in the form of a sphere. To do this it is enough to consider a sharp change of the permittivity but the integral equation method allows one to take into account arbitrary time changes in the permittivity, for example, by using a step-by-step approach. In this case, an arbitrary time change of the medium parameters is approximated by a sequence of step functions. The relation between the solution and the equation free term obtained at the first step is transferred to the other terms with the corresponding changes of the permittivity value.

Therefore, it is further assumed that the permittivity abruptly changes from the background value ε to the value ε_1 in the region $V(t)$ at the zero moment. It yields the operator $\hat{V}_e(t) = h(t)\hat{I}(\varepsilon_1 - \varepsilon)/\varepsilon$ where $\theta(t)$ is the Heaviside unit function. If the region V does not depend on time, the application of the Laplace transform $\hat{L} = \int_0^\infty dt e^{-pt}$ to equation (4.123) gives the equation in the spatial domain only

$$\mathbf{E}(p,\mathbf{r}) = \mathbf{E}_0(p,\mathbf{r}) + \frac{1}{4\pi}\left(\nabla\nabla - \frac{p^2}{v^2}\right)\int_V d\mathbf{r}'\frac{\exp(-p|\mathbf{r}-\mathbf{r}'|/v)}{|\mathbf{r}-\mathbf{r}'|}\hat{I}\frac{\varepsilon_1 - \varepsilon}{\varepsilon}\mathbf{E}(p,\mathbf{r}') \tag{4.124}$$

This equation is similar to the known one (for example, as in [11]) if $p = ik$ is substituted into Eq. (4.124).

In the following treatment, a sphere of radius a is considered, and it is assumed that the sphere center is located at the origin point of a system of axes (Fig. 4.19).

To solve Eq. (4.124), the well-known method of spherical vector functions is used [14]. These functions are given by the following formulae:

$$\mathbf{L}_{mn}(pr,\Omega) = \frac{1}{p}\frac{d}{dr}[j_n(pr)]\mathbf{P}_{mn}(\theta,\varphi) + \sqrt{n(n+1)}\frac{1}{pr}j_n(pr)\mathbf{B}_{mn}(\theta,\varphi) \tag{4.125}$$

$$\mathbf{M}_{mn}(pr,\Omega)=\sqrt{n(n+1)}j_n(pr)\mathbf{C}_{mn}(\theta,\varphi) \tag{4.126}$$

$$\mathbf{N}_{mn}(pr,\Omega)=n(n+1)\frac{1}{pr}j_n(pr)\mathbf{P}_{mn}(\theta,\varphi)+\sqrt{n(n+1)}\frac{1}{pr}\frac{d}{dr}[rj_n(pr)]\mathbf{B}_{mn}(\theta,\varphi), \tag{4.127}$$

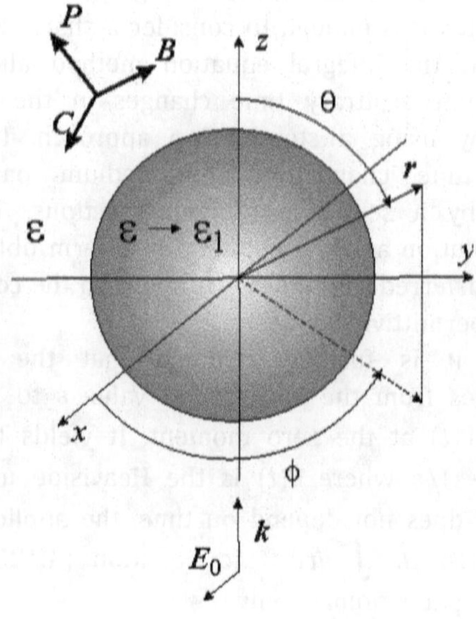

Figure 4.19 The geometry of the problem statement.

where $\mathbf{P}_{mn}=\hat{r}P_n^m(\cos\theta)e^{im\varphi}$, $\mathbf{B}_{mn}=[\hat{\theta}\frac{d}{d\theta}P_n^m(\cos\theta)+\hat{\varphi}\frac{im}{\sin\theta}P_n^m(\cos\theta)]$ $e^{im\varphi}\sqrt{n(n+1)}$, $\mathbf{C}_{mn}=-\hat{r}\times\mathbf{B}_{mn}$ are the longitudinal and the transverse angular functions of the solid angle $\Omega=(\theta,\varphi)$, $P_n^m(\cos\theta)$ is the associated Legendre function and $j_n(pr)$ is the spherical Bessel function. The use of these functions allows us to represent the kernel of Eq. (4.124) in the form of the series

$$\frac{\exp(-p|\mathbf{r}-\mathbf{r}'|/v)}{|\mathbf{r}-\mathbf{r}'|}\hat{I}=-\frac{p}{v}\sum_{n=0}^{\infty}\sum_{m=-n}^{n}\frac{2n+1}{n(n+1)}\frac{(n-m)!}{(n+m)!}[\mathbf{M}_{mn}(-ipr_{<}/v,\Omega)\mathbf{M}_{mn}^{h}(-ipr_{>}/v,\Omega')+\mathbf{N}_{mn}(-ipr_{<}/v,\Omega)\mathbf{N}_{mn}^{h}(-ipr_{>}/v,\Omega')+n(n+1)\mathbf{L}_{mn}(-ipr_{<}/v,\Omega)\mathbf{L}_{mn}^{h}(-ipr_{>}/v,\Omega')], \tag{4.128}$$

where $r_<$ ($r_>$) denotes the smaller (larger) r, r' and the functions $\mathbf{L}^h_{mn}$, $\mathbf{M}^h_{mn}$, $\mathbf{N}^h_{mn}$ differ from (4.125)–(4.127) by the function $h_n^{(2)}(pr)$ instead of $j_n(pr)$.

We consider now a plane wave $\mathbf{E}_0(t, \mathbf{r}) = \mathbf{e}_x E_0 e^{i\omega(t - z/v)}$ propagating along the z axis with an electric field vector directed along the x axis (Fig. 4.19). Then, the Laplace transform of the initial field and its expansion over the spherical vector functions gives

$$\mathbf{E}_0(p,\mathbf{r}) = \frac{E_0}{p-i\omega}\sum_{n=0}^{\infty}\frac{(2n+1)i^{n+1}}{2n(n+1)}\left[\mathbf{M}_{-1n}(\omega r/v,\Omega) - \mathbf{M}_{1n}(\omega r/v,\Omega) - \mathbf{N}_{-1n}(\omega r/v,\Omega) - \mathbf{N}_{1n}(\omega r/v,\Omega)\right] \quad (4.129)$$

This provides a basis on which to seek the solution to (4.124) in the form of two expansions $\mathbf{E}(p, \mathbf{r}) = \mathbf{E}^M(p, \mathbf{r}) + \mathbf{E}^N(p, \mathbf{r})$ over two kinds of spherical vector functions:

$$\mathbf{E}^M(p,\mathbf{r}) = \sum_{n=0}^{\infty}\sum_{m=-n}^{n}[A_{mn}(p)\mathbf{M}_{mn}(\omega r/v, \Omega) + B_{mn}(p)\mathbf{M}_{mn}(-ipr/v_1, \Omega)] \quad (4.130)$$

$$\mathbf{E}^N(p,\mathbf{r}) = \sum_{n=0}^{\infty}\sum_{m=-n}^{n}\left[\bar{A}_{mn}(p)\mathbf{N}_{mn}(\omega r/v, \Omega) + \bar{B}_{mn}(p)\mathbf{N}_{mn}(-ipr/v_1, \Omega)\right], \quad (4.131)$$

where $v_1 = c/\sqrt{\varepsilon_1}$ is the phase velocity in the changed medium inside the sphere and A_{mn}, B_{mn} are unknown coefficients. The substitution of these expressions into (4.124), integration over the sphere region and algebraic manipulations yields these coefficients

$$A_{-1n} = -A_{1n} = -\bar{A}_{-1n} = -\bar{A}_{1n} = \frac{(2n+1)i^{n+1}}{2n(n+1)}\frac{v_1^2}{v^2}E_0\frac{p+i\omega}{p^2+\omega_1^2} \quad (4.132)$$

$$B_{-1n} = -B_{1n} = \frac{(2n+1)i^{n+1}}{2n(n+1)}\frac{v^2-v_1^2}{v^2}\frac{E_0}{p-i\omega}\frac{p^2}{p^2+\omega_1^2}\frac{U_n(\omega,p)}{U_n(-ivp/v_1, p)} \quad (4.133)$$

$$\bar{B}_{-1n} = \bar{B}_{1n} = -\frac{(2n+1)i^{n+1}}{2n(n+1)}\frac{v^2-v_1^2}{v^2}\frac{E_0}{p-i\omega}\frac{p^2}{p^2+\omega_1^2}\frac{\bar{U}_n(\omega,p)}{\bar{U}_n(-ivp/v_1,p)}. \quad (4.134)$$

Here $\omega_1 = v_1\omega/v$ is the frequency transformed by the change of the permittivity inside the sphere. The functions in (4.133) and (4.134) are determined by the formulae

$$U_n(s,p) = \frac{s}{v} j_{n+1}(sa/v)h_n^{(2)}(-ipa/v) + \frac{ip}{v} j_n(sa/v)h_{n+1}^{(2)}(-ipa/v) \quad (4.135)$$

and

$$\bar{U}_n(s,p) = \frac{n+1}{a}\left(\frac{ip}{s} - \frac{s}{ip}\right) j_n(sa/v)h_n^{(2)}(-ipa/v)$$
$$-\frac{ip}{v} j_{n+1}(sa/v)h_n^{(2)}(-ipa/v) - \frac{s}{v} j_n(sa/v)h_{n+1}^{(2)}(-ipa/v) \quad (4.136)$$

All other coefficients in (4.130) and (4.131) are equal to zero.

The similarity of the spatial structure of the solutions (4.130), (4.131) to the structure of the free term (4.129) and the relations (4.132)–(4.134) allows one to consider an arbitrary time variation of the permittivity by approximating it with step functions in time.

4.6.2 Analysis of the Inner Field

The application of the inverse Laplace transform to (4.130), (4.131) together with the formulae (4.132)–(4.134) gives the field in time domain. It can be seen that instead of one time harmonic dependence of the initial wave with the frequency ω equal for all vector terms in (4.129), the transformed field gains a whole spectrum which is given by the poles of the functions (4.130)–(4.134). The first terms in the square brackets in (4.130) and (4.131) have the poles $p_\pm = \pm i\omega_1$ in the functions $A_{\pm 1n}(p)$ (4.132) that give two sets of waves with the transformed frequencies $\pm\omega_1$ for each spherical vector: $\sim e^{\pm i\omega_1 t}\mathbf{M}_{\pm 1n}(\omega r/v, \Omega)$ and $\sim e^{\pm i\omega_1 t}\mathbf{N}_{\pm 1n}(\omega r/v, \Omega)$. One set of these waves represents the spherical waves that diverge from the sphere center whereas the other represents the waves that converge to the center. The spatial structure of these waves remains unchanged compared to the initial wave but the frequency changes in such a way that the wave phase velocity is equal to the velocity in the changed medium v_1.

The second terms in (4.130) and (4.131) have a more complex spectrum. First of all there are waves with the initial wave frequency ω determined by the pole $p = i\omega$ in (4.133) and (4.134). These waves maintain the frequency of the initial wave but they change the spatial dependence: $\sim e^{\pm i\omega t}\mathbf{M}_{\pm 1n}(\omega r/v_1, \Omega)$ and $\sim e^{i\omega t}\mathbf{N}_{\pm 1n}(\omega r/v_1, \Omega)$. These waves are excited by the wave impinging onto the sphere after the permittivity change when the phase velocity inside the sphere becomes equal to v_1. The range of existence for these waves is defined by the asymptotic behavior of the whole integrand in the inverse Laplace transform at infinity. It gives the broadened sphere layer $a - v_1 t \leq r \leq a$ which is created at the sphere surface at the zero moment and begins to broaden until its inner boundary collapses at the sphere center at the moment $t_0 = a/v_1$ (Fig. 4.20). After this moment, the inner waves with frequency ω exist in the whole bulk of the sphere.

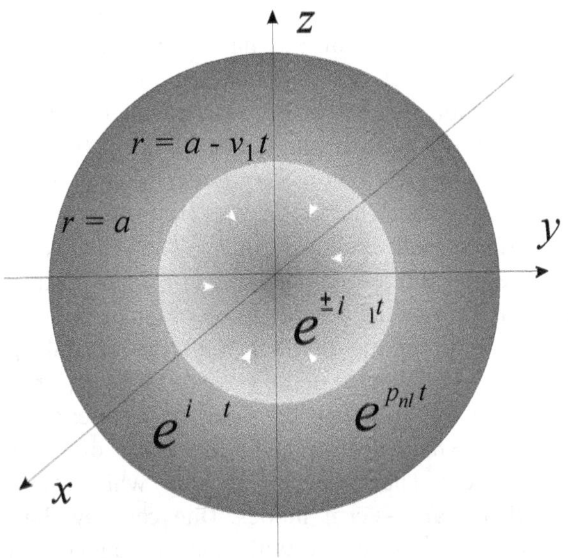

Figure 4.20 Structure of the waves distribution in the time interval $0 \leq t \leq a/v_1$.

The poles $\pm p_1 = \pm i\omega_1$ of the second terms in (4.130) and (4.131) give the waves with the same frequency as in the first terms but they exist only in the broadened sphere layer $a - v_1 t \leq r \leq a$ during the interval $0 < t < a/v_1$. These waves annihilate the similar waves $\sim e^{\pm i\omega_1 t}$ from the first terms via the equality $\mathbf{M}_{mn}\ (-ipr/v_1,\ \Omega)U_n(-ipv/v_1,\ p)|_{p\ =\ \pm i\omega_1} = \mathbf{M}_{mn}(\omega r/v,\ \Omega)U_n(\omega,\ p)$. Thus, the waves $\sim e^{\pm i\omega_1 t}\mathbf{M}_{\pm 1n}(\omega r/v,\ \Omega)$ and $\sim e^{\pm i\omega_1 t}\mathbf{N}_{\pm 1n}(\omega r/v,\ \Omega)$ exist only in the contractive sphere region $0 \le r \le a - v_1 t$ until the moment $t_0 = a/v_1$. This differs substantially from the case of an unbounded medium where the initial wave with frequency ω is substituted by the waves with the transformed frequency ω_1 completely in the whole space.

The sphere layer $a - v_1 t \le r \le a$ is occupied also by waves of indefinite spectra determined by the zeros of the denominators (4.135) and (4.136) in (4.133) and (4.134) respectively. These spectra are different for various spherical vectors $\mathbf{M}_{\pm 1n}$ and $\mathbf{N}_{\pm 1n}$ which are represented by various polarizations of the partial spherical terms of the initial wave. This is similar to the effect of various polarizations of an initial wave in the case of a 2D problem. In that case, the permittivity change occurs in a cylindrical resonator and the spectrum for the E-polarization of the initial wave (i.e., the vector **E** is parallel to the cylinder axes) is given by the zeros of the denominator in (4.133), whereas for the H-polarization (i.e., the vector **H** is parallel to the cylinder axis) it is given by the zeros of the denominator in (4.134). These two cases of polarization can be independently realized in practice in the 2D case, whereas in the 3D problem considered both polarizations exist simultaneously and they cannot be separated as they are the components of a single plane wave.

The spectra given by the second group of terms in Eqs. (4.130) and (4.131) depend on the number n of the partial spherical vector and create an infinite set of the waves $\sim e^{p_{nl}t}\mathbf{M}_{\pm 1n}(-ip_{nl}r/v_1, \Omega)$ and $\sim e^{\overline{p}_{nl}t}\mathbf{N}_{\pm 1n}(-i\overline{p}_{nl}r/v_1, \Omega)$ with the complex frequencies p_{nl} and $\overline{p}_{nl}$, $l = 0, 1, 2,...$ which are equal to the zeros of (4.135) and (4.136), respectively. These spectra include whispering gallery modes as well as Fabry–Perot modes. One can say that at the moment $t_0 = a/v_1$, the waves with the frequencies $\pm\omega_1$ are replaced by the waves of these spectra completely in the whole sphere.

4.6.3 Analysis of the Exterior Field

The substitution of the inner field found into Eq. (4.124) and integration gives the field outside the sphere. It consists of the initial field and the scattered one, $\mathbf{E}_{ex} = \mathbf{E}_0 + \mathbf{E}_{sc}$. The latter field separates into two polarizations, $\mathbf{E}_{sc} = \mathbf{E}_{sc}^M + \mathbf{E}_{sc}^N$:

$$\mathbf{E}_{sc}^M(p,\mathbf{r}) = \frac{v^2 - v_1^2}{v^2}\frac{E_0}{p - i\omega}\frac{p^2}{p^2 + \omega_1^2}\sum_{n=0}^{\infty}\frac{(2n+1)i^{n+1}}{2n(n+1)}$$

$$[\mathbf{M}_{-1n}^h(-ipr/v,\Omega) - \mathbf{M}_{1n}^h(-ipr/v,\Omega)]\frac{U_n^b(\omega,p)}{U_n(-ivp/v_1,p)} \quad (4.137)$$

$$\mathbf{E}_{sc}^N(p,\mathbf{r}) = \frac{v^2 - v_1^2}{v^2}\frac{E_0}{p - i\omega}\frac{p^2}{p^2 + \omega_1^2}\sum_{n=0}^{\infty}\frac{(2n+1)i^{n+1}}{2n(n+1)}$$

$$[\mathbf{N}_{-1n}^h(-ipr/v,\Omega) - \mathbf{N}_{1n}^h(-ipr/v,\Omega)]\frac{\bar{U}_n^b(\omega,p)}{\bar{U}_n(-ivp/v_1,p)}, \quad (4.138)$$

where the numerators are determined by the functions

$$U_n^b(s,p) = \frac{s}{v}j_{n+1}(sa/v)j_n(-ipa/v_1) + \frac{ip}{v_1}j_n(sa/v)j_{n+1}(-ipa/v_1) \quad (4.139)$$

and

$$\bar{U}_n^b(s,p) = \frac{n+1}{a}\left(\frac{ip}{s} - \frac{s}{ip}\right)j_n(sa/v)j_n(-ipa/v_1)$$

$$-\frac{ip}{v_1}j_{n+1}(sa/v)\,j_n(-ipa/v_1) - \frac{s}{v}j_n(sa/v)j_{n+1}(-ipa/v_1) \quad (4.140)$$

and the denominators are as in the interior field, the formulae (4.135) and (4.136). The identity

$$U_n^b(\omega,p)U_n(-ivp/v_1,p) + U_n(\omega,p)U_n^b(-ivp/v_1,p) = \frac{v}{pa^2}U_n^b(\omega,pv/v_1) \quad (4.141)$$

is used for obtaining (4.137) and (4.138).

One can see that the exterior field consists of the wave with the frequency ω provided by the pole $p = i\omega$ in (4.137), (4.138) and the waves of the spectra provided by the zeros of the functions $U_n(-ivp/v_1, p)$ and $\bar{U}_n(-ivp/v_1, p)$ which are the same as in the expressions for the interior field. The waves with the frequencies $\pm\omega_1$ are absent in the exterior field as the points $p_\pm = \pm i\omega_1$ are not singular for the expressions (4.137) and (4.138). This is because the numerators $U_n^b(\omega,p)$, $\bar{U}_n^b(\omega,p)$ in these expressions are equal to zero at these points, $U_n^b(\omega, \pm i\omega_1) = 0$, $\bar{U}_n^b(\omega, +i\omega_1) = 0$.

Calculation of the Poynting vector $\mathbf{S} = \varepsilon_0 c^2\ \mathrm{Re}([\mathbf{E} \times \mathbf{B}^*])/2$ and integration over all angles gives the cross-section for the partial wave of the complex frequency $p_{nk} = i\omega_{nk} + \alpha_{nk}$

$$\sigma_{nk} = |a_n|^2 \frac{1+n^2(n+1)^2}{n(n+1/2)(n+1)} f(p_{nk})V_n(p_{nk})f^*(p_{nk})$$

$$\times V_n^*(p_{nk})u_n(p_{nk},r)w_n^*(p_{nk},r)e^{-2\alpha_{nk}t} \tag{4.142}$$

It is seen that the spatial dependence is given by the functions

$$u_n = \sqrt{n(n+1)}\frac{iv}{pr}h_n^{(2)}(-ipr/v),\ w_n = \sqrt{n(n+1)}\frac{iv}{pr}\frac{d}{dr}\left[rh_n^{(2)}(-ipr/v)\right] \tag{4.143}$$

and the temporal dependence has a decaying character. It follows from (4.142) that de-excitation of the different partial waves is going on with different rates which are determined by the real parts of the complex frequency $p_{nk} = i\omega_{nk} + \alpha_{nk}$.

So, the temporal analysis of the electromagnetic field transformed by the time jump change of the permittivity in the dielectric sphere illuminated by a harmonic plane wave shows that the evolution of the interior field is different in various time intervals, which are defined by the time needed for the wave to move from the sphere surface to its center. In the interval $0 \le t \le a/v_1$ the region of the sphere consists of two sub-regions $0 \le r \le a - v_1t$ and $a - v_1t < r \le a$. The waves with the frequency $\omega_1 = \omega\sqrt{\varepsilon/\varepsilon_1}$ exist only in the first collapsing sub-region. The field evolution in the second sub-region is determined by the creation of "wide spectra" waves that are different for various vector components of the field. These spectra waves occupy the whole sphere after the moment a/v_1. The field scattered outside the sphere contains the wave with the frequency of the initial wave and the waves of the spectra that are the same as in the interior field. This phenomenon is a consequence of the finite size of the region where the permittivity changes in time.

4.7 Evolution of Waves after Plasma Ignition in a Sphere

The transformation of an incident plane harmonic wave caused by the instantaneous creation of a plasma sphere is investigated in this section. As discussed in the previous section, the exact expressions for the transformed field are obtained by virtue of the solution of an initial and boundary value electromagnetic problem for Maxwell's equations based on the longitudinal and transverse spherical vector functions. The section is devoted to the investigation of the key processes in the interaction of an electromagnetic field with a spherical plasma particle after its creation. An unconditioned ratio of the sphere radius and the spatial scale of the electromagnetic field is assumed. The investigation takes into account the full vector nature of the electromagnetic field by using the expansions over the set of orthogonal vector spherical functions.

4.7.1 Solution to the Problem

A 3D initial spherical symmetric problem for Maxwell's equations is formulated for the case when in a field of a plane monochromatic wave $E_0(t, \mathbf{r}) = e_x E_0 e^{i\omega(t - z/v)}$ a plasma sphere appears at zero moment of time (Fig. 4.20). An electromagnetic problem for a field is formulated as an initial and boundary value problem in the form of the integral equation, the Laplace transform of which is (4.124). A solution is constructed by using the spherical vector functions (4.125)–(4.127). An interior field can be represented in the form of two expansions corresponding to TE and TM polarizations:

$$\mathbf{E}(t,\mathbf{r}) = \mathbf{E}_{\text{early}}(t,z) + \hat{L}^{-1}(\mathbf{E}^M(p,\mathbf{r}) + \mathbf{E}^N(p,\mathbf{r})), \tag{4.144}$$

where

$$\mathbf{E}_{\text{early}} = \frac{v_1}{2v^2}[(v_1 + v)e^{i\omega_1 t} + (v_1 - v)e^{-i\omega_1 t}]E_0 \mathbf{e}_x e^{-i\omega z/v} \tag{4.145}$$

is the field at the early stage of the evolution. This field consists of two waves with transformed frequency

$$\omega_1 = \sqrt{\varepsilon\omega^2 + \omega_e^2} \Big/ \sqrt{\varepsilon / \varepsilon_1} \tag{4.146}$$

Other two terms in Eq. (4.144) are sums of waves represented by the vector modes as in (4.130) and (4.131)

$$\mathbf{E}^M(p, \mathbf{r}) = \sum_{n=1}^{\infty} B_{-1n}(p)[\mathbf{M}_{-1n}(-ipr/v_1, \Omega) - \mathbf{M}_{1n}(-ipr/v_1, \Omega)] \tag{4.147}$$

$$\mathbf{E}^N(p, \mathbf{r}) = \sum_{n=1}^{\infty} \bar{B}_{-1n}(p)\left[\mathbf{N}_{1-n}(-ipr/v_1, \Omega) + \mathbf{N}_{1n}(-ipr/v_1, \Omega)\right], \tag{4.148}$$

where the coefficients are given by the formulae

$$B_{-1n} = \frac{(2n+1)i^{n+1}}{2n(n+1)} \frac{v^2 - v_1^2}{v^2} \frac{E_0}{p - i\omega} \frac{p^2}{p^2 + \omega_1^2} \frac{U_n(\omega, p)}{U_n(-ivp/v_1, p)} \tag{4.149}$$

$$\bar{B}_{-1n} = -\frac{(2n+1)i^{n+1}}{2n(n+1)} \frac{v^2 - v_1^2}{v^2} \frac{E_0}{p - i\omega} \frac{p^2}{p^2 + \omega_1^2} \frac{\bar{U}_n(\omega, p)}{\bar{U}_n(-ivp/v_1, p)}. \tag{4.150}$$

The spatial-temporal behavior of the waves (4.147) and (4.148) is determined by the functions

$$U_n(s, p) = \frac{s}{v} j_{n+1}\left(\frac{sa}{v}\right) h_n^{(2)}\left(\frac{-ipa}{v}\right) + \frac{ip}{v} j_n\left(\frac{sa}{v}\right) h_{n+1}^{(2)}\left(\frac{-ipa}{v}\right) \tag{4.151}$$

$$\begin{aligned}\bar{U}_n(s, p) = {} & \frac{n+1}{a}\left(\frac{ip}{s} - \frac{s}{ip}\right) j_n\left(\frac{sa}{v}\right) h_n^{(2)}\left(\frac{-ipa}{v}\right) \\ & - \frac{ip}{v} j_{n+1}\left(\frac{sa}{v}\right) h_n^{(2)}\left(\frac{-ipa}{v}\right) - \frac{s}{v} j_n\left(\frac{sa}{v}\right) h_{n+1}^{(2)}\left(\frac{-ipa}{v}\right).\end{aligned} \tag{4.152}$$

The field outside the sphere after its creation is given by similar expressions but with different coefficients

$$\mathbf{E} = \mathbf{E}_0 + \mathbf{E}_{sc}^M + \mathbf{E}_{sc}^N \tag{4.153}$$

$$\mathbf{E}_{sc}^M(p, \mathbf{r}) = \sum_{n=1}^{\infty} B_{-1n}^b [\mathbf{M}_{-1n}(-ipr/v, \Omega) - \mathbf{M}_{1n}(-ipr/v, \Omega)] \tag{4.154}$$

$$\mathbf{E}^{N}_{sc}(p,\mathbf{r})=\sum_{n=1}^{\infty}\bar{B}^{b}_{-1n}[\mathbf{N}_{-1n}(-ipr/v,\,\Omega)+\mathbf{N}_{1n}(-ipr/v,\,\Omega)] \tag{4.155}$$

$$B^{b}_{-1n}=\frac{(v^2-v_1^2)E_0}{v^2(p-i\omega)}\frac{p}{p^2+\omega_1^2}\frac{(2n+1)i^{n+1}}{2n(n+1)}\frac{U^{b}_{n}(\omega,p)}{U_n(-ivp/v_1,p)} \tag{4.156}$$

$$\bar{B}^{b}_{-1n}=\frac{(v^2-v_1^2)E_0}{v^2(p-i\omega)}\frac{p}{p^2+\omega_1^2}\frac{(2n+1)i^{n+1}}{2n(n+1)}\frac{\bar{U}^{b}_{n}(\omega,p)}{\bar{U}_n(-ivp/v_1,p)} \tag{4.157}$$

$$U^{b}_{n}(s,p)=\frac{s}{v}j_{n+1}\left(\frac{sa}{v}\right)j_n\left(\frac{-ipa}{v_1}\right)+\frac{ip}{v_1}j_n\left(\frac{sa}{v}\right)j_{n+1}\left(\frac{-ipa}{v_1}\right) \tag{4.158}$$

$$\bar{U}^{b}_{n}(s,p)=\frac{n+1}{a}\left(\frac{ip}{s}-\frac{s}{ip}\right)j_n\left(\frac{sa}{v}\right)j_n\left(\frac{-ipa}{v_1}\right)$$
$$-\frac{ip}{v_1}j_{n+1}\left(\frac{sa}{v}\right)j_n\left(\frac{-ipa}{v_1}\right)-\frac{s}{v}j_n\left(\frac{sa}{v}\right)j_{n+1}\left(\frac{-ipa}{v_1}\right) \tag{4.159}$$

The expressions obtained allow analysis of the evolution of the electromagnetic field inside, as well as outside, the plasma sphere after its creation.

4.7.2 The Evolutionary Process

Analysis shows a process which is similar to that observed for the dielectric sphere considered in the previous section. Instead of one time harmonic dependence of the initial wave with the frequency ω the transformed field acquires a whole spectrum. First of all, the field evolution process is divided into two stages. The first, early, stage lasts from the process beginning (when the plasma sphere appears) until the moment when this process is qualitatively replaced by another one. The time interval for this stage is equal to $0<t<a\sqrt{\varepsilon_1}/c$ and the whole spatial region of the sphere divides into two sub-regions, the inner sphere $0 \le r \le a - ct/\sqrt{\varepsilon_1}$ and the sphere layer $a - ct/\sqrt{\varepsilon_1} \le r \le a$, similar to that in Fig. 4.20. Inside the inner sphere, the initial wave is transformed entirely into two waves given by the formula (4.145).

These waves have the new frequency (4.146) and this process is analogous to the simple process in an unbounded medium. In a similar manner to the unbounded process, only the frequency changes; the spatial distribution of the field remains unchanged and is the same as in the initial plane wave. The phase velocity of these secondary waves is equal to $c/\sqrt{\varepsilon_1}$. This process is restricted spatially by the contractive inner sphere $0 \le r \le a - ct/\sqrt{\varepsilon_1}$ which collapses in the sphere center at the moment $t_0 = a/\sqrt{\varepsilon_1}/c$. It differs substantially from the case of an unbounded medium where the initial wave with frequency ω is substituted by the waves with the transformed frequency ω_1 instantaneously in the whole space. In the case considered there is the second sub-region in the same time interval—the sphere layer $a - ct/\sqrt{\varepsilon_1} \le r \le a$. This layer is created at the sphere surface at the zero moment and begins to broaden until its inner boundary collapses at the sphere center at the moment $t_0 = a\sqrt{\varepsilon_1}/c$.

In the sphere layer there are the waves with the initial wave frequency ω. These waves maintain the frequency of the initial wave but they change the spatial dependence: $\sim e^{i\omega t}\mathbf{M}_{\pm 1n}(\omega r/v_1, \Omega)$ and $\sim e^{i\omega t}\mathbf{N}_{\pm 1n}(\omega r/v_1, \Omega)$. These waves are excited by the wave impinging onto the sphere after the plasma appearing. After the moment $t_0 = a\sqrt{\varepsilon_1}/c$ the inner waves with the frequency ω exist in the whole bulk of the sphere. The waves with frequencies ω_1 transform into indefinite spectra. These spectra are different for various spherical vectors $\mathbf{M}\pm_{1n}$ and $\mathbf{N}\pm_{1n}$, depend on the number n of the partial spherical vector and create an infinite set of waves: TE waves

$$\mathbf{E}^M(t,\mathbf{r}) = \mathbf{E}_1^M(\mathbf{r})e^{i\omega t} + \sum_{n=1}^{\infty}\sum_{l} E_{nl}^M(p_{nl})[\mathbf{M}_{-1n}(k_{nl},\Omega) - \mathbf{M}_{1n}(k_{nl},\Omega)]e^{p_{nl}t} \tag{4.160}$$

and TM waves

$$\mathbf{E}^N(t,\mathbf{r}) = \mathbf{E}_1^N(\mathbf{r})e^{i\omega t} + \sum_{n=1}^{\infty}\sum_{l} E_{nl}^N(\bar{p}_{nl})[\mathbf{N}_{-1n}(\bar{k}_{nl},\Omega) + \mathbf{N}_{1n}(\bar{k}_{nl},\Omega)]e^{\bar{p}_{nl}t} \tag{4.161}$$

Here p_{nl} and $\bar{p}_{nl}$, $l = 0, 1, 2, \ldots$ are the zeros of dispersion equations for the eigen-frequencies for TE and TM waves

$$U_n(-ivp/v_1, p) = 0 \text{ and } \bar{U}_n(-ivp/v_1, p) = 0 \tag{4.162}$$

respectively, Table 4.1 given at the end of the section. One can see that amongst the excited waves there are waves with very high Q-factors (bold in Table 4.1).

Table 4.1a The eigen-frequency for TE waves, $\mathbf{M}_{\pm 1n}$

n	Re(p_k)	Im(p_k)	$Q(p_k)$	n	Re(p_k)	Im(p_k)	$Q(p_k)$
2	-0.36027	5.795801	8.043621	9	7.37E-05	5.191384	7.04E+04
2	-0.3829	19.63995	25.64623	**9**	**-0.00568**	**5.19225**	**9.14E+02**
7	**-0.00139**	**5.131524**	**1845.876**	9	-0.3578	19.80761	5.54E+01
7	-0.30145	7.516539	12.46723	**10**	**0.000122**	**5.233264**	**4.27E+04**
7	-0.36687	18.3031	24.94527	10	-0.3463	18.70175	5.40E+01
8	**-3.98E-05**	**5.145795**	**1.29E+05**	**11**	**-0.00016**	**5.271872**	**3.28E+04**
8	-0.36243	19.05795	5.26E+01	11	-0.07614	9.686253	1.27E+02

Table 4.1b The eigen-frequency for TM waves, $\mathbf{N}_{\pm 1n}$

n	Re(p_k)	Im(p_k)	$Q(p_k)$
2	-0.37278	6.033678	8.0929
2	-0.35546	16.90632	23.78085
7	0.000279	4.959192	8877.004
7	0.010958	5.414122	247.0322
7	-0.26289	19.00854	36.15278

The $\mathbf{M}_{\pm 1n}$ waves are pure lateral ones but the $\mathbf{N}_{\pm 1n}$ waves have radial components for the electric field.

The exterior field given by (4.154) and (4.155) also consists of waves with different polarizations. This field contains the initial wave with the frequency ω and the new waves of the spectra provided by the zeros of the functions $U_n(-ivp/v_1, p)$ and $\bar{U}_n(-ivp/v_1, p)$ which are the same as in the expressions for the interior field. The temporal behavior of some of these waves is shown in Fig. 4.21. The waves with the frequencies ω_1 are absent in the exterior field.

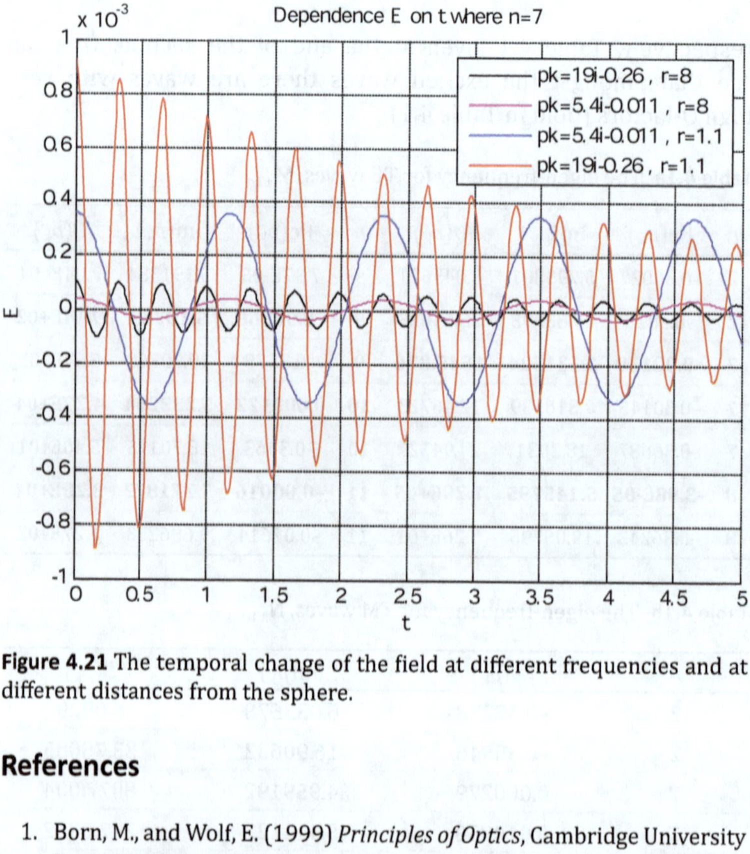

Figure 4.21 The temporal change of the field at different frequencies and at different distances from the sphere.

References

1. Born, M., and Wolf, E. (1999) *Principles of Optics*, Cambridge University Press, Cambridge, 952 p.
2. Mittra, R., and Lee, S. W. (1971) *Analytical Techniques in the Theory of Guided Waves*, The Macmillan Company, New York, 324 p.
3. Bateman, H., and Erdely, A. (1954) *Tables of Integral Transforms*, New York, Toronto, London: McGraw-Hill Book Comp., Inc., 344 p.
4. Prudnikov, A. P., Brychkov, Yu A., and Marichev, O. I. (1992) *Integrals and Series: Special Functions*, New York: Gordon and Breach Science Publishers, 750 p.
5. Kuo, S., and Ren, A. (1993) Experimental study of wave propagation through a rapidly created plasma, *IEEE Trans. Plasma Sci.*, **21**(1), 53–56.

6. Hawkins, R. J., and Kallman, J. S. (1994) Lasing in tilted-waveguide semiconductor laser amplifiers, *Opt. Quantum Electronics*, **26**, S207–S217.

7. Hagness, S. C., Joseph, R. M., and Taflove, A. (1996) Subpicosecond electrodynamics of distributed Bragg reflector microlasers: Results from Finite Difference Time Domain simulations, *Radio Sci.*, **31**(4), 931–941.

8. Bakunov, M. I., and Maslov, A. V. (1999) Frequency upshifting of electromagnetic radiation via oblique incidence on an ionization front, *IEEE Trans. Plasma Sci.*, **27**(3), 655–663.

9. Shifman, Y., and Leviatan, Y. (2001) On the use of spatio-temporal multiresolution analysis in method of moments solutions of transient electromagnetic scattering, *IEEE Trans. Antenn. Propag.*, **49**(8), 1123–1129.

10. Taflove, A. (1995) *Computational Electrodynamics: The Finite-Difference Time-Domain Method*, Boston, Artech House.

11. Jackson, J. D. (1999) *Classical Electrodynamics*, John Wiley and Sons, Inc.

12. Kokkorakis, G. C., and John, G. F. (2007) EM fields induced in homogeneous dielectric spheres by external sources, *IEEE Trans. Antenn. Propag.*, **55**(11), 3178–3190.

13. Kalluri, D. K. (1998) *Electromagnetics of Complex Media (Frequency Shifting by a Transient Magnetoplasma Medium)*. CRC Press: Boca Raton, Boston, New York.

14. Jackson, D. J. (1998) *Classical Electrodynamics*, John Wiley & Sons, New York.

Chapter 5

Non-Stationarity of Electromagnetic Waves Caused by the Movement of a Medium Boundary

5.1 Transformation of an Electromagnetic Wave by a Uniformly Moving Boundary of a Medium

Medium movement as a whole—movement of medium parameters, movement of a medium boundary together with movement of a medium or without it—leads to various non-stationary phenomena. Apart from the classical uniform movement of a medium or its boundaries, other forms of movement are of interest [1, 2]. Rotational movement has been investigated most often. It leads to the scattering [3–9] and amplification [10, 11] of electromagnetic waves and to the appearance of peculiar ponderomotive effects [12, 13]. The modulation and spectral transformation of electromagnetic waves being reflected from vibrating surfaces have also been investigated [14–17]. Nevertheless, uniform movement is also interesting as, in addition to some fundamental problems, a whole series of new problems is concerned with a complex form of moving boundaries [2, 18] or with the radiation of sources crossing through a moving complex boundary [19, 20].

Non-Stationary Electromagnetics: An Integral Equations Approach (Second Edition)
Alexander Nerukh and Trevor Benson

ISBN 978-981-4774-95-6 (Hardcover), 978-0-429-02773-4 (eBook)
www.panstanford.com

The problem of "paradoxes" of moving boundaries [21–23] is particularly interesting; these paradoxes occur because of a discrepancy in the number of secondary waves and the number of boundary conditions. Abandoning a sharp boundary idealization, and considering a structure with transition layer boundary, is proposed in [21] to resolve these paradoxes. However, it will be shown in this chapter that the Volterra integral equation approach allows these "paradoxes" to be resolved directly.

In the statement of a problem with a moving boundary there is one more idealization, namely, movement stationarity assuming that the movement has begun at infinite past time. Abandoning this idealization, by considering a movement that begins at a finite moment of time, leads to the appearance of new peculiarities in the wave transformation on a moving boundary. In this chapter, such peculiarities are considered with an abrupt uniform movement of a boundary beginning at zero moment of time, as well as with a smooth "turning on" of a boundary movement according to a relativistic uniformly accelerated law. In the latter approach, the continuity of a boundary velocity change allows the development of the evolution of the wave transformation process to be traced.

Resolution of all of these questions is possible by considering a problem that is one-dimensional (1D) with respect to spatial coordinates. If one assumes that fields have only components normal to a longitudinal axis, and are uniform with respect to the transverse axes, then the problem becomes a scalar one and is described by the equation

$$E = E_0 - \frac{1}{2\varepsilon v}\int_{-\infty}^{\infty}\int dt'dx'\frac{\partial^2}{\partial t^2}\theta\left(t - t' - \frac{|x - x'|}{v}\right)$$
$$\times\chi(t',x')[\varepsilon_1 - \varepsilon + (\varepsilon_2 - \varepsilon_1)\varphi(t')]E(t',x') \quad (5.1)$$

over the whole time axis (-∞, ∞). If the permittivity of the medium inside the region $\chi = 1$ changes by jump from ε_1 to ε_2 at zero moment that is $\varphi(t) = \theta(t)$ then it follows from (5.1) that

$$E = F - \frac{\varepsilon_2 - \varepsilon}{2\varepsilon v}\int d\vec{x}'\frac{\partial^2}{\partial t^2}\theta\left(t - t' - \frac{|x - x'|}{v}\right)\chi^{(+)}(t',x')E(t',x'), \quad (5.2)$$

where

$$F = E_0 - \frac{\varepsilon_1 - \varepsilon}{2\varepsilon v}\int_{-\infty}^{0} dt' \int_{-\infty}^{\infty} dx' \frac{\partial^2}{\partial t^2}\theta\left(t - t' - \frac{|x - x'|}{v}\right)\chi^{(-)}(t', x')E_1(t', x'). \quad (5.3)$$

These equations describe the field in the second and the third stages of the evolution of interaction of an electromagnetic field with a moving medium or boundary.

5.1.1 Discrepancy of Secondary Waves and Boundary Conditions Numbers

Interaction of an electromagnetic wave with the moving interface between two media depends sufficiently on the way the moving boundary is created. A moving boundary can separate moving and immovable media, and a moving boundary can be created as a result of the propagation of some pumping wave that changes the properties of a medium but does not set it in a motion. The character of a scattered field will be different in these cases. Solution of the problem in a differential statement can lead to a vagueness determined by a discrepancy between the number of secondary waves and the number of boundary conditions.

In the first case described above, the movement is created by external sources (generally of non-field origin). Therefore, medium velocity and boundary velocity can take arbitrary values between zero and the velocity of light c. It means that the boundary velocity can be in an arbitrary ratio with respect to phase (group) velocities of waves in the immovable medium. Ambiguity of determination of a number of scattering waves does not occur in this case because of the medium moves on one side of the boundary and the phase (group) velocities of waves that have passed into the moving medium depend on the medium velocity and are determined by the formula [24]

$$v_{\text{ph}}^{\pm} = v_{\text{gr}}^{\pm} = c\frac{1 \pm n_1\beta}{\beta \pm n_1}, \quad (5.4)$$

where n_1 is the refractive index of the moving medium in its frame of rest, $\beta = u/c$ and u is the velocity of the movement of the medium boundary and as well as the medium itself.

It follows from the formula that there exists only one wave $v_{\text{ph}}^{\pm}$ in the field that passes into the medium which satisfies a condition of taking energy from the boundary for an arbitrary value of $|\beta| < 1$. Indeed, from the inequality $(1 + n_1\beta)/(n_1 + \beta) > \beta$ it follows that $\beta^2 < 1$ for this wave. The other wave v_{ph}^- annot take energy from the boundary as $v_{ph}^- < 0$ if $1 - n_1\beta > 0$. In the opposite case $1 - n_1\beta < 0$ the phase velocity is positive, $v_{ph}^- > 0$, but it can be greater the boundary velocity, $v_{ph}^- > \beta$, only if $\beta^2 > 1$; that is impossible.

It is worth noting that the frequency and the amplitude

$$\omega_1 = \omega\frac{(1-n\beta)(1+n_1\beta)}{1-\beta^2},\quad A_1 = A_0\frac{2n}{n+n_1}\frac{(1-n\beta)(1+n_1\beta)}{1-\beta^2} \tag{5.5}$$

of the wave transmitted into the moving medium have no peculiarities for permissible values of β, whereas the wave reflected into the immovable medium has a peculiarity when $\beta = -1/n$

$$\omega_0 = \omega\frac{1-n\beta}{1+n\beta},\quad A_r = A_0\frac{n-n_1}{n+n_1}\frac{1-n\beta}{1+n\beta}. \tag{5.6}$$

In the second case of moving boundary considered, the medium movement is absent on the both sides of the boundary, the phase (group) velocities of waves do not depend on the boundary velocity, and the characteristics of the transmitted wave are determined by the expressions

$$\omega_1 = \omega\frac{1-n\beta}{1+n_1\beta},\quad A_1 = A_0\frac{2n}{n+n_1}\frac{1-n\beta}{1+n_1\beta}. \tag{5.7}$$

It follows from these expressions that the transmitted wave also has a peculiarity (when $\beta = 1/n_1$). Moreover in interim cases, when $v < u < v_1$ and $v > u > v_1$, it is possible for three secondary waves to exist that satisfy the condition of taking energy from the boundary; there can be one reflected wave and two transmitted ones, or two reflected waves and one transmitted one. Ambiguity appears because there are only two boundary conditions, as was been acknowledged for the first time by L. A. Ostrovsky [21–23]. To resolve this question, it was proposed in refs. [21–23] to consider a structure with a transition layer and to take into account a point of synchronism where wave velocity coincides with the boundary velocity. This gives a missing condition.

5.1.2 Resolution of Moving Boundaries "Paradoxes"

An extension of the boundary must be taken into account when wavelength is comparable with the width of the transition layer [24]. In the approximation of an ideal thin boundary, a solution of the problem in a differential statement is ambiguous. Such ambiguity does not occur in the integral statement of the problem. Before showing the resolution of the "paradoxes" by virtue of the integral method, it is necessary to make some further remarks.

A moving boundary can be considered as a source of electromagnetic waves, so it makes sense to compare the formulae (5.6) and (5.7) with the formulae for the frequency and amplitude of an electromagnetic waves radiated by an oscillator that is moving in vacuum. It is known that when the oscillator is coming nearer to an observer then the frequency and the amplitude of the radiated field tend to infinity as $\beta \to 1$. It is one of the arguments of the restriction of the light velocity in vacuum in the special relativity theory. The tendency to infinity of the frequency and the amplitude described by the formulae (5.6) and (5.7) has the same connotation in the case when the movement of the medium is absent and the movement of the medium parameter jump is created by the field propagating in the same medium. It means that some phase velocity has an extreme character. So, if the boundary moves to meet an incident wave then its velocity is restricted by the wave phase velocity in the background $|\beta| < 1/n$. If the boundary moves away the incident wave then the inequality $\beta < 1/n_1$ must be fulfilled.

So, there is the restriction $|u| < v$ with the oncoming movement and $u < v_1$ with the accompanying one. Ambiguity in a definition of wave number will also take place with the oncoming movement of the boundary in the case of $n_1 > n$ ($v_1 < v$). In this case, it is possible for two waves to exist with phase velocities $\pm v_1$.

Let us consider now the solution to the problem of the reflection of the electromagnetic wave $E_0 = \exp[i\omega(t - x/v)]$ by the boundary moving uniformly in the immovable medium. This problem is described by Eq. (5.1), into which we must substitute $\varepsilon_1 = \varepsilon_2$, $\varphi(t) = 0$, $\chi(t, x) = \theta(x - ut)$. To solve this equation, we construct the resolvent of the third stage of the interaction by generalizing the results obtained for the case of the uniform movement of the boundary with velocity u. The presence of the characteristic function $\chi = \theta(x - ut)$ changes the kernel of the equation for the resolvent in the following way:

$$\langle \mathbf{p} | \hat{K}_3 | \mathbf{p}' \rangle = \langle \mathbf{p} | \chi \hat{K}_2 \chi | \mathbf{p}' \rangle$$

$$= \frac{2\lambda v}{p^2 + v^2 k^2} \cdot \frac{p^2}{p - p' + iu(k - k')} \frac{i}{k - k - i0}$$

$$+ \alpha v^2 \frac{p + iuk}{p - ivk} \cdot \frac{1}{p - p' + iu(k - k')} \frac{1}{p + iuk - i(v + u)k'}, \quad (5.8)$$

where $\alpha = \frac{v^2 - v_2^2}{2vv_2^2}$.

The matrix elements for the product $\chi \hat{R}_2 \chi$ will take a form that is analogous to (5.8) with the corresponding substitution $v \to v_2$:

$$\langle \mathbf{p} | \chi \hat{R}_2 \chi | \mathbf{p}' \rangle = 2\alpha \left\{ \frac{v_2^2}{v} \frac{p^2}{p^2 + v_2^2 k^2} \frac{i}{k - k' - i0} \right.$$

$$\left. + \frac{v_2^3}{2v} \frac{p + iuk}{p - iv_2 k} \frac{1}{p + iuk - i(v_2 + u)k'} \right\} \cdot \frac{1}{p - p' + iu(k - k')}. \quad (5.9)$$

To construct $\hat{R}_2$ let us add the item determined by the reflection of the characteristic

$$\langle \mathbf{p} | \hat{R}_3 | \mathbf{p}' \rangle = 2\alpha \left\{ \frac{v_2^2}{v} \frac{p^2}{p^2 + v_2^2 k^2} \frac{i}{k - k' - i0} \right.$$

$$\left. + \frac{v_2^3}{2v} \frac{p + iuk}{p + iuk - i(v_2 + u)k'} \left[\frac{1}{p - iv_2 k} - \frac{R}{p + iv_2 k} \right] \right\} \cdot \frac{1}{p - p' + iu(k - k')}, \quad (5.10)$$

where $\mathrm{Re}(p - p') > 0$.

Substitution of this expression into Eq. (5.8) gives the same value for R as in the case of the immovable medium: $R = \frac{v - v_2}{v + v_2}$. It is worth noting that the boundary movement is taken into account in the resolvent entirely by the characteristic function. This may be seen from the comparison of the expressions for the resolvent for a motionless boundary and (5.10).

The expression (5.10) gives the matrix elements for the resolvent in the coordinate representation

$$\langle \mathbf{x} | \hat{R}_3 | \mathbf{x}' \rangle = \langle \mathbf{x} | \chi \hat{R}_2 \chi | \mathbf{x}' \rangle - \lambda \frac{v_2}{v} \chi(\mathbf{x}) \frac{\partial}{\partial t} R_u \times \frac{\partial}{\partial t} \theta \left(v_2 t - x - \frac{v_2 - u}{v_2 + u} (v_2 t' + x') \right) \chi(\mathbf{x}'), \tag{5.11}$$

where R_u is the reflectance for a plane wave from a moving boundary [24],

$$R_u = \frac{v - v_2}{v + v_2} \frac{v_2 - u}{v_2 + u}. \tag{5.12}$$

The second item in (5.11) corresponds to the reflected characteristic (see Fig. 5.1). As in the case of the immovable boundary, the slope of the reflected characteristic is determined by the medium parameters not the boundary. Let us analyze the case of the oncoming movement of the incident wave and the boundary when the ambiguity in the definition of the secondary waves number remains in spite of the remarks mentioned above.

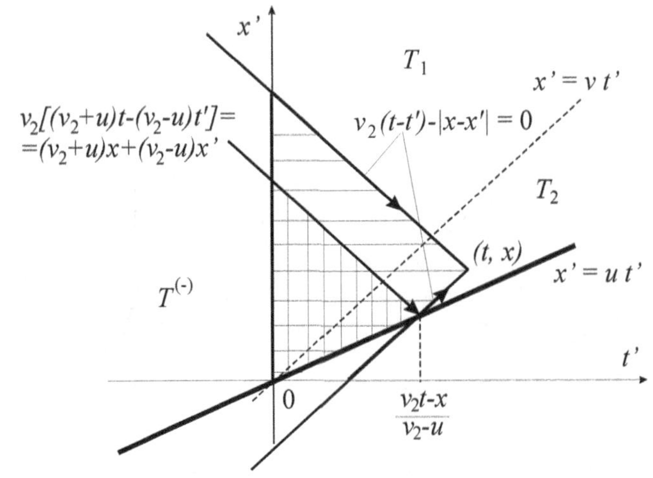

Figure 5.1 Characteristics for the construction of the resolvent (5.11).

Thus, there are possible four cases for relations between the velocities of waves and boundaries: completely sub-light $u < v < v_2$; $u < v_2 < v$; completely super-light $u > v > v_2$; $u > v_2 > v$; interim $v < u < v_2$; and interim $v_2 < u < v$. The third case could be excluded if one takes into account the remarks made above.

The solution of the problem in the integral method is implemented in two stages: In the first stage, the inner field is determined and then, in the second stage, the external field is determined by virtue of the inner one. It allows one to split the four cases above into two groups depending on the relation between the boundary velocity and the phase velocity of the transmitted wave.

As the problem considered is stationary, that is the medium boundary moves uniformly during the while from time axis, $-\infty < t < \infty$, then the expression describing the transmitted as well as reflected field is Eq. (5.1) with the characteristic function $\chi(\mathbf{x}) = \theta(x - ut)$. The solution to this equation is constructed by virtue of the resolvent (5.11), which is extended on the negative time half-axis also. Based on the arguments given in Chapter 3, Section 3.1, one can conclude that the solution given by the resolvent will be unique and unambiguous as the region of the field definition and the region of the equation definition coincide. Furthermore, the regions $T^{(-)}$ and T_1 go to infinity and only the region T_2 remains.

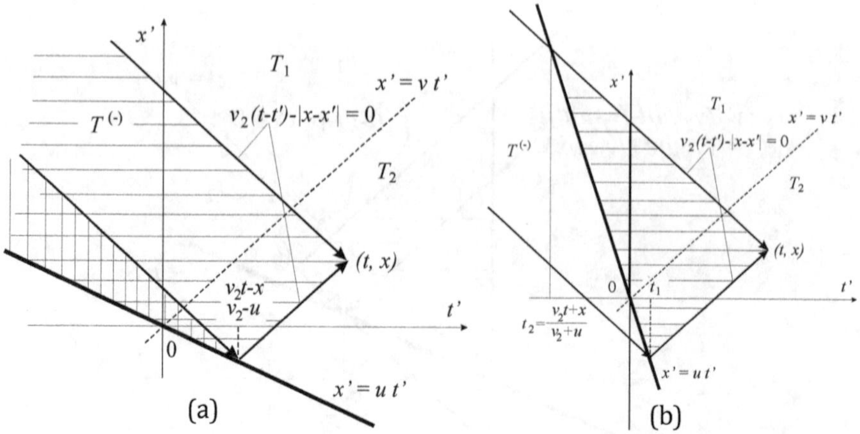

Figure 5.2 Influence of the boundary velocity on the formation of the resolvent.

As the movement is oncoming the resolvent (5.11) has different forms depending on the relationship between the velocities u and v_2. In the case $|u| < v_2$ (Fig. 5.2a) the reflected characteristic of the resolvent is inside the world region of the reflecting medium, $\chi = 1$. Consequently, the resolvent must be taken in the whole form (5.11) in which the term $\langle \mathbf{x} | \hat{R}_2 | \mathbf{x}' \rangle$ is determined by the expression

$$\langle \mathbf{x} | \hat{R}_2 | \mathbf{x}' \rangle = -\lambda \frac{v_2}{v} \frac{\partial^2}{\partial t^2} \theta\left(t - t' - \frac{|x - x'|}{v_2} \right) \theta(t - t').$$

This term corresponds to the integration over the region denoted by the horizontal hatching in Fig. 5.2a. The term with the coefficient R_u is denoted by the double hatching.

The solution of Eq. (5.1) found by virtue of this resolvent coincides with known expression for the transmitted wave [24]

$$E = E_0 + \hat{R}_3 E_0 = \frac{\omega_1^{(+)}}{\omega} A e^{i\omega_1^{(+)}\left(t - \frac{x}{v_2}\right)}, \tag{5.13}$$

where $\omega_1^{(+)} = \omega \frac{v_2}{v} \frac{v - u}{v_2 - u}$, $A = \frac{2v_2}{v + v_2}$.

In the case $|u| > v_2$ (Fig. 5.2b) the reflected characteristic of the resolvent will be outside the world region $\chi = 1$. Therefore, one has to take $R_u = 0$ in the expression for the resolvent. Calculation of the transmitted field by virtue of such a resolvent shows that the field consists of two waves

$$E = E_0 + \hat{R}_2 E_0 = \frac{\omega_1^{(+)}}{\omega} A^{(+)} e^{i\omega_1^{(+)}\left(t - \frac{x}{v_2}\right)} + \frac{\omega_1^{(-)}}{\omega} A^{(-)} e^{i\omega_1^{(-)}\left(t + \frac{x}{v_2}\right)}, \tag{5.14}$$

where $\omega_1^{(-)} = \omega \frac{v_2}{v} \frac{v - u}{v_2 + u}$, $A^{\pm} = \frac{v \pm v_2}{2v}$. It also coincides with known results.

The transmitted field obtained allows one to calculate the reflected field if one considers the observation point in the formula (5.1) to be located outside of the world region of the diffraction object. In the completely sub-light case, $|u| < v$, v_2, the following known expression for the reflected wave appears:

$$E_r = -\frac{v - v_2}{v + v_2}\frac{v - u}{v + u}e^{i\omega\frac{v-u}{v+u}\left(t+\frac{x}{v}\right)}. \qquad (5.15)$$

In the completely super-light case, $|u| > v, v_2$, the reflection is absent as the intersection of the kernel region of the formula (5.1) and the region $\chi = 1$ is equal to zero (Fig. 5.3a).

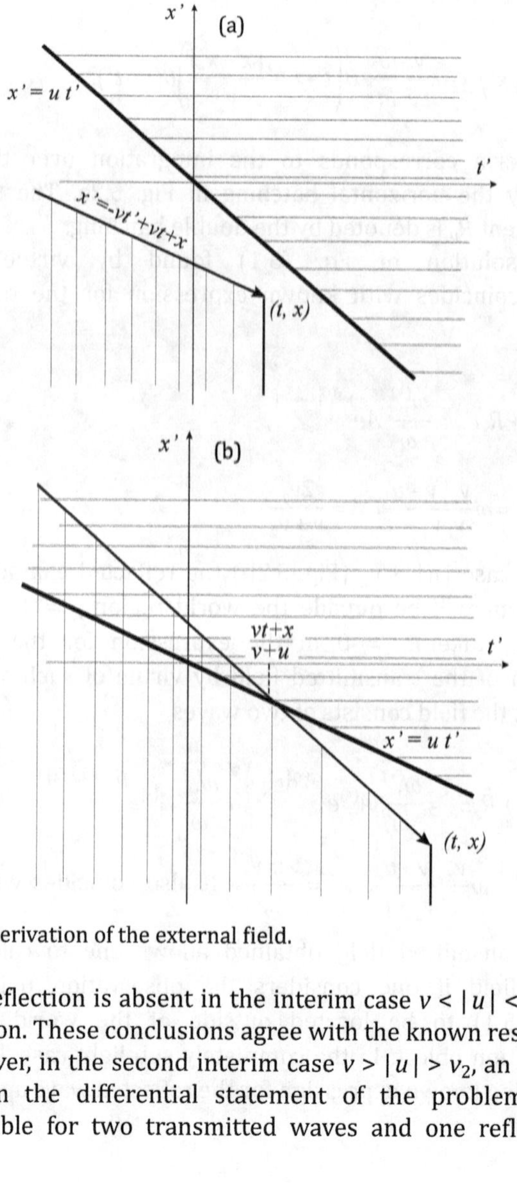

Figure 5.3 Derivation of the external field.

The reflection is absent in the interim case $v < |u| < v_2$ for the same reason. These conclusions agree with the known results.

However, in the second interim case $v > |u| > v_2$, an ambiguity appears in the differential statement of the problem because it is possible for two transmitted waves and one reflected one to exist (Fig. 5.3b). In the integral statement of the problem, we obtain a unique answer for the reflected field by substituting the field (5.14) in the integral| (5.1):

$$\frac{\partial^2}{\partial t^2}\int_{-\infty}^{\infty}\int dt'dx'\theta(x'-ut')\theta\left(t-t'-\frac{x-x'}{v}\right)\left\{A^{(+)}\frac{\omega_1^{(+)}}{\omega}e^{i\omega_1^{(+)}\left(t'-\frac{x'}{v_2}\right)}\right.$$

$$\left.+A^{(-)}\frac{\omega_1^{(-)}}{\omega}e^{i\omega_1^{(-)}\left(t'+\frac{x'}{v_2}\right)}\right\}=\frac{v_2(v-u)}{2(v+u)}-\frac{v_2(v-u)}{2(v+u)}\equiv 0 \tag{5.16}$$

In this case, the contributions from the two transmitted waves (5.14) are cancelled mutually, therefore in this case the reflected wave is absent. It is easy to make sure that the boundary conditions are fulfilled in this case, that is, the values $E-\frac{u}{c}B$ and $H-\frac{u}{c}D$ are continuous on the boundary. Indeed, the magnitudes of these values on the both sides of the boundary are given by the expressions

$$E_2-\frac{u}{c}B_2=E_0-\frac{u}{c}B_0=\frac{v-u}{v}e^{i\omega\frac{v-u}{v}t},$$

$$B_2-\frac{u}{c}\varepsilon_2E_2=B_0-\frac{u}{c}\varepsilon E_0=\frac{c}{v}\,\frac{v-u}{v}e^{i\omega\frac{v-u}{v}t}, \tag{5.17}$$

respectively. It is taken into account that the media are nonmagnetic, that is, $B=\mu_0H$.

It is worth noting one other peculiarity of the wave interaction with a boundary that is moving uniformly. The expressions for the fields have a meaning for a sub-light, as well as for super-light, velocity of the boundary. However, when the boundary velocity is equal to the wave phase velocity, then the frequency of one of the secondary waves and its amplitude turn to infinity. It is the idealization of an infinitely powerful source of the boundary movement and it shows that a real movement does not have such a velocity. It will be shown below that the approximation of an infinitely powerful source of the boundary movement leads to a creation of infinite discontinuities in the electromagnetic fields in the case of non-stationary movement.

In the more general case, a structure of the transition boundary layer should be considered that takes into account additional factors. These include medium dispersion, because of a limitless increase of wave frequencies [22, 24, 25], and medium non-linearity because of a limitless increase of wave amplitudes [26].

5.2 Evolution of an Electromagnetic Wave after the Beginning of Medium Boundary Movement

The problem considered in the previous section shows that a uniform movement of an interface between two media leads to the existence of one or two waves in the transmitted field. This situation changes drastically with a non-uniform movement. The simplest case is when the boundary movement changes as a jump from zero to the value u.

In the general case, the boundary movement can be regarded as non-uniform if the boundary was at rest before some zero moment of time ($\chi^{(-)} = \theta(x)$) and is moving by some law $\chi_s(t)$ after this zero moment. If the medium inside the half-space $\chi^{(+)} = 1$ does not change its properties ($\varepsilon_2 = \varepsilon_1$, $\varphi(t) = 0$), then after $t \geq 0$ the problem will be described by the expressions (5.2) and (5.3) with $\varepsilon_2 = \varepsilon_1$.

Let a monochromatic wave $E_0 = \exp[i\omega(t - \frac{x}{v})]$ fall upon the boundary. Then the solution of Eq. (5.1) determining the transmitted field in the region $x > 0$ before zero moment is well known and equal to $E^{(-)} = A\ \exp[i\omega\tau^{-}]$, where $A = \frac{2}{n+1}$, $n = \frac{v}{v_2}$, $\tau^{\pm} = t \pm \frac{x}{v_2}$. Substituting the expression for $E^{(-)}$ into (5.3), we find the free term for Eq. (5.2) as

$$F = \theta(x - vt)ne^{i\omega_0\left(t - \frac{x}{v}\right)} + nF_0 e^{-i\omega_0\left(t + \frac{x}{v}\right)} + \theta(vt - x)e^{i\omega\left(t - \frac{x}{v}\right)}, \qquad (5.18)$$

where $F_0 = \dfrac{n-1}{n+1}$, $\omega_0 = n\omega$.

The resolvent (5.11) determines two regions in the half-plane $t \geq 0$ in which the interaction of the field with the boundary has sufficiently different character (see Fig. 5.4).

The region $T_1(0 \leq t < x/v_2)$ is characterized by the absence of the influence of the interface because of a restricted velocity of signal propagation. In the region $T_2(ut \leq n < v_2t)$, the boundary has an influence. In the region $T_1(v_2t \leq x)$ the field is not influenced by the changing of the boundary state and is determined by the same expression as before zero moment, when the boundary was motionless, that is $E_1 = E^{(-)}$.

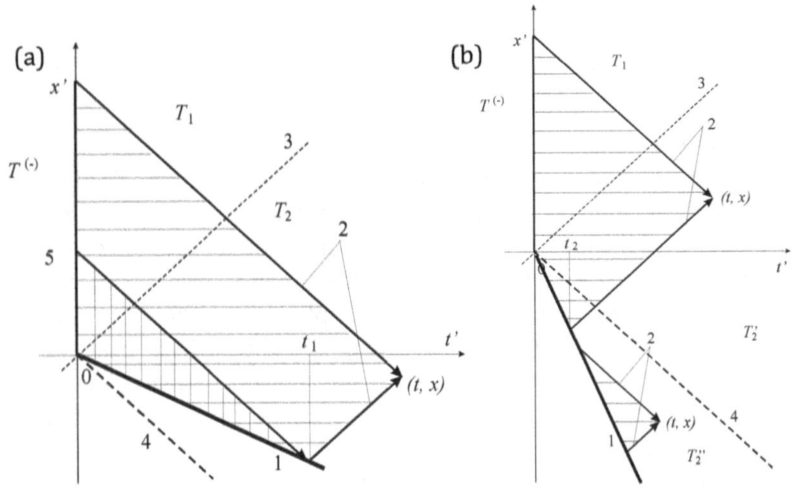

Figure 5.4 Dependence of the structure of the field on the relation between the boundary velocity and the phase one: (a) corresponds to $u > v_2$, (b) corresponds to $u < -v_2$. The lines on the figures are described by the following equations: (1) $x' = ut'$; (2) $v_2(t - t') - |x - x'| = 0$; (3) $x' = v_2t$; (4) $x' = -v_2t'$; (5) $x' = -v_2t' + v_2\frac{v_2+u}{v_2-u}\tau^{(-)}$. The time points are $t_1 = \frac{v_2+u}{v_2-u}\tau^{(-)}$; $t_2 = \frac{v_2}{v_2+u}\tau^{(+)}$.

The field structure in the region T_2 depends on the relation between the boundary velocity u and the wave phase velocity v_2. If $u > -v_2$ (Fig. 5.4a), then the expression for the field is determined by the free term by virtue of the resolvent (5.11) and coincides with the formula (5.12) for the stationary case. If $u < -v_2$, then the region T_2 is divided onto two sub-regions T_2' $(v_2t \geq x \geq -v_2t)$ and $T_2''(-v_2t \geq x \geq ut)$ (Fig. 5.4b). In the sub-region T_2' the field consists of two waves

$$E_2' = F + \hat{R}F = \frac{\omega_1^{(+)}}{\omega}A^{(+)}e^{i\omega_1^{(+)}\tau^-} + F_0\frac{n-1}{2n}\frac{v+u}{v_2-u}e^{-i\omega_2^{(+)}\tau^-}, \tag{5.19}$$

where $\omega_2^{(\pm)} = \omega(v+u)(v_2 \mp u)^{-1}$. In comparison with the stationary case, the change is that one more direct wave is added to the direct wave described by the first term in (5.19). This new wave is determined by the term F_0 in the free term in (5.18). This wave frequency $\omega_2^{(\pm)}$ satisfies the condition of the equality of its phase and the F_0 phase on the boundary:

$$-\omega_0\left(t+\frac{x}{v}\right)=\omega_2^{(+)}\left(t-\frac{x}{v_2}\right)\Bigg|_{x=ut} \tag{5.20}$$

A mechanism for the creation of this wave is as follows. A free term F is the result of the interaction between the field and the medium during all the time before zero moment. This term should be considered as an initial condition for the field after zero moment. A sudden change of boundary velocity at zero moment evidently causes the same splitting of the past wave, as in the case of an abrupt change in permittivity examined in Chapter 3. However, as the splitting takes place on the boundary, the possibility of realizing several waves in the medium depends on the ratio between the boundary velocity and the waves' velocities in the medium. With $u > -v_2$ there is no splitting, as it is impossible for the waves moving towards the boundary to exist in the transmitted field. In this case, with $-u < v$, a reflected wave exists, defined by the expression (5.14). With $u < -v_2$ there is a possibility of realizing splitting, that is reflected waves appear in the medium ε_2 except the direct past waves. Indeed, calculation of the field in region T_2'' gives us the waves

$$E_2''=\frac{\omega_1^{(+)}}{\omega}A^{(+)}e^{i\omega_1^{(+)}\tau^-}+\frac{\omega_1^{(-)}}{\omega}A^{(-)}e^{i\omega_1^{(-)}\tau^+}$$
$$+F_0\frac{n-1}{2n}\frac{\omega_2^{(-)}}{\omega}e^{-i\omega_2^{(+)}\tau^-}+F_0\frac{n+1}{2n}\frac{\omega_2^{(-)}}{\omega}e^{-i\omega_2^{(-)}\tau^+}. \tag{5.21}$$

The correctness of these conclusions can be easily verified by substituting expressions for E_1, E'_2, and E''_2 into the integral Eq. (5.2). With this, the areas of integration take the forms shown in Fig. (5.5).

It is worth noting that the frequencies and amplitudes of the both reversed waves become infinite when $u = -v_2$, as in the stationary case.

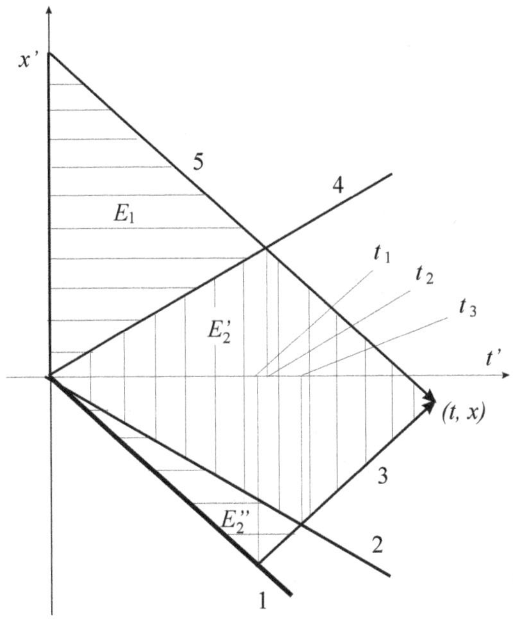

Figure 5.5 The splitting of the wave at moment the velocity jump. Here the lines are described by the following equations: (1) $x' = ut'$; (2) $x' = -v_1t'$; (3) $x' = vt' - v\tau_0^{(-)}$; (4) $x' = v_1t'$; (5) $x' = -vt' + v\tau_0^{(+)}$. The time points are $t_1 = v\tau_0^{(-)}/(v-u)$; $t_2 = n\tau_0^{(+)}/(n+1)$; $t_3 = n\tau_0^{(-)}/(n+1)$.

Let us now examine the boundary conditions in the case $u > -v_2$ when the existence of reflected waves is possible. Substituting (5.21) into (5.1) we find after some calculation that the contributions of the first two waves compensate one another, as in the stationary case, while the second pair of waves, caused by the initial wave F_0 in the free term in (5.18), define the following reflected wave:

$$E_r = \frac{n-1}{n+1} n e^{-i\omega n\left(t+\frac{x}{v}\right)}. \tag{5.22}$$

The frequency of this wave coincides with the frequency of the wave F_0. The amplitudes of the waves on the moving boundary satisfy the boundary conditions, although their magnitudes are different from the stationary case, which is defined by the expression (5.17):

$$E_2 - \frac{u}{c}B_2 = (E_0 + E_r) - \frac{u}{c}(B_0 - B_r)$$
$$= \frac{n-\beta_2}{n} e^{i\omega\frac{n-\beta_2}{n}t} + \frac{n+1}{n+1}(n+\beta_2)e^{-i\omega(n+\beta_2)t}, \tag{5.23}$$

$$B_2 - \frac{u}{c}\varepsilon_2 E_2 = (B_0 + B_r) - \frac{u}{c}\varepsilon(E_0 - E_r)$$
$$= c\frac{n-\beta_2}{nv} e^{i\omega\frac{n-\beta_2}{n}t} - c\frac{n-1}{n+1}\frac{n+\beta_2}{v} e^{-i\omega(n+\beta_2)t}, \tag{5.24}$$

where $\beta_2 = u/v_2$.

Thus, non-stationarity of the boundary movement breaks the balance which guaranteed the absence of the reflected field in the stationary case, with idealization of the "sharp boundary," even if the relations $v_2 < |u| < v$ are true.

5.3 Relativistic Uniform Accelerated Movement of a Medium Boundary

Special features of wave transformations, which appear in the case of a non-stationary boundary movement examined above, also become apparent in the case of smooth non-stationarity of the moment when the boundary velocity reaches the magnitude of the wave phase velocity continuously. Smooth non-stationarity is realized by non-uniform movement of the boundary. Direct examining of irregular forward movement in macroscopic electrodynamics was conducted in connection with examining waves propagating in an indefinite homogeneous linearly accelerated medium [27, 28]. The presence of a non-uniform moving boundary was implied also in the papers [1, 2, 18]; however, the influence on scattered fields of just non-uniform movement was not examined in these papers. This is why it is interesting to consider the scattering of electromagnetic waves by a boundary, which makes irregular movement.

Let us examine relativistic uniformly accelerated movement, which is the simplest and most natural non-inertial movement. With such a movement, the velocity of a boundary is changing constantly from zero to relativistic magnitudes and thus passing constantly through all special points. The latter are concerned with conversion of scattered field amplitudes into infinity as well as with the ambiguity in the definition of the number of scattered waves. It is also essential that with such a movement the problem becomes non-stationary. Studying such a problem by selecting solutions based on qualitative reasoning was conducted in the paper [29]; however, not all the solutions were taken into account.

The law of boundary movement towards the impinging wave takes the form [30] $x_s(t) = -v\left(\sqrt{\xi^2 + t^2} - \xi\right)$, where $\xi = c/w$; w is a value of the acceleration in a boundary reference system. The boundary movement velocity $u(t) = -vt(\xi^2 + t^2)^{-1/2}$ with $t \to \infty$ tends to its limiting value $-v$ that is to the wave velocity in the background medium.

As the medium properties are not changed at the zero moment in the region $T_2(\chi > v_2t)$ the transmitted field is the same as before this zero moment, $E_2 = E_1$, that is as in the case of a uniform movement. In the region $(\chi > v_2t)$, the resolvent and the nature of interaction are different. If $n = v/v_2 < 1$ (Fig. 5.6a) then the resolvent characteristic reflected from the boundary world line is in the world region $\chi^{(+)} = \theta(x - x_s(t)) = 1$ during all time.

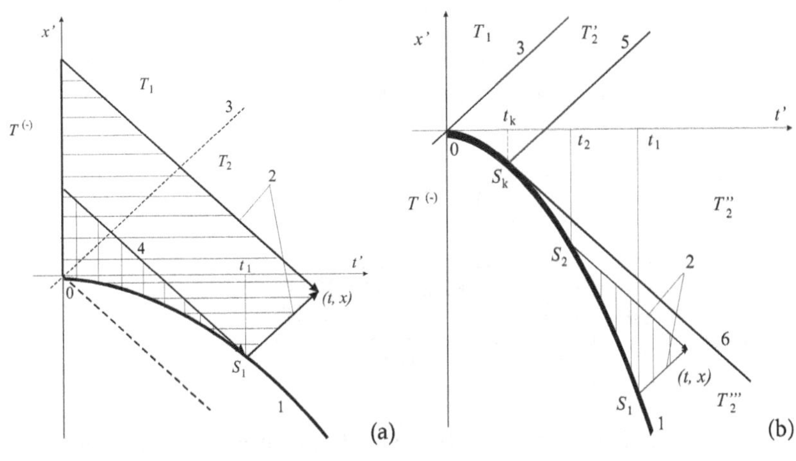

Figure 5.6 Resolvent characteristics. Here: (1) $x' = -v(\sqrt{\xi^2 + t^2} - \xi)$; (2) $v_2(t - t') - |x - x') = 0$; (3) $x' = v_2t'$; (4) $t' - x'/v_2 = \varphi(\tau^{(-)})$; (5) $t' - x'/v_2 = \tau_{ew}$; (6) $t' + x/v_2 = \tau_{th}$.

Therefore the resolvent takes the following form:

$$\langle \mathbf{x}|\hat{R}_3|\mathbf{x}'\rangle = -\theta(x-x_s(t))\frac{\partial}{\partial t}\frac{\lambda}{n}\left\{\frac{\partial}{\partial t}\theta\left(t-t'-\frac{|x-x'|}{v_2}\right)\right.$$

$$\left.+R_u(\tau^-)\frac{\partial}{\partial t}\theta\left(\varphi(\tau^-)-t'-\frac{x'}{v_2}\right)\right\}\theta(x'-x_s(t')), \qquad (5.25)$$

where $\varphi(\tau) = 2t_1 - \tau$; $t_{1,2}$ is the point of intersection of the lower and upper characteristics of the resolvent $\hat{R}_3$ with the world boundary line:

$$t_{1,2}(\tau) = \frac{n}{n^2-1}\{\pm\xi - \tau/n + \sqrt{\xi^2 \pm 2n\xi\tau + \tau^2}\}. \qquad (5.26)$$

Coefficient R_u in (5.25), which is now a function of the observation point, is defined by the same expression as in (5.11):

$$R_u(\tau) = \frac{n-1}{n+1}\frac{v_2 - u_1(\tau)}{v_2 + u_1(\tau)}$$

$$= \frac{n-1}{n+1}\frac{\sqrt{\xi^2+2n\xi\tau+\tau^2}}{\frac{2n^2}{n^2-1}\left(\xi+\frac{\tau}{n}\right)-\frac{n^2+1}{n^2-1}\sqrt{\xi^2+2n\xi\tau+\tau^2}} \qquad (5.27)$$

where the boundary's velocity is taken at the moment of the boundary "meeting" with the resolvent characteristic $u_1(t) = u(t_1)$:

$$u(t_1) = -\frac{vt_1}{\sqrt{t_1^2+\xi^2}} = \frac{-vt_1}{\xi+\frac{\tau^-}{n}-\frac{t_1}{n}}$$

$$= vn\frac{\xi+\frac{\tau^-}{n}-\sqrt{\xi^2+2n\xi\tau^- -(\tau^-)^2}}{n^2\left(\xi+\frac{\tau^-}{n}\right)-\sqrt{\xi^2+2n\xi\tau^- +(\tau^-)^2}}. \qquad (5.28)$$

If $n > 1$ (Fig. 5.6b), then the reflected characteristic belongs to the region $\chi^{(+)} = 1$ only until the moment of contact $t_{tn} = \xi(n^2-1)^{-1/2}$. At this moment, the boundary's velocity becomes equal to the wave velocity v_2. A beam $\tau^- = \tau_{lw} = \xi[(n^2+1)(n^2-1)^{-\frac{1}{2}} - n]$ on Fig. 5.6b separates the sub-region T_3'' from the region T_3.

The reflected characteristic does not contribute into the resolvent any more in T_3''. A tangential beam $\tau^+ = \tau_{tn} = \xi(n - \sqrt{n^2 - 1})$ separates the sub-region T_3''' in which both resolvent characteristics cross the boundary world line. In the sub-regions T_3'' and T_3''' it must be assumed that $R_u \equiv 0$ in the expression (5.25) for the resolvent. These differences become apparent in the structure of the transmitted field. With $n < 1$ in the whole region T_3 and with $n > 1$ only in the sub-region T_3' $(\tau^- < \tau_{lw} = \xi((n^2+1)(n^2-1)^{-\frac{1}{2}} - n))$, the field which is defined by the resolvent (5.25) consists of one forward wave

$$E_3 = \frac{2v}{v + v_2} \frac{-i}{\omega} \frac{\partial}{\partial t} e^{i\omega\psi_1^{(+)}(\tau^-)}, \tag{5.29}$$

where $\psi_{1(\tau)}^{(\pm)} = \dfrac{1}{n \pm 1}\{\tau - \xi \pm \sqrt{n^2 \pm 2n\xi\tau + \tau^2}\}.$ (5.30)

The frequency of this wave is a function of the observation point

$$\bar{\omega}_1^{(+)} = \frac{\partial}{\partial t}(\omega\psi_1^{(+)}(\tau^-)) = \frac{\omega}{n+1}\left\{1 + \frac{n\xi + \tau^-}{\sqrt{\xi^2 + 2n\xi\tau^- + (\tau^-)^2}}\right\} \tag{5.31}$$

On the line $\tau^- = 0$ separating the regions T_2 and T_3', the frequency is equal to the frequency of the wave $E_2 = E_1$, that is to the frequency of the impinging wave. It is natural since this line corresponds to zero velocity of the boundary.

If we substitute the expression for the boundary velocity (5.28) into the formula for the transmitted wave frequency in the stationary case (5.13), then we find that the expression for the frequency $\bar{\omega}_1^{(+)}$ (5.31) can be expressed through the boundary velocity in a standard way:

$$\bar{\omega}_1^{(+)} = \omega \frac{v_2}{v} \frac{v - u_1(\tau^-)}{v_2 - u_1(\tau^-)} \tag{5.32}$$

Because the amplitude of the wave given by (5.29) is proportional to the frequency, $|E_3| = \frac{2v_2}{v+v_2}\left|\frac{\bar{\omega}_1^{(+)}}{\omega}\right|$, the behavior of both these wave parameters is defined by the behavior of the boundary velocity and it is changing in time, as shown in Fig. 5.7.

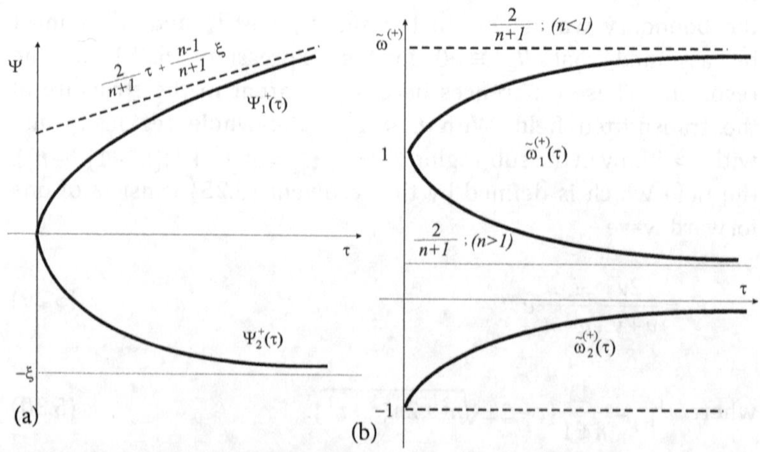

Figure 5.7 The phase characteristics.

With $n < 1$, the transmitted field description is exhausted by the given correlations. The same results are obtained via a different approach used in the paper [29]. In this paper the solution is constructed by matching, and it coincides with that obtained by the resolvent method for $n > 1$ and also in the interval $0 < \tau^- < \tau_{lw}$. However, over other intervals, the picture becomes more complex and the matching of solutions becomes more difficult as uncertainties appear in the results given in [29] as the solution is built only in some sub-regions of the problem domain.

There are no such uncertainties when we solve this problem by the resolvent method. Let us find the transmitted field for the case of n>1 in the sub-regions T_3'' and T_3'''. In these sub-regions the resolvent is defined by the expression (5.25) with $R_u = 0$. In the sub-region $T_3''(t^- > \tau_{iw}, \tau^- > \tau_{tn})$ the transmitted field consists of two forward waves

$$E_3'' = -\frac{n+1}{2n}\frac{1}{\omega}\frac{\partial}{\partial t}e^{i\omega_1^{(+)}(\tau^-)} + \frac{n-1}{2n}F_0\frac{i}{\omega}\frac{\partial}{\partial t}e^{i\omega_1\psi_2^{(+)}(\tau^-)}, \qquad (5.33)$$

where $\omega_1 = n\omega$ and $\psi_1^{(+)}(\tau-)$ is from (5.30)

$$\psi_2^{(\pm)}(\tau) = \frac{1}{n \mp 1}\{\tau + \xi \mp \sqrt{\xi^2 \pm 2n\xi\tau + \tau^2}\}. \quad (5.34)$$

The first term on the right hand side of Eq. (5.33) is the same wave as in (5.29) but with a changed amplitude. This change occurs because the boundary velocity becomes a "super-light" one for the transmitted waves, $|u(\tau^-)| > v_2$, and coincides with a similar change of amplitude observed for a movement with constant velocity, see (5.13).

The second item on the right hand side of (5.33) represents the forward wave with the frequency

$$\begin{aligned}\bar{\omega}_2^{(+)} &= \frac{\partial}{\partial t}(\omega_1 \psi_0^{(+)}(\tau^-)) \\ &= \frac{\omega_1}{n-1}\left\{1 - \frac{n\xi + \tau^-}{\sqrt{\xi^2 + 2n\xi\tau^- + (\tau(-))^2}}\right\} \\ &= -\omega_1 \frac{v_2}{v}\frac{v + u_1(\tau^-)}{v_2 - u_1(\tau^-)}. \end{aligned} \quad (5.35)$$

This frequency dependence on velocity is fully analogous to the corresponding expression for the frequency of the second forward wave (5.18) for the uniform boundary movement that starts at zero moment. Accordingly, the origin of two waves (5.33) also is connected with the splitting on the boundary of the initial waves represented by the free term F. The phase and frequency of the second forward waves are represented in Fig. 5.7.

Splitting leads to the formation of the reflected wave, as in the case of uniform movement. The calculation of the field in the sub-regions T_3''' gives four waves

$$\begin{aligned} E_3''' = &-\frac{n+1}{2n}\frac{i}{\omega}\frac{\partial}{\partial t}e^{i\omega\psi_1^{(+)}(\tau^-)} + \frac{n-1}{2n}F_0\frac{i}{\omega}\frac{\partial}{\partial t}e^{i\omega_1\psi_2^{(+)}(\tau^-)} \\ &-\frac{n-1}{2n}\frac{i}{\omega}\frac{\partial}{\partial t}e^{i\omega\psi_1^{(-)}(\tau^+)} + \frac{n+1}{2n}F_0\frac{i}{\omega}\frac{\partial}{\partial t}e^{i\omega_1\psi_1^{(-)}(\tau^+)}. \end{aligned} \quad (5.36)$$

The behavior of the phases and frequencies of the reflected waves is shown in Fig. 5.8.

Summarizing, we can draw the conclusion that at the instant when the boundary velocity becomes equal to the wave phase velocity, $|u| = v_2$, two discontinuities are formed in the transmitted field region. They spread with the velocity v_2 in opposite directions and they are of the same nature as in the case of a stationary uniform movement.

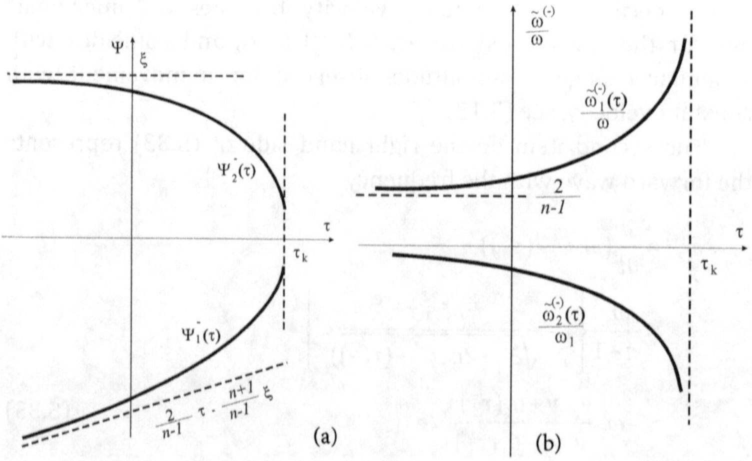

Figure 5.8 Phase characteristics.

The line $\tau^- = \tau_{lw}$ is a line of discontinuity of a finite quantity, while the reflected waves fields have an infinite jump on the line $\tau^+ = \tau_{tn}$. In the case of a uniform movement, it follows from (5.19) that the reflected waves have a finite jump because the boundary velocity immediately takes the value $|u| > v_2$. With $|u| \to v_2$ this jump also tends to infinity. The infinite jump is a result of the idealization of the medium model. Really, to solve the problem correctly we have to take into account the dispersion of the medium because in the vicinity of the line of infinite discontinuity the reflected wave frequencies take infinite values.

We find the reflected field by integrating in the formula (5.1) the expressions obtained for the transmitted field and assuming that the observation point is outside the object of diffraction. In the region $\tau_0^+ < 0$ $\tau_0^\pm = t \pm x/v$ the boundary does not yet have an influence, and the reflected wave is defined by the Fresnel formulae. In the region $0 < \tau_0^+ < \xi$ a wave reflects from the moving boundary. Thus, with $n < 1$ a general expression for the field takes the form:

$$E_r = -\frac{n-1}{n+1}\frac{i}{\xi}\frac{\partial}{\partial t}e^{i\omega\psi_r(\tau_0^+)}, \tag{5.37}$$

where $\psi_r(\tau) = \begin{cases} \tau, \tau < 0 \\ \dfrac{\xi\tau}{\xi - \tau}, 0 < \tau < \xi \end{cases}$.

The line $\tau_0^+ = \xi$ is an asymptote for the boundary world line therefore $\tau_0^+ \to \xi$ means that the boundary goes to infinity. With this the frequency

$$\omega_r = \frac{\partial}{\partial t}(\omega\psi_r) = \omega\frac{\xi^2}{(\xi - \tau_0^+)^2}, \tag{5.38}$$

and the amplitude of the reflected wave tend to infinity; this is a result of the tendency of the boundary velocity to the limiting value v. The increase of the field amplitude can be interpreted as an increasing compression of the field before the accelerated moving boundary.

In the case of the leaving movement, the boundary world line is defined by the law

$$x_s(t) = v_2(\sqrt{\xi^2 + t^2} - \xi). \tag{5.39}$$

The velocity of the boundary movement asymptotically tends to the value of the phase velocity v_2. If $n > 1$ ($v > v_2$), then interaction between the impinging wave and the boundary takes place at any moment. The transmitted field defined by the full resolvent (5.25) and consists only of the forward wave in the form of (5.29) with the phase

$$\Phi(\tau^-) = \frac{2\xi n - (n+1)\tau^-}{2(\xi - \tau^-)}\tau^- \tag{5.40}$$

and the frequency

$$\omega_3 = \frac{\partial}{\partial t}(\omega\Phi(\tau^-)) = \frac{2n\xi^2 + (n+1)\tau^-(\tau^- - 2\xi)}{2n(\xi - \tau^-)^2}. \tag{5.41}$$

The frequency and amplitude of the transmitted wave tend to infinity with $\tau^{-} \to \xi$ that is with $u \to v_2$. This effect is analogous to the one observed in the area of the reflected field for the oncoming movement. In this case, the asymptote of the medium boundary world line is a straight line $\tau^{-} = \xi$ (Fig. 5.9).

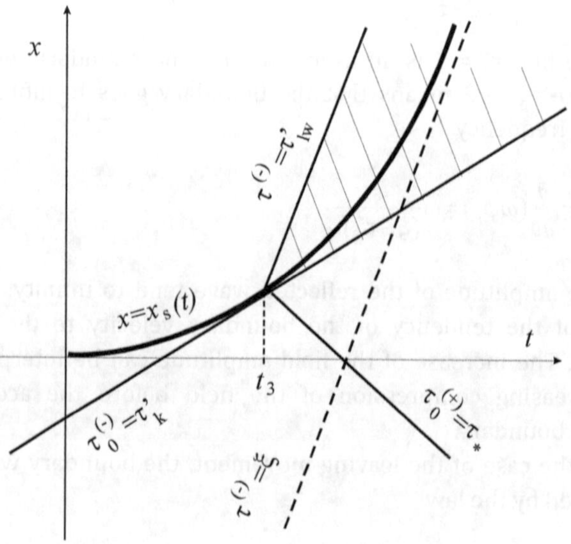

Figure 5.9 Time-spatial zones for the diffracted field (5.29).

If $n < 1$ then the impinging wave interacts with the boundary only until the moment $t_3 = \xi n(n^2 - 1)^{-1/2}$. In the future period, the impinging wave remains behind the running away boundary, and as a result, a dilative shadow region appears on both sides of the boundary (the cross-hatched region on Fig. 5.9). This effect is analogous to the one described in the paper [31]. The transmitted field is described by the function (5.40); however, its development is completely different compared to the case with $n > 1$. The behavior of the phase and the frequency and consequently the amplitude is shown in Fig. 5.10.

With the beginning of the accelerated movement of the boundary for the case $n < 1$ the amplitude of the transmitted wave decreases and becomes equal to zero with $\tau^{-} = \tau_H^1 = \xi(1 - \sqrt{(1-n)(1+n)^{-1}})$, which corresponds to the shadow boundary. The second shadow boundary coincides with the straight line $\tau_0^- = \tau'_\kappa = \frac{\xi}{n}(1-\sqrt{1-n^2})$.

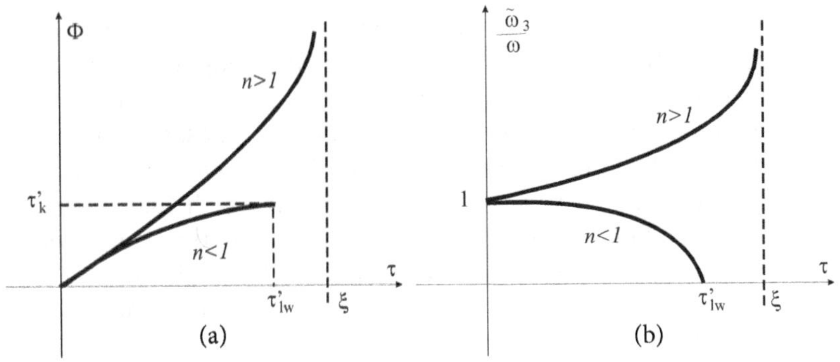

Figure 5.10 Phase characteristics.

The reflected wave in the case of the accompanied movement is described by the phase function

$$\psi_r = \begin{cases} & \tau_0^+, \tau_0^+ < 0 \\ \dfrac{n}{n^2-1}\left[\dfrac{n^2+1}{n}\tau_0^+ + 2\left(\xi - \sqrt{\xi^2 + 2\dfrac{\xi}{n}\tau_0^+ + (\tau_0^+)^2}\right)\right], & 0 < \tau_0^+ < \tau_* \end{cases}. \quad (5.42)$$

Restriction of the reflected field domain by the line $\tau_0^+ = \tau_* = \frac{\xi}{n}\left[(1+n^2)(1-n^2)^{-1/2} - 1\right]$ makes sense only in the case $n < 1$ and corresponds to the moment of the impinging wave separation from the boundary. The amplitude of the reflected field in this case falls down to zero with $\tau_0^+ \to \tau_*$. In the case $n > 1$, the reflected field exists for all τ_0^+ its amplitude remains finite during all time and tends to $|E_r| = (\frac{1-n}{1+n})^2$ as $\tau_0^+ \to \infty$.

Thus, temporary asymmetry caused by non-stationarity can generate a number of effects, such as the formation of propagating discontinuities in the field and the formation of the new transmitted and reflected waves in the case of the super-light movement of the boundary with respect to one of the media. The latter are absent in the stationary case.

5.4 Electromagnetic Field Energy Accumulation in a Collapsing Dielectric Layer

The temporal variations of medium parameters result in alterations in both the frequency and the amplitude of the electromagnetic wave propagating in the medium. The wave reflection from a moving medium boundary causes a similar effect. A combination of both these two phenomena may produce qualitatively new effects, the consequences of which can be both the accumulation of electromagnetic energy and the generation of electromagnetic pulses. An investigation of these phenomena is performed in this chapter. The interaction of an electromagnetic wave with a plane dielectric layer will be studied when the dielectric boundaries are moving to meet each other after starting at the zero moment of time, so that after a certain time interval the collapse of the layer happens. It will be demonstrated that this collapse causes an excitation of electromagnetic wave packets with an unlimited increase in frequency (under the approximation of an infinitely powerful mechanism which stimulates the boundary motion).

5.4.1 Increase of the Wave Amplitudes in the Collapsing Layer

Let us examine now the evolution of the electromagnetic field in the layer which, first, appears at the zero time moment in the homogeneous medium background and then collapses at the moment $t = a/2u$ because of the uniform motion of its boundaries, each with the velocity u in opposite directions, towards each other. The resolvent is derived in the same way as in Chapter 3, but the uniform boundary movement means that a transformation of the resolvent described by the expression (5.11) has to be taken into account. The multiple reflections of the resolvent characteristics divide the region of the layer in spatial-temporal coordinates, described by the characteristic function $\chi = \theta(x - ut)\theta(a - ut - x)$, into the zones shown in Fig. 5.11.

These zones are not equidistant now and the distance between them decreases with their number

$$t_n - t_{n-1} = p^{1-n} \frac{a}{v_2 + u}, \qquad (5.43)$$

where the parameter is $p = \frac{v_2 + u}{v_2 - u}$. (5.44)

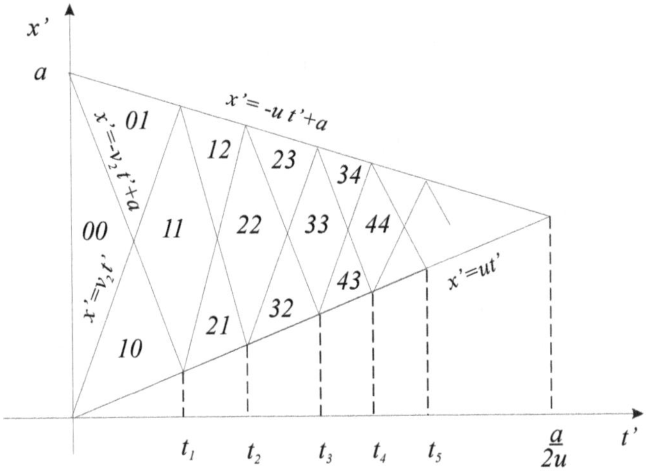

Figure 5.11 Spatial-temporal zones corresponding to certain number of reflections from each boundary of the layer with the collapsing boundaries.

It follows that an infinitely large number of zones exists in the finite time interval whose duration is equal to $a/2u$. In the case when $u > v_2$ the whole triangle formed by the world lines of the layer boundaries is completely placed within the zone 00 and the field has no time to experience the influence of the layer boundaries, and is thus the same as that in the infinite medium. This is the reason for considering only the case $u < v_2$ to investigate the field behavior features due to the layer collapsing.

If the observation point is in the mn-th zone then the resolvent has the form

$$\hat{R} = (\hat{R}_1 + \hat{R}_{mn})\theta(x' - ut')\theta(a - ut' - x'), \tag{5.45}$$

where the matrix elements of the operator $\hat{R}_{mn}$ are

$$\left\langle \mathbf{x} \,|\, \hat{R}_{mn} \,|\, \mathbf{x}' \right\rangle = -\zeta \frac{v_2}{v} \frac{\partial^2}{\partial t^2} \left\{ \sum_{k=1}^{m} R_u^k \theta\left[-v_2 t' + (-1)^k x' \right.\right.$$
$$\left.\left. + p^{i_k} \left(1 + \frac{u}{v_2} q^2 \frac{1 - p^{k-i_k}}{1 - p^2} \right) v_2 \tau^{(-)} - p^{i_k} \frac{1 - p^{k-i_k}}{1 - p^2} qa \right] \right.$$

$$+\sum_{k=1}^{n} R_u{}^k \theta\Big[-v_2 t' - (-1)^k x' + p^{i_k}\left(1+\frac{u}{v_2}q^2\frac{1-p^{k-i_k}}{1-p^2}\right)v_2\tau^{(+)} - p^{k-i_k}\frac{1-p^{k+i_k}}{1-p^2}qa\Big]\Big\} \tag{5.46}$$

with $q = \frac{2v_2}{v_2 - u}$, $\tau^{(\pm)} = t \pm x/v_2$ and $i_k = [1 + (-1)^k]/2$.

The reflection coefficient is given by $R_u = \frac{v-v_2}{v+v_2}\frac{v_2-u}{v_2+u}$. (5.47)

Let an initial wave $E_0(t, x) = E_0 e^{i\omega(t - x/v)}$ propagate in the background. After the layer appears the field in the zone 00 is the same as in the case of the layer with stationary boundaries, i.e., it is determined by (3.123). The field in the zone *mm* is described by the following formula:

$$E_{mm} = \frac{\omega^{(+)}}{\omega}C_0\Bigg\{e^{i\omega^{(+)}(t-x/v_2)} + \sum_{k=1}^{m-1} R_1{}^k \exp\left[i\omega^{(+)}p^k(t-(-1)^k x/v_2) - i\frac{1-p^{\frac{k+i_k}{2}}}{1-p^2}q\frac{\omega^{(+)}}{v_2}a\right]\Bigg\}$$
$$+R_1{}^m C_1 \exp\left[i\omega_2 p^m(t-(-1)^m x/v_2) - i\frac{1-p^{\frac{m+i_m}{2}}}{1-p^2}q\frac{\omega_2}{v_2}a\right]$$
$$+R_1{}^m C_2 \exp\left[-i\omega_2 p^m(t+(-1)^m x/v_2) + i\frac{1-p^{\frac{m+i_m}{2}}}{1-p^2}pq\frac{\omega_2}{v_2}a\right], \tag{5.48}$$

where $\omega^{(\pm)} = \omega\frac{v_2}{v}\frac{v-u}{v_2 \pm u}$, $C_0 = \frac{2v_2}{v+v_2}$, $C_{1,2} = \frac{v_2(v_2 \pm v)}{2v^2}$ and $R_1 = pR = \frac{v_2+u}{v_2-u}\frac{v-v_2}{v+v_2}$.

In this case, the field structure is essentially more complicated compared to the case of the field in the layer with stationary boundaries. There are also two waves excited by the waves marked by the multipliers C_1 and C_2 in each zone resulting from splitting the initial wave due to the permittivity jump. However, the frequencies of these waves increase with increasing the zone number because the collapsing layer, which reflects the internal field, becomes thinner with time. In addition to these waves, there is a set of m waves being proportional to C_0 in each zone that is the waves generated by the wave impinging on the layer. This set has a spectrum which consists of a discrete set of frequencies $\omega^{(+)}p^k$ where k = 0, 1, ..., m - 1. As a result of the multiple reflections the frequencies of all the waves increase with increasing the zone number, i.e., they increase in time. The similar effect can be observed in the case of a cavity with a uniformly moving wall as reported in [31–40].

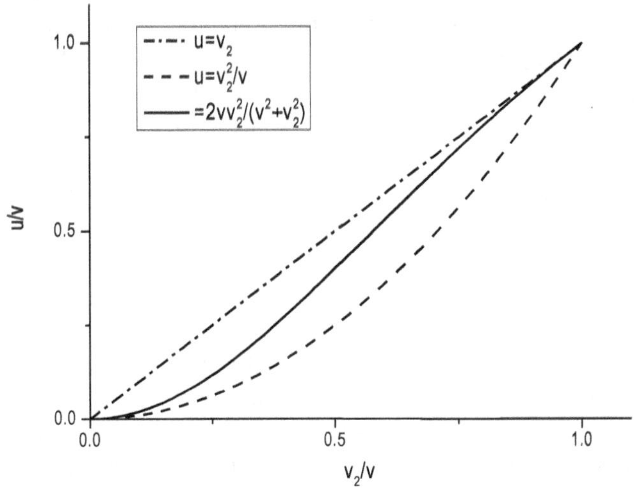

Figure 5.12 The lines of the critical relations between the layer boundary velocity and wave phase velocity: u/v = $(v_2/v)^2$—the dashed line, $u/v > 2v_2^2/(v^2 + v_2^2)$—the solid line, $u/v = v_2/v$—the dot-dashed line.

The coefficient of the frequency multiplication is equal to p given by (5.44) It increases if $u \to v_2$. The behavior of the wave amplitudes is determined by relation between the wave phase velocities and the velocity of the layer boundary motion. It follows from (5.48) that R_1 > 1 if $vu - v_2^2 > 0$, i.e., $u > c\sqrt{\varepsilon/\varepsilon_2}$ and that the wave amplitudes increase infinitely in this case. It is essential that this process takes place during a finite time interval. The jump of the permittivity inside the layer at the zero moment of time yields a discrete set of steps, while the layer boundary motion makes the number of steps infinitely large. An infinite number of elementary reflections take place within the finite time interval here. The region between the curve $u/v = (v_2/v)^2$ (the dotted curve in Fig. 5.12) and the curve $u/v = v_2/v$ (the dot-dashed line in Fig. 5.12) is characterized by an increasing of the internal wave amplitudes. Consequently, the increase of the internal wave amplitudes is possible only in an optically dense layer, i.e., when $v_2/v < 1$.

5.4.2 The Energy Accumulation in the Layer

The process considered above can also accumulate energy in the layer owing to the transformation of boundary kinetic energy into the energy of the electromagnetic field. To investigate this problem let us examine the energy relations for the waves being excited by the split waves C_1 and C_2, as described by the last terms in (5.48).

The energy balance equation has the following form:

$$\frac{\partial w}{\partial t} = -\frac{\partial}{\partial x} S, \tag{5.49}$$

where, in Gaussian units, $w = (\varepsilon_2 E^2 + B^2)/8\pi$ is the energy density and $S = cE_y B_z/8\pi$ stands for the density of the energy flux. It is assumed that the electric field vector only has the component E_y.

Let us integrate both sides in Eq. (5.49) over the layer thickness. It must be taken into account that jumps in the field exist in the layer on the wave fronts whose propagation world lines are parallel to the boundaries of the zones, as shown in Fig. 5.13.

The coordinates of these jumps in the interval $[t_{m-1}, t_m]$ are

$$x_m^{(1)} = -v_2 t + \frac{1}{p^{m-1}} \frac{1-p^m}{1-p} a, \quad x_m^{(2)} = v_2 t + \frac{1}{p^{m-1}} \frac{1-p^{m-1}}{1-p} a. \tag{5.50}$$

The integration of (5.49), taking into account the field discontinuities, yields the energy balance equation in the layer within the time interval $[t_{m-1}, t_m]$:

$$\begin{aligned} \frac{dW}{dt} = &-u\{w(x_2 - 0) + w(x_1 + 0)\} + v_2\{\langle w \rangle_{x_m^{(1)}} - \langle w \rangle_{x_m^{(2)}}\} \\ &-\{S(x_2 - 0) - S(x_1 + 0)\} + \langle S \rangle_{x_m^{(1)}} + \langle S \rangle_{x_m^{(2)}}, \end{aligned} \tag{5.51}$$

where $W = \int_{x_1}^{x_2} w dx$ is the wave total energy in the layer, $x_1 = ut$, $x_2 = -ut + a$ are the coordinates of the layer boundaries, $\langle S \rangle_{x_m} = S(x_m + 0) - S(x_m - 0)$ and $\langle w \rangle_x = w(x + 0) - w(x - 0)$ are the jumps in the corresponding values.

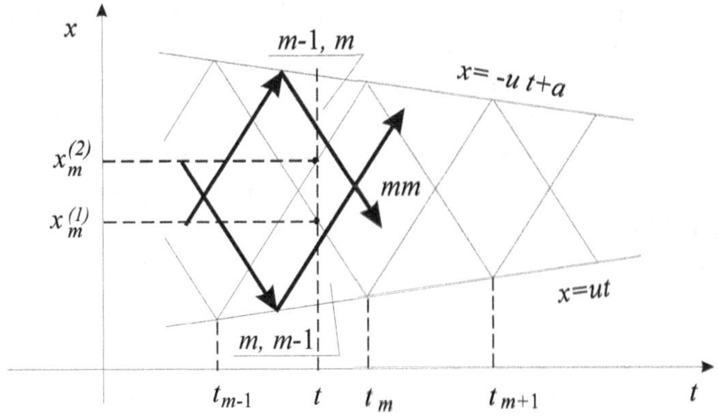

Figure 5.13 The time-spatial picture for the determination of the energy flux.

The group of waves in the zone mm and the adjacent zones are given by the following expressions:

$$E_{mm} = C_1 R_1^m f_m^{(-)}(t,x) + C_2 R_1^m f_m^{(+)}(t,x) \tag{5.52}$$

$$E_{m-1m} = C_1 R_1^{m-1} [f_{m-1}^{(-)}(t,x) + R_1 f_m^{(-)}(t,x)] \tag{5.53}$$

$$E_{mm-1} = C_2 R_1^{m-1} [f_{m-1}^{(+)}(t,x) + R_1 f_m^{(+)}(t,x)], \tag{5.54}$$

where

$$f_m^{(-)}(t,x) = \cos\left[\omega_2 p^m (t - (-1)^m x / v_2) - \frac{1 - p^{(m+i_m)/2}}{1 - p^2} q \frac{\omega_2 a}{v_2}\right] \tag{5.55}$$

$$f_m^{(+)}(t,x) = \cos\left[\omega_2 p^m (t + (-1)^m x / v_2) - \frac{1 - p^{(m-i_m)/2}}{1 - p^2} pq \frac{\omega_2 a}{v_2}\right] \tag{5.56}$$

Substitution of (5.52)–(5.56) into (5.51) yields the following equation for the group of waves under consideration:

$$\frac{dW}{dt}=\varepsilon_2(v_2-u)(R_1^2-p)R_1^{2(m-1)}\times\{C_1^2\cos^2(\omega_m t-\varphi_1)+C_2^2\cos^2(\omega_m t-\varphi_2)\},\tag{5.57}$$

where $\omega_m=p^{m-1}\dfrac{v_2+u}{v}\omega$.

In the stationary case ($u=0$) one obtains $dW/dt<0$ meaning that the wave energy in the layer decreases and the waves "fade out." However, if the layer boundaries move, the energy balance allows such a relation between the parameters that provides $dW/dt>0$, which provides for the possibility of energy accumulation. It follows from (5.57) that this energy accumulation can happen when $R_1^2-p>0$. Taking into account that

$$R_1^2-p=p^2\frac{2v}{v_2+u}\frac{1}{(v+v_2)^2}\left[\frac{u}{v}(v^2+v_2^2)-2v_2^2\right],\tag{5.58}$$

the criterion for the energy accumulation is obtained as follows:

$$\frac{u}{v}>\frac{2v_2^2}{v^2+v_2^2}\tag{5.59}$$

The critical boundary of the energy accumulation region is indicated by the solid curve in Fig. 5.12. The region of the energy accumulation, unlike the region of the wave amplitude increase, can exist in an optically dense layer as well as in a layer whose optical density is less than that of the environment.

5.4.3 Generation of Electromagnetic Pulses by the Collapsing Layer

Substituting the internal field determined in the previous section into (5.1) and integrating it the field reflected from the collapsing layer is obtained; it has the following form of a packet localized between the planes vt_{m+1} and vt_m:

$$E_m=\frac{2v^2}{v_2(v+v_2)}\frac{v_2+u}{v+u}C_1^{i_m}C_2^{1-i_m}R_1^m e^{-i\omega p^m\frac{v_2+u}{v+u}(t+x/v)+(-1)^m i\omega\eta_m\frac{a}{v}},\tag{5.60}$$

where $\eta_m=-\frac{v_2-u}{2u}p^{1-i_m}(1-p^{m+i_m})$. Figure 5.14 illustrates the dependence of the reflected wave frequency $\omega_3=\omega p^3\dfrac{v_2+u}{v+u}$ in the third pulse on the layer relative permittivity (meaning the ratio $v_2/v = \sqrt{\varepsilon/\varepsilon_2}$) for different boundary velocities. The frequency is much larger near the internal phase velocity then near the boundary of the amplitude growth.

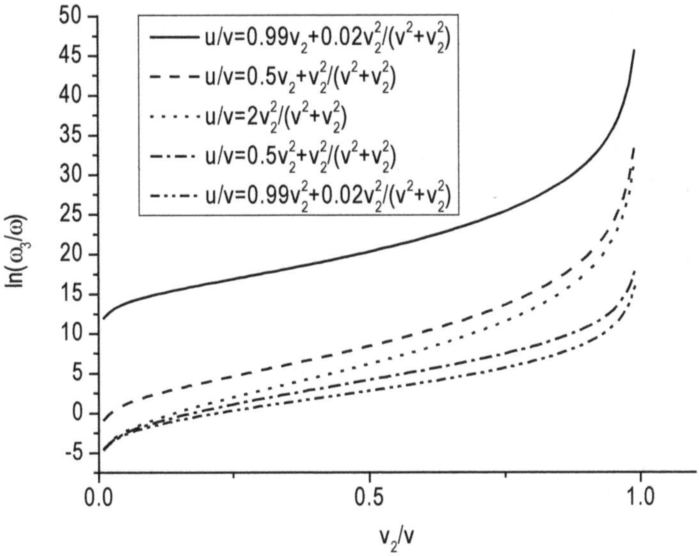

Figure 5.14 Dependence of the frequency in the 3D reflected pulse on the layer relative permittivity for various boundary velocities: near the internal phase velocity—solid line; the central velocity of the energy accumulation region—dash line; near the boundary of the energy accumulation region—dotted line; near the boundary of the amplitude growth—dash-dot-dot line.

The dependence of the wave pulse duration on the permittivity in the layer is shown in Fig. 5.15. The behavior of the duration is the inverse of the frequency behavior.

The field energy in the packet is proportional to $\left(\frac{v-v_2}{v+v_2}\right)^{2m} p^m$ because the packet length is determined by $v(t_{m+1}-t_m)=p^{-m}\frac{va}{v_2+u}$. Hence, for the parameter values providing the energy accumulation in the layer, the field energy in the pulse coming outside this layer infinitely increases when $m \to \infty$, i.e., for the time moment approaching the moment of the layer collapse. Note that the wave frequency (and so the quantum energy) increases more slowly than the pulse length decreases. This leads to the quantum number within the pulse tending to zero with $m \to \infty$, as follows:

$$\int_{vt_{m+1}}^{vt_m} \frac{wdl}{\omega_m} \approx \frac{w}{p^m}\frac{v}{p^m} \Rightarrow \frac{R_1^{2m}}{p^{2m}} = \left(\frac{v-v_2}{v+v_2}\right)^{2m} \to 0. \tag{5.61}$$

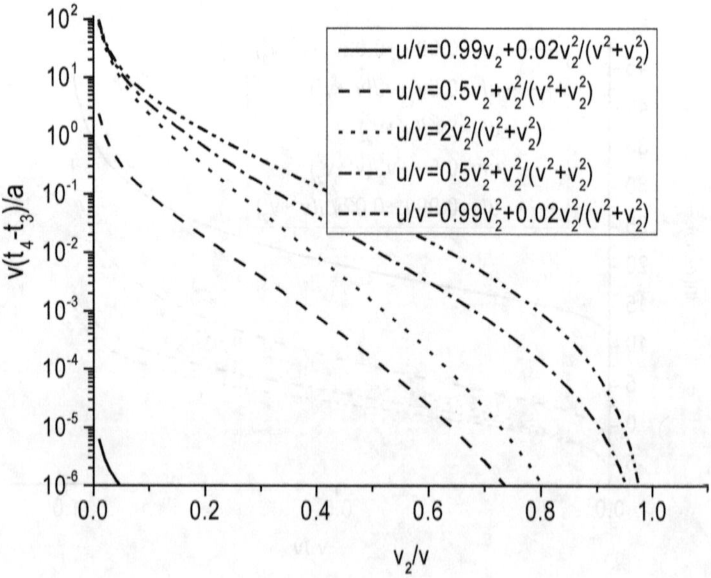

Figure 5.15 Dependence of the duration in the 3D reflected pulse on the permittivity in the layer: near the internal phase velocity—solid line; the central velocity of the energy accumulation region—dashed line; near the boundary of the energy accumulation region—dotted line; near the boundary of the amplitude growth—dash-dot-dot line.

The dependence the volume energy density on the relative permittivity inside the layer for different boundary velocities is shown in Fig. 5.16. The closer the boundary velocity is to the inside phase velocity the greater the volume energy density.

The above investigations show that a combined action of two mechanisms, such as a temporal variation of the permittivity and the motion of the medium boundaries, whose separate actions induce the phenomenon of wave amplitude and frequency alterations, result in the accumulation of electromagnetic energy in the collapsing dielectric layer created at a certain time moment. This process is accompanied by the excitation of electromagnetic pulses outside the layer, the duration of which decreases to zero at the moment of the layer collapse, whereas the field frequency within this pulse infinitely increases. The pulse energy density also increases infinitely, if one assumes an infinitely powerful process which causes the movement of the boundaries.

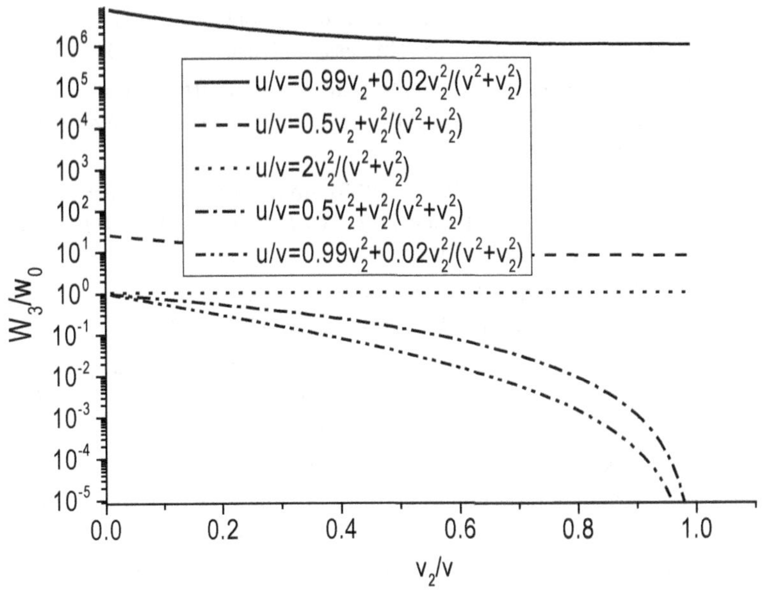

Figure 5.16 Dependence of the volume energy density in the third reflected pulse on the permittivity inside the layer: near the internal phase velocity—solid line; the central velocity of the energy accumulation region—dashed line; near the boundary of the energy accumulation region—dotted line; near the boundary of the amplitude growth—dash-dot-dot line.

References

1. Krasilnikov, V. N., Lutchenko, L. N. (1973) Principle of seeming location of an interface and generalization of V. A. Fock's, reflection formulae on the case of moving boundaries, *Problems of waves diffraction and propagation,* (in Russian) Leningrad University Publishing House, **12**, pp. 150–157.
2. Buldyrev, V. S., Maslova, G. N. (1982) Reflection of modulated wave from moving boundary of arbitrary shape, Wave and diffraction, *Proc. VIII USSR Symp. Waves Diffraction Propagation,* (in Russian) Moscow, **3**, pp. 280–283.
3. Petrov, B. M. (1972) Spectral characteristics of a field scattered by uniformly moving and revolving impedance cylinder, *Radiotechnika i Electronika,* (in Russian) **17**(9), pp. 1799–1804.

4. Van Bladel, J. (1973) Relativity theory of revolving disks, *Proc. IEEE,* **63**(3), pp. 3–9.
5. Bolotovskiy, B. M., Zhemanova, E. O. (1985) Solution to a wave equation with boundary conditions given on revolving cylinder surface, *Computing Math. Math. Phys.*, (in Russian) Moscow, **25**, pp. 143–149.
6. Tanaka, K. (1985) A general relation of scattered waves by a rotating object, *Proc. 1985 Int. Symp. Antennas Propag., Kyoto, Aug. 20–22,* **1**, pp. 249–252.
7. Janjieri, G. V., Pikulin, V. D. (1985) About asymmetrical waves in revolving dielectric cylinder, *Proc. IX USSR Symp. Waves Diffraction Propagation, Telavi,* (in Russian) Tbilisi, **2**, pp. 313–315.
8. Rao, T. C. K. (1986) Electromagnetic wave propagation over rotating dielectric cylinder, *Can. J. Phys.,* **64**(11), pp. 1517–1521.
9. Nerukh, A. G., Khizhnyak, N. A. (1978) Electromagnetic wave diffraction on moving regions of uniform medium, *Ukrainian Phys. J.*, (in Russian) **23**(12), pp. 2009–2014.
10. Bolotovski, B. M., Stolyarov, S. N. (1980) Electromagnetic waves amplification with presence of moving boundaries, *Einstein Collection 1977,* (in Russian) Moscow, "Nauka", pp. 73–130.
11. Zeldovich Ya. B., Starobinskiy, A. A. (1976) Superreflection and quantum birth of particles, *Proceedings on Mathematical Physics,* (in Russian) Leningrad, "Nauka", pp. 25–48.
12. Kokodiy, N. G. (1978) Transverse ponderomotor effect in rotating dissipative cylinder, *Optika i spektroskopiya,* (in Russian) **45**(2), pp. 213–218.
13. De Zutter, D. (1986) Radiation pressure and torque exerted by a plane wave on rotating circular cylinder, *URSI Symp. Inf. Electromagn Theory,* Budapest, Aug. 25–29, 1986, Part A, Budapest, pp. 206–208.
14. Freilikher, V. D., Fuks, I. M. (1969) Spectrum of signal scattered by vibrating interface, *Izv. vuz. Radiofizika,* (in Russian) **12**(1), pp. 114–119.
15. Van Bladel, J., De Zutter, D. (1981) Reflection from linear vibrating objects, *IEEE Trans. Antennas Propag.*, **29**(4), pp. 629–637.
16. Andreev, A. M., Ginzburg, V. M. (1985) refraction and scattering of electromagnetic waves on a surface water-air strained by radiation pressure of ultrasound, *Radiotechnika i Electronika,* (in Russian) **30**(5), pp. 881–885.
17. Censor, D. (1985) Electromagnetic scattering by harmonic expanding surfaces and related complex resonances, *Radio Sci.,* **20**(1), pp. 25–34.

18. Pogorzhelski, R. J. (1973) Electromagnetic scattering from a radially moving spherical discontinuity, *J. Appl. Phys.*, **44**(1), pp. 168–173.

19. Laziev, E. M. (1984) Spontaneous radio-frequency radiation of charged particles, *Thesis on Doc. of Science*, (in Russian) Erevan, **174** p.

20. Davydov, V. A., Kolesov, V. V. (1982) Radiation of charged filament on moving degraded boundary between two media, *Izv. vuz. Radiofizika*, (in Russian) **25**(7), pp. 815–820.

21. Ostrovski, L. A. (1971) Some general relations for waves on moving boundary between two media, *Zhurnal experimentalnoy i teoreticheskoy fiziki (ZhETPh)*, (in Russian) 61, **2**(8), pp. 551–561.

22. Ostrovski, L. A. (1975) Some "paradoxes of moving boundaries" in electrodynamics, *Uspekhi fizicheskikh nauk (UPhN)*, (in Russian) **116**(2), pp. 313–325.

23. Ostrovski, L. A., Stepanov, N. S. (1971) Nonresonance parametric phenomena in distributed systems, *Izv. vuz. Radiofizika*, (in Russian) **14**(4), pp. 489–529.

24. Stolyarov, S. N. (1978) Boundary value problems in electrodynamics of moving media, *Einstein collection 1975–1976*, (in Russian) Moscow, "Nauka", pp. 152–215.

25. Stolyarov, S. N. (1976) About limitation on efficacy of frequencies transformation on moving boundaries, *Pisma v Zhurnal of technicheskoy fiziki*, (in Russian) **2**(19), pp. 880–882.

26. Ostrovski, L. A., Stepanov, N. S. (1967) Correctness of problems on waves interaction with moving jump of parameter, *Izv. vuz. Radiofizika*, (in Russian) **10**(8), pp. 1183–1188.

27. Mo, T. C. (1971) Electromagnetic wave propagation in a uniformly accelerated simple medium, *Radio Sci.*, **6**(6), pp. 673–679.

28. Tanaka, K. (1974) Electromagnetic fields in linearly accelerated medium, *Trans. Inst. Electron. Commun. Eng. Jap.*, **57**(2), pp. 75–82.

29. Nerukh, A. G., Khizhnyak, N. A. (1979) Wave scattering by uniformly accelerated boundary between media, *Zhurnal technicheskoy fiziki*, (in Russian) **49**(2), pp. 225–230.

30. Landau, L. D., Lifshitz, E. M. (1967) *Field Theory*, Moscow, (in Russian) "Nauka", 460 p.

31. Stetsenko, O. A. (1963) About pressing of electromagnetic field between two planes, *Izv. vuz. Raditehnika*, **6**(6), pp. 695–700 (in Russian).

32. Krasilnikov, V. N. (1975) Regions of "running" half-shadow, *Izv. vuz. Radiofizika*, (in Russian) **18**(6), pp. 894–900.

33. Nerukh, A. G. (1989) On the transformation of electromagnetic waves by a nonuniformly moving boundary between two media, *Soviet Phys. Tech. Phys.*, **34**(3), pp. 281–284.

34. Borisov, V. (1987) *Transient Electromagnetic Waves*, Leningrad State University Press, Leningrad (in Russian).

35. Nerukh, A. G., Khizhnjak, N. A. (1991) *Modern Problems of Transient Macroscopic Electrodynamics*, Test-Radio Publ., Kharkov (in Russian).

36. Kalluri, D. K. (1988) On reflection from a suddenly created plasma half-space: Transient solution, *IEEE Trans. Plasma Sci.*, **16**, pp. 11–16.

37. Kalluri, D. K., Goteti, V. R. (1992) Frequency shifting of electromagnetic radiation by sudden creation of a plasma slab, *J. Appl. Phys.*, **72**, pp. 4575–4580.

38. Bakunov, M. I., Maslov, A. V. (1998) Trapping of an electromagnetic wave by the boundary of a time-varying plasma, *Phys. Rev. E*, **57**, p. 5978–5986.

39. Bakunov, M. I., Maslov, A. V., Zhukov, S. N. (1999) Scattering of a surface plasmon polarisation by rapid plasma creation in a semiconductor slab, *J. Opt. Soc. Am. B*, **16**(11), p. 1942.

40. Nerukh, A. G., Yemelyanov, K. M. (2000) An interaction of electromagnetic field with a collapsing plasma layer, *EUROEM 2000–Euro Electromagnetics*, Edinburgh, UK.

41. Bateman, H., and Erdely, A. (1954) *Tables of Integral Transforms* (New York, Toronto, London: McGraw-Hill Book Comp., Inc.) 344 p.

42. Villermaux, E., Bossa, B. (2009) Single-drop fragmentation determines size distribution of raindrops, *Nat. Phys.*, **5**(9), pp. 697–702.

43. Mahmud, G., Campbell, C. J., Bishop, K. J. M., Komarova, Y. A., Chaga, O., Soh, S., Huda, S., Kandere-Grzybowska, K., Grzybowski, B. A. (2009) Directing cell motions on micropatterned ratchets, *Nat. Phys.*, **5**(8), pp. 606–612.

44. Lin, Y.-H., Lin, W.-Y., Lee, G.-B. (2009) Image-driven cell manipulation, *IEEE Nanotechnol. Mag.*, **3**(3), pp. 6–11.

45. Humar, M., Ravnik, M., Pajk, S., Musevic, I. (2009) Electrically tunable liquid crystal optical microresonators, *Nat. Photonics*, **3**(10), pp. 595–600.

46. Sretenskiy, L. N. (1946) *Theory of Newton Potential*, Gostekhizdat, Moscow.
47. Khizhnyak, N. A. (1986) *Integral Equations of Macroscopic Electrodynamics*, "Naukova dumka" Publ., Kiev, 280 p. (in Russian).
48. Khizhnyak, N. A., Lysenko, O. E. (1970) Diffraction of electromagnetic wave on an object of a complex form, *Zhurnal tekhnicheskoy fiziki*, **40**(12), p. 2581 (in Russian).

Chapter 6

An Electromagnetic Field in a Metallic Waveguide with a Moving Medium

The first part of the book were devoted to the basic electromagnetic transient phenomena caused by time varying properties of a homogeneous medium or a medium with a very simple discontinuity in the form of a plane boundary. The time variation can be caused by changing in time of the medium properties or by moving the medium boundaries. As an extension of this work, transients in waveguide and resonator structures are now investigated. Waveguides with perfectly conducting walls and dielectric waveguides will be considered.

The influence of a moving medium on a wave guided in a rectangular waveguide with perfectly conducting walls is considered in this chapter, along with the wave evolution. The transformation of a guided wave in a waveguide filled with a dielectric medium which uniformly moves with relativistic velocity is investigated. Based on this investigation, it is shown possible to model various phenomena connected with the interaction of electromagnetic waves with a boundary of a relativistic moving medium in the presence of waveguide dispersion. The interaction of the electromagnetic field with plasma "flashed" at some zero moment of time in the waveguide, and then uniformly expanding, is considered.

Non-Stationary Electromagnetics: An Integral Equations Approach (Second Edition)
Alexander Nerukh and Trevor Benson

ISBN 978-981-4774-95-6 (Hardcover), 978-0-429-02773-4 (eBook)
www.panstanford.com

The investigations are carried out by using the evolution approach developed in this book. A chain of evolution equations for a non-stationary electromagnetic field in a waveguide is derived and solutions to these equations are obtained by virtue of the resolvent method.

6.1 Expansion of an Electromagnetic Field by Non-Stationary Eigen-Functions of a Waveguide

The general integral Eq. (1.59)

$$\mathbf{E} = \mathbf{E}_0 - \frac{1}{v^2}\frac{\partial^2 G_0}{\partial t^2} * \chi \frac{1}{\varepsilon_0 \varepsilon}(\mathbf{P}_1 - \mathbf{P}_{\text{ex}})$$
$$- \frac{\partial G_0}{\partial t} * \{\text{curl}\chi\mu_0\mu(\mathbf{M}_1 - \mathbf{M}_{\text{ex}}) + \mu_0\mu(\chi\mathbf{j}_1 + \mathbf{i}\delta(S))\} \qquad (6.1)$$

will be considered for the case of a field in a rectangular metallic waveguide when Green's function has the dyadic form (1.49)

$$G_{ij} = \frac{v^2}{4\pi}\left(\frac{\partial^2}{\partial x_i \partial x_k} - \frac{1}{v^2}\frac{\partial^2}{\partial t^2}\delta_{ik}\right)\tilde{\tilde{f}}^{E,kj} \qquad (6.2)$$

The elements of this tensor are defined by the functions (1.54) for the rectangular waveguide whose longitudinal axis is directed along the x axis and transverse dimensions are equal to a along the y axis and b along the z axis:

$$\tilde{\tilde{f}}_{ij}^{E} = \frac{8\pi v}{ab}\delta_{ij}\sum_{m,n}\tilde{\tilde{\Gamma}}_{mn} f_i^{E,mn} f_j^{\prime E,mn}, \qquad (6.3)$$

where the coefficients $\tilde{\tilde{\Gamma}}_{mn} = \int_{-\infty}^{t} d\tau \int_{-\infty}^{\tau} d\tau \, \Gamma_{mn}(\tau' - t', x - x')$ (6.4)

are determined by the functions

$$\Gamma_{mn}(t,x) = J_0\left(\omega_{mn}\sqrt{t^2 - \frac{x^2}{v^2}}\right)\theta\left(t - \frac{|x|}{v}\right). \qquad (6.5)$$

Here $\omega_{mn} = v\pi\sqrt{(m/a)^2 + (n/b)^2}$ is an eigen-frequency of a mn-mode.

Substitution of (6.2) into (6.1) gives the specific form of the integral equation for a waveguide

$$E_i = E_{0i} + \frac{1}{\varepsilon}\int_{-\infty}^{\infty} dt' \int_{-\infty}^{\infty} dx' \int_0^a dy' \int_0^b dz' \left(\frac{\partial^2}{\partial x_i \partial x_j} - \frac{1}{v^2}\frac{\partial^2}{\partial t^2}\delta_{ij} \right) \tilde{\hat{f}}_{ij}^E$$
$$\times \left\{ \frac{\partial^2}{\partial t'^2}\chi\left(P_j - \frac{\varepsilon - 1}{4\pi}E_j\right) + ce_{jkl}\frac{\partial^2}{\partial x_k' \partial t'}\left(P_l - \frac{\mu - 1}{4\pi\mu}B_l\right) + \frac{\partial}{\partial t'}(\chi j_j - i_j \delta(S)) \right\},$$

(6.6)

where e_{jkl} is the completely anti-symmetric tensor of rank 3.

As the integrals with respect to t' and x' represent convolutions the differentiation by these variables can be brought from the field functions to Green's function. After differentiation and some manipulations we obtain the Eq. (6.6) in the form

$$\mathbf{E} = \mathbf{E}_0$$
$$+ \frac{8\pi v}{ab}\sum_{m,n}\int d\mathbf{x}' \left\{ \frac{1}{\varepsilon}\left((\hat{f}^{E,mn}\hat{\kappa}^*)\hat{\kappa} - \hat{f}^{E,mn}\frac{1}{v^2}\frac{\partial^2}{\partial t^2} \right) \right\} \hat{f}'^{E,mn}\left\{ \Gamma_{mn}\chi\left(\mathbf{P}_{in} - \frac{\varepsilon - 1}{4\pi}\mathbf{E}\right) \right.$$
$$\left. + \tilde{\Gamma}_{mn}(\chi \mathbf{j}_{in} + \mathbf{i}\delta(S)) - \hat{f}^{E,mn}\frac{\mu}{c}\frac{\partial}{\partial t}\left[\hat{\kappa} \times \Gamma_{mn}\chi\hat{f}'^{E,mn}\left(\mathbf{M}_{in} - \frac{m - 1}{4\pi\mu}\mathbf{B}\right) \right] \right\},$$

(6.7)

where $\hat{f}^{E,mn}$ is the tensor with the elements $f_{ij}^{E,mn} = \delta_{ij}\Gamma_{mn} f_i^{E,mn} f_j'^{E,mn}$, $\hat{\kappa} = (\partial/\partial x_1, -\kappa_m, -\kappa_n)$, $\hat{\kappa}^* = (\partial/\partial x_1, \kappa_m, \kappa_n)$, $\kappa_m = m\pi/a$, $\kappa_n = n\pi/b$ and $\hat{f}\mathbf{A}$ means a product of a diagonal tensor $\hat{f}$ with a vector $\mathbf{A}$.

The kernel in Eq. (6.7) is a series of the waveguide eigen-functions. It is natural to represent all field values in the form of an expansion on waveguide eigen-functions which have to be vectorial ones. It is common to derive these functions under the assumption that fields can be represented in the form of an expansion in terms of modes which are monochromatic plane inhomogeneous waves with a certain frequency and a certain wave number. In the general case of non-monochromatic fields, with inseparable dependence on time and spatial coordinates, the expressions for these vector eigen-functions have to be corrected. The derivation of these eigen-functions follows the scheme presented in the book [1].

It is well known that in a regular waveguide all waves are divided into two classes: E- or TM-waves (transverse magnetic waves) and H- or TE-waves (transverse electric waves). For the TM-waves the longitudinal components of the electric field satisfy the first boundary-value problem which consists of Eq. and the boundary condition

$$\frac{\partial^2 E}{\partial y^2}+\frac{\partial^2 E}{\partial z^2}+\kappa^2 E=0, \tag{6.8}$$

$$E|_{L}=0,$$

where L is the rectangular contour of the waveguide cross-section. The eigen-functions of this problem are

$$E_x=\sin\kappa_m y\sin\kappa_n z,\ \kappa_m=m\pi/a,\ \kappa_n=n\pi/b. \tag{6.9}$$

Expansion for the transverse electric components is given by the vector eigen-functions $\mathbf{e}_{mn\perp}$ which are expressed through the scalar function (6.9) by the following

$$\begin{aligned}\mathbf{e}_{mn\perp}&=\nabla_\perp E_x=\kappa_m f_{22}^{E,mn}\mathbf{e}_y+\kappa_n f_{33}^{E,mn}\mathbf{e}_z\\&=\kappa_m\cos\kappa_m y\sin\kappa_n z\mathbf{e}_y+\kappa_n\sin\kappa_m y\cos\kappa_n z\mathbf{e}_z\end{aligned} \tag{6.10}$$

Here the scalar functions are given by (1.51)

$$\left.\begin{aligned}f_{11}^{E,mn}&=\sin\kappa_m y\sin\kappa_n z\\f_{22}^{E,mn}&=\cos\kappa_m y\sin\kappa_n z\\f_{33}^{E,mn}&=\sin\kappa_m y\cos\kappa_n z\end{aligned}\right\} \tag{6.11}$$

The axes are directed as shown in Fig. 1.3 where the coordinates x, y, z will be numerated as 1, 2, 3. It is easy to check that these functions satisfy the first vector boundary-value problem

$$\nabla_\perp^2\mathbf{e}_{mn\perp}+\kappa_{mn}^2\mathbf{e}_{mn\perp}=0;\ \ \mathbf{e}_{mnt}\big|_L=0;\ \ \operatorname{div}_\perp\mathbf{e}_{mnt}\big|_L=0,$$

where $\kappa_{mn}^2=\kappa_m^2+\kappa_n^2$ are the eigen-values.

The vectors (6.10) give the system of the vector functions for a magnetic field

$$\mathbf{e}^*_{mn} = [\mathbf{e}_x, \nabla_\perp E_x] = -\kappa_n f_{33}^{E,mn}\hat{\mathbf{e}}_y + \kappa_m f_{22}^{E,mn}\hat{\mathbf{e}}_z \tag{6.12}$$

The system of the vector functions for an electric field is given by the Maxwell's equation $\mathbf{E} = \mathrm{curl}\int^t \mathbf{B}dt$. Omitting integration by time that is inessential in this case we obtain

$$\mathbf{e}_{mn} = \kappa^2_{mn} f_{11}^{E,mn}\hat{\mathbf{e}}_x + \kappa_m f_{22}^{E,mn}\hat{\mathbf{e}}_y \frac{\partial}{\partial x} + \kappa_n f_{33}^{E,mn}\hat{\mathbf{e}}_z \frac{\partial}{\partial x} \tag{6.13}$$

Thereby the electrical and magnetic fields of the TM-waves are represented by the series on the systems of the vector eigen-functions (6.12) and (6.13):

$$\mathbf{E} = \sum_{m,n} \mathbf{e}_{mn} E_{mn}(t,x) \tag{6.14}$$

$$\mathbf{B} = \sum_{m,n} \mathbf{e}^*_{mn} B_{mn}(t,x) \tag{6.15}$$

The vector eigen-functions for TE-waves generated by the scalar functions

$$H_x = \cos\kappa_m y \cos\kappa_n z \tag{6.16}$$

are equal to

$$\mathbf{b}_{mn\perp} = \nabla_\perp H_x = -\kappa_m f_{22}^{B,mn}\hat{\mathbf{e}}_y - \kappa_n f_{33}^{B,mn}\hat{\mathbf{e}}_z$$

$$= -\kappa_m \sin\kappa_m y \cos\kappa_n z\hat{\mathbf{e}}_y - \kappa_n \cos\kappa_m y \sin\kappa_n z\hat{\mathbf{e}}_z, \tag{6.17}$$

where $f_{11}^{B,mn} = \cos\kappa_m y \cos\kappa_n z,\ f_{22}^{B,mn} = f_{33}^{E,mn},\ f_{33}^{B,mn} = f_{22}^{E,mn}$ (6.18)

This system of functions satisfies the second vector boundary-value problem

$$\nabla^2_\perp \mathbf{b}_{mn\perp} + \kappa^2_{mn}\mathbf{b}_{mn\perp} = 0;\ (\mathbf{b}_{mn\perp}, \mathbf{n})\big|_L = 0;\ (\mathrm{curl}_\perp, \mathbf{b}_{mn})_\perp\big|_L = 0.$$

The function (6.16) gives the system of functions for an electric field

$$\mathbf{b}^{*}_{mn} = [\mathbf{e}_x, \nabla_{\perp} H_x] = \kappa_n f_{33}^{B,mn} \hat{\mathbf{e}}_y - \kappa_m f_{22}^{B,mn} \hat{\mathbf{e}}_z \tag{6.19}$$

and the Maxwell's equation $\mathbf{B} = -\text{curl}\int^{t} \mathbf{E}dt$ gives the system of the functions for a magnetic field

$$\mathbf{b}_{mn} = \kappa^2_{mn} f_{11}^{B,mn} \hat{\mathbf{e}}_{\mathbf{x}} - \kappa_m f_{22}^{B,mn} \hat{\mathbf{e}}_y \frac{\partial}{\partial x} - \kappa_n f_{33}^{B,mn} \hat{\mathbf{e}}_z \frac{\partial}{\partial x}. \tag{6.20}$$

Thereby the fields for the TE-waves are given by the series

$$\mathbf{E} = \sum_{m,n} \mathbf{b}^{*}_{mn} E_{mn}(t,x), \tag{6.21}$$

$$\mathbf{B} = \sum_{m,n} \mathbf{b}_{mn} B_{mn}(t,x). \tag{6.22}$$

The series for the fields (6.14), (6.15), (6.21) and (6.22) allows one to reduce the vector integral Eq. (6.7) to a system of scalar integral equations with respect to the expansion coefficients. If a material object in the waveguide has an arbitrary shape then the field contains the waves of both classes and the system of integral equations is very intricate. As the main interest in this book is the non-stationarity of electrodynamics processes we confine ourselves further to the case when the object is a plane-parallel insertion filling the whole waveguide cross-section. In this case, the object characteristic function $\chi = \chi(t, x)$ depends on one spatial coordinate only.

Let us consider the case when the initial field consists of H- or TE-waves. It is evident that the geometric properties of the object eliminate the appearance of waves of the other class than the initial one. So, the vectors **E**, **P**, and **j** can be represented in the form of the expansion (6.21) and the vectors **B** and **M** in the form of the expansion (6.22). These expansions allow integrating over the transverse cross-section of the waveguide in the integral Eq. (6.7):

$$\mathbf{E} = \mathbf{E}_0 - 2\pi v \sum_{m,n} \mathbf{b}^{*}_{mn} \int_{-\infty}^{\infty} dt' \int_{-\infty}^{\infty} dx' \left\{ \frac{1}{\varepsilon v^2} \frac{\partial^2}{\partial t^2} \left\{ \Gamma_{mn} \chi \left(P_{mn} - \frac{\varepsilon - 1}{4\pi} E_{mn} \right) \right. \right.$$
$$\left. \left. + \tilde{\Gamma}_{mn} (\chi j_{mn} + i_{mn} \delta(S)) \right\} + \frac{\mu}{c} \frac{\partial}{\partial t} \left(\frac{\partial \Gamma_{mn}}{\partial x} \frac{\partial}{\partial x'} - \kappa^2_{mn} \Gamma_{mn} \right) \chi \left(M_{mn} - \frac{\mu - 1}{4\pi\mu} B_{mn} \right) \right\} \tag{6.23}$$

Equating coefficients of the terms with the same vector-functions reduces (6.23) to the temporal and 1-D spatial integral scalar equation for an arbitrary mode of the electric field whose index is omitted

$$E = E_0 - \frac{2\pi}{\varepsilon v}\int_{-\infty}^{\infty} dt' \int_{-\infty}^{\infty} dx' \left\{ \frac{\partial^2}{\partial t^2}\left[\Gamma_k \chi \left(P_{in} - \frac{\varepsilon - 1}{4\pi} E \right) + \tilde{\Gamma}_k (\chi j_{\text{in}} + i\delta(s)) \right] \right.$$
$$\left. + c\frac{\partial}{\partial t}\left(\frac{\partial \Gamma_k}{\partial x}\frac{\partial}{\partial x'} - \kappa_k^2 \Gamma_k \right) \chi \left(\mathbf{M}_{\text{in}} - \frac{\mu}{4\pi\mu} B \right) \right\}, \quad (6.24)$$

$$\text{where } \Gamma_k = J_0\left(\omega_k \sqrt{(t-t')^2 - \frac{(x-x')^2}{v^2}} \right) \theta\left(t - t' - \frac{|x-x'|}{v} \right) \quad (6.25)$$

$$\tilde{\Gamma}_k = \int_{-\infty}^{t} d\tau \Gamma_k(\tau' - t', x - x'),\ \omega_k = v\kappa_k. \quad (6.26)$$

and the index k designates the oscillation mode considered, $\kappa_k = \kappa_{mn}$. The differentiation operations in this equation can be brought out the integrals as the latter are convolutions.

If the background medium is not magnetic (the relative permeability $\mu = 1$), and the object consists of such a medium that the magnetization vector **M** has transverse components only, then the integral equation can be simplified. In this case the magnetic field **B** does not enter directly in the equation and the magnetization **M** can be expanded as a series of the transverse eigen-vectors

$$\mathbf{b}_{mn\perp} = -\kappa_m f_{22}^{B,mn} \hat{\mathbf{e}}_y - \kappa_n f_{33}^{B,mn} \hat{\mathbf{e}}_z \quad (6.27)$$

in which, in comparison with vectors (6.20), the derivative $\partial/\partial x$ is brought to the coefficients M_{mn}. In this case, the coefficients of the new expansion $\mathbf{M} = \sum_{m,n} \mathbf{b}_{mn\perp} \bar{M}_{mn}(t,x)$ are connected with the old ones by the relation $\bar{M}_{mn} = \frac{\partial}{\partial x} M_{mn}$. One has to take into account also that the equality κ_{mn} would be stated in (6.20). After manipulations the Eq. (6.24) takes the form

$$E = E_0 - 2\pi \int_{-\infty}^{\infty} dt' \int_{-\infty}^{\infty} dx' \left\{ \frac{1}{\varepsilon v}\frac{\partial^2}{\partial t^2}\left[\Gamma_k \left(P_{in} - \frac{\varepsilon - 1}{4\pi} E \right) \chi + \tilde{\Gamma}_k (\chi j_{in} + i\delta(s)) \right] \right.$$
$$\left. + \frac{1}{\sqrt{\varepsilon}}\frac{\partial^2}{\partial t \partial x} \Gamma_k \chi \bar{\mathbf{M}}_{in} \right\} \quad (6.28)$$

As the integrals in this equation represent convolutions, the differentiation operators can be brought out the integrals.

Thereby the integral equation for the electric field in the rectangular metallic waveguide containing a dielectric object filling the whole cross-section of the waveguide and restricted by the plane-parallel boundaries which are normal to the waveguide longitudinal axis is given by (6.28).

Equation (6.28) is the starting one for the investigation of the non-stationarity in the waveguide which will be carried out further. It is the Volterra integral equation and it breaks up into a chain of evolutionary linked equations which can be solved by the resolvent method as in the case of an unbounded medium considered in the previous chapters. The existence of waveguide dispersion gives effects of theoretical and practical interest in the case of such simple dynamics as uniform movement of a restricted dielectric, even if the object inside the waveguide does not change its properties. Such phenomena will be considered in the following sections.

6.2 Equations for a Field in the Waveguide with a Non-Stationary Insertion

The influence of the waveguide dispersion on the electromagnetic transients is revealed in the stationary movement of the medium or its boundary [3–6]. The stationary movement is understood as the movement which does not change its character since the infinitely remote past. This investigation is interesting because it reveals peculiarities brought about by the waveguide dispersion. The results can be compared with results for strictly non-stationary phenomena, for example, when the movement begins at some finite moment of time.

We consider a dielectric non-stationary insertion filling the whole cross-section of the waveguide and moving along it. This insertion is restricted by planes $x_1(t)$ and $x_2(t)$ which are perpendicular to the waveguide longitudinal axes (Fig. 6.1). The waveguide itself is filled by a motionless background nonmagnetic, $\mu = 1$, medium. A guided wave of a given mode E_0 is falling on the layer from the side $x < x_1(t)$. One must distinguish two cases: (i) the medium inside the insertion is motionless and (ii) the medium inside the insertion moves together with its boundary. In both cases, we consider only the most practical case when the movement is directed along the waveguide.

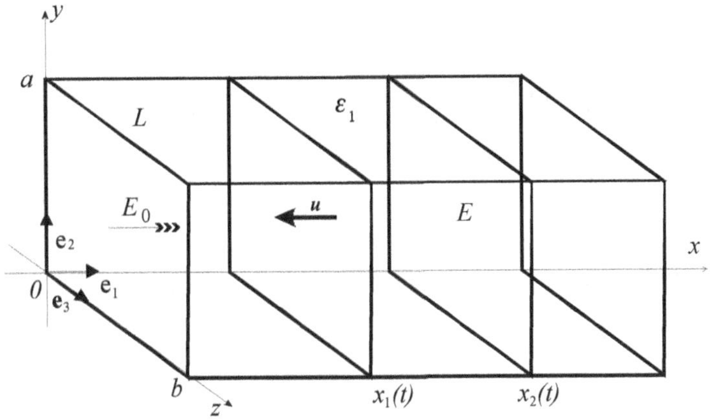

Figure 6.1 Movement of the insertion in the waveguide.

If the medium in rest is non-conducting and nonmagnetic and as a dielectric polarizability α'_{ij}, then the general constituent relations [3] give the expressions for polarization vectors for the medium uniformly moving with the velocity **u**

$$P_i = \gamma^2 \gamma_{ij} \alpha'_{jm} (\gamma_{mn} E_n + e_{mns} \beta_n B_s), \tag{6.29}$$

$$M_i = -\gamma^2 e_{ijk} \beta_j \alpha'_{kn} (e_{mns} \beta_n B_s + \gamma_{mn} E_n). \tag{6.30}$$

Here $\beta = \mathbf{u}/c$, $\gamma^2 = 1/(1 - \beta^2)$, $\gamma_{ij} = \delta_{ij} + \frac{1-\gamma}{\gamma\beta^2}\beta_i\beta_j$, e_{mns} is the complete antisymmetric tensor of the third rank. For an isotropic medium the polarizability is a scalar one, $\alpha'_{ij} = \alpha'\delta_{ij}$, so the constituent relations (6.29), (6.30) are simplified

$$\mathbf{P} = \alpha'\gamma^2 \{\mathbf{E} - \beta(\beta\mathbf{E}) + [\beta\mathbf{B}]\}, \mathbf{M} = [\mathbf{P}\beta]. \tag{6.31}$$

These relations provide for transverse **E**, **P**, and **M** fields in the case of TE-waves as the mode transformation is absent under the chosen geometry of the insertion.

The stationarity of the problem allows one to use a representation of all field values in the form of a Fourier transform with respect to the time variable. In this representation the fields inside the insertion consist of two waves

$$E=\sum_{s=1}^{2}\int_{-\infty}^{\infty}\frac{d\nu}{2\pi}\tilde{E}_s(\nu)e^{i\nu t-ik_s(\nu)x}$$

$$P-\frac{\varepsilon-1}{4\pi}E=\sum_{s=1}^{2}\int_{-\infty}^{\infty}\frac{d\nu}{2\pi}\tilde{\bar{P}}_s(\nu)e^{i\nu t-ik_s(\nu)x},$$

$$\bar{M}=\sum_{s=1}^{2}\int_{-\infty}^{\infty}\frac{d\nu}{2\pi}\tilde{\bar{M}}_s(\nu)e^{i\nu t-ik_s(\nu)x}. \tag{6.32}$$

If the kernel (6.25) is represented as a Fourier transform also

$$\Gamma_k=\int_{-\infty}^{\infty}\frac{d\omega}{2\pi}e^{i\omega(t-t')}\frac{1}{i\nu k_0(\omega)}e^{-ik_0(\omega)|x-x'|}, \tag{6.33}$$

then the integral Eq. (6.24) in the previous section, after integration by x' over the interval $[x_1(t),\ x_2(t)]$ with x belonging to this interval, takes the form

$$\sum_{s=1}^{2}\int_{-\infty}^{\infty}\frac{d\omega}{2\pi}\tilde{E}_s(\omega)e^{i\omega t-ik_s(\omega)x}$$

$$=E_0+\frac{2\pi}{c^2}\sum_{s=1}^{2}\int_{-\infty}^{\infty}\frac{d\omega}{2\pi}e^{i\omega t-ik_s(\omega)x}\frac{\omega^2}{k_s^2-k_0^2}\left(\tilde{\bar{P}}_s(\nu)-\frac{ck_s(\omega)}{\omega}\tilde{\bar{M}}_s(\nu)\right)$$

$$-\frac{2\pi}{c^2}\sum_{s=1}^{2}\int_{-\infty}^{\infty}\frac{d\omega}{2\pi}e^{i\omega t}\frac{i\omega^2}{k_s}\left\{e^{-ik_0(\omega)x}\int_{-\infty}^{\infty}\frac{d\nu}{2\pi}\alpha_s^+(\omega,\nu)\left(\tilde{\bar{P}}_s(\nu)-\frac{ck_0(\omega)}{\omega}\tilde{\bar{M}}_s(\nu)\right)\right.$$

$$\left.+e^{ik_0(\omega)x}\int_{-\infty}^{\infty}\frac{d\nu}{2\pi}\alpha_s^-(\omega,\nu)\left(\tilde{\bar{P}}_s(\nu)+\frac{ck_0(\omega)}{\omega}\tilde{\bar{M}}_s(\nu)\right)\right\}. \tag{6.34}$$

Here

$$\alpha_s^{\pm}(\omega,\nu)=\frac{i}{k_0(\omega)\mp k_s(\nu)}\int_{-\infty}^{\infty}dt'e^{i(\nu-\omega)t'-i[k_s(\nu)\mp k_0(\omega)]x_{1,2}(t')}, \tag{6.35}$$

$x_1(t)$ corresponds to α_s^+, $x_2(t)$ corresponds to α_s^- and $\nu k_0=\sqrt{\omega^2-\omega_k^2}$, where ω_k is the cutoff frequency for the mode considered.

The integrands to the left and to the right in (6.34) are equal to each other as the integrals are the Fourier transforms. Equating in this equality the terms with the same exponents with respect to x gives the system of two equations reflecting the extinction theorem of Ewald–Oseen [7]. One equation gives the dispersion equation

$$(k_s^2(\omega)-k_0^2(\omega))\tilde{E}_s = 4\pi\frac{\omega^2}{c^2}\left(\tilde{\tilde{P}}_s - \frac{ck_s(\omega)}{\omega}\tilde{\tilde{M}}_s\right) \qquad (6.36)$$

and the other terms compensate the falling wave that gives the following equation of the first kind for the amplitudes:

$$e^{-ik_0(\omega)x}\sum_{s=1}^{2}\int_{-\infty}^{\infty}\frac{d\nu}{2\pi}\alpha_s^+(\omega,\nu)\left(\tilde{\tilde{P}}_s(\nu)-\frac{ck_0(\omega)}{\omega}\tilde{\tilde{M}}_s(\nu)\right)$$
$$+e^{ik_0(\omega)x}\sum_{s=1}^{2}\int_{-\infty}^{\infty}\frac{d\nu}{2\pi}\alpha_s^-(\omega,\nu)\left(\tilde{\tilde{P}}_s(\nu)+\frac{ck_0(\omega)}{\omega}\tilde{\tilde{M}}_s(\nu)\right)=\frac{c^2}{2\pi}\int_{-\infty}^{\infty}dte^{-i\omega t}\frac{k_0(\omega)}{i\omega^2}E_0(t)$$
$$(6.37)$$

If the observation point is out of the insertion then we have the formula for finding the external field through the internal one which can be found by solving Eq. (6.37). These formulae have the following forms:

For the wave transmitted through the layer ($x > x_2$)

$$E = E_0 + \frac{2\pi}{c^2}\sum_{s=1}^{2}\int_{-\infty}^{\infty}\frac{d\omega}{2\pi}\int_{-\infty}^{\infty}\frac{d\nu}{2\pi}e^{i\omega t-ik_0(\omega)x}4\pi\frac{\omega^2}{k_0(\omega)}N_s^+(\omega,\nu)\left(\tilde{\tilde{P}}_s(\nu)-\frac{ck_0(\omega)}{\omega}\tilde{\tilde{M}}_s(\nu)\right)$$
$$(6.38)$$

and for the wave reflected from the layer ($x \le x_2$)

$$E = E_0 + \frac{2\pi}{c^2}\sum_{s=1}^{2}\int_{-\infty}^{\infty}\frac{d\omega}{2\pi}\int_{-\infty}^{\infty}\frac{d\nu}{2\pi}e^{i\omega t+ik_0(\omega)x}4\pi\frac{\omega^2}{k_0}N_s^-(\omega,\nu)\left(\tilde{\tilde{P}}_s(\nu)+\frac{ck_0(\omega)}{\omega}\tilde{\tilde{M}}_s(\nu)\right)$$
$$(6.39)$$

In these formulae, the kernels are determined by the integrals

$$N_s^{\pm}(\omega,\nu)=\int_{-\infty}^{\infty}dt'\frac{e^{i(\nu-\omega)t'}}{k_s(\nu)\pm k_0(\omega)}\{e^{-i[k_s(\nu)\mp k_0(\omega)]x_2(t')}-e^{-i[k_s(\nu)\mp k_0(\omega)]x_1(t')}\}.$$
$$(6.40)$$

More detailed investigations will be carried out in the next sections for the undisturbed field in the form of the guided wave $E_0 = E_0 \exp[i(\omega_0 t) - k_0(\omega_0)x]$. Substitution of this expression into (6.37) breaks it into the system of two equations

$$\sum_{s=1}^{2}\int_{-\infty}^{\infty}\frac{dv}{2\pi}\alpha_s^{+}\left(\tilde{\tilde{P}}_s(v)-\frac{ck_0(\omega)}{\omega}\tilde{\tilde{M}}_s(v)\right)=\frac{c^2k_0(\omega)}{i\omega^2}E_0\delta(\omega-\omega_0) \qquad (6.41)$$

$$\sum_{s=1}^{2}\int_{-\infty}^{\infty}\frac{dv}{2\pi}\alpha_s^{-}\left(\tilde{\tilde{P}}_s(v)+\frac{ck_0(\omega)}{\omega}\tilde{\tilde{M}}_s(v)\right)=0, \qquad (6.42)$$

where $\delta(\omega)$ is the Dirac delta-function.

Thereby Eqs. (6.36)–(6.42) determine completely the fields inside the insertion as well as outside it.

6.3 Vibration of a Boundary of a Plane Dielectric Resonator

An important instance of a time-varying medium is the vibrating movement of the medium boundary which has been investigated in some papers [8–11]. We investigate the field in a vibrating dielectric resonator which is formed by the insertion considered in the previous section. One of the plane boundaries of this resonator which is normal to the waveguide longitudinal axes is vibrating with the frequency Ω, $x_1 = \xi_0 \sin(\Omega t)$. The opposite boundary of the resonator is motionless, $x_2 = l$ (see Fig. 6.2). It is assumed that the vibration amplitude is small in comparison with characteristic spatial scales of the problem. If the Fourier images for the polarizations of the medium inside the dielectric insertion with permittivity ε_1 have simple forms $\tilde{\tilde{P}}_s = (\varepsilon_1 - \varepsilon)\tilde{E}_s/4\pi$, $\tilde{\tilde{M}}_s = 0$, s = 1, 2 then the dispersion Eq. (6.36) gives the ordinary dispersion relation $k_s^2 = (\omega^2 - \omega_{k1}^2)/v_1^2$ with $\omega_{k1} = v_1\kappa_{mn}$, $v_1 = \sqrt{\varepsilon_1}$ for the waves inside the insertion. If the undisturbed wave TE-wave is a monochromatic wave $E_0 = E_0 \exp[i(\omega_0 t) - k_0(\omega_0)x]$ then the equations for the field (6.41), (6.42) take the form

$$\int_{-\infty}^{\infty}\frac{dv}{2\pi}\{K_1^{(1)}\tilde{E}_1(v)+K_2^{(1)}\tilde{E}_2(v)\}=\frac{4\pi k_0(\omega)}{i\omega^2(\varepsilon_1-\varepsilon)}E_0\delta(\omega-\omega_0) \qquad (6.43)$$

$$\int_{-\infty}^{\infty} \frac{dv}{2\pi} \{K_1^{(2)} \tilde{E}_1(v) + K_2^{(2)} \tilde{E}_2(v)\} = 0 \tag{6.44}$$

with the kernels

$$K_s^{(1)} = \varphi_s^{(1)} \sum_{n=-\infty}^{\infty} \delta(v - \omega - n\Omega) \tag{6.45}$$

$$\varphi_s^{(1)} = \frac{i}{k_0(\omega) + (-1)^s k(v)} \int_0^{2\pi} e^{-i\xi_0 [k_0(\omega) + (-1)^s k(v)] \sin x + i\frac{v-\omega}{\Omega} x} dx \tag{6.46}$$

$$K_s^{(2)} = \varphi_s^{(2)} \delta(v - \omega), \ \varphi_s^{(2)} = \frac{2\pi i}{k_0(\omega) + (-1)^{s-1} k(v)} e^{-i[k_0(\omega) + (-1)^{s-1} k(v)]l}, \tag{6.47}$$

where $k = \sqrt{\omega^2 - \omega_{k1}^2} / v_1$. For calculation of the kernel $K_s^{(1)}$ the integration interval in (6.41) is divided into sections of 2π length and the equality $\sum_{n=-\infty}^{\infty} e^{i2\pi n\omega} = \sum_{n=-\infty}^{\infty} \delta(\omega - n)$ is used [12].

Equation (6.44) gives links between amplitudes

$$\tilde{E}_2 = \psi(\omega) \tilde{E}_1 \tag{6.48}$$

with $\psi(\omega) = \dfrac{k(\omega) - k_0(\omega)}{k(\omega) + k_0(\omega)} e^{-2ik(\omega)l}$. (6.49)

This link leads to the functional equation for the amplitude $\tilde{E}_1(\omega)$ following from (6.43)

$$\sum_{n=-\infty}^{\infty} \varphi(\omega + n\Omega, \omega) \tilde{E}_1(\omega + n\Omega) = \frac{4\pi k_0(\omega) c^2}{i\omega^2 (\varepsilon_1 - \varepsilon)} E_0 \delta(\omega - \omega_0), \tag{6.50}$$

where $\varphi(v, \omega) = \varphi_1^{(1)}(v, \omega) + \psi(v) \varphi_2^{(1)}(v, \omega)$.

The solution to this equation is given by the formula

$$\tilde{E}_1(\omega) = \sum_{m=-\infty}^{\infty} a_m(\omega_0) \delta(\omega - \omega_0 + m\Omega) \tag{6.51}$$

whose coefficients satisfy the system

$$\sum_{m+n=0} \varphi(\omega_0 - m\Omega, \omega_0) a_m = \frac{4\pi k_0 c^2 E_0}{i\omega_0^2 (\varepsilon_1 - \varepsilon)} \tag{6.52}$$

$$\sum_{m+n=p} \varphi(\omega_0 - m\Omega, \omega_0 - p\Omega) a_m = 0, \quad p = \pm 1, \pm 2, \ldots \tag{6.53}$$

The small value of the vibration amplitude allows one to truncate this system. If one confines oneself to the first three coefficients then the solution to the system (6.52)–(6.53) has the form

$$\left.\begin{aligned} a_{-1} &= \frac{A}{\Delta}(\varphi_{01}\varphi_{-1-1} - \varphi_{0-1}\varphi_{1-1}) \\ a_0 &= \frac{A}{\Delta}(\varphi_{1-1}\varphi_{-11} - \varphi_{11}\varphi_{-1-1}) \\ a_1 &= \frac{A}{\Delta}(\varphi_{0-1}\varphi_{11} - \varphi_{01}\varphi_{1-1}) \end{aligned}\right\}, \tag{6.54}$$

where $\varphi_{mn} = \varphi(\omega_0 + m\Omega, \omega_0 + n\Omega)$; $A = \dfrac{4\pi k_0 c^2 E_0}{i\omega_0^2(\varepsilon_1 - \varepsilon)}$ (6.55)

and the denominator is

$$\Delta = \varphi_{-10}\begin{vmatrix} \varphi_{11} & \varphi_{01} \\ \varphi_{1-1} & \varphi_{0-1} \end{vmatrix} + \varphi_{00}\begin{vmatrix} \varphi_{-11} & \varphi_{11} \\ \varphi_{-1-1} & \varphi_{1-1} \end{vmatrix} + \varphi_{10}\begin{vmatrix} \varphi_{01} & \varphi_{0-1} \\ \varphi_{-11} & \varphi_{-1-1} \end{vmatrix}. \tag{6.56}$$

For obtaining the explicit expression for (6.55) one can integrate in (6.46) using the known formula [13]

$$\frac{1}{\kappa_{mn}^{(s)}}\int_0^{2\pi} e^{i\xi_0\kappa_{mn}^{(s)}\sin x + i(m-n)x}dx = (-1)^{m-n}\frac{i}{\kappa_{mn}^{(s)}}J_{m-n}(\xi_0\kappa_{mn}^{(s)}),$$

where $\kappa_{mn}^{(s)} = k_0(\omega_0 + n\Omega) + (-1)^s k(\omega_0 + m\Omega)$. It gives the expressions for the coefficients φ_{mn} through the Bessel functions

$$\varphi_{mn} = \frac{(-1)^{m-n}i}{\kappa_{mn}^{(1)}}J_{m-n}(\xi_0\kappa_{mn}^{(1)}) + \psi(\omega_0 + m\Omega)\frac{(-1)^{m-n}i}{\kappa_{mn}^{(2)}}J_{m-n}(\xi_0\kappa_{mn}^{(2)}) \tag{6.57}$$

The problem considered is that of the forced oscillations of a dielectric resonator with a vibrating wall that obeys the harmonic law with a frequency Ω. If the exiting wave has the frequency ω_0 then oscillations with the combined frequencies $(\omega_0 + p\Omega)$, $p = 0, \pm1, \pm2, \ldots$ are established. The amplitudes of the first three oscillations are given by the formulae (6.54) with an accuracy to the first power of the small parameter $|\xi_0 \kappa_{mn}^{(s)}| << 1$. As one would anticipate the amplitude of the main oscillation a_0, corresponding to the frequency ω_0, does not depend on the wall vibration. The oscillation amplitudes of the first combination order $\omega_0 \pm \Omega$ are proportional to the amplitude of the wall vibrations.

The truncation of the system (6.52), (6.53) is possible as the small perturbation corresponds to a solution expansion with respect to a small parameter which is $|\xi_0 \kappa_{mn}^{(s)}| << 1$. Further simplification can be done by using the expansions of Bessel's functions using this small parameter. It yields the following formulae for the coefficients

$$a_{-1} = \xi_0 \frac{A}{2A_{00}A_{11}}; \quad a_0 = \frac{A}{A_{00}}; \quad a_1 = \xi_0 \frac{A}{2A_{00}A_{-1-1}}, \tag{6.58}$$

$$\text{where } A_{nn} = \frac{1}{\kappa_{nn}^{(1)}} \left\{ 1 - \left(\frac{\kappa_{nn}^{(1)}}{\kappa_{nn}^{(2)}} \right)^2 e^{-2ik(\omega_0 + n\Omega)l} \right\}. \tag{6.59}$$

The resonances in the insertion are determined by the resonance condition, which is provided by equating the denominator of (6.56) to zero. This condition can be achieved by equating to zero any of the multipliers (6.59), giving two equations for the resonance frequencies

$$ik(\tilde{\omega})l = \ln \frac{\pm k_0(\tilde{\omega}) \mp k(\tilde{\omega})}{k_0(\tilde{\omega}) + k(\tilde{\omega})}, \quad \tilde{\omega} = \omega_0 + n\Omega, \; n = 0, \pm 1. \tag{6.60}$$

In addition to the ordinary group of resonances for a dielectric resonator with motionless walls that corresponds to $n = 0$ in (6.60), there are two groups of resonances shifted by the frequency $\pm\Omega$. The equation has a solution for both signs in (6.60) if the insertion is more electrically dense than the background that is if $\varepsilon_1/\varepsilon > 1$. As the initial wave frequency must be greater than the cutoff frequency, $\omega_0 > \omega_k$ then the resonance can be come only for the initial wave frequency $\omega_0 = \sqrt{\varepsilon/\varepsilon_1}\,\omega_k + \Omega$ if the vibrating frequency is $\Omega > (1 - \sqrt{\varepsilon/\varepsilon_1})\omega_k$.

To compare the amplitudes of the exciting oscillations we calculate numerically the ratios of the satellite amplitudes to the central one in dimensional variables

$$\frac{a_{-s}}{a_0} = \bar{\xi}_0 \frac{(h_{0s} - h_s/p)(h_{0s} + h_s/p)^2}{(h_{0s} + h_s/p)^2 - (h_{0s} - h_s/p)^2 e^{-ih_s b_k}}, \tag{6.61}$$

where $s = \pm 1$, $h_{0s} = \sqrt{\eta_s^2 - 1}$, $h_s = \sqrt{\eta_s^2 - p^2}$, $\eta_s = (\omega_0 + s\Omega)/\omega_k$

$$\bar{\xi}_0 = \xi_0 \omega_k / v, \; p = \sqrt{\varepsilon/\varepsilon_1}, \; b_k = (l\omega_k)/(vp) = \pi l \sqrt{m^2/a^2 + n^2/b^2} \tag{6.62}$$

and the infinitesimally small condition has the form

$$|\bar{\xi}_0 (h_{0s} \pm h_s)| << 1 \tag{6.63}$$

The relative satellite amplitudes a_{-1}/a_0 are shown in Figs. 6.2 and 6.3. It is seen that the amplitude of the satellite a_{-1} (Fig. 6.2a) has finite minima, whereas the other satellite a_{+1} (Fig. 6.2b) has discontinuities at some values of the vibration frequency which correspond to the resonances. The dependence of the amplitude on the layer width (Fig. 6.3) has an oscillatory character as might be anticipated.

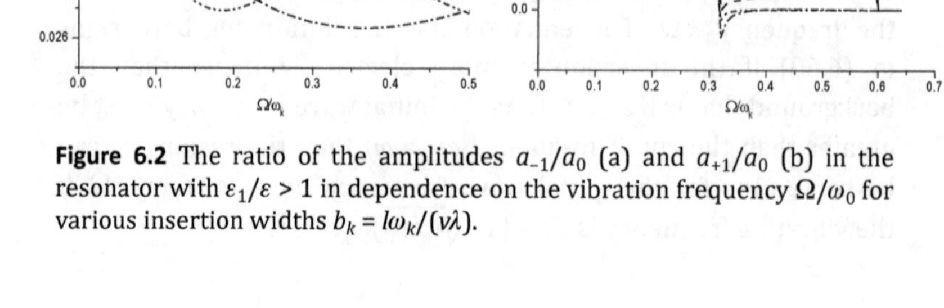

Figure 6.2 The ratio of the amplitudes a_{-1}/a_0 (a) and a_{+1}/a_0 (b) in the resonator with $\varepsilon_1/\varepsilon > 1$ in dependence on the vibration frequency Ω/ω_0 for various insertion widths $b_k = l\omega_k/(v\lambda)$.

In the case when the insertion is less electrically dense than the background, $\varepsilon_1/\varepsilon < 1$, then the equation in (6.60) has solutions only for the upper sign. The same condition $\omega_0 > \omega_k$ allows the resonance to be at the frequency $\omega_0 = \sqrt{\varepsilon/\varepsilon_1}\,\omega_k + \Omega$ for arbitrary vibrating frequency Ω and on the frequency $\omega_0 = \sqrt{\varepsilon/\varepsilon_1}\,\omega_k - \Omega$ for $\Omega < (\sqrt{\varepsilon/\varepsilon_1} - 1)\omega_k$.

The dependence of the relative satellite amplitudes on the frequency has a similar character to the previous case (Fig. 6.4) as well as for dependence on the layer width (Fig. 6.5).

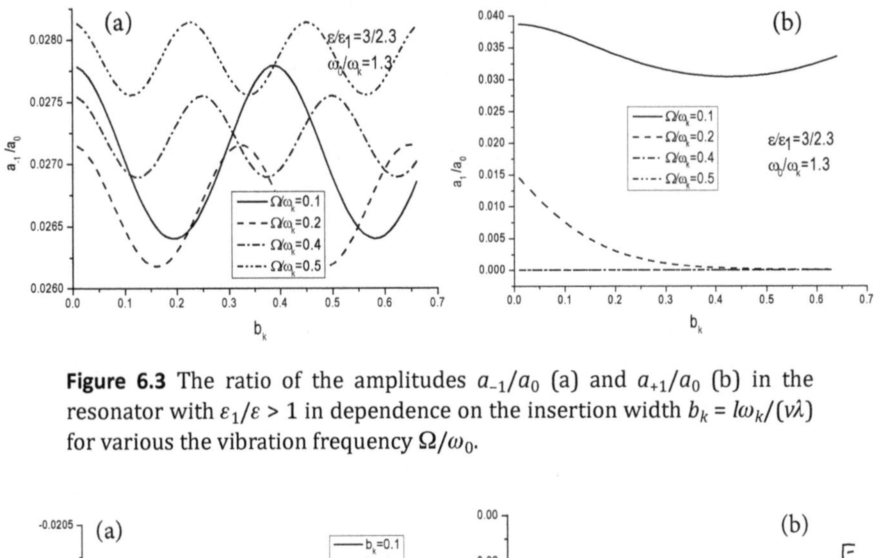

Figure 6.3 The ratio of the amplitudes a_{-1}/a_0 (a) and a_{+1}/a_0 (b) in the resonator with $\varepsilon_1/\varepsilon > 1$ in dependence on the insertion width $b_k = l\omega_k/(v\lambda)$ for various the vibration frequency Ω/ω_0.

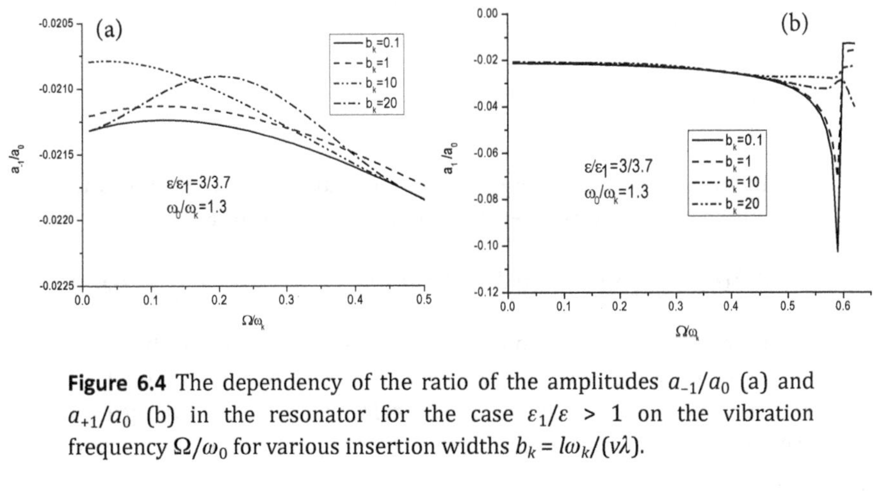

Figure 6.4 The dependency of the ratio of the amplitudes a_{-1}/a_0 (a) and a_{+1}/a_0 (b) in the resonator for the case $\varepsilon_1/\varepsilon > 1$ on the vibration frequency Ω/ω_0 for various insertion widths $b_k = l\omega_k/(v\lambda)$.

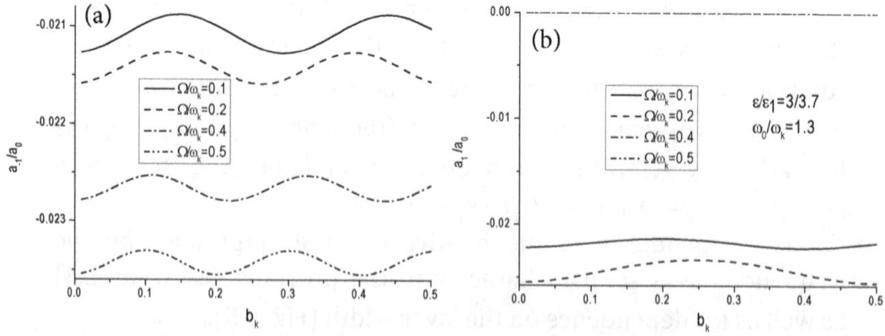

Figure 6.5 The dependency of the ratio of the amplitudes a_{-1}/a_0 (a) and a_{+1}/a_0 (b) in the resonator on the insertion width $b_k = l\omega_k/(\nu\lambda)$ for the case when $\varepsilon_1/\varepsilon > 1$ and for various vibration frequencies Ω/ω_0.

6.4 Uniform Movement of a Dielectric Layer in Presence of Waveguide Dispersion

The uniform relativistic movement of a dielectric medium together with its boundary [14] represents another example of a time-varying medium in a waveguide. We consider the interaction of a TE-wave with a dielectric layer moving in the waveguide. As layer boundaries are planes located normally to the waveguide axes the mode transformation is absent and the transformed fields can be represented in the form $\mathbf{E} = \sum_{m,n} \mathbf{b}^*_{mn} E_{mn}(t,x)$ (6.21) where $\mathbf{b}^*_{mn} = \kappa_n f^{B,33}_{mn}\hat{\mathbf{e}}_y - \kappa_m f^{B,22}_{mn}\hat{\mathbf{e}}_z$ is the vector (6.19) and

$$E(t,x) = \sum_{s=1}^{2} \int_{-\infty}^{\infty} \frac{d\nu}{2\pi} \tilde{E}_s(\nu) e^{i\nu t - ik_s(\nu)x} \tag{6.64}$$

In this case, the medium movement cannot be neglected and one has to use the constituents for a moving medium (6.31)

$$\mathbf{P} = \alpha'\gamma^2 \{\mathbf{E} - \boldsymbol{\beta}(\boldsymbol{\beta}\mathbf{E}) + [\boldsymbol{\beta}\mathbf{B}]\},\ \mathbf{M} = [\mathbf{P}\boldsymbol{\beta}], \tag{6.65}$$

where α' is a medium polarizability in the frame of the rest, $\boldsymbol{\beta} = \mathbf{u}/c$, $\mathbf{u}$ is the medium velocity, c is the velocity of light in vacuum and $\gamma = (1 - \beta^2)^{-1/2}$ is the relativistic factor. Taking into account that $\boldsymbol{\beta}$ and $\mathbf{E}$ are mutually perpendicular we obtain from (6.65)

$$\mathbf{P} = \sum_{m,n} \mathbf{b}^*_{mn} \sum_{s=1}^{2} \int_{-\infty}^{\infty} \frac{d\nu}{2\pi} \alpha'(\nu)\gamma^2 \left(1 - \frac{uk_s(\nu)}{\nu}\right) \tilde{E}_{mns} e^{i\nu t - ik_s(\nu)x} \tag{6.66}$$

$$\mathbf{M} = \sum_{m,n} \mathbf{b}_{mn\perp} \sum_{s=1}^{2} \int_{-\infty}^{\infty} \frac{d\nu}{2\pi} \alpha'(\nu)\gamma^2 \beta \left(1 - \frac{uk_s(\nu)}{\nu}\right) \tilde{E}_{mns} e^{i\nu t - ik_s(\nu)x}, \tag{6.67}$$

where $\mathbf{b}_{mn\perp} = -\kappa_m f_{mn}^{B,22} \hat{\mathbf{e}}_y - \kappa_n f_{mn}^{B,33} \hat{\mathbf{e}}_z$ is the vector (6.27). It follows immediately from these relations that

$$\left.\begin{aligned} \tilde{\bar{P}}_s &= \left\{ \gamma^2 \alpha'(\nu) \frac{\nu - uk_s(\nu)}{\nu} - \frac{\varepsilon - 1}{4\pi} \right\} \tilde{E}_s \\ \tilde{M}_s &= \gamma^2 \alpha'(\nu) \beta \frac{\nu - uk_n(\nu)}{\nu} \tilde{E}_s \end{aligned}\right\}. \tag{6.68}$$

Substitution of these formulae to the dispersion Eq. (6.36) gives the dispersion equation for the waves inside the layer

$$[1 - (1 + 4\pi\alpha'(\omega))\beta^2] k_s{}^2 + 8\pi\alpha'(\omega)\beta \frac{\omega}{c} k_s - [1 + 4\pi\alpha'(\omega) - \beta^2] \frac{\omega^2}{c^2} + \frac{\varepsilon\omega_k^2}{c^2\gamma^2} = 0 \tag{6.69}$$

This equation is the generalization of the dispersion equation for plane waves in a medium moving in a waveguide [15].

A medium with waveguide dispersion (waveguides, cold isotropic plasma) can be described by the effective permittivity $\varepsilon_{\text{eff}} = \varepsilon_1(1 - \omega_{k1}^2/\omega^2)$ where ω_{k1} is a cutoff frequency and ε_1 is the medium permittivity at the rest. If this expression is substituted in the general formula of the paper [15] then we obtain the dispersion equation in the medium moving in the waveguide

$$\frac{1}{\gamma_{\varepsilon 1}^2} k_s^2 + 2(\varepsilon_1 - 1)\beta \frac{\omega}{c} k_s - [\varepsilon_1 - \beta^2] \frac{\omega^2}{c^2} + \frac{\varepsilon\omega_k^2}{c^2\gamma^2} = 0, \tag{6.70}$$

where $\gamma_{\varepsilon 1}^2 = (1 - \varepsilon_1\beta^2)^{-1}$. The same equation is given by substitution of $4\pi\alpha' = \varepsilon_1 - 1$ in (6.69).

Solutions to this equation determine wave numbers for waves passed into moving dielectric

$$k_{1,2} = \frac{\omega}{c}\left\{-\beta\gamma_{\varepsilon 1}^2(\varepsilon_1 - 1) \pm \sqrt{\varepsilon_1}\frac{\gamma_{\varepsilon 1}^2}{\gamma^2}\sqrt{1 - \frac{\gamma^2}{\gamma_{\varepsilon 1}^2}\frac{\omega_{k1}^2}{\omega^2}}\right\}. \tag{6.71}$$

The same expression for dispersion relation is derived in [16] by virtue of the Lorentz transformation.

One can see from (6.71) that the cutoff frequency for the waveguide filled by the moving dielectric makes sense for the under-Cherenkov-movement only ($\gamma_{\varepsilon 1}^2 > 0$). In this case, the cutoff frequency is equal to $\omega_{\text{cut relativ}} = \gamma\gamma_{\varepsilon 1}^{-1}\omega_k$. In the case of the Cherenkov movement ($\gamma_{\varepsilon 1}^2 < 0$) the waveguide loses the beyond-cutoff property. Disappearance of the beyond-cutoff waveguide property occurs when the medium velocity is equal to the phase velocity of a wave in an unbounded medium. The reason for that is the effect of the field drag by the medium, and it corresponds very well to Brillouin's interpretation for the running wave in the waveguide according to which the phenomenon of a wave cutoff sets in when Brillouin waves propagate transverse to a waveguide [17].

The dependence on frequency of the angle between the wave propagation direction and the waveguide axis is $\cos\theta' = \sqrt{1 - (\omega_k'/\omega')^2}$ at rest. It gives $\omega' = \omega_k'/\sqrt{1 - (\cos\theta')^2}$. In the laboratory frame, where the medium moves along the waveguide with the velocity $u = \beta c$, the wave frequency is given by the expression $\omega = \omega'\gamma(1 + u\cos\theta'/v')$. It leads to $\omega' = \omega[\gamma(1 + u\cos\theta'/v')]^{-1}$ where v' is the wave phase velocity in the motionless medium. These expressions give the link between the frequency in the laboratory frame and the angle

$$\omega = \omega_k'\gamma(1 + u\cos\theta'/v')/\sqrt{1 - (\cos\theta')^2} \tag{6.72}$$

Wave aberration caused by the field drag by the medium changes the angle of the wave propagation direction [18] as

$$\tan\theta = \frac{v'\sin\theta'}{\gamma(v'\cos\theta' + u)} \tag{6.73}$$

The cutoff phenomenon comes when the Brillouin waves propagate transversely to the waveguide. Assuming $\theta = \pi/2$ in (6.73) leads to $\cos\theta' = -u/v'$ that gives from (6.72)

$$\omega_k = \omega_k' \gamma \sqrt{1-(u/v')^2}. \tag{6.74}$$

Substitution of the phase velocity in the motionless medium $v' = c/\sqrt{\varepsilon}$ into (6.74) gives the same expression for the critical frequency as follows from (6.71) Thereby, the waveguide allows waves of any frequency to propagate in the case of the Cherenkov movement ($\gamma_{\varepsilon 1}^2 < 0$).

The dispersion relation (6.71) allows one to construct the solution of the equation for the amplitudes. Using the designation following from (6.68)

$$L_s^{\pm}(\nu,\omega) = \frac{4\pi}{\tilde{E}_s}\left\{\tilde{\tilde{P}}_s \mp \frac{ck_0(\omega)}{\omega}\tilde{M}_s\right\}$$

$$= \gamma^2 4\pi\alpha'(\nu)\frac{[\nu - uk_s(\nu)][\omega \mp uk_0(\omega)]}{\nu\omega} - (\varepsilon - 1) \tag{6.75}$$

the equations for the amplitudes take the form

$$\sum_{s=1}^{2}\int_{-\infty}^{\infty}\frac{d\nu}{2\pi}K_s^{(1)}(\nu,\omega)L_s^{(+)}(\nu,\omega)\tilde{E}_s(\nu) = \frac{4\pi c^2 k_0(\omega)}{i\omega^2}E_0\delta(\omega-\omega_0) \tag{6.76}$$

$$\sum_{s=}^{2}\int_{-\infty}^{\infty}\frac{d\nu}{2\pi}K_s^{(2)}(\nu,\omega)L_s^{(-)}(\nu,\omega)\tilde{E}_s(\nu) = 0 \tag{6.77}$$

These equations are suitable for consideration of the interaction of an electromagnetic wave with a dielectric domain moving in a waveguide with any velocity not exceeding the light velocity in vacuum, that is constrained only by the inequality $u < c$. To reveal the peculiarities of waveguide dispersion in the presence of moving medium and boundaries we consider the semi-infinite dielectric restricted by a transverse plane boundary and moving uniformly in the waveguide. In this case, $4\pi\alpha' = \varepsilon_1 - 1$, $x_1 = ut$, $x_2 = \infty$ that gives from (6.34)

$$K_s^{(1)}(\nu,\omega) = \frac{2\pi i\delta(\nu - uk_s(\nu) - \omega_u)}{k_0(\omega) - k_s(\nu)}, K_s^{(2)}(\nu,\omega) = 0, \tag{6.78}$$

where $\omega_u = \omega - uk_0(\omega)$.

So, only one equation (6.76) for the amplitudes of two possible waves leaves

$$\sum_{s=1}^{2}\int_{-\infty}^{\infty}\frac{d\nu}{2\pi}\frac{\delta(\nu-uk_s(\nu)-\omega_u)}{k_0(\omega)-k_s(\nu)}L_s^{(+)}(\nu,\omega)\hat{E}_s(\nu)=\frac{-2c^2k_0(\omega)}{\omega^2}E_0\delta(\omega-\omega_0) \tag{6.79}$$

The frequencies for the passed waves are given by the dispersion equation

$$\frac{\gamma_{\varepsilon1}^2}{\gamma^2}\nu\pm\sqrt{\varepsilon_1}\beta\frac{\gamma_{\varepsilon1}^2}{\gamma^2}\nu\sqrt{1-\frac{\gamma^2}{\gamma_{\varepsilon1}^2}\frac{\omega_{k1}^2}{\nu^2}}=\omega_u, \tag{6.80}$$

which follows from equating to zero the argument of the δ-function in (6.79). The same dispersion equation for the frequency ν can be derived from the condition for the equality of phases at the boundary $\nu t-k_{1,2}(\nu)x\,|_{x=ut}=\omega t-k_0(\omega)x\,|_{x=ut}$ if $k_{1,2}(\nu)$ is substituted by the expression from (6.71).

One or two solutions may correspond to every sign in (6.80) depending on the frequency ω:

$$\nu^{\pm}(\omega)=\gamma^2(\omega_u\pm\sqrt{\varepsilon_1}\beta\sqrt{\omega_u^2-\gamma^{-2}\omega_{k1}^2}) \tag{6.81}$$

The choice of the sign is determined by the relation between the boundary movement velocity and the group velocity for every wave passed into the moving dielectric:

$$v_{gr}^{\pm}=\frac{u}{|\beta|}\frac{\sqrt{\varepsilon_1}|\beta|\omega_u\pm\sqrt{\omega_u^2-\gamma^{-2}\omega_{k1}^2}}{\sqrt{\varepsilon_1}\omega_u\pm|\beta|\sqrt{\omega_u^2-\gamma^{-2}\omega_{k1}^2}} \tag{6.82}$$

Analysis of Eq. (6.82) shows that the condition for the energy withdrawal from the boundary, $v_{gr}^{\pm}>u$, in the whole ranges of frequencies ω and boundary velocities u can be satisfied only for one wave with the group velocity $v_{gr}^{\pm}$. It is worth noting that neglecting of the medium movement and taking into account the boundary movement only leads in the Cherenkov case to an ambiguity in the problem solution owing to possibility of the existence of two passed waves. The solution to Eq. (6.79) with these peculiarities gives the field passed through the boundary into moving dielectric. This field has different characteristics over various frequencies intervals, as will now be discussed.

In the case of runaway movement ($\beta>0$) the incident wave interacts with a boundary when the boundary velocity is less than the group velocity of the incident wave only. This is satisfied only when $\gamma_\varepsilon^2 > 0$ for the frequencies of the incident wave $\omega_0 > \gamma_\varepsilon \omega_k$. The passed field is absent for other frequencies for $\gamma_\varepsilon^2 > 0$ and for all frequencies for $\gamma_\varepsilon^2 < 0$.

The passed field is the running wave with the frequency $\nu_0 = \nu^+(\omega_0)$ if $\omega_0 > \omega_2$ where

$$\omega_2 = \gamma_\varepsilon^2 \left\{ \frac{\omega_{k1}}{\gamma} + \sqrt{\varepsilon}\beta \sqrt{\frac{\omega_{k1}^2}{\gamma^2} - \frac{\omega_k^2}{\gamma_{\varepsilon 1}^2}} \right\} \tag{6.83}$$

The transmission coefficient for this wave is equal to

$$T = \frac{2\nu_0}{\beta\omega_0} \frac{(\nu_0 - \omega_0)\,|ck_0 - \varepsilon\beta\omega_0|}{(\varepsilon_1 - 1)\gamma^2\omega_u^2 - (\varepsilon - 1)\nu_0\omega_0} \tag{6.84}$$

In the frequency interval $\gamma_\varepsilon \omega_k < \omega_0 < \omega_2$ the passed field is represented by the exponential wave

$$E = TE_0 e^{\omega_1(t - x/u) + i\gamma^2\omega_u(t - \beta x/c)}, \tag{6.85}$$

where $\omega_1 = \sqrt{\varepsilon_1}\beta\gamma\sqrt{\omega_{k1}^2 - \gamma^2\omega_u^2}$. The physical mechanism of the exponential wave (6.85) formation links with the waveguide cutoff feature and is explained by the double Doppler effect. This effect is that a moving boundary can be considered as a source of electromagnetic waves excited by the incident wave. This source re-radiates waves into a motionless or moving medium.

In the case of the opposed movement ($\beta < 0$) the passed field exists for all above-critical frequencies ($\omega_0 > \omega_k$) and it is the running wave with parameters ν_0 and T if the frequency satisfies $\omega_0 > \omega_2$. The field is the exponential wave (6.85) in the interval $\omega_k < \omega_0 < \omega_2$. It is worth noting that the interval $[\omega_k, \omega_2]$ exists in all cases if $\gamma_\varepsilon^2 > 0$ and in the case $\gamma_\varepsilon^2 \omega_{k1} > \gamma\omega_k$ only for $\gamma_\varepsilon^2 < 0$. The exponential wave will decrease when the boundary moves away from the observation point.

The reflected field is defined as

$$E_r = \frac{-1}{2c^2} \int_{-\infty}^{\infty} \frac{d\nu}{2\pi} \frac{\nu^2 \delta(\nu + uk_0(\nu) - \omega_u)}{k_0(\nu)[k_0(\nu) + k_1(\nu_0)]} TL_1^{(-)}(\nu, \nu_0) e^{i\nu t + ik_0(\nu)x} \tag{6.86}$$

The equality of the δ-function argument under this integral is the equality of the phases of the incident and reflected waves and it determines the reflected wave frequency

$$\omega_r = \gamma_\varepsilon^2(\omega_u \pm \sqrt{\varepsilon}\,\beta\sqrt{\omega_u^2 - \gamma_\varepsilon^{-2}\omega_k^2}) \tag{6.87}$$

The group velocity for the reflected wave is

$$v_{gr}^{\pm} = \frac{u}{|\beta|\sqrt{\varepsilon}}\frac{\sqrt{\varepsilon}|\beta|\gamma_\varepsilon^2\omega_u \pm |\gamma_\varepsilon^2|\sqrt{\omega_u^2 - \gamma_\varepsilon^{-2}\omega_k^2}}{\gamma_\varepsilon{}^2\omega_u \pm \sqrt{\varepsilon}\,\beta|\gamma_\varepsilon^2|\sqrt{\omega_u^2 - \gamma_\varepsilon^{-2}\omega_k^2}} \tag{6.88}$$

Comparison of this velocity with the boundary velocity shows that for the opposed motion the reflected field exists for all frequencies for under-Cherenkov movement only, $\gamma_\varepsilon^2 > 0$, and for the runaway boundary for the frequencies $\omega_0 > \gamma_\varepsilon\omega_k$. In the Cherenkov case, $\gamma_\varepsilon^2 > 0$, the reflection is absent for the runaway boundary as the incident wave does not run down the boundary and it passed completely for the opposed motion. This analysis gives the final expression for the reflected wave frequency

$$\omega_r = \omega_0\gamma_\varepsilon^2(1 + \varepsilon\beta^2 - 2\sqrt{\varepsilon}\,\beta\sqrt{1 - \omega_k^2/\omega_0^2}) \tag{6.89}$$

and the wave-number

$$k_0(\omega_r) = \omega_0\gamma_\varepsilon^2\frac{\sqrt{\varepsilon}}{c}((1 + \varepsilon\beta^2)\sqrt{1 - \omega_k^2/\omega_0^2} - 2\sqrt{\varepsilon}\,\beta) \tag{6.90}$$

Calculation of the integral (6.86) gives the reflectivity

$$R = \left(\frac{\omega_r}{\omega_0}\right)^2\frac{v_0 - \omega_{0r}}{v_0 - \omega_r}S, \tag{6.91}$$

where S is the factor determined by the form of the constituent equations in the case of the medium movement together with the boundary

$$S = \frac{\omega_0}{\omega_r}\frac{(\varepsilon_1 - 1)\gamma^2\omega_u{}^2 - (\varepsilon - 1)v_0\omega_r}{(\varepsilon_1 - 1)\gamma^2\omega_u{}^2 - (\varepsilon - 1)v_0\omega_0} \tag{6.92}$$

If the medium is motionless, and only the boundary moves, then $S \equiv 1$. The formula (6.91) coincides with the expression for the reflectivity derived in a differential formulation in [6] for the problem of the wave reflection from a moving semi-infinite medium in a waveguide.

The expression (6.91) is rather general because it is the solution to the problem which represents a model describing the processes of the reflection of the electromagnetic waves for various situations.

(i) The reflection of the electromagnetic wave from the dielectric half-space moving through another dielectric. Assuming $\omega_k = \omega_{k1} = 0$ in (6.91) and substituting v_0 from (6.81) one obtains the reflectivity as in [19, 20]

$$R = \frac{1-\sqrt{\varepsilon_1}\beta}{1+\sqrt{\varepsilon}\beta}\frac{\sqrt{\varepsilon_1}-\sqrt{\varepsilon}}{\sqrt{\varepsilon_1}+\sqrt{\varepsilon}} \tag{6.93}$$

(ii) In the case of a movement, only a boundary which divides two motionless dielectric media [21] the same formula (6.93) is obtained if one takes $\omega_k = \omega_{k1} = 0$, $S = 1$ and $\nu = \gamma_{\varepsilon 1}^2(\omega_u \pm \sqrt{\varepsilon_1}\beta\sqrt{\omega_u^2 - \gamma_{\varepsilon 1}^{-2}\omega_k^2})$.

(iii) To obtain the reflectivity for the reflection of waves from the moving plasma half-space in vacuum as well as through a dielectric with the permittivity ε one has to substitute in (6.91) $\varepsilon_1 = \varepsilon$, $\omega_k = 0$, $S = \omega_0/\omega_r$ and v from (6.81). As a result, one obtains

$$R = \frac{1-\sqrt{\varepsilon_1}\beta}{1+\sqrt{\varepsilon_1}\beta}\frac{1-\sqrt{Q}}{1+\sqrt{Q}}, \quad Q = 1 - \frac{\omega_{k1}^2}{\omega_0^2}\frac{1+\sqrt{\varepsilon_1}\beta}{1-\sqrt{\varepsilon_1}\beta}, \tag{6.94}$$

which coincide with expressions in [22–24].

The frequency of the reflected wave is determined by the velocity of the moving boundary and does not depend on medium properties behind the boundary. The group velocity for the reflected wave

$$v_{gr} = c\frac{1-\varepsilon\beta^2}{|1-\varepsilon\beta^2|}\frac{\left|(1+\varepsilon\beta^2)\sqrt{1-\omega_k^2/\omega_0^2} - 2\sqrt{\varepsilon}\beta\right|}{1-2\sqrt{\varepsilon}\beta\sqrt{1-\omega_k^2/\omega_0^2}+\varepsilon\beta^2} \tag{6.95}$$

remains distinct from zero even if $\omega_0 \to \omega_k$ that is when the group velocity of the incident wave tends to zero. This fact is very important for energetics in the reflection from a moving boundary.

The expression for the reflectivity (6.91) is simplified using the identity $\varepsilon\omega_k^2 = \varepsilon_1\omega_{k1}^2$:

$$R = \frac{\omega_r}{\omega_0}\frac{1-\sqrt{A}}{1+\sqrt{A}}, \quad A = \frac{\varepsilon_1}{\varepsilon}\frac{\omega_u^2 - \omega_{k1}^2/\gamma^2}{\omega_u^2 - \omega_k^2/\gamma_\varepsilon^2} \tag{6.96}$$

It follows from (6.89) and (6.96) that the frequency of the reflected wave and the reflectivity therewith grows infinitely when $\varepsilon\beta^2 \to 1$, not only for the meeting movement (as in the case of infinite space) but for the running away movement also. However in the case of the running away movement the minimal frequency $\gamma_\varepsilon\omega_k$ provides for the appearance of the reflected field growth also. In connection with this, the following situation appears before our eyes. Let an initial field exist in the whole waveguide before the boundary movement begins, and that the boundary movement has turned on adiabatically in the infinite remote past. If the velocity of this movement is $u > (c/\sqrt{\varepsilon})\sqrt{1-(\omega_k/\omega)^2}$ then the energy of the incident wave lags behind the boundary ($u < v_{g0}$), that is the leakage of the energy cumulated in the waveguide before the movement began occurs. The velocity of the phase movement of the incident wave exceeds the boundary velocity if $(\sqrt{\varepsilon}\beta)^{-1} > \sqrt{1-(\omega_k/\omega)^2}$; this happens for all frequencies in the case of $\gamma_\varepsilon^2 > 0$ and for $\omega < \varepsilon_1\beta^2|\gamma_\varepsilon|\omega_k$ in the case of Cherenkov movement. For other values of the velocity (or the frequency) the phase movement as well as the energy of the incident wave is opposite to the movement of the boundary and the medium. It means that the phase of the incident wave becomes negative. It is evident that for correct investigations of all of these processes one needs a statement of an initial value problem when the beginning of the boundary or medium movement is taken into account.

In addition to this effect, the wave dispersion leads also to the fact that the movement of the part of the homogeneous dielectric in the waveguide provokes the wave reflection. Indeed, if $\varepsilon_1 = \varepsilon$ then the reflectivity does not turn into zero. In an unbounded space, a similar effect occurs only in the case of oblique incidence of the wave onto the boundary. If one takes into account the interpretation of the running wave in the waveguide as the superposition of two crisscrossing plane waves (Brillouin waves) then the analogy between these two effects becomes complete [25].

6.5 Penetration of an Electromagnetic Wave through Plasma Boundary after Its Start in a Waveguide

The interaction of an electromagnetic wave with a moving boundary considered in the previous section shows that continuity of the movement, that is, its beginning at the infinite remote past, can lead to ambiguity in the problem statement. The movement beginning can occur earlier or later than the field switching. As a consequence, all transient phenomena remain out of view. Information losses grow if there are non-stationarities of another kind, for example a change in time of medium properties.

To withdraw this ambiguity we consider the evolution of an electromagnetic wave interaction with a restricted plasma by a plane whose boundary begins the uniform movement at zero moment of time in the waveguide. The time jump of the plasma density at this moment is taken into account by assuming that an electromagnetic field interacts with a half-restricted plasma formed at zero moment of time, and then expanded in a waveguide. In this case there are two contributions to dispersion, waveguide and plasma dispersion. The electromagnetic wave propagates from the empty side of the waveguide. A wave source works from the defined moment of time and is located at some distance from the point where the plasma boundary was at zero moment. The problem is solved by an evolutionary approach based on the integral Volterra equation.

It is assumed that an homogeneous cold plasma is created at zero moment of time in the waveguide half-space, that is fills the whole cross-section of the waveguide and is restricted on one side by a plane that is normal to the waveguide axis x. The plasma density is characterized by a plasma frequency ω_e. After its creation the plasma expands such that its boundary moves with the velocity $u = \beta c$ (Fig. 6.6).

A signal source

$$\mathbf{j}_{\text{ex}} = \sum_k \mathbf{b}_k \delta(x - x_0)\theta(t - t_0) j_k(t) \tag{6.97}$$

operates from the point in time t_0 and is located at a waveguide cross-section x_0, where x_0 is the distance from the plasma boundary at a moment when it comes into being. In (6.97) $\mathbf{b}_k$ is a waveguide vector eigen-function (6.19) and k denotes a waveguide mode with a corresponding waveguide critical frequency ω_k. With the geometry of the problem considered, there is no mode transformation.

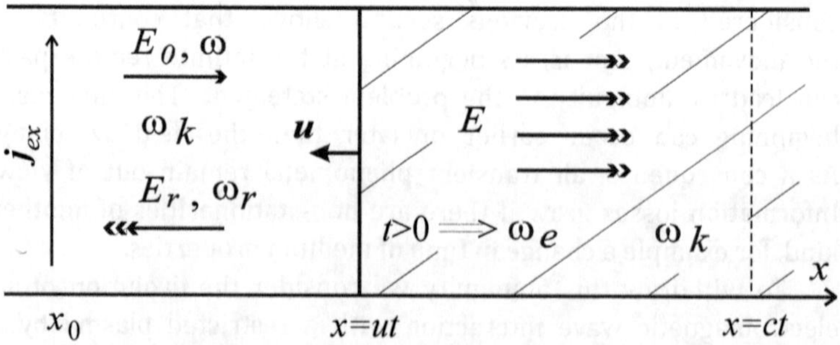

Figure 6.6 Penetration of a wave into plasma.

An electromagnetic signal generated by the current $\mathbf{j}_{ex}$ propagates from the empty side of the waveguide and is described by the expression

$$E_0(t,x) = -\frac{2\pi}{c}\sum_k\left[\Gamma_k(t-t_0,x-x_0)j_k(t_0) + \int_{t_0}^{\infty}\Gamma_k(t-t',x-x_0)\frac{\partial j_k(t')}{\partial t'}dt'\right], \tag{6.98}$$

where, $\Gamma_k(t-t',x-x') = J_0\left(\omega_k\sqrt{(t-t')^2 - \frac{(x-x')^2}{c^2}}\right)\theta\left(t-t'-\frac{|x-x'|}{c}\right),$ (6.99)

J_0 is the Bessel function, θ is the Heaviside unit function, c is the light velocity in vacuum.

The electromagnetic field penetrating into the half-restricted plasma after zero moment of time satisfies an evolutionary Volterra integral equation:

$$E = F - \frac{2\pi}{c}\int_0^{\infty}dt'\int_0^{\infty}dx'\frac{\partial^2}{\partial t^2}\Gamma_k(t-t',x-x')\theta(x'-ut')P(t',x'). \tag{6.100}$$

Here, $F = \theta(x - ut)E_0(t, x)$ is a free term of the integral equation and a constitutive relation for cold plasma is given by

$$P = \frac{\omega_e^2}{4\pi}\int_0^t (t - t')\theta(x - ut')E(t', x')dt'. \tag{6.101}$$

The solution to Eq. (6.100) is found by the resolvent, which satisfies Eq. (3.7)

$$\langle \mathbf{p} | \hat{R} | \mathbf{p}' \rangle = \langle \mathbf{p} | \hat{K} | \mathbf{p}' \rangle + \int d\mathbf{p}_1 \langle \mathbf{p} | \hat{K} | \mathbf{p}_1 \rangle \langle \mathbf{p}_1 | \hat{R} | \mathbf{p}' \rangle. \tag{6.102}$$

The kernel of this equation $\langle \mathbf{p} | \hat{K} | \mathbf{p}' \rangle = \left\langle \mathbf{p} \middle| \frac{2\pi}{c}\frac{\partial^2}{\partial t^2}\Gamma_k(t - t', x - x')\chi\hat{P} \middle| \mathbf{p}' \right\rangle$, where $\hat{P}$ is the operator determined by (6.101), depends on the direction of the boundary velocity. If $u > 0$ then

$$\langle \mathbf{p} | \hat{K} | \mathbf{p}' \rangle = \omega_e^2 \left\{ \frac{p^2}{p^2 + c^2k^2 + \omega_k^2} \frac{i}{k - k' - i0} + \frac{c}{2\varphi_k} \frac{(q - \beta\varphi_k)^2}{(\varphi_k - \beta p - ick)(\varphi_k - \beta q - ick'\gamma^{-2})} \right\} \frac{1}{(q - iuk')^2 (q - p')}, \tag{6.103}$$

where $q = p + iuk$, $\gamma = 1/\sqrt{1 - \beta^2}$, $\varphi_k = \sqrt{q^2 + \omega_k^2\gamma^{-2}}$, and that branch of the square root is taken which has the positive real part, $\mathrm{Re}\varphi_k > 0$.

If $u > 0$ then

$$\langle \mathbf{p} | \hat{K} | \mathbf{p}' \rangle = \omega_e^2 \left\{ \frac{p^2}{p^2 + c^2k^2 + \omega_k^2} \frac{i}{k - k' - i0} + \frac{c}{2\gamma^4\varphi_k} \frac{1}{(\varphi_k - \beta p - ick)(\varphi_k - \beta q - ick'\gamma^{-2})} \right\} \frac{1}{q - p'} \tag{6.104}$$

The resolvent, as the solution to Eq. (6.102) according to the algorithm used in the previous sections, consists of two terms. The first one is determined by a plasma frequency change

$$\left\langle \mathbf{p} \middle| \hat{R}_1^{(+)} \middle| \mathbf{p}' \right\rangle = \frac{\omega_e^2}{q - p'} \frac{p^2}{p^2 + c^2k^2 + \omega_{ke}^2} \frac{i}{k - k' - i0} \frac{1}{(q - iuk')^2} \tag{6.105}$$

for $u > 0$ and

$$\left\langle \mathbf{p}\left|\hat{R}_1^{(-)}\right|\mathbf{p}'\right\rangle = \frac{\omega_e^2}{q-p'}\frac{p^2}{p^2+c^2k^2+\omega_{ke}^2}\frac{i}{k-k'-i0} \quad (6.106)$$

for $u > 0$. In these equations $\omega_{ke}^2 = \omega_k^2 + \omega_e^2$.

The second term takes into account the boundary and its movement. For $u > 0$

$$\left\langle \mathbf{p}\left|\hat{R}_2^{(+)}\right|\mathbf{p}'\right\rangle = \omega_e^2\left\{\frac{p^2}{p^2+c^2k^2+\omega_{ke}^2}\frac{i}{k-k'-i0} + \frac{c}{2\varphi_{ke}}\left[\frac{1}{\varphi_{ke}-\beta p-ick}+\frac{R(q)}{\varphi_{ke}+\beta p+ick}\right]\frac{(q-\beta\varphi_{ke})^2}{\varphi_{ke}-\beta q-ick'\gamma^{-2}}\right\}$$
$$\frac{1}{(q-iuk')^2(q-p')} \quad (6.107)$$

and for $u < 0$

$$\left\langle \mathbf{p}\left|\hat{R}_2^{(-)}\right|\mathbf{p}'\right\rangle = \omega_e^2\left\{\frac{p^2}{p^2+c^2k^2+\omega_{ke}^2}\frac{i}{k-k'-i0} + \frac{c}{2\gamma^4\varphi_{ke}}\left[\frac{1}{\varphi_{ke}-\beta p-ick}+\frac{R(q)}{\varphi_{ke}+\beta p+ick}\right]\frac{1}{\varphi_{ke}-\beta q-ick'\gamma^{-2}}\right\}$$
$$\frac{1}{q-p'} \quad (6.108)$$

In these formulae, $\varphi_{ke} = \sqrt{q^2+\omega_{ke}^2\gamma^{-2}}$ and

$$R(q) = \frac{\varphi_k(q)-\varphi_{ke}(q)}{\varphi_k(q)+\varphi_{ke}(q)}. \quad (6.109)$$

Integration in (6.102) must be done in such a way that the condition $\mathrm{Re}[\varphi_{ke}(q) \mp \beta q] > 0$ is fulfilled.

The expression (6.109) is the analog to the reflectivity of an electromagnetic wave from a plane motionless plasma boundary in a waveguide [26, 27]

$$R(\omega) = \frac{\sqrt{\omega^2-\omega_{ke}^2}-\sqrt{\omega^2-\omega_k^2}}{\sqrt{\omega^2-\omega_{ke}^2}+\sqrt{\omega^2-\omega_k^2}} \quad (6.110)$$

If one takes into account the boundary movement by transitioning to a moving coordinate system with the substitution $\omega \to \gamma(\omega - iuk)$, changes the Fourier transform for the Laplace transform by $p = i\omega$ and takes into account that reflection is going on the plasma side, that means $R(\omega) \to -R(\omega)$, then (6.110) is equal to (6.109) exactly.

The resolvents found allow us to write the solution to Eq. (6.102) in the form (for $u > 0$)

$$E = F(t,x) + \int_0^\infty dt' \int_0^\infty [R_1^{(+)}(t,t',x,x') + R_2^{(+)}(t,t',x,x')]F(t',x')dx'. \tag{6.111}$$

Here $R_1^{(+)}(t,t',x,x') = \int_{\eta-i\infty}^{\eta+i\infty} \frac{dp}{2\pi i} \frac{1}{2c\psi(p)} e^{p(t-t')-\psi(p)\frac{|x-x'|}{c}} \theta(x'-ut')$, (6.112)

where $\psi(p) = \sqrt{p^2 + \omega_{ke}^2}$ and

$$R_2^{(+)}(t,t',x,x') = \int_{\eta-i\infty}^{\eta+i\infty} \frac{dp}{2\pi i} \int_{-\infty}^{\infty} \frac{ds}{2\pi} R(q) \frac{\exp\left[p(t-t') + is\left(x - \frac{\beta}{1-\beta^2}t'\right) - \frac{\varphi_{ke}(q) - \beta p}{c(1-\beta^2)}(x'-ut')\right]}{2\varphi_{ke}(q)\left[\varphi_{ke}(q) + \beta p + ics\right]} \theta(x'-ut') \tag{6.113}$$

Let us now consider the transformation in the expanded plasma of the initial wave generated by the harmonic source $j_k = e^{i\omega t}$ in (6.97). According to (6.98) and (6.99) this wave is given by the expression

$$E_0(t,x) = E_0\theta\left(t - t_0 - \frac{|x-x_0|}{c}\right)\left\{J_0\left(\omega_k\sqrt{(t-t_0)^2 - \frac{(x-x_0)^2}{c^2}}\right) + i\omega \int_{\frac{|x-x_0|}{c}}^{t-t_0} J_0\left(\omega_k\sqrt{\tau^2 - \frac{(x-x_0)^2}{c^2}}\right)e^{i\omega(t-\tau)}d\tau\right\}. \tag{6.114}$$

This expression can be simplified in extreme cases. If the source location and the moment of its turning on are matched by the condition $t_0 - x_0/c \approx 0$ then the incident wave of the mode considered might be presented approximately in the form of a plane wave:

$$E_0(t,x) = E_0\theta\left(t - \frac{x}{c}\right)e^{i[\omega t - k_0(t,x)x]}, \quad (6.115)$$

where $k_0(t, x) = \omega/c$ if $x \approx ct$ and $k_0(t,x) = \sqrt{\omega^2 - \omega_k^2}/c$ if $x << ct$ and $t \to \infty$.

To reveal important features of the phenomenon considered we confine ourselves to the simplest expression for the incident field (6.115) assuming that k_0 = const is suitable to the corresponding values in various time-spatial regions. It gives such an expression for the free term in (6.115)

$$F(t,x) = E_0\theta(x - ut)\theta\left(t - \frac{x}{c}\right)e^{i[\omega t - k_0 x]}. \quad (6.116)$$

The transformed field is obtained after integrating in (6.115)

$$E(t,x) = E_0\left[\theta(x - ut) - \frac{\omega_e^2}{\omega_{ke}^2 + c^2k_0^2 - \omega^2}\right]\theta(t - x/c)e^{i(\omega t - k_0 x)}$$

$$+ \frac{\omega_e^2 E_0}{\omega_{ke}^2 + c^2k_0^2 - \omega^2}\theta(t - x/c)e^{i(\omega_3 t - k_3 x)} + I_1(t,x) + I_2(t,x), \quad (6.117)$$

where

$$I_1(t,x) = -\omega_e^2(1-\beta)\int_{\eta - i\infty}^{\eta + i\infty}\frac{dp}{2\pi i}\frac{\exp[pt - \psi(p)x/c]}{2\psi(p)[p - \beta\psi(p) - i\omega_1][p - \psi(p) - i\omega_2]}, \quad (6.118)$$

$$I_2(t,x) = \frac{\omega_e^2}{\gamma^2}\int_{\eta - i\infty}^{\eta + i\infty}\frac{dp}{2\pi i}\frac{[\varphi_k - \varphi_{ke}]\exp\left[\frac{p + \beta\varphi_{ke}}{1 - \beta^2}t - \frac{\beta p + \varphi_{ke}}{1 - \beta^2}x/c\right]}{2\varphi_{ke}[\varphi_k + \varphi_{ke}][p - i\omega_1][p + \varphi_{ke} - i(1+\beta)\omega_2]}, \quad (6.119)$$

$\omega_1 = \omega - uk_0, \omega_2 = \omega - ck_0$, $\omega_3 = (\omega_{ke}^2 + \omega_2^2)/(2\omega_2)$, $k_3 = (\omega_{ke}^2 - \omega_2^2)/(2c\omega_2)$, and $\psi(p) = \sqrt{p^2 + \omega_{ke}^2}$.

At first let us analyze the integral (6.118). Its integrand has two branch points $\pm pb = \pm i\omega_{ke}$ and three poles $p_{1,2} = i\gamma^2(\omega_1 \pm \beta\sqrt{\omega_1^2 - \varphi_{ke}^2\gamma^2})$ and $p_3 i\omega_3$. The single-valued choice of the root branch $\psi(p) = \sqrt{p^2 + \omega_{ke}^2}$ is determined by the condition Re $\psi(p) > 0$ provides a decay of $e^{-\psi x/c}$ with $x \to \infty$. To implement such a decay condition for the incident wave imaginary additions are introduced: $\omega_k \to \omega_k + i\varepsilon$ and $\omega_e \to \omega_e + i\delta$. Taking into account these additions removes degeneracy from the branch points and the poles which move away from the imaginary axes (Fig. 6.7a). To provide the required choice of the root branch of $\psi(p)$ the branch-cuts are drawn as shown in Fig. 6.7a; that gives Re $\psi(p) > 0$ on the upper part of the Riemann surface [28, 29]. The equations for the branch-cuts have the forms: $\xi\eta = -\varepsilon\omega_k - \delta\omega_e$ if $p = \xi + i\eta$. Analysis of the equations determining the singular points in (6.118) shows that the poles p_1 and p_3 located on the upper part of the Riemann surface but the pole p_2 on the lower part.

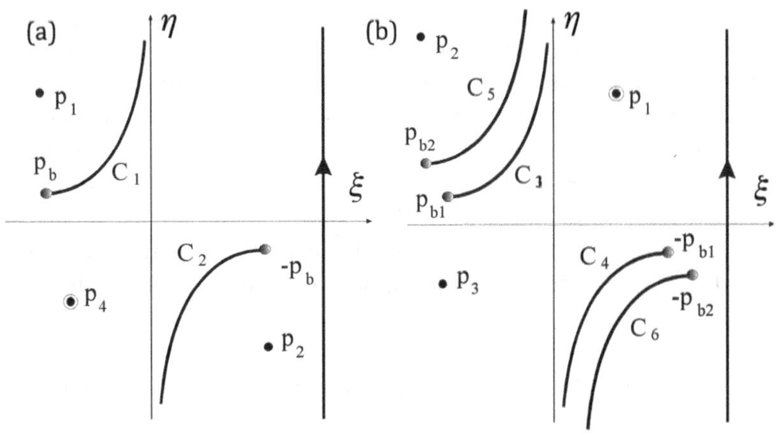

Figure 6.7 Branch-cuts for calculation of the integrals in (6.118) and (6.119).

Calculation of the residuals at the poles on the upper part of the Riemann surface gives for the integral I_1, (6.118),

$$I_1(t,x)=\left[-\frac{\omega_e^2 E_0}{\omega_{ke}^2+c^2k_0^2-\omega^2}e^{i(\omega_3 t-k_3 x)}+E_2^{(1)}(t,x)+S(t,x)\right]\theta(t-x/c). \tag{6.120}$$

Here $E_2^{(1)}(t,x)=\dfrac{-\omega_e^2}{2\gamma^2\varphi_{ke}^*\left[\beta\omega-ck_0+\varphi_{ke}^*\right]}e^{i(\nu t-\kappa x)}$, (6.121)

where $\nu=\gamma^2(\omega_1+\beta\varphi_{ke}^*)$, $\kappa=\gamma^2(\beta\omega_1+\varphi_{ke}^*)/c$, $\varphi_{ke}^*=\sqrt{\omega_1^2-\omega_{ke}^2/\gamma^2}$ and $S(t, x)$ is the sum of the I_1-kind-integrals over the branch-cuts $C_{1,2}$ in Fig. 5.7a. The frequency ν and the wave number κ coincide with corresponding characteristics of the penetrated wave in the case when the boundary movement begins in the infinite past. The frequency ν satisfies the condition of the equality of the incident and penetrated wave phases on the moving plasma boundary

$$(\omega t-k_0 x)|_{x=ut}=(\nu t-x\sqrt{\nu^2+\omega_{ke}^2}/c)|_{x=ut}. \tag{6.122}$$

Let us now pay attention to the integral $I_2(t, x)$ (6.119) which determines the contribution of the interaction with the moving boundary to the penetrated field. The integrand in (6.119) has four branch points: $\pm p_{b1}=i\omega_{ke}\gamma^{-1}$ and $\pm p_{b1}=i\omega_k\gamma^{-1}$. The branch cuts drawn as shown in Fig. 5.7b provide fulfillment of the conditions $\mathrm{Re}\varphi_k>0$ and $\mathrm{Re}\varphi_{ke}>0$ on the upper list of the Riemann surface. These lines are given by the equations $\xi\eta=-(\varepsilon\omega_k+\delta\omega_e)\gamma^{-2}$ for $C_{3,4}$ and $\xi\eta=-\varepsilon\omega_k\gamma^{-2}$ for $C_{5,6}$. Both singular points of the integrand $p_1=i[(1-\beta)\omega_{ke}^2+(1+\beta)\omega_2^2]/(2\omega_2)$, $p_2=i\omega_1$ are located on the upper part of the Riemann surface. Calculation of the residuals at the poles on this upper part gives for the integral I_2, (6.119),

$$I_2(t,x)=[E_2^{(2)}(t,x)+E_3(t,x)+S_R(t,x)]\theta(t-x/c). \tag{6.123}$$

Here

$$E_2^{(2)}(t,x)=\frac{\omega_e^2 R(i\omega_1)}{2\gamma^2\varphi_{ke}^*[\beta\omega-ck_0-\varphi_{ke}^*]}e^{i(\nu t-\kappa x)} \tag{6.124}$$

$$E_3(t,x) = \frac{\{\sqrt{[(1+\beta)\omega_2^2 + (1-\beta)\omega_{ke}^2]^2 - 4\omega_2^2\omega_k^2\gamma^{-2}} + (1+\beta)\omega_2^2 - (1-\beta)\omega_{ke}^2\}^2}{4\omega_2^2[\omega_{ke}^2 + c^2k_0^2 - \omega^2]\gamma^{-2}}$$

$$e^{i(\bar{\omega}_3 t - \bar{k}_3 x)}, \tag{6.125}$$

$$\text{where } \bar{\omega}_3 = \frac{(1-\beta)^2\omega_{ke}^2 + (1+\beta)^2\omega_2^2}{2\omega_2\gamma^{-2}}, \quad \bar{k}_3 = \frac{(1+\beta)^2\omega_2^2 - (1-\beta)^2\omega_{ke}^2}{2c\omega_2\gamma^{-2}} \tag{6.126}$$

and $S_R(t, x)$ is the sum of the I_2-kind-integrals over the branch-cuts $C_{3,4}$ and $C_{5,6}$ in Fig. 5.8b.

The frequency ω_4 delivering the singularity to the integrand in $I_2(t, x)$ satisfies the condition of equality of the incident wave phase on the moving plasma boundary and the penetrated wave phase translated into the rest frame of the boundary

$$(\omega t - k_0 x)|_{x=ut} = \gamma^{-1}(\omega_1 t' - \varphi_{ke}^* x'/c)|_{x=ut}, \tag{6.127}$$

where $t' = \gamma^{-1}(t - xu/c^2)$, $x' = \gamma^{-1}(x - ut)$.

Summation of all expressions in (6.117) gives for the penetrated field in the region between the plasma boundary and the plane $x = ct$:

$$E(t, x) = E_1 + E_2 + E_3 + E_{tr}. \tag{6.128}$$

The first term

$$E_1(t,x) = F(t,x) - \frac{\omega_e^2}{\omega_{ke}^2 + c^2k_0^2 - \omega^2} e^{i(\omega t - k_0 x)}, \quad \omega_{ke}^2 = \omega_k^2 + \omega_e^2 \tag{6.129}$$

is a relic of the incident wave in the penetrated field. Near the signal front ($x \approx ct$) this wave frequency and the wave number are as in the incident wave but its amplitude changes in the ratio of ω_k^2/ω_{ke}^2. This relic disappears with distance from the wave front as the wavenumber $k_0 \to \frac{1}{c}\sqrt{\omega^2 - \omega_k^2}$ and the two terms in (6.129) denominator cancel each other. In an open space ($\omega_k = 0$) this relic is absent. In the empty waveguide ($\omega_e = 0$) this relic is the whole incident wave as the other items in (6.128) are equal to zero.

The second term in (6.128) is the sum of (6.121) and (6.124). It represents the stationary wave as in the case when plasma movement begins in the infinite remote past:

$$E_2 = \frac{\omega_e^2(\varphi_k^* - \beta\omega + ck_0)}{(\varphi_k^* + \varphi_{ke}^*)(\omega_{ke}^2 + c^2k_0^2 - \omega^2)} e^{i(\Omega_2 t - k_2 x)}, \tag{6.130}$$

where $\varphi_k^* = \sqrt{\Omega_1^2 - (1-\beta^2)\omega_k^2}$, $\varphi_{ke}^* = \sqrt{\Omega_1^2 - (1-\beta^2)\omega_{ke}^2}$.

In the open space ($\omega_k = 0$) the wave E_2 represents the whole penetrated field. It vanishes in the absence of the plasma ($\omega_e = 0$). Far away from the signal front ($x << ct$), this wave amplitude becomes equal to

$$E_2 = \frac{2(ck_0 - \beta\omega)}{(\varphi_{ke}^* + ck_0 - \beta\omega)}, \tag{6.131}$$

and coincides with the corresponding one in the case of the open space ($\omega_k = 0$).

The third term in (6.128) is absent near the signal front ($x \approx ct$) and far away from it ($x << ct$); this term fulfills the function of the relic

$$E_3 = -4\gamma^{-2}\omega_2^2\omega_e^2\Big[(1-\beta)\omega_{ke}^2 - (1+\beta)\omega_2^2 + \sqrt{[(1+\beta)\omega_2^2 + (1-\beta)\omega_{ke}^2]^2 - 4\gamma^{-2}\omega_{ke}^2\omega_2^2}\Big]^{-2} e^{i(\bar{\omega}_3 t - \bar{k}_3 x)}, \tag{6.132}$$

where

$$\bar{\omega}_3 = (1+\beta^2)\omega - 2\beta ck_0 + \frac{(1-\beta)\omega_e^2}{2\omega_2}, \quad \bar{k}_3 = (1+\beta^2)k_0 - 2\beta\omega/c - \frac{(1-\beta)\omega_e^2}{2c\omega_2}. \tag{6.133}$$

This wave amplitude turns to zero with $\omega_k = 0$ as well as with $\omega_e = 0$. That is this wave exists only with the availability of a double dispersion mechanism, namely waveguide and plasma dispersion. When the plasma boundary moves with a relativistic velocity ($\beta \to -1$) this wave also turns to zero.

The fourth term in (6.128) is determined by the integrals over branch cuts in the complex plane and yields the continuous wave spectrum that describes the transition process. This field is absent near the wave front $x = ct$. The transient is damped out when $t \to \infty$. The distribution of the secondary waves in the interval ($x = ut$, $x = ct$) that is between the plasma boundary and the wave front is shown in Fig. 6.8.

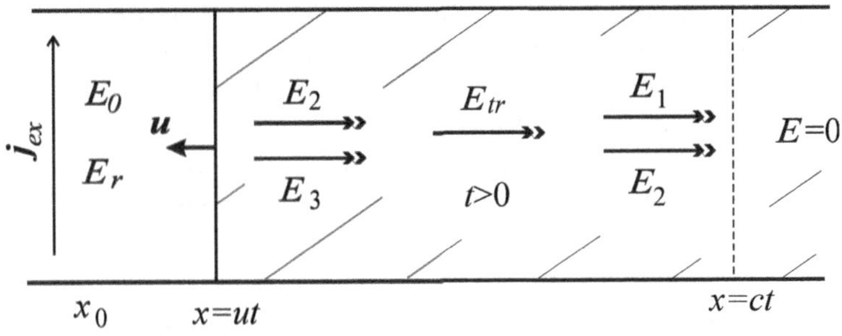

Figure 6.8 The distribution of the secondary waves.

Thus, the penetration of the electromagnetic wave through the moving plasma boundary excites a complex field in the plasma. Near the wave front, this field consists of two waves—the relic of the incident wave and the stationary wave that is the analogue of the wave for the case when the movement of the plasma boundary begins at the infinite remote past. In an open space, the relic disappears and the stationary wave represents the whole penetrated field. When the plasma is absent, the stationary wave disappears and the relic represents the whole penetrated field. Far away from the wave front, the field consists of two waves also. The new wave performs the function of the relic but this wave exists only with the double dispersion mechanism—the waveguide and the plasma dispersion. The second wave is the stationary wave that converts to the incident wave if the plasma is absent. The whole interval between the plasma boundary and the wave front is filled by the transient field that is asymptotically damped out in time.

6.6 Interaction of an Electromagnetic Wave with a Plasma Bunch Moving in a Metallic Waveguide

Investigations of electromagnetic wave interaction with moving plasma has been carried out for a long time since the first publications by Landecker [30] and Lampert [31], and it remains of interest up to now [7, 20, 23, 24, 26, 30–35]. It is well known that the electromagnetic wave exhibits simultaneously the amplitude and frequency change and duration compression under such an interaction [36–38]. It has been shown in previous studies that the electromagnetic wave amplitude and frequency can be enhanced appreciably after its reflection from a moving medium boundary. In Refs. [20, 23, 26–38], the method for multiplication of a wave frequency when a wave reflects from plasma moving in the opposite direction has been described in detail, and the experimental realization of such a structure was shown in Refs. 26 and 32. As reflection is most effective when the wave phase velocity is near the plasma velocity, a slow-wave structure has been used in this work. Frequency multiplication of 11–20% at the initial frequency 24.75 MHz has been achieved with strong wave slowdown (200–300 times) and the plasma velocity near 6×10^4 m/sec. However, the gain and a frequency multiplication factors in this method are confined by difficulties in the production of a slow-wave structure, the technology for which increases in complexity with frequency growth and wave slowdown. Furthermore, this approach does not allow one to adjust the gain and frequency multiplication factors independently as they change simultaneously with variations in the ratio of wave phase velocity to plasma velocity. These disadvantages can be avoided if one takes into account the influence of a waveguide on the interaction of an electromagnetic wave with a medium and another mechanism turns out in action. It appears that an artificial slowdown of the wave is not needed; slowing of the group velocity is natural for a waveguide. In this case, adjustment of a gain factor and a frequency multiplication factor are in practice implemented separately. The efficiency of these changes is characterized by the ratio of the boundary velocity to the wave group velocity; this is of vital importance in those waveguides in which a double dispersion mechanism occurs [33, 34]. The importance of this mechanism for the energy characteristics of the reflected electromagnetic waves will be shown in this section. Usually the power characteristics are considered; however, if the plasma arises and starts to move at a certain time moment then the wave interaction with plasma is accompanied by the appearance of a transient electromagnetic field. The spectral structure of the secondary electromagnetic waves is of interest even in the case of a steady uniform movement of the medium boundary [39]. If the movement is not steady, the evolution of the electromagnetic wave has to be considered as in the previous section [40].

6.6.1 Main Ratios for Electromagnetic Waves in a Waveguide with a Relativistic Moving Plasma Bunch

To investigate the influence of the double dispersion mechanism on the energy characteristics of waves reflected from a moving plasma bunch we consider a bunch of homogeneous cold plasma moving in a waveguide. It is assumed to be restricted by two planes being normal to the waveguide axis x. This bunch moves along the waveguide, as shown in Fig. 6.9. For the geometry of the problem considered there is not any mode transformation and only a single guided wave of TE type, characterized by the waveguide frequency ω_k, is considered. The incident electromagnetic TE wave $\mathbf{E}_0(t, x) = \mathbf{b}_\perp E_0 \exp(i(\omega t - k_0 x))$ propagates from the empty left side of the waveguide, $k_0(\omega) = \frac{1}{c}\sqrt{\omega^2 - \omega_k^2}$ is the wave number in the empty waveguide.

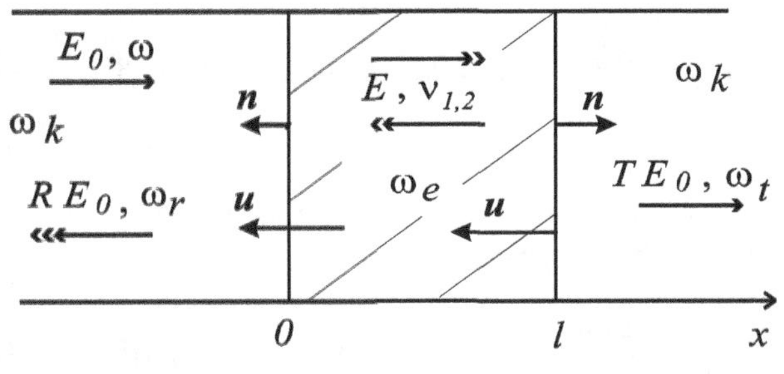

Figure 6.9 The geometry of the problem.

If the electric field is represented in the form

$$E(t,x)=\sum_{s=1}^{2}\int_{-\infty}^{\infty}\frac{d\nu}{2\pi}\tilde{E}_s(\nu)e^{i\nu t-ik_s(\nu)x} \tag{6.134}$$

then, following on from (6.41) and (6.42), its Fourier component must satisfy

$$\sum_{s=1}^{2}\int_{-\infty}^{\infty}\frac{d\nu}{2\pi}\alpha_s^{(+)}(\nu,\omega)L_s^{(+)}(\nu,\omega)\tilde{E}_s(\nu)=\frac{4\pi c^2k_0(\omega)}{i\omega^2}E_0\delta(\omega-\omega_0) \tag{6.135}$$

$$\sum_{s=1}^{2}\int_{-\infty}^{\infty}\frac{d\nu}{2\pi}\alpha_s^{(-)}(\nu,\omega)L_s^{(-)}(\nu,\omega)\tilde{E}_s(\nu)=0, \tag{6.136}$$

where

$$L_s^{\pm}(\nu,\omega)=\frac{4\pi}{\tilde{E}_s}\left\{\tilde{P}_s\mp\frac{ck_0(\omega)}{\omega}\tilde{M}_s\right\}$$
$$=\gamma^2 4\pi\alpha'(\nu)\frac{\left[\nu-uk_s(\nu)\right]\left[\omega\mp uk_0(\omega)\right]}{\nu\omega}, \tag{6.137}$$

$\alpha'(\nu)$ is the plasma polarizability at rest, u is the plasma bunch velocity and the character of the movement is taken into account by the functions

$$\alpha_s^{\pm}(\omega,\nu)=\frac{i}{k_0(\omega)\mp k_s(\nu)}\int_{-\infty}^{\infty}dt'e^{i(\nu-\omega)t'-i\left[k_s(\nu)\mp k_0(\omega)\right]x_{1,2}(t')}, \tag{6.138}$$

with $x_1(t)$ corresponding to α_s^+ and $x_2(t)$ corresponding to α_s^-. The wave numbers for the inner waves must satisfy the dispersion Eq. (6.69)

$$[1-(1+4\pi\alpha'(\omega))\beta^2]k_s^2+8\pi\alpha'(\omega)\beta\frac{\omega}{c}k_s-[1+4\pi\alpha'(\omega)-\beta^2]\frac{\omega^2}{c^2}+\frac{\varepsilon\omega_k^2}{c^2\gamma^2}=0 \tag{6.139}$$

The wave transmitted through the bunch, $x \geq x_2(t)$, is determined by the integral

$$E_t=\frac{2\pi}{c^2}\sum_{s=1}^{2}\int_{-\infty}^{\infty}\frac{d\omega}{2\pi}\int_{-\infty}^{\infty}\frac{d\nu}{2\pi}e^{i\omega t-ik_0(\omega)x}\frac{\omega^2}{k_0(\omega)}N_s^+(\omega,\nu)L_s^+(\omega,\nu)\tilde{E}(\nu) \tag{6.140}$$

and the wave reflected from the bunch, $x < x_1(t)$, by the integral

$$E_r = -\frac{2\pi}{c^2}\sum_{s=1}^{2}\int_{-\infty}^{\infty}\frac{d\omega}{2\pi}\int_{-\infty}^{\infty}\frac{dv}{2\pi}e^{i\omega t+ik_0(\omega)x}\frac{\omega^2}{k_0(\omega)}N_s^-(\omega,v)L_s^-(\omega,v)\tilde{E}(v) \quad (6.141)$$

In these formulae the kernels are determined by the integrals

$$N_s^{\pm}(\omega,v) = \int_{-\infty}^{\infty}dt'\frac{e^{i(v-\omega)t'}}{k_s(v)\mp k_0(\omega)}\{e^{-i[k_s(v)\mp k_0(\omega)]x_2(t')} - e^{-i[k_s(v)\mp k_0(\omega)]x_1(t')}\} \quad (6.142)$$

The polarizability of motionless cold isotropic plasma $4\pi\alpha'(v) = -\omega_e^2/[\gamma^2(v-uk_s(v))^2]$ where ω_e is the plasma frequency allows one to obtain roots of the dispersion Eq. (6.139). These roots $k_{1,2}(v) = \pm\sqrt{v^2-\omega_{ke}^2}$ where $\omega_{ke}^2 = \omega_k^2+\omega_e^2$ define the number of the inner waves and do not depend on the plasma velocity as the dispersion properties of a cold plasma and a waveguide are the same.

Assuming the coordinates of the bunch boundaries $x_1 = ut$, $x_2 = l + ut$, we find from (6.138)

$$\alpha_s^{\pm}(\omega,v) = \frac{2\pi i\delta(v-uk_s(v)-\Omega_{1,2})}{k_0(\omega)\mp k_s(v)}\begin{cases}1\\ e^{-i[k_s(v)+k_0(\omega)]l}\end{cases}, \quad (6.143)$$

where $\Omega_{1,2} = \omega\pm uk_0(\omega)$. This expression gives for (6.76) and (6.136)

$$\sum_{s=1}^{2}\int_{-\infty}^{\infty}\frac{dv}{2\pi}\frac{\delta(v-uk_s(v)-\Omega_1)}{k_0(\omega)-k_s(v)}\frac{\omega_e^2\Omega_1}{\omega v(v-uk_s(v))}\tilde{E}_s(v) = \frac{2c^2k_0(\omega)}{\omega^2}E_0\delta(\omega-\omega_0) \quad (6.144)$$

$$\sum_{s=1}^{2}\int_{-\infty}^{\infty}\frac{dv}{2\pi}\frac{\delta(v-uk_s(v)-\Omega_2)}{k_0(\omega)+k_s(v)}\frac{\omega_e^2\Omega_2}{\omega v(v-uk_s(v))}e^{-i[k_0(\omega)+k_s(v)]l}\tilde{E}_s(v) = 0 \quad (6.145)$$

The structure of solutions of these equations is determined by the zeros of the δ-functions arguments. This equality of the arguments to zero is equivalent to equality of the phases of inner and outer waves on the bunch boundaries. Indeed,

$\Omega_{10} = \omega - uk_0(\omega)$ is the phase of the incident wave with the frequency ω_0 on the bunch boundary $x_1 = ut$, $\Omega_2 = \omega_r + uk_0(\omega_r)$ is the phase of the reflected wave with the frequency ω_r on the same boundary. Equality of these phases gives the frequency of the reflected wave

$$\omega_r = \gamma^2[(1+\beta^2)\omega_0 - 2uk_0(\omega_0)] \tag{6.146}$$

However, Fig. 6.10 shows that the phase equality $\Omega_{10} - \Omega_2$ is possible with following movement (i.e., one for which the plasma bunch phase velocity $\beta > 0$ such that the electromagnetic wave follows it) only for $\omega_0 > \omega_k' = (1+\beta^2)\gamma^2\omega_k$. Therefore, for $\omega_0 > \omega_k'$, the phase of the reflected wave is provided by the second branch of the function $\Omega_1(\omega)$ that is $\Omega_1(\omega_0) = \Omega_1(\omega_r)$. For the case when the electromagnetic wave and plasma velocities oppose each, which we hitherto refer to as 'opposite movement', (plasma bunch phase velocity $\beta > 0$), the graphs for $\Omega_1(\omega)$ and $\Omega_2(\omega)$ change places, so the unique correspondence between Ω_{10} and ω_0 exists in the whole range $\omega_0 > \omega_k$.

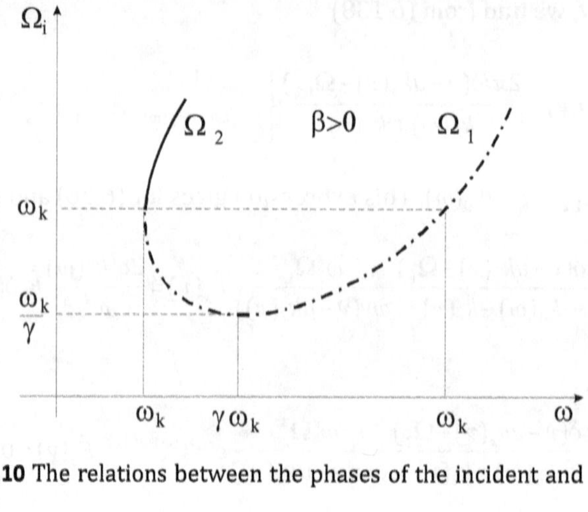

Figure 6.10 The relations between the phases of the incident and reflected waves.

Interaction between the electromagnetic field and the moving plasma bunch depends on the ratio between the group velocities of the incident and reflected waves and the bunch velocity. In the case of the following movement ($\beta > 0$) the group velocity of the incident wave

$$v_{g0} = c\sqrt{1-\omega_k^2/\omega_0^2} \tag{6.147}$$

becomes less than the bunch velocity if $\omega_0 > \gamma\omega_k$ that is when the incident wave does not come up the bunch, therefore the interaction is absent.

Behind the bunch, only the reflected wave stays. This wave can be named as the wave skipped by the bunch overtaking the incident wave. The group velocity of the reflected wave is

$$\begin{aligned} v_{gR} &= c\frac{d\omega_r}{dk_0} = c\frac{\sqrt{\omega_r^2-\omega_k^2}}{\omega_r} \\ &= c\frac{2\beta\omega_0-(1+\beta^2)ck_0(\omega_0)}{(1+\beta^2)\omega_0-2uk_0(\omega_0)} \end{aligned} \tag{6.148}$$

and its dependency on ω_0 is shown in Fig. 6.11.

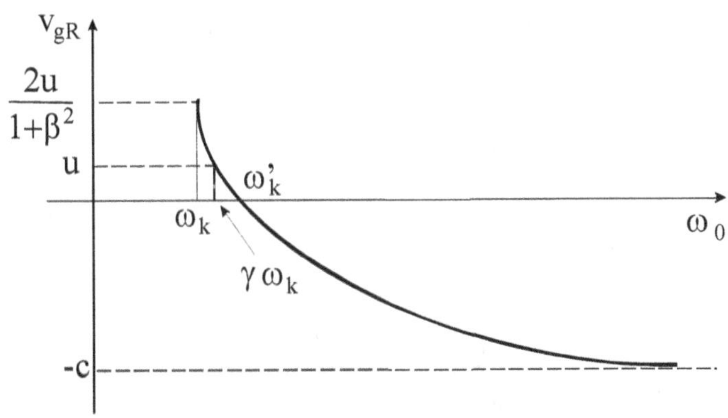

Figure 6.11 The group velocity of the reflected wave.

Figure 6.11 shows that if $\omega_0 > \gamma\omega_k$ then the "reflected" wave outruns the bunch that is it does not exist behind the bunch. It may exist ahead the bunch only and it frequency has to be ω_r. Thus, in the frequency range $\omega_0 < \omega_0 > \gamma\omega_k$ and the following movement, the bunch can interact with the overtaking incident wave only. In this process, it re-radiates forward a wave of frequency ω_r. In the case $\omega_0 > \gamma\omega_k$ the incident wave comes up the bunch. The reflected wave lags behind the bunch but moves in the direction towards it in the frequency range $\gamma\omega_k < \omega_0 < \omega_k' = (1+\beta^2)\gamma^2\omega_k$. Only when $\omega_0 < \omega_k'$ does the reflected wave propagate towards the incident one. In the case of opposite movement ($\beta < 0$) the interaction between the bunch and the incident wave exists for any frequency and the reflected wave has negative velocity.

Consider now the structure of the solution to Eqs. (6.144), (6.145) Equating the arguments of the δ-functions in these equations to zero

$$\nu - uk_s(\nu) = \Omega_i,\ i = 1,2 \tag{6.149}$$

gives the frequencies for the inner waves

$$\nu_s = \gamma^2(\Omega_i \pm \beta\sqrt{\Omega_i^2 - \omega_{ke}^2/\gamma^2}) \tag{6.150}$$

It follows from the equality of phases $\Omega_1(\omega_0) = \Omega_2(\omega_r)$ that the inner waves are functions of the phase $\Omega_{10} = \Omega_1(\omega_0)$. The wave numbers of these waves are determined by the expressions following from (6.149) and (6.150)

$$k_s = \frac{1}{c}\gamma^2(\beta\Omega_{10} \pm \sqrt{\Omega_{10}^2 - \omega_{ke}^2/\gamma^2}),\ s = 1,2 \tag{6.151}$$

Equations (6.150) and (6.151) allow one to determine the group velocities

$$v_{gs} = \frac{d\nu_s}{dk_s} = c\frac{\beta\Omega_{10} \pm \frac{\beta}{|\beta|}\sqrt{\Omega_{10}^2 - \omega_{ke}^2/\gamma^2}}{\Omega_{10} \pm |\beta|\sqrt{\Omega_{10}^2 - \omega_{ke}^2/\gamma^2}},\ s = 1,2 \tag{6.152}$$

The dependency of the group velocity on the phase $\Omega_{10} = \Omega_1(\omega_0)$ is shown in Fig. 6.12.

As the bunch has two reflecting boundaries there exists two inner waves. The wave with v_{g1} satisfies the condition of energy withdrawal on the boundary $x_1(t)$ whereas the wave v_{g2} satisfies this condition on $x_1(t)$.

According to this the structure of the solutions to (6.144), (6.145) can be set in the form

$$\tilde{E}_s(\nu) = a_s \delta(\nu - \nu_{s0}), \tag{6.153}$$

where $\nu_s = \nu_s(\Omega_{10})$. Substitution of (6.153) into (6.144), (6.145) gives

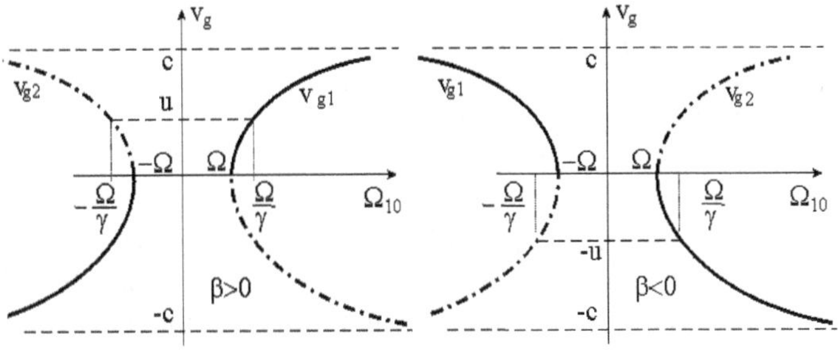

Figure 6.12 The group velocities of the inner waves.

$$\sum_{s=1}^{2}\int_{-\infty}^{\infty}\frac{d\nu}{2\pi}\frac{\delta(\Omega_{10}-\Omega_1)}{k_0(\omega)-k_s(\nu_{s0})}\frac{\omega_e^2\Omega_1}{\omega\nu_{s0}\Omega_{10}}a_s = \frac{2c^2k_0(\omega)}{\omega^2}E_0\delta(\omega-\omega_0) \tag{6.154}$$

$$\sum_{s=1}^{2}\int_{-\infty}^{\infty}\frac{d\nu}{2\pi}\frac{\delta(\Omega_{10}-\Omega_2)}{k_0(\omega)+k_s(\nu_{s0})}\frac{\omega_e^2\Omega_2}{\omega\nu_{s0}\Omega_{10}}e^{-i[k_0(\omega)+k_s(\nu_{s0})]l}a_s = 0 \tag{6.155}$$

Equation (6.155) gives

$$\frac{a_1}{\nu_{10}(k_0' + k_1)}e^{-i[k_0+k_1]l} + \frac{a_2}{\nu_{20}(k_0' + k_2)}e^{-i[k_0+k_2]l} = 0, \tag{6.156}$$

where $k_0' = k_0(\omega_r)$. After rearranging (6.154) according to

$$\delta(\Omega_{10} - \Omega_1) = \delta(\omega_0 - \omega)/\left|\partial\Omega_{10}/\partial\omega_0\right| = ck_0\delta(\omega_0 - \omega)/\left|ck_0 - \beta\omega\right|$$

and taking into account (6.156) we obtain

$$a_{1,2} = \frac{2c\nu_{1,20}}{\omega_0\omega_e^2}\frac{(ck_0 - \beta\omega_0)(k_0 - k_{1,2})E_0}{1 - \dfrac{k_0 - k_{1,2}}{k_0 - k_{2,1}}\dfrac{k_0' + k_{2,1}}{k_0' + k_{1,2}}e^{\pm i[k_2 - k_1]l}} \tag{6.157}$$

These formulae are true for any frequency in the case of a following movement and for the frequencies

$\omega_0 > \omega_k' = (1+\beta^2)\gamma^2\omega_k$ with the opposite movement. In the case $\gamma\omega_k < \omega_0 < \omega_k'$ Eq. (6.155) is an identity for any a_s as $\Omega_{10} - \Omega_2 \neq 0$. In this case the δ-function in (6.154) rearranges to $\delta(\Omega_{10} - \Omega_1) = ck_0\delta(\omega - \omega_0)/|ck_0 - \beta\omega| + \delta(\omega - \omega_r)/|\partial\Omega_1/\partial\omega_r|$. As the multiplier $1/|\partial\Omega_1/\partial\omega_r|$ must be equal to zero we obtain for a_s'

$$a_{1,2}' = \frac{2cv_{1,20}}{\omega_0\omega_e^2}\frac{(ck_0 - \beta\omega_0)(k_0 - k_{1,2})E_0}{1 - \dfrac{k_0 - k_{1,2}}{k_0 - k_{2,1}}\dfrac{k_r - k_{2,1}}{k_r - k_{1,2}}}, \tag{6.158}$$

where $k_r = k_0' = k_0(\omega_r) = \gamma^2[(1+\beta^2 ck_0(\omega_0)) - 2\beta\omega_0]/c$. Therefore the amplitudes of the inner waves do not depend on the bunch length in the frequency range $\gamma\omega_k < \omega_0 < \omega_k'$ when the wave movement is one that follows the plasma.

That fact that the amplitudes of the inner waves do not depend on the bunch length is a consequence of so-called relativistic effect of cutoff [18]. It is shown in this work that transformation of running harmonic wave parameters ω', $k' = \sqrt{n^2(\omega'/c)^2 - \Gamma^2}$ into a moving reference frame gives new parameters $\omega = \gamma(\omega' + uk')$, $k = \gamma(k' + u\omega'/c^2)$. Here Γ is determined by the lateral sizes of the waveguide and n is refractive index. If $uk' < 0$ then it is possible to satisfy the equality $k = 0$ with the frequency $\omega' = c\Gamma/\sqrt{n^2 - \beta^2}$ in the waveguide frame of rest. This frequency exceeds the waveguide cutoff frequency $c\Gamma/n$ in the waveguide frame of rest. Therefore, a running wave of frequency ω appears in the moving reference frame, and this wave does not depend on the longitudinal coordinate; that is all components of the field have the same phase on the longitudinal axis. It is natural that in this case the field does not react with an inhomogeneity placed along the waveguide. However, as the frequencies on both sides of the moving parameter discontinuity are different, the amplitudes on both sides of the moving parameter discontinuity are different also; this is caused by the equality of tangential field components on the discontinuity. This is confirmed by Eq. (6.158). Simple analysis of overcritical frequencies in the waveguide confirms that. Indeed, in the frame of rest of the waveguide the propagating waves can exist only if $\omega' > \omega_k$ as $k' = \sqrt{\omega'^2 - \omega_k^2}/c$.

Transition to the frame moving opposite to these waves gives $\omega = \gamma(\omega' + u\sqrt{\omega'^2 - \omega_k^2}/c)$, $k = \sqrt{\omega^2 - \omega_k^2}/c = \gamma(k' + u\omega'/c^2)$. It is seen that if $\omega' > \omega_k$ then $\omega > \gamma\omega_k$ whereas the dispersion relation admits the existence of the running waves with frequencies $\omega_k < \omega \le \gamma\omega_k$. To provide such frequencies requires that the frequency in the frame of rest $\omega' = \gamma(\omega - u\sqrt{\omega^2 - \omega_k^2}/c)$ takes values in the interval $\omega_k/\gamma < \omega' > \omega_k$. These values are less than the critical frequency.

Thus, the analysis shows that the presence of a moving medium in dispersive systems cannot satisfy the assumptions of the idealized problem statement. Such a statement requires differentiation of moments at the beginning of the field excitement and at the beginning of the medium movement beginning (or upon switching on any other non-stationarity).

The frequencies and the wave numbers of the inner waves are real if the satisfy the condition $\omega_0 > \omega^* = \gamma(\omega_{ke} + \beta\omega_e) > \gamma\omega_k$. In the case when the electromagnetic wave and plasma movements are in oppose directions the running waves exist in the range $\omega_0 > \omega_k$ if $\omega_{ke} > \gamma\omega_k$, and in the range $\omega_0 > \omega^*$ if $\omega_{ke} > \gamma\omega_k$.

In the case $\Omega_1^2 < \omega_{ke}^2/\gamma^2$ the frequencies ν_s as well as the wave numbers k_s become complex, that is the regime of running waves becomes impossible. It means physically that the frequency of the boundary-source $\omega_s = \gamma\Omega_{10}$ (see Sect. 6.4) is less the critical frequency of the loaded part of the waveguide ω_{ke}. This condition is fulfilled if $\omega_0 > \omega^*$ for the following movement. In the case of the opposite movement, the regime of cutoff is met when $\omega_0 > \omega_k$ if $\omega_{ke} > \gamma\omega_k$ and $\omega_0 > \omega^*$ if $\omega_{ke} > \gamma\omega_k$.

The inner field found allows one to calculate the field outside the bunch using the formulae (6.140) and (6.141). These fields contain one running wave for all frequencies $\omega_0 > \omega_k$ in the opposite movement and for $\omega_0 > \gamma\omega_k$ in the following movement. Calculation of (6.140) and (6.141) with the explicit expressions for $N_s^{\pm}(\omega,\nu)$ (6.40) gives for the transmitted wave

$$E_t = \frac{\omega_0\omega_e^2}{c(ck_0 - \beta\omega_0)}\left\{a_1\frac{e^{i[k_0-k_1]l} - 1}{\nu_1(k_0 - k_1)} + a_2\frac{e^{i[k_0-k_2]l} - 1}{\nu_2(k_0 - k_2)}\right\} \tag{6.159}$$

and for the wave reflected from the bunch

$$E_r = \frac{\omega_0 \omega_e^2}{c(ck_0 - \beta\omega_0)} \left\{ a_1 \frac{1 - e^{-i[k_0 + k_1]l}}{v_1(k_0 + k_1)} + a_2 \frac{1 - e^{-i[k_0 + k_2]l}}{v_2(k_0 + k_2)} \right\}. \quad (6.160)$$

The transmitted wave has the frequency ω_0, and the reflected wave has the frequency ω_r, (6.146). The formulae (6.159) and (6.160) are true for $\omega_0 > \omega_k$ in the case of opposite movement and for $\omega_0 > \omega_k'$ in the case of following movement. To obtain the formulae for the frequency range $\gamma\omega_k < \omega_0 < \omega_k'$, one must substitute a_s by a_s' (6.158) in (6.159), (6.160) and $k_0 \to -k_0$ as the reflected wave propagates towards the running away bunch.

It is worth noting the analogy between the interaction of a waveguide wave with a plane-parallel plasma bunch in a waveguide and interaction of a plane wave impinging on the same layer at an angle θ in free space. Indeed, cosine of the angle for propagation of the Brillouin waves in the waveguide is equal to $\Lambda = \sqrt{1 - \omega_k^2 / \omega^2}$. If one substitutes Λ for cos θ in (6.150), (6.151), (6.157), (6.159) and (6.160) these formulae transform into formulae determining waves scattered by a plasma layer in the free space [41].

6.6.2 Characteristic Matrix for Waves in a Waveguide with a Plasma Layer

The interaction of an electromagnetic wave with a plasma bunch moving in a waveguide considered above can be used for amplitude amplification and frequency shift of an electromagnetic wave reflected from a moving medium boundary. It is a common practice to characterize the efficiency of this reflection by the ratio of a boundary velocity to a wave phase velocity. However, in a dispersive structure the ratio of a boundary velocity to a wave group velocity is of prime importance. It appears most clearly in a waveguide when a double dispersion mechanism exists. The importance of this mechanism for the energy characteristics of reflected electromagnetic waves will be shown further.

To investigate the interaction of an electromagnetic wave with a plasma bunch moving in a waveguide in more detail we use another approach based on a characteristic matrix. For that we consider the problem statement given in Sect. 6.1 when a single guided TE wave $\mathbf{E}_0(t,x)=\mathbf{b}_\perp E_0\exp(i(\omega t-k_0x))$, $k_0(\omega)=\frac{1}{c}\sqrt{\omega^2-\omega_k^2}$, which is characterized by the waveguide frequency ω_k, impinges from the empty left side of the waveguide onto a plasma bunch moving in a waveguide. The homogeneous cold plasma in the bunch is characterized by a plasma frequency ω_e. With the geometry of the problem considered there is no mode transformation.

An electromagnetic field in the waveguide can be written as follows:

$$\mathbf{E}=E(t,x)\mathbf{b}_\perp,\ \mathbf{B}_\perp=B(t,x)[\mathbf{e}_x,\mathbf{b}_\perp] \tag{6.161}$$

where $\mathbf{b}_\perp$ is the transverse eigen-vector for the mode considered, $\mathbf{e}_x$ is the x-axis unit vector directed along the waveguide, and $B=-c\int_{-\infty}^{t}\frac{\partial E}{\partial x}dt$ follows from Maxwell's equations. It is emphasized here that the incident wave interacts with the bunch at any frequency ω if the wave and the bunch move in opposite directions, $u<0$. If $u>0$ then this interaction takes place for $\omega>\gamma\omega_k$ only. In the case $\omega<\gamma\omega_k$ and $u>0$ the incident wave does not catch up the bunch as its group velocity is less than the bunch velocity. In the following discussion only the movement in opposite directions will be considered.

The fields must satisfy the boundary conditions on the moving boundary

$$\mathbf{n}\times[\mathbf{E}]=\beta[\mathbf{B}_\perp],\qquad \mathbf{n}\times[\mathbf{H}]=-\beta[\mathbf{D}_\perp], \tag{6.162}$$

where $\mathbf{H}=\mathbf{B}-4\pi\mathbf{M}$, $\mathbf{D}=\mathbf{E}+4\pi\mathbf{P}$, $\mathbf{P}$ and $\mathbf{M}$ are electrical and magnetic polarizations of the medium, $\boldsymbol{n}$ is the normal unit vector to the moving boundary of the plasma bunch; the subscript $\perp$ denotes a component that is normal to the vector $\boldsymbol{n}$, $\boldsymbol{\beta}=\mathbf{u}/c$, $\boldsymbol{u}$ is the velocity of the plasma and its boundaries and the brackets $[\cdot]$ denote a jump of the discontinuous vector function on the boundary. Constitutive relations for the moving medium, which is an isotropic nonmagnetic one at rest, are:

$$\mathbf{P}=\gamma^2\alpha'(\mathbf{E}-\beta(\beta,\mathbf{E})+\beta+\mathbf{B}),\ \mathbf{M}=-\beta\times\mathbf{B}, \tag{6.163}$$

where $\gamma^2 = (1 - \beta^2)^{-1}$ is the relativistic factor, and α' is the susceptibility of the medium at rest.

It follows from the boundary conditions that the field combinations $E - \beta B$ and $B - \beta E$ are continuous across the moving bunch boundaries.

For determining the field in the bunch, and on its boundaries, the characteristic matrix method is used [20, 21]. The field in the bunch consists of two waves $E \sim e^{i(\nu_1 t - k_1 x)}$, $e^{i(\nu_2 t + k_2 x)}$ from which the continuous combinations mentioned above have the following forms:

$$\mathbf{Q}_1 = \begin{pmatrix} B - \beta E \\ E - \beta B \end{pmatrix} = \frac{1}{\nu_1} \begin{pmatrix} ck_1 - \beta \nu_1 \\ \nu_1 - uk_1 \end{pmatrix} e^{i(\nu_1 t - k_1 x)},$$
$$\mathbf{Q}_2 = \begin{pmatrix} B - \beta E \\ E - \beta B \end{pmatrix} = \frac{1}{\nu_2} \begin{pmatrix} -ck_2 - \beta \nu_2 \\ \nu_2 + uk_2 \end{pmatrix} e^{i(\nu_2 t + k_2 x)} \tag{6.164}$$

These vector-functions are linearly independent. It is more convenient to use other linearly independent vector-functions which can be constructed as follows:

$$\mathbf{Q}_1^* = a_{11}\mathbf{Q}_1 + a_{12}\mathbf{Q}_2, \ \mathbf{Q}_2^* = a_{21}\mathbf{Q}_1 + a_{22}\mathbf{Q}_2. \tag{6.165}$$

These two vectors set up a fundamental matrix

$$N(t,x) = (\mathbf{Q}_1^* \ \mathbf{Q}_2^*), \tag{6.166}$$

in which the constants a_{ij} are chosen so that the fundamental matrix links the field in the arbitrary bunch cross-section with the field on its boundary $x = ut$:

$$\mathbf{Q}(t,x) = N(t,x)\mathbf{Q}(t,ut) \tag{6.167}$$

It means that $N(t,ut) = \begin{pmatrix} 1 & 0 \\ 0 & 1 \end{pmatrix}$. From this and (6.166) it follows that

$$
\begin{aligned}
N_{11} &= \gamma^2\left(\frac{(ck_1-\beta v_1)(v_2+uk_2)}{c(v_1k_2+v_2k_1)}e^{-ik_1\xi}+\frac{(v_1-uk_1)(ck_2+\beta v_1)}{c(v_1k_2+v_2k_1)}e^{ik_2\xi}\right),\\
N_{12} &= \gamma^2\frac{(ck_1-\beta v_1)(ck_2+\beta v_1)}{c(v_1k_2+v_2k_1)}(e^{-ik_1\xi}-e^{ik_2\xi}),\\
N_{21} &= \gamma^2\frac{(v_1-uk_1)(v_2+uk_2)}{c(v_1k_2+v_2k_1)}(e^{-ik_1\xi}-e^{ik_2\xi}),\\
N_{22} &= \gamma^2\left(\frac{(v_1-uk_1)(ck_2+\beta v_1)}{c(v_1k_2+v_2k_1)}e^{-ik_1\xi}+\frac{(ck_1-\beta v_1)(v_2+uk_2)}{c(v_1k_2+v_2k_1)}e^{ik_2\xi}\right),
\end{aligned}
\tag{6.168}
$$

where $\xi = x - ut$.

The matrix M that is the inverse of the matrix N is called the characteristic matrix [20, 21]. It links the field values on the boundary $x = ut$ with the ones in the arbitrary bunch cross-section

$$\mathbf{Q}(t, ut) = M(\xi)\mathbf{Q}(t,x), \tag{6.169}$$

where $M(t,x) = \begin{pmatrix} M_{11} & M_{12} \\ M_{21} & M_{22} \end{pmatrix} = N^{-1}(t,x) = \begin{pmatrix} N_{22} & -N_{12} \\ -N_{21} & N_{11} \end{pmatrix}$. (6.170)

The elements of the matrix M are

$$
\left.
\begin{aligned}
M_{11} &= \gamma^2\left(\frac{(v_1-uk_1)(ck_2+\beta v_1)}{c(v_1k_2+v_2k_1)}e^{-ik_1\xi}+\frac{(ck_1-\beta v_1)(v_2+uk_2)}{c(v_1k_2+v_2k_1)}e^{ik_2\xi}\right)\\
M_{12} &= -\gamma^2\frac{(ck_1-\beta v_1)(ck_2+\beta v_1)}{c(v_1k_2+v_2k_1)}(e^{-ik_1\xi}-e^{ik_2\xi})\\
M_{21} &= -\gamma^2\frac{(v_1-uk_1)(v_2+uk_2)}{c(v_1k_2+v_2k_1)}(e^{-ik_1\xi}-e^{ik_2\xi})\\
M_{22} &= \gamma^2\left(\frac{(ck_1-\beta v_1)(v_2+uk_2)}{c(v_1k_2+v_2k_1)}e^{-ik_1\xi}+\frac{(v_1-uk_1)(ck_2+\beta v_1)}{c(v_1k_2+v_2k_1)}e^{ik_2\xi}\right)
\end{aligned}
\right\},
\tag{6.171}
$$

where $\xi = x - ut$. Note that $\det N - \det M = 1$.

The frequencies and the wave numbers of the waves in the bunch are subject to the dispersion relation for the plasma moving in a waveguide. This relation can be derived from Maxwell's equations if one takes into account that for fields which are proportional to $\exp(i(\omega t - \mathbf{kr}))$ the constitutive relation (6.163) has the following form:

$$\mathbf{P} = \gamma 2\alpha'(\omega,\mathbf{k})\left\{\left(1-\frac{c}{\omega}(\mathbf{k},\beta)\right)\mathbf{E} - \mathbf{b}(\beta,\mathbf{E}) + \frac{c}{\omega}\mathbf{k}(\beta,\mathbf{E})\right\},\quad \mathbf{M} = -\beta\times\mathbf{P}, \tag{6.172}$$

where the equation $\mathbf{B} = (c/\omega)\mathbf{k} \times \mathbf{E}$ has been used. For a cold plasma at rest, it holds that $\alpha'(\omega, k) = \alpha'(\omega') = -\omega_e^2/(4\pi\omega'^2)$ where $\omega' = \gamma(\omega - (\mathbf{u},\mathbf{k}))$ is the wave frequency in the coordinates of resting medium, and ω_e is the Lorentz-covariant plasma frequency. If a transverse plane wave propagates along the direction of the medium velocity then these constitutive relations become simpler

$$\mathbf{P} = -\frac{\omega_e^2}{4\pi\omega(\omega-uk)}\mathbf{E},\qquad \mathbf{M} = -\beta\times\mathbf{P}. \tag{6.173}$$

These relations enable one to obtain the dispersion equation for the waves in the plasma moving in a waveguide

$$\left[1+(1+4\pi\alpha')\beta^2\right]k^2 + 2\pi\alpha'\beta\frac{\nu}{c}k - \left[1+4\pi\alpha' - \beta^2\right]\frac{\nu^2}{c^2} + \frac{\omega_k^{\ 2}}{c^2\gamma^2} = 0, \tag{6.174}$$

where $4\pi\alpha' = -\omega_e^2\gamma^{-2}(\nu - uk(\nu))^{-2}$. The last expression yields

$$k_{1,2}(\nu) = \pm\frac{1}{c}\sqrt{\nu^2 - \omega_{ke}^2}, \tag{6.175}$$

where $\omega_{ke}^2 = \omega_k^2 + \omega_e^2$.

Wave phase equalities $\nu - uk_i(\nu) = \Omega_2(\nu)$ on the bunch boundaries can be used to find the frequencies and the wave numbers for the internal waves:

$$\begin{aligned} \nu_{1,2} &= \omega(\gamma^2(1-\beta\Lambda) \pm \beta\gamma\delta\sqrt{q^2-1}), \\ k_{1,2} &= \frac{\omega}{c}(\pm\gamma^2\beta(1-\beta\Lambda) + \gamma\delta\sqrt{q^2-1}), \end{aligned} \tag{6.176}$$

where $\delta = \omega_e/\omega$ is the plasma factor, $q = \gamma(\Lambda - \beta)/\delta$ and $\Lambda^{-1} = (\sqrt{1-(\omega_k/\omega)^2})^{-1}$ is a waveguide factor for the mode considered. It follows from (6.176) that the plasma bunch is non-transparent if $q < 1$.

Taking into account these expressions one can derive the characteristic matrix from (6.171)

$$M = e^{-i\frac{\gamma^2\beta\Omega l}{c}} \begin{pmatrix} \cos\varphi & -i\dfrac{c\kappa}{\Omega_2}\sin\varphi \\ -i\dfrac{\Omega_2}{c\kappa}\sin\varphi & \cos\varphi \end{pmatrix}, \tag{6.177}$$

where $\kappa = c^{-1}\sqrt{\Omega_2^2 - \gamma^{-2}\omega_{ke}^2}$, $\varphi = \gamma^2\kappa l = \gamma\delta\omega c^{-1} l\sqrt{q^2-1}$.

The matrix (6.144) allows one to obtain expressions for characteristics of the scattered waves. If R is the reflectance of a plane wave from the bunch and T is the external transmittance of a plane wave through the bunch then the vector $\mathbf{Q}$ on the outer bunch boundaries will be represented by formulae:

$$\mathbf{Q}_R = \frac{1}{\omega}\begin{pmatrix} ck_0 - \beta\omega \\ \omega - uk_0 \end{pmatrix} E_0 e^{i(\omega - uk_0(\omega))t} + \frac{R}{\omega_r}\begin{pmatrix} -ck_r - \beta\omega_r \\ \omega_r + uk_r \end{pmatrix} E_0 e^{i(\omega_r + uk_r)t},$$

$$\mathbf{Q}_T = T\begin{pmatrix} ck_t - \beta\omega_t \\ \omega_t - uk_t \end{pmatrix} E_0 e^{i(\omega_2 - uk_t)t - ik_t l} \tag{6.178}$$

where $k_{r,t} = k_0(\omega_{r,t})$, $\omega_{r,t}$ are the frequencies of the reflected and the transmitted waves, respectively, and l is the bunch length. As the vector $\mathbf{Q}$ is continuous across the bunch boundaries so the frequencies of the reflected and transmitted waves and the reflectance and the external transmittance can be derived from (6.169) which will be written in this case as $\mathbf{Q}_R = M(l)\mathbf{Q}_T$. This gives the expressions for reflectance and external transmittance

$$R = \frac{\omega_r}{\omega}\frac{\gamma^2\beta\Omega(\omega-\gamma^2\Omega)(N_{11}-N_{22}) - (\omega-\gamma^2\Omega)^2 N_{21} + \gamma^4\beta^2\Omega^2 N_{12}}{\gamma^2\beta\Omega(\omega-\gamma^2\Omega)(N_{11}+N_{22}) - (\omega-\gamma^2\Omega)^2 N_{21} - \gamma^4\beta^2\Omega^2 N_{12}},$$

$$T = \frac{2\gamma^2\beta\Omega(\omega-\gamma^2\Omega)}{\gamma^2\beta\Omega(\omega-\gamma^2\Omega)(N_{11}+N_{22}) - (\omega-\gamma^2\Omega)^2 N_{21} - \gamma^4\beta^2\Omega^2 N_{12}}, \tag{6.179}$$

where $\Omega = \omega - uk_0(\omega)$.

The frequency of the reflected wave $\omega_r = \gamma^2[(1+\beta^2)\omega - 2uk_0(\omega)]$ allows one to introduce the frequency multiplication coefficient (FMC) F for the reflected wave as

$$F = \omega_r/\omega = (1-2\beta\Lambda+\beta^2)/(1-\beta^2), \tag{6.180}$$

where $\Lambda^{-1} = (\sqrt{1-(\omega_k/\omega)^2})^{-1}$ is a waveguide factor for the mode considered.

The frequency of the external transmitted wave is equal to the frequency of the initial wave, $\omega_t = \omega$, when $u < 0$.

6.7 Frequency Multiplication and Amplitude Amplification

At first we analyze the frequency multiplication coefficient (FMC) for the reflected wave which is determined by (6.180)

$$F = \omega_r/\omega = (1-2\beta\Lambda+\beta^2)/(1-\beta^2) \tag{6.181}$$

The frequency of the external transmitted wave is equal to the frequency of the initial wave, $\omega_t = \omega$, when $u < 0$, i.e., when the plasma bunch moves towards the incident wave propagation.

The frequency multiplication coefficient (FMC) for the reflected wave $F = (1-2\beta\Lambda+\beta^2)/(1-\beta^2)$ does not depend on the interior parameters of the bunch (length and plasma frequency). It is determined by the bunch velocity u and the waveguide factor Λ^{-1}. The magnitude $\Lambda^{-1} = 1$ corresponds to an infinite space and the maximum influence of the waveguide corresponds to $\Lambda^{-1} \to \infty$. Figure 6.13 demonstrates the dependence of FMC on the bunch velocity for various values of the waveguide factor.

Unlike the infinite medium the FMC in the waveguide is determined only for bunch velocities that satisfy the inequalities $-1 < \beta \le \beta_0$ where $\beta_0 = \Lambda$ is the velocity of the initial-wave-bunch synchronism when the initial wave catches up with the bunch. The FMC inside the interval $(0, \beta_0)$ is less than unity reaching its minimum, $F_{min} = \sqrt{1-\Lambda^2}$, at the point $\beta_{min} = (1-\sqrt{1-\Lambda^2})\Lambda^{-1}$. In the case of the meeting movement ($\beta < 0$) the FMC is greater than one, tending to infinity when $\beta \to -1$. Such behavior of the

FMC agrees with the behavior of the group velocity of the reflected wave,

$$v_{gR} = c(2\beta - (1+\beta^2)\Lambda)(1-2\beta\Lambda+\beta^2)^{-1} \tag{6.182}$$

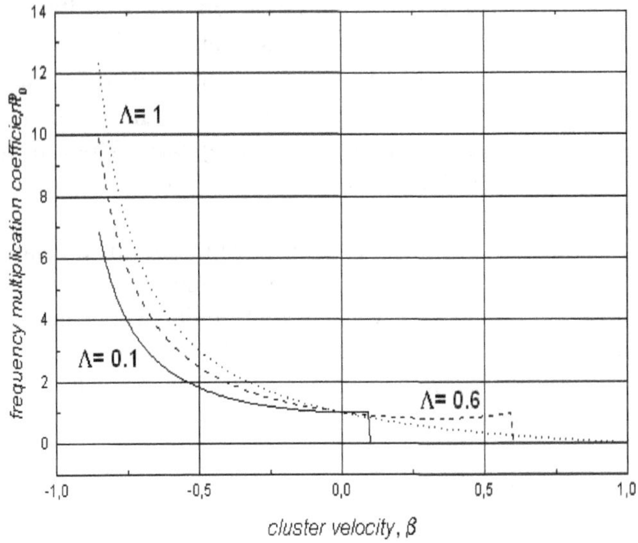

Figure 6.13 Frequency multiplication coefficient versus the bunch velocity.

In the case when $\beta = \beta_{\min}$ the group velocity is equal to zero, $v_{gR} = 0$, meaning that the reflected field represents itself a standing wave which is left by the bunch in the region it passed through. For $\beta = \beta_{\min}$ the group velocity is positive, $v_{gR} > 0$. It means that the reflected wave propagates in the direction of the outgoing bunch boundary. The reflected wave moves together with the boundary if $\beta = \beta$, and consequently $v_{gR} = u$. In the case of a meeting movement of the bunch and the wave, $\beta < 0$, the reflected wave is always propagating away from the bunch boundary, $v_{gR} < -c\Lambda$, where $c\Lambda$ is the group velocity of the reflected wave when the bunch movement is absent.

Expressions for the reflectance R and the external transmittance T follow from (6.179) after substitution of N_{ij} (6.168)

$$R = \frac{Pf2i\sin\varphi}{(1-f^2)\cos\varphi + i(1+f^2)\sin\varphi}, \quad T = \frac{(1-f^2)\exp(i\gamma\delta\omega c^{-1}ql)}{(1-f^2)\cos\varphi + i(1+f^2)\sin\varphi}, \tag{6.183}$$

where $f = (q - \sqrt{q^2 - 1})^2$. It must be emphasized here that the value

$$R_0 = f = (q - \sqrt{q^2 - 1})^2. \tag{6.184}$$

presents the reflectance of a semi-infinite plasma bunch.

Let us now focus our attention upon the velocity interval $(-1, \beta_0)$, particularly to the sub-interval (β_1, β_0) where $\beta_1 = (\Lambda - \delta\sqrt{\delta^2 - \Lambda^2 + 1})(\delta^2 + 1)^{-1}$. The interior field consists of the damped waves in this subinterval because the frequency and wavenumber in (6.176) acquires imaginary parts. The reflectance of the semi-infinite bunch $R_0 = F(q - \sqrt{q^2 - 1})^2$ is complex with $(R_0) = F$ in the sub-interval (β_1, β_0).

The value of $|R_0|$ peaks at $\beta \approx \beta_1$ which is not a relativistic value. One can see from Fig. 6.14, which shows the reflectance of the semi-infinite bunch for various waveguide factors and plasma densities, that the maximum values of $|R_0|$ are reached at $\beta \approx \beta_1$ when this velocity can be far below relativistic values that is shown in Table 6.1.

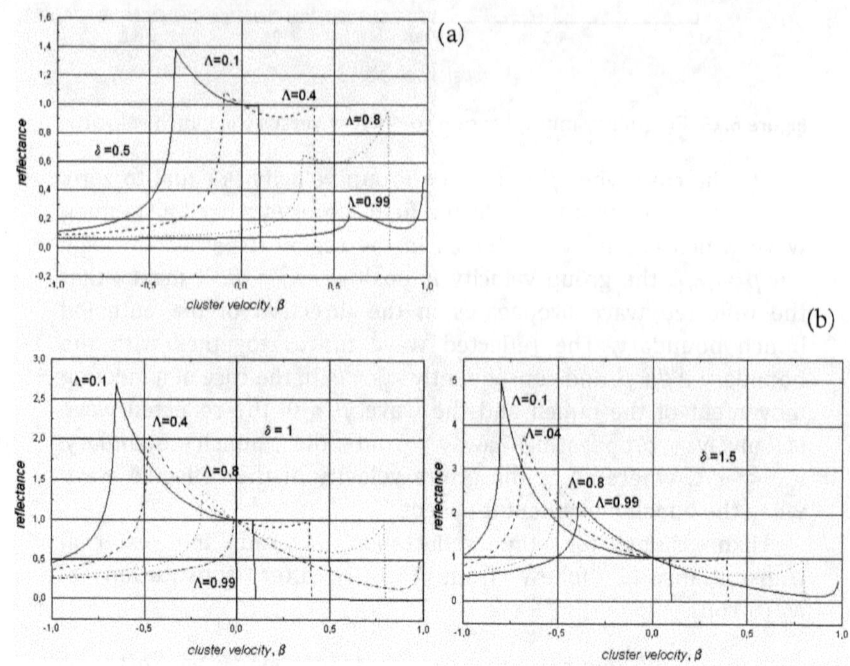

Figure 6.14 Reflectance of the semi-infinite bunch for various waveguide factors and plasma densities: (a) small plasma density, (b) large plasma density.

Table 6.1 The β_1 values for various plasma densities δ

	$\delta = 0.5$	$\delta = 1.0$	$\delta = 1.5$
$\Lambda^{-1} = 10$	0.365	0.655	0.800
$\Lambda^{-1} = 2.5$	−0.097	−0.472	−0.688
$\Lambda^{-1} = 1.25$	0.327	−0.183	−0.499
$\Lambda^{-1} = 1.001$	0.584	−0.090	−0.390

The reflectance of the finite length bunch differs from the semi-infinite one by a factor whose character strongly depends on the position of the bunch velocity with respect to the indicated subinterval. Inside this subinterval the reflectance is determined by the expression

$$R = \frac{R_0 2\sinh\varphi^*}{e^{\varphi^*} - (q + i\sqrt{1-q^2})^4 e^{-\varphi^*}}, \tag{6.185}$$

where $\varphi^* = \gamma\delta\omega c^{-1} l\sqrt{1-q^2}$.

Outside this subinterval the reflectance is given by

$$R = \frac{R_0 2i\sin\varphi}{e^{l\varphi} - (q - \sqrt{q^2-1})^4 e^{-l\varphi}}, \tag{6.186}$$

where $\varphi = \gamma\delta\omega c^{-1} l\sqrt{q^2-1}$. Dependence of the reflectance on the bunch velocity is shown in Fig. 6.15.

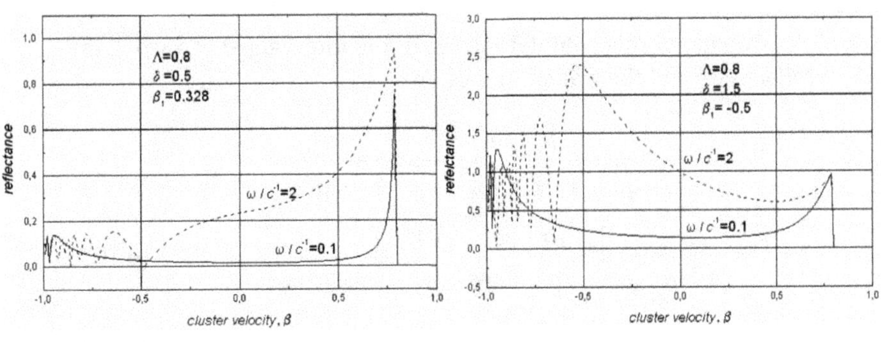

Figure 6.15 Reflectance versus the bunch velocity for a finite length bunch.

6.7.1 Enhanced Reflectivity from the Moving Plasma Bunch

The reflectivity $\overline{R}$ and the transmittance $\overline{T}$ of a bunch are determined by the relations

$$\overline{R} = S_R / S_0 = RR^* v_{gR} / v_{g0}, \quad \overline{T} = S_T / S_0 = TT^*, \tag{6.187}$$

where S_0, S_R, S_T are the energy fluxes of incident, reflected and transmitted waves, respectively, and v_{g0}, v_{gR}, v_{gT} are the group velocities of these waves. Here, it has been taken into account that $|v_{gT}| = |v_{g0}|$. The energy fluxes are determined by the component of the Poynting vector directed along the waveguide (in Gaussian units),

$$S = \frac{c}{8\pi}\mathrm{Re}[\mathbf{EH}^*]_x = \frac{c^2 k}{8\pi\omega}EE^* = \frac{v_g}{8\pi}EE^* \tag{6.188}$$

where k is the longitudinal wave number and v_g is the group velocity of the wave considered.

The reflectivity and the transmittance in the subinterval of non-transparency ($\beta_1 \le \beta \le \beta_0$) are:

$$\overline{R} = \frac{\overline{R}_0 \sinh^2\varphi^*}{4q^2(1-q^2)+\sinh^2\varphi^*}, \quad \overline{T} = \frac{4q^2(1-q^2)}{4q^2(1-q^2)+\sinh^2\varphi^*}, \tag{6.189}$$

where $\overline{R}_0 = \dfrac{(1+\beta^2)\Lambda - 2\beta}{\Lambda(1-2\beta\Lambda+\beta^2)}P^2$ (6.190)

is the reflectivity of the semi-infinite bunch in the subinterval considered.

Outside of this subinterval, i.e., in the interval ($-1 < \beta < \beta_1$), the reflectivity and the transmittance are:

$$\overline{R} = \frac{\overline{R}_0 4\sin^2\varphi}{[1-(q-\sqrt{q^2-1})^4]^2 + 4(q-\sqrt{q^2-1})^4\sin^2\varphi},$$

$$\overline{T} = \frac{[1-(q-\sqrt{q^2-1})^4]^2}{[1-(q-\sqrt{q^2-1})^4]^2 + 4(q-\sqrt{q^2-1})^4\sin^2\varphi}, \tag{6.191}$$

where $\bar{R}_0 = \dfrac{(1+\beta^2)\Lambda - 2\beta}{\Lambda(1-2\beta\Lambda+\beta^2)}P^2(q-\sqrt{q^2-1})^4$ (6.192)

is the reflectivity of the semi-infinite bunch outside this subinterval.

The direction of the reflected energy flux is determined by the sign of $\bar{R}_0$ or, namely, by the direction of the group velocity of the reflected wave. For $\beta > \beta_{\min}$ the sign of $\bar{R}_0$ is negative which corresponds to the movement of the reflected energy flux in the direction of the bunch. A synchronism appears and the reflected energy moves together with the bunch if $\beta = \beta_0 = \Lambda$. The movement of the reflected energy is absent and a standing wave is created if $\beta = \beta_{\min}$. For $\beta < \beta_{\min}$ the sign of $\bar{R}_0$ is positive and the reflected energy moves outwards from the bunch. If $\beta_{\min} > 0$, i.e., in the case where the wave and bunch movements are in meeting directions, the reflected energy flux is always directed backwards from the bunch.

The value of $\bar{R}_0$ changes from –1 to 0 if the bunch velocity changes from the synchronism value β_0 to the standing wave value $\beta_{\min}$. The left boundary of the non-transparent subinterval β_1 is positive for a small plasma density and for the waveguide factor Λ differing sufficiently from zero. Then, the value of $\bar{R}_0$ is very small on the left side of β_1, reaching the values $\bar{R}_0|_{\beta=0} = (\Lambda - \sqrt{\Lambda^2-\delta^2})^4\delta^{-4}$ and $\bar{R}_0|_{\beta=-1} = \delta^4/(4\Lambda(1+\Lambda)^2)$. For instance, $\bar{R}_0 \approx 10^{-4}$ if $\delta \approx 0.1$, $\Lambda \approx 0.5$. With an increase in the waveguide factor Λ^{-1}, even for a small plasma density, it may appear that β_1 becomes negative. In this case, $\bar{R}_0|_{\beta=0} = 1$, the value of $\bar{R}_0|_{\beta=0} = -1$ is as it was above, and the reflectivity peaks near β_1. At this point $\bar{R}_0|_{\beta_1} = 4\delta\Lambda^{-1}\sqrt{1-\Lambda^2} - 3$. For a bunch of a high plasma density, meaning that $\delta \geq 1$, the influence of the waveguide becomes greater, which is illustrated in Fig. 6.16b. The strong influence of the waveguide parameters can be explained by the fact that the group velocity of the incident wave tends to zero with $\Lambda^{-1} \to \infty$; however, the group velocity of the reflected wave does not tend to zero in this case.

The reflectivity of a finite-length bunch possesses similar properties to that of a semi-infinite one, and its behavior is shown in Fig. 6.17.

The above investigations have revealed that the maximum reflectivity of the electromagnetic wave in the waveguide is observed for smaller values of the bunch velocity than relativistic ones. The value of this plasma bunch velocity depends on the parameters of the plasma and the waveguide, and can be very small. The ratio of the frequency of the transformed field to the incident wave frequency is more than unity, and so this effect of the frequency change can be called an effective multiplication of the frequency, if the velocity of the bunch movement in the waveguide is slow. The values of the significant parameters mentioned above are presented in Table 6.2.

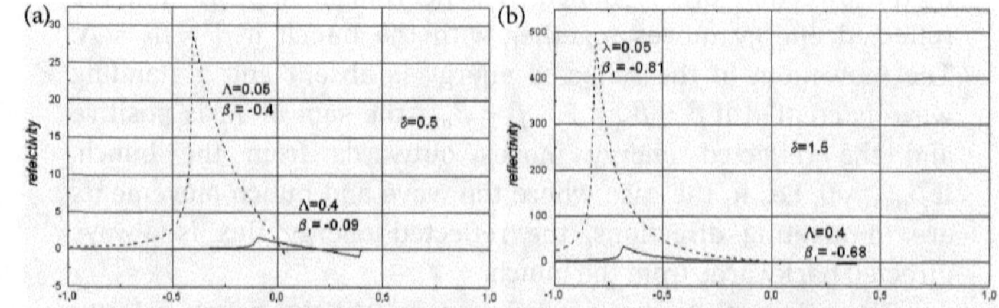

Figure 6.16 Reflectivity of the semi-infinite bunch for the various waveguide factors and various plasma densities: (a) small plasma density, (b) large plasma density.

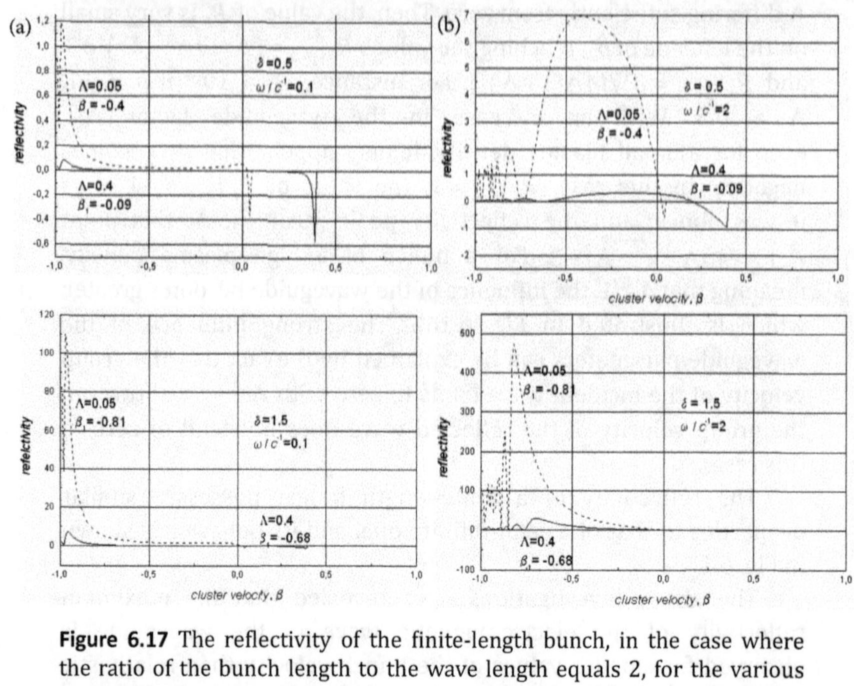

Figure 6.17 The reflectivity of the finite-length bunch, in the case where the ratio of the bunch length to the wave length equals 2, for the various waveguide factors and the various plasma density: (a) the small plasma density, (b) the large plasma density.

Table 6.2 The reflectivity of the half-infinite bunch for different parameters of the plasma

Frequency multiplication coefficient, P	The relativistic factor, γ	The waveguide factor, Λ^{-1}	The plasma factor, δ	Reflectivity of the semi-infinite cluster, $\overline{R}_0$
2	1.05	1.25	1.25	4.5
2	1.12	2.5	1.1	7.2
2	1.21	20	.073	70.0
10	1.84	1.25	3.0	125.0
10	1.96	2.5	2.55	225.0
10	2.3	20	2.18	2000.0

The reflectivity attains its maximum when $q \approx 1$, which yields the formula $\beta_{\max} = (f\sqrt{f^2-1} - D\sqrt{1+D^2})/(D^2+f^2)$, $f = \omega/\omega_k$, $D = \omega_e/\omega_k$ for the plasma velocity. This condition can hold for a small value of the velocity and a suitable adjustment of the plasma and waveguide parameters can provide a negative value for $\beta_{\max}$, Fig. 6.15, when the frequency multiplication occurs. As in this case ω_r/ω is greater than one, then this can be considered as an effective multiplication of the frequency. A comparison with the case of the (non-dispersive) dielectric moving cluster, Section 6.4, shows that the shift of the reflectivity maximum to small values of the cluster velocity owes only to the existence of the double dispersion mechanism, the plasma dispersion and the waveguide dispersion, as in the case of a dielectric cluster.

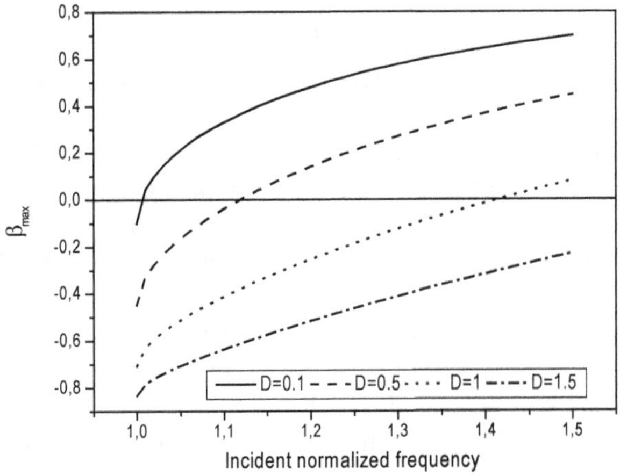

Figure 6.18 The plasma velocity providing the maximum value for the reflectivity.

Comparing these results with the case of a dielectric (nondispersive) bunch, where the shift of plasma bunch velocity, β_1, corresponding to the maximum reflectivity of the electromagnetic wave in the waveguide is absent, one can conclude that the shift of the reflectivity maximum towards smaller values of the bunch velocity is due to the presence of a double dispersion mechanism, namely, the plasma dispersion and the waveguide dispersion.

6.8 Resonance Effects in a Stratified Plasma Cluster Moving in a Waveguide

The effectiveness of wave reflection from a plasma bunch moving in a waveguide can be increased using the resonance effects due to the stratified structure of the cluster. Investigation of such a phenomenon using the characteristic matrix method is presented in this section. The matrix for one plasma slab moving in a waveguide is generalized for a double plasma slab. Such a composite slab is characterized by two thicknesses and two values of plasma frequency. A stratified cluster is modeled by a package of such double slabs. The reflectance of this cluster depends on the cluster velocity and the structure coefficient, which is determined by slab thickness, its plasma density and a package function which is determined by the number of the double plasma slabs. In the case of quarter-wavelength slabs the modulus of the structure coefficient tends to unity with an increase in the number of the slabs and a sharpening of the stratification. The structure coefficient is equal to zero when the stratification is smooth as well as in the case of half-wavelength slabs.

It will be shown that for a weak stratified cluster the structure coefficient gets its maximum, which is equal to unity, if a resonance condition for an effective thickness is satisfied. This effective thickness can be controlled by varying the slab number.

For a strongly stratified cluster, the reflectance depends on the ordering of the slabs, namely which slab has the denser plasma, the first one or the second one. If the first slab density is smaller than the second slab one then the reflectance of the package is greater than the reflectance of the double slab. In the reverse case, the resonance character of reflection becomes more apparent. The resonance thickness of the plasma slabs depends on the plasma density and its velocity. It will be shown that the frequency multiplication of an electromagnetic wave can be enhanced sufficiently as compared to a homogeneous bunch by virtue of a variation in the slab parameters.

6.8.1 The Characteristic Matrix for Stratified Plasma Cluster

The results obtained in the previous section were derived under the assumption that the medium moving along with the boundary is homogeneous. If the medium has a layered structure then one can expect the appearance of resonance effects. Such effects are investigated in this section for the case of the restricted stratified plasma cluster moving in a rectangular regular waveguide (see Fig. 6.19).

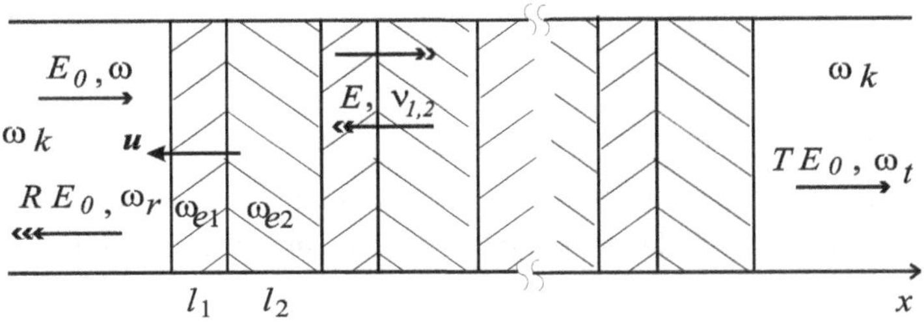

Figure 6.19 The stratified plasma cluster moving in the waveguide.

The incident electromagnetic wave is of the TE type, $\mathbf{E}_0(t,x) = \mathbf{b}_\perp E_0 \exp(i(\omega t - k_0 x))$, where $k_0(\omega) = \sqrt{\omega^2 - \omega_k^2}/c$ is the wave number in an empty waveguide. It propagates to meet the plasma cluster which moves with the velocity $u < 0$.

If the cluster is restricted by planes that are perpendicular to the waveguide axis x then the interaction of the field with the cluster can be described by the characteristic matrix (6.177):

$$M = e^{-i\frac{\gamma^2 \beta \Omega l}{c}} \begin{pmatrix} \cos\varphi & -i\dfrac{c\kappa}{\Omega}\sin\varphi \\ -i\dfrac{\Omega}{c\kappa}\sin\varphi & \cos\varphi \end{pmatrix}, \tag{6.193}$$

where $\kappa = c^{-1}\sqrt{\Omega^2 - \gamma^{-2}\omega_{ke}^2}$, $\varphi = \gamma^2 \kappa l$, $\gamma^{-2} = 1 - \beta^2$, $\beta = u/c$, $\Omega = \omega - uk_0(\omega)$.

Using this expression one can easily find the matrix of a double layer with the parameters ω_{e1}, ω_{e2}, l_1, l_2 using the multiplication of matrices [20, 21]

$$M_{21} = M_1(l_1)M_2(l_2) = e^{-i\frac{\gamma^2\beta\Omega(l_1+l_2)}{c}}\begin{pmatrix} m_{11} & m_{12} \\ m_{21} & m_{22} \end{pmatrix}, \tag{6.194}$$

where

$$m_{11} = \cos\varphi_1 \cos\varphi_2 - \frac{\kappa_2}{\kappa_1}\sin\varphi_1 \sin\varphi_2,$$

$$m_{12} = i\frac{\Omega}{c}\left(\frac{1}{\kappa_2}\cos\varphi_1 \sin\varphi_2 + \frac{1}{\kappa_1}\sin\varphi_1 \cos\varphi_2\right),$$

$$m_{21} = i\frac{c}{\Omega}(\kappa_1 \sin\varphi_1 \cos\varphi_2 + \kappa_2 \cos\varphi_1 \sin\varphi_2),$$

$$m_{11} = \cos\varphi_1 \cos\varphi_2 - \frac{\kappa_1}{\kappa_2}\sin\varphi_1 \sin\varphi_2, \tag{6.195}$$

where $\kappa_i = c^{-1}\sqrt{\Omega^2 - \gamma^{-2}\omega_{kei}^2}$, $\omega_{kei}^2 = \omega_k^2 + \omega_{ei}^2$ and $\varphi_i = \gamma^2 \kappa_i l_i$.

In the case when the medium inside the cluster is a packet of N double plasma layers the characteristic matrix is equal to

$$M_{2N} = e^{-i\frac{\gamma^2\beta\Omega N(l_1+l_2)}{c}}\begin{pmatrix} M_{11} & M_{12} \\ M_{21} & M_{22} \end{pmatrix}, \tag{6.196}$$

where the matrix elements are defined by the expressions:

$$M_{ii} = -U_{N-2}(a) + U_{N-1}(a)m_{ii}, \quad M_{ij} = U_{N-1}(a)m_{ij}. \tag{6.197}$$

The functions $U_N(a) = \dfrac{\sin(N+1)\psi}{\sin\psi}$ are the Chebyshev polynomials and their argument is

$$a = \cos\psi = \cos\varphi_1 \cos\varphi_2 - \frac{1}{2}\left(\frac{\kappa_1}{\kappa_2} + \frac{\kappa_2}{\kappa_1}\right)\sin\varphi_1 \sin\varphi_2. \tag{6.198}$$

6.8.2 Resonance Effects

The parameters of the reflected and transmitted electromagnetic waves can be easily determined using the characteristic matrix. It turns out that the frequency of the transmitted wave does not change and the frequency of the reflected wave is equal to $\omega_r = \gamma^2[(1 + \beta^2)\omega - 2uk_0(\omega)$ similarly to the case of the homogeneous cluster discussed in Section 6.6. The reflectance is determined by the expression

$$R_{2N} = F\frac{A_1 + iB_1}{A_2 + D\Phi_N + iB_2}, \tag{6.199}$$

where

$$\begin{aligned}
A_1 &= (p-s)[h^2R_1(R_2^2-1)+R_2(R_1^2-1)]\sin\varphi_1\sin\varphi_2,\\
A_2 &= [ph^2R_1(R_2^2-1)+sR_2(R_1^2-1)]\sin\varphi_1\sin\varphi_2\\
&\quad -[sh^2R_1(R_2^2-1)+pR_2(R_1^2-1)]\cos\varphi_1\cos\varphi_2,\\
B_1 &= 4R_1R_2(p\cos\varphi_1\sin\varphi_2+sh^2\sin\varphi_1\cos\varphi_2),\\
B_2 &= 2\left[pR_2(R_1^2+1)\sin\varphi_1\cos\varphi_2+sh^2R_1(R_2^2+1)\cos\varphi_1\sin\varphi_2\right],\\
D &= pR_2(R_1^2-1)+sh^2R_1(R_2^2-1)
\end{aligned} \tag{6.200}$$

The following notations are used here:

$R_1 = (q-\sqrt{q^2-1})^2$, $R_2 = \frac{(q-\sqrt{q^2-h^2})^2}{h^2}$, $p = \sqrt{q^2-h^2}$, $s = \sqrt{q^2-1}$, $q = \gamma(ck_0 - \beta\omega)\omega_{e1}^{-1}$; $h = \omega_{e2}/\omega_{e1}$ is the cluster stratification parameter and $F = \omega_r\omega^{-1}$ is the frequency multiplication coefficient (FMC). The product FR_1 is the reflectance for the semi-infinite cluster and the parameter q characterizes the transparency of the cluster. The internal waves become fading ones when $q < 1$. The function

$$\Phi_N = \sin(N-1)\psi/\sin N\psi \tag{6.201}$$

represents a packet factor.

The reflectivity is then equal to

$$S_{2N} = \frac{v_{gR}}{v_{g0}}R_{2N}R_{2N}^* = \frac{v_{gR}}{v_{g0}}F^2\overline{S}_{2N}, \tag{6.202}$$

where $v_{gR,0}$ stand for the group velocities of the reflected and incident waves. The first factor $\frac{v_{gR}}{v_{g0}}F^2$ depends only on the waveguide parameters and the whole cluster velocity, reaching a maximum in the order of 10^2–10^3 if the condition $q \approx 1$ is fulfilled. The last condition can be satisfied for small values of the cluster velocity for a certain combination of plasma and waveguide parameters as was shown in the previous section. In fact, the internal structure of the cluster contributes to the reflectivity expression through the second factor, $\overline{S}_{2N}$. Even though this factor depends on the parameters q, $\varphi_{1,2}$ for a motionless cluster the presence of the motion adds another possibility to vary these parameters by choosing the cluster velocity variations. In the next treatment the influence of the cluster stratification on the second multiplier $\overline{S}_{2N}$ will be considered for $q \geq 1$ because such a value allows one to attain the greatest effectiveness of the first factor in the reflectivity.

For arbitrary layer thicknesses, but for the weak cluster stratification, $h \approx 1$, the expression for $\overline{S}_{2N}$ has the simple form,

$$\overline{S}_{2N} = [1 + 4s^2 \coth^2 N(\varphi_1 + \varphi_2)]^{-1} \tag{6.203}$$

It follows from (6.203) that $\overline{S}_{2N}$ reaches a maximum value of 1 under the condition $2\tan(N\psi) = \tan\psi$.

For a strongly stratified cluster $\overline{S}_{2N}$ depends on the order of the slabs, namely, whether the first or the second slab possesses the larger or smaller plasma density. If the first slab density is smaller than the second one, i.e., $h >> 1$, all coefficients in (6.200) become complex and the value a takes the following form:

$$a = \cos\varphi_1 \cosh\varphi_2' + \frac{h}{2s}\sin\varphi_1 \sinh\varphi_2' \tag{6.204}$$

where $\varphi_2' = h\gamma(\omega_{e1}/c)l_2$. Denoting by prime and double prime respectively the real and imaginary parts for each coefficient in (6.200) we get

$$\overline{S}_{2N} = \frac{|A_1|^2 + |B_1|^2 + 2(B_1'A_1'' - A_1'B_1'')}{|D|^2 \Phi_N^2 + 2(F_1 D' + F_2 D'')\Phi_N + F_1^2 + F_2^2}, \tag{6.205}$$

where $F_1 = A_2' - B_2''$, $F_2 = A_2'' + B_2'$.

The denominator in (6.205) takes its minimum value (which is equal to $|D|^{-2}(F_1D'' - F_2D')^2)$ for the following value of the packet factor:

$$\Phi_{\text{res}} = -|D|^{-2}(F_1D' + F_2D'') \tag{6.206}$$

The reflectivity of the packet of layers exceeds the reflectivity of the double layer when the number of the double layers N satisfies the condition $\Phi_N = 0$. This excess can be characterized by the excess coefficient $G = \bar{S}_{2N\max}\bar{S}_{21}^{-1}$ being determined by the following equation in the case of the stratification considered:

$$G = \bar{S}_{2N\max}\bar{S}_{21}^{-1} = |D|^2 \frac{F_1^2 + F_2^2}{(F_1D'' - F_2D')^2} \tag{6.207}$$

An example of dependence of this coefficient on the layer parameters contained in the values φ_1 and φ_2 is presented in Fig. 6.20 for a cluster comprised of three double layers moving with the velocity $\beta = 0.7$. It is seen that there are peak values of the excess coefficient for certain parameters of the layers in comparison with a one double layer.

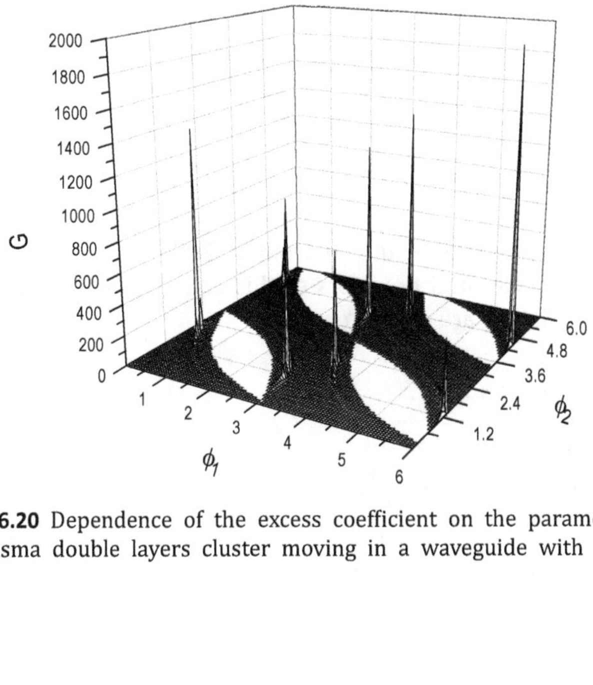

Figure 6.20 Dependence of the excess coefficient on the parameters of the plasma double layers cluster moving in a waveguide with velocity $\beta = 0.7$.

In order to elucidate the principle of the possible influence of the cluster stratification on the frequency multiplication effectiveness of the reflected wave we investigate in detail the case of a strongly stratified cluster when the first slab is more dense then the second one (h << 1). The investigation can be performed analytically in this case, because all the coefficients in (6.205) are real and the expression for $\overline{S}_{2N}$ has the following form:

$$\overline{S}_{2N} = \frac{A_1^2 + B_1^2}{(A_2 + D\Phi_N)^2 + B^2} \qquad (6.208)$$

The resonance condition is determined by the equality

$$A_2(\varphi_1, \varphi_2) + D\Phi_N(a) = 0 \qquad (6.209)$$

The excess coefficient is then equal to

$$G = 1 + \left(\frac{A_2}{B_2}\right)^2 = 1 + \frac{a^2}{\left(\cos\varphi_1 \sin\varphi_2 + \dfrac{1}{2s}\sin\varphi_1 \cos\varphi_2\right)^2} \qquad (6.210)$$

One can write down an explicit dependence on a for the packet factor because ψ = arccos a:

$$\Phi_1 = 0, \Phi_2 = \frac{1}{2a}, \Phi_3 = \frac{2a}{4a^2 - 1}, \Phi_4 = \frac{4a^2 - 1}{4a(2a^2 - 1)},$$

$$\Phi_5 = \frac{4a(2a^2 - 1)}{16(a^2 - 1)^2 + 20(a^2 - 1) + 5}, \ldots \qquad (6.211)$$

The values h << 1, s << 1 provide $A_2 \approx -aD$ that leads to the condition: $\Phi_N(a) - a = 0$ from (6.209). Using the formula, (6.211)) one can find the roots of this equation. For instance, for N = 2, 3, 4, 5, they are equal to

$$N = 2, \;\; a_{1,2} = \pm 1/\sqrt{2},$$

$$N = 3, \;\; a_{1=0,} \; a_{2,3} = \pm\sqrt{3}/2,$$

$$N = 4, \;\; a_{1-4} = \pm\sqrt{1/2 \pm 1/2\sqrt{2}},$$

$$N = 2, \;\; a_1 = 0, \; a_{2-5} = \pm\sqrt{1 + \frac{-6 \pm \sqrt{20}}{16}}. \qquad (6.212)$$

The values of the magnitudes of the roots of (6.209) above determine the lines on the plane (φ_1, φ_2). The extreme values of the excess coefficient G on these lines, i.e., under the condition (6.209) are equal to

$$G_1 = \frac{1}{1-a^2} \text{ when } \varphi_1 = k\pi, \quad \cos\varphi_2 = (-1)^k a,$$

and

$$G_2 = \frac{1}{1-4s^2a^2} \text{ when } \varphi_1 = \frac{\pi}{2} + k\pi, \quad \sin\varphi_2 = (-1)^{k+1} 2sa, \quad (6.213)$$

where $k = 0, 1$.

In the first case, $G_1 \to \infty$ for $a \to \infty$. It occurs with an increase in the number of layers, i.e., in accordance with the tendency following from (6.212). However, the multiplier $\bar{S}_{2N}$ has a small value. Indeed, substituting $\varphi_1 = k\pi$, $\cos\varphi_2 = (-1)^k a$ in (6.208) yields $\bar{S}_{2N} \cong \frac{1}{4}h^4$ while $\bar{S}_{21} = \frac{1}{4}h^4(1 - a^2)$. The second case, corresponding to the excess coefficient value G_2, is more effective. In this case, $\bar{S}_{2N} \cong (1-4s^2a^2)^{-1}$, $\bar{S}_{21} \cong 1$, thereby increasing the excess coefficient, i.e., increasing its relative value, $\bar{S}_{2N}$, as well as its absolute value, with an increase in the value $4s^2a^2$.

It is convenient to retrace the dependence of $\bar{S}_{2N}$ on the cluster velocity, if the geometric thickness of the layers in the rest frame of the cluster $l_{01,2} = \gamma l_{1,2}$ is introduced. The optical thicknesses then take the form $\varphi_1 = s\frac{\omega_{e1}}{c}l_{01}$, $\varphi_2 = p\frac{\omega_{e1}}{c}l_{02}$. It follows from $h << 1$, $q \approx 1$ that $p \approx 1$ and $s = \sqrt{q^2-1} = \frac{1}{\omega_{e1}}\sqrt{(\omega')^2 - \omega_{ke1}^2}$, where $\omega' = \gamma(\omega - uk)$ is the frequency of the incident wave in the rest frame of the cluster. In accordance with the second result in (6.213) it can be found that $\bar{S}_{2N}$ has a resonance maximum for the parameters combinations:

$$\frac{1}{c}\sqrt{(\omega')^2 - \omega_{ke1}^2}\, l_{01} = \frac{\pi}{2} + k\pi,$$

$$\sin\frac{\omega_{e1}}{c}l_{02} = (-1)^{k+1}\frac{2a}{c\omega_{e1}}\sqrt{(\omega')^2 - \omega_{ke1}^2}, \quad k = 0,1 \quad (6.214)$$

These conditions reveal that the resonance thickness of the first layer is determined by the plasma density of this layer and by the cluster velocity while the thickness of the second layer is determined by the resonance value of the parameter a as well. When the multiplier $\bar{S}_{2N}$ exceeds unity one can see that the resonance character of the cluster stratification influences the reflectivity. A great increase of the reflectivity effectiveness and, consequently, of the frequency multiplication effectiveness, can be reached by a proper choice of the cluster parameters.

References

1. Nikolsky, V. V. (1989) *Electrodynamics and Radio Waves Propagation* (in Russian), Moscow, "Nauka" 544 p.
2. Jackson, J. D. (1999) *Classical Electrodynamics*, John Wiley & Sons 808 p.
3. Pauli, W. (1958) *Theory of Relativity*, Pergamon Press, 936 p.
4. Nerukh, A. G., and Khizhnyak, N. A. (1976) Movement of a boundary of two dielectrics in a waveguide, *Zhurnal technicheskoy fiziki* (in Russian), **46**(1), 21–30.
5. Nerukh, A. G., and Khizhnyak, N. A. (1978) Interaction of electromagnetic waves with a boundary of relativistic moving medium in presence of waveguide dispersion, *Izv. Vuzov. Radiofizika* (in Russian), **21**(10), 492–1497.
6. Gazaryan, E. D., Laziev, E. M., and Ter-Pogosyan, A. D. (1978) Interaction of electromagnetic waves with a moving semi-infinite medium in a waveguide, *Izv. Vuzov. Radiofizika* (in Russian), **21**(10), 1517–1519.
7. Born, M., and Wolf, E. (1964) *Principles of Optics*, Pergamon Press, Oxford, 856 p.
8. Freilikher, V. D., Fuks, I. M. (1969) Signal spectrum scattered by oscillating boundary, *Izv. Vuzov. Radiofizika* (in Russian), **12**(1), 114–119.
9. Krasilnikov, V. N., and Magomadova, T. A. (1974) About parametric generation of electromagnetic oscillations in spherical resonator with radius changing in time, *Problems of Diffraction and Wave Scattering* (in Russian), 13, 155–159.
10. Nerukh, A. G. (1975) *Scattering of Electromagnetic Waves by Moving Objects*, Dissertation on PhD in Radio Physics, Kharkov, 150 p. (in Russian).

11. Van Bladel, J., and de Zutter, D. (1981) Reflection from linear vibrating objects: Plane mirror at normally incidence, *IEEE Trans. Antennas Propag.*, **AP-29**(4), 629–637.
12. Vladimirov, V. S. (1967) *Equations of Mathematical Physics* (in Russian), "Nauka", Moscow, 436 p.
13. Bateman, H., and Erdely, A. (1954) *Tables of Integral Transforms*, New York, Toronto, London: McGraw-Hill Book Comp., Inc., 344 p.
14. Nerukh, A. G., and Khizhnyak, N. A. (1978) Interaction of electromagnetic waves with a boundary of relativistic moving medium in presence of waveguide dispersion, *Izv. vuzov. Radiofizika* (in Russian), **21**(10), 1492–1497.
15. Bolotovski, B. M., and Stolyarov, S. N. (1974) Boundary value problems in electrodynamics of moving media (in Russian), *Uspekhi fizicheskih nauk*, **114**(4), 569–608.
16. Aalund, M., and Johansen, G. (1971) Generalized plane waves and waveguide modes in a moving isotropic medium, *J. Appl. Phys.*, 1971, **2**(7), 2669–2674.
17. Katzelenbaum, B. E. (1966) *High Frequency Electrodynamics* (in Russian), "Nauka", Moscow, 237 p.
18. Landau, L. D., and Lifshitz, E. M. (1975) *The Classical Theory of Fields*, vol. 2, 4th ed., Butterworth-Heinemann.
19. Stolyarov, S. N. (1962) Interaction of electromagnetic waves with moving interface between media, *Izv. Vuzov. Radiofizika* (in Russian), **5**(4), 671–676.
20. Yeh, C. (1966) Reflection and transmission of electromagnetic waves by a moving plasma medium, *J. Appl. Phys.*, **37**(8), 3079–3084.
21. Ostrovskiy, L. A., and Stepanov, N. S. (1971) Nonresonance parametric phenomena in distributed systems, *Izv. Vuzov. Radiofizika* (in Russian), **14**(4), 489–529.
22. Lampert, M. A. (1959) Meeting of an electromagnetic wave with "Cherenkov electron gas, *Millimeter and Submillimeter Waves*, Moscow, 145–149.
23. Fainberg, Ya. B., and Tkalich, V. S. (1959) Reflection of an electromagnetic wave from plasma moving through a dielectric medium in presence of a static magnetic field, *Zhurnal teknicheskoy fiziki* (in Russian), **29**(4), 491–496.
24. Stolyarov, S. N. (1978) *Boundary Value Problems in Electrodynamics of Moving Medium, Einstein Collected Articles* (in Russian), 1975–1976, Moscow, "Nauka", 152–215.

25. Nerukh, A. G., and Khizhnyak, N. A. (1978) Diffraction of electromagnetic waves on moving domains of homogeneous medium (in Russian), *Ukrainien fizicheskiy zhurnal*, **23**(12), 2009–2014.

26. Kurilko, V. I. (1961) Electromagnetic wave reflection from plasma moving in a waveguide of slow waves, *Zhurnal teknicheskoy fiziki* (in Russian), **31**(8), 899–905.

27. Nerukh, A. G., and Khizhnyak, N. A. (1980) Reflection of radio waves from a plasma bunch moving in a waveguide, *Izv. vuzov. Radiofizika* (in Russian), **23**(5), 517–522.

28. Felsen, L. B., and Marcuvitz, N. (1973) *Radiation and Scattering of Waves*, Prentice-Hall, Inc., New Jersey.

29. Mittra, R., and Lee, S. W. (1974) *Analytical Techniques in the Theory of Guided Waves*, The Macmillan Company, New York.

30. Landecker, K. (1952) Possibility of frequency multiplication and wave amplification by means of some relativistic effects, *Phys. Rev.*, **86**(6), 852–855.

31. Lampert, M. A. (1956) Reflection of electromagnetic waves by Cerenkov electron gas, *Phys. Rev.*, **102(**2), 299–304.

32. Zagorodnov, O. G., Fainberg, J. B., and Egorov, A. M. (1960) About reflection of an electromagnetic wave from plasma moving in a waveguide of slow waves (in Russian), *Zhurnal tekhnicheskoy fizifki*, **30**(1), 7–9.

33. Zagorodnov, O. G., Fainberg, J. B., Egorov, A. M., and Bolotovsky, B. M. (1961) Frequency multiplication by means of a plasma "collapse" *Soviet Physics-Techn. Physics (Engl. Transl.)*, **6,** 297–302.

34. Kurilko, V. I., and Miroshnichenko, V. I. (1962) Electromagnetic wave reflection by a moving plasma, *Soviet Phys. Techn. Phys. (Engl. Transl.)*, **7,** 803–808.

35. Nerukh, A. G., and Khizhnjak, N. A. (1983) Energy relations for the interaction of an electromagnetic wave with a plasma cluster moving in a waveguide, *Radiophys. Quantum Electronics (Engl. Transl.)*, **26,** 12–20.

36. Nerukh, A. G., Khizhnjak, N. A., and Minko, P. E. (1990) Electromagnetic wave reflection from stratified plasma cluster moving in a waveguide, *Radiophys. Quantum Electronics (Engl. Transl.)*, **33,** 5–12.

37. Mori, W. B. (1991) Generation of tunable radiation using an underdense ionization front, *Phys. Rev.* **A**, **44(**8), 5118–5121.

38. Pauli, W. (1921) *Relativitivitatstheorie*, Encyclopedie der math. Wiss., Band vol. 2, Heft 4, Leipzig §35. 359 p.

39. Landau, L. D., and Lifshitz, E. M. (1959) *Electrodynamics of Continuous Media*, Moscow, GIFML (in Russian), 462 p.

40. Tolman, R. C. (1969) *Relativity, Thermodynamics and Cosmology*, Oxford, Clarendon Press, 490 p.

41. Nerukh, A. G., and Khizhnyak, N. A. (1998) Enhanced reflection of an electromagnetic wave from a plasma cluster moving in a waveguide, *Microwave Opt. Technol. Lett.*, **17**(4), 267–273.

42. Nerukh, A. G., Scherbatko, I. V., and Sakhnenko, N. K. (1999) Penetration of electromagnetic wave into plasma expanding in a waveguide, *J. Appl. Electromagnetism*, **2**(1), 63–70.

43. Nerukh, A. G., and Khizhnyak, N. A. (1978) Diffraction of electromagnetic waves on moving domains of homogeneous medium (in Russian), *Ukrainien fizicheskiy zhurnal*, **23**(12), 2009–2014.

44. Abeles, F. (1950) Recherches sur la propagation des ondes electromagnetiques sinusoidales dans les milieux stratifies. Application aux couches minces, *Ann. Phys.* **12**(5), 708–762.

Chapter 7

Non-Stationary Electromagnetic Processes in Time-Varying Dielectric Waveguides

Analytical investigations of the transient behavior of electromagnetic waves in waveguides have a long history when considering only empty waveguides with conductive walls [1–4]. In these early works, straight waveguides were considered in which the phenomenon is described by a 1D equation in the time or frequency domain because the transverse field components are determined via the longitudinal ones in the general formulation. Using a Fourier integral transform, some aspects of the general nature of transient propagation were investigated. It was shown that the normal steady-state structure does not exist in the transient phase since the different field components propagate in a non-identical manner that is characterized by the group velocity.

The theory of pulse distortion and transmission in perfect and imperfect metallic waveguides was constructed in [4] using Laplace transforms, where it was shown that with the development of pulse techniques, and possibilities for waveguide trunk communication, the pulse distortion by a waveguide became one of its principal features. This feature sets an upper limit to repeater spacing, and becomes an important parameter governing the

Non-Stationary Electromagnetics: An Integral Equations Approach (Second Edition)
Alexander Nerukh and Trevor Benson

ISBN 978-981-4774-95-6 (Hardcover), 978-0-429-02773-4 (eBook)
www.panstanford.com

design of numerous components. It was admitted in this work that somewhat allied problems were investigated by Sommerfeld and Brillouin, namely the propagation of plane waves in unbounded homogeneous dispersive media. In waveguides, on the other hand, the proximity of physical boundaries gives rise to modal dispersion. Furthermore, the problem obtains further new features when the boundaries are transparent, i.e., when the field inside the waveguide core depends on a medium outside of it.

There are many modern applications in quantum electronics, quantum computing, and optical technology that rely on the interaction of electromagnetic waves and media with time-varying, non-linear, and/or anisotropic parameters. Examples are the direct up-conversion of microwave signals into the optical domain by an electro-optical modulator or a micro-photonic modulator [5] which are the main phenomena in photonic front-end receiver architectures [6, 7] for microwave and millimeter wave signal processing. To protect electronic systems against high power microwaves, new radiofrequency front-end technology has been constructed based on a dielectric antenna coupled to an electro-optic resonator which converts the received electromagnetic signal to a modulated optical signal carried away from the antenna front-end via an optical fiber. All-dielectric photonic-assisted receivers have been proposed for this application; the complete lack of metal and electronics in the front-end offers immunity against damage from intense electromagnetic radiation [8].

In all these devices, there is, in one way or another, an interaction between an electromagnetic field and a time-varying medium. In practice, temporal variation of the material's refractive index can be caused by an external source current or a field source. For many applications, the frequency of the external source is much lower than that of the signal, which permits the signal behavior to be modeled using a slowly varying envelope to incorporate the effects of the time-varying material. For example, modern modulators typically use an external 40 GHz microwave signal, while the frequency of the propagating optical signal is of the order of few THz. However, the steadily increasing demand for higher bit rates requires the use of higher modulation frequencies and recently multiple quantum wells driven by terahertz fields have been investigated both analytically and experimentally for this purpose [9, 10]. At such high modulation and switching speeds, the modeling approaches based upon the slowly varying envelope paradigm, such as the finite difference beam propagation method [5], are less appropriate and hence more rigorous modeling techniques need to be used. In particular, transient effects cannot be neglected with such high modulation rates and need to be properly accounted for [11].

Dielectric waveguide tuning can be achieved by changing the material refractive index which can be practically realized by varying the input signal in a non-linear waveguide [12], by voltage control [13], or by free carrier plasma injection [14]. An enhanced frequency shift occurs in a slow wave photonic structure with modulated permittivity [15]. The reflection of a light pulse from a shock-like dielectric modulation propagating through a photonic crystal demonstrates the reverse Doppler effect [16] and a frequency shift which is tunable by adjusting the band gap size [17]. A dynamic photonic crystal whose properties are modulated while a pulse is inside it can be used to stop, store, and time-reverse light pulses [18].

Surface plasmon polaritons have great potential as information carriers for next-generation, highly integrated nano-photonic devices. A number of techniques for controlling the propagation of guided surface plasmon polariton signals have been demonstrated. However, with sub-microsecond or nano-second response times at best, these techniques are likely to be too slow for future applications in such fields as data transport and processing. In [19] it is reported that femtosecond optical frequency plasmon pulses can propagate along a metal–dielectric waveguide and that they can be modulated on the femtosecond timescale by direct ultrafast optical excitation of the metal, thereby offering unprecedented terahertz modulation bandwidth—at a speed at least five orders of magnitude faster than existing technologies.

Maxwell's equations are self-consistent only for electromagnetic fields in a vacuum. In a general medium, the constitutive equations and boundary conditions significantly complicate both the formulation and solution of electromagnetic problems. Such problems become even more complex when the media are not only inhomogeneous, but are also time varying. Such a situation can be met when considering the propagation of electromagnetic signals in dielectric or semiconductor waveguides, in particular in the context of modulators, pulsed lasers, and frequency conversion. The proper description and investigation of the physics of these phenomena is motivated by their significant importance to optical communication technology, in particular interactions between microwave and optical pulses and semiconductor active media in waveguides. The solution of such electromagnetic problems has demanded accurate time-domain techniques.

Pure analytical approaches reveal and allow investigation of the main features of transient phenomena in waveguide structures with non-transparent boundaries. However, the possibilities of such approaches are constrained when dielectric waveguides are investigated. Difficulties grow in the case of a complex medium in the core, especially in the case of a non-linear medium.

More general cases are considered by semi-analytical methods when investigations have a preliminary analytic stage, in which a solution to the original equations is represented in the premeditated form of a slowly varying amplitude or a sum of definite harmonic waves in the case of a non-linear process [20–23]. In these investigations, the analytical approach was supported by numerical calculations. For example, a rigorous integral formulation is used in [24] to provide a numerical approach to the problem of an inhomogeneous dispersive slab illuminated by an incident TEM plane wave with arbitrary time dependence but a specific electromagnetic field direction. The integral equation is derived from the wave equation via a Fourier transform using Green's function for an unbounded homogeneous medium and slab inhomogeneity is taken into account by introducing a co-ordinate dependent permittivity. The numerical process consists of a space-time discretization and the solution is determined step by step from simple recurrence formulae. Numerical investigation of various aspects of non-linear phenomena in optical waveguides described by the paraxial wave equation is developed by Romanova and Melnikov [25] via a classical mathematical method. The modern method of a spatio-temporal multi-resolution analysis in the method of moments is used in [26] for the solution of transient electromagnetic scattering. However, the possibilities of semi-analytical methods are restricted by their problem statement itself which does not allow complex phenomena to be considered, for example when considering ultra-short pulses or a medium with complex features.

The development of a computer technique widens greatly when using a numerical analysis, such as the FDTD method, the TLM method or beam propagation methods [27–30] and numerous other works. However, such pure numerical methods are restricted from the beginning by a special statement of the problem, specific conditions and parameters and they do not give a complete qualitative picture of a phenomenon. Moreover, they make great computational demands. The development of analytical-numerical methods by the application of the integral equation method developed in this book is considered in this chapter.

It was shown in Chapter 2 that a temporal change in the permittivity of an unbounded medium transforms a primary harmonic plane wave to new secondary ones having different frequencies but the same wavenumber as the primary wave, as originally described in [31, 32]. This general feature is also observed when a plane wave is normally incident onto a plane interface between two media, the permittivity of one of which changes abruptly in time; this has been considered in Sections 3.2–3.4 and earlier in [33, 34]. However, in this case the spatial structure of the waves also becomes more complex. Nevertheless, the monochromatic character of the secondary waves is not disturbed if the medium is non-dissipative. In the case, when an electromagnetic wave is obliquely incident onto a plane boundary with a time-varying medium, not only does the structure of the system of monochromatic waves become more complex, but a continuous wave spectrum also appears as was shown in Sections 3.9–3.11, [34, 35]. All these circumstances arise in a dielectric waveguide with time-varying media as will be considered in this chapter. A time domain integral equation technique is presented to take into account, in one formulation, a complex combination of boundary and initial conditions while also permitting the medium parameters to change in time.

7.1 Wave Equations for Longitudinal and Transverse Components in Generalized Functions

According to the approach developed in Chapter 1, in order to obtain the Volterra integral equations for the electromagnetic fields in a time-varying dielectric waveguide, it is necessary to formulate the problem in terms of generalized functions. As a waveguide is a longitudinally uniform structure, the electromagnetic fields can be expressed in terms of their longitudinal and transverse components $\mathbf{E} = \mathbf{E}_\perp + E\mathbf{e}_3$, $\mathbf{B} = \mathbf{B}_\perp + B\mathbf{e}_3$, where it is assumed that the z axis (axis number 3) is directed along the structure. Using the following representation of the fields in Maxwell's equations, formulated in generalized functions and derivatives [36, 37]

$$\text{curl}\mathbf{E} = -\partial_t \mathbf{B}, \quad \text{curl}\mu_0^{-1}\mathbf{B} = \partial_t \varepsilon_0 \hat{\varepsilon}\mathbf{E} + \hat{\sigma}\mathbf{E} \tag{7.1}$$

we consider a longitudinally invariant waveguide, formed by a core medium described by the operators $\hat{\varepsilon}_r$, $\hat{\sigma}_r$ and a cladding, or background, medium, described by the parameters ε, σ. All media are non-magnetic, $\hat{\mu} = 1$. Then the medium in the whole space is described by the generalized operators $\hat{\varepsilon} = \varepsilon_0\varepsilon + \chi\varepsilon_0(\hat{\varepsilon}_r - \varepsilon)$, $\hat{\sigma} = \chi(\hat{\sigma}_r - \sigma)$. Here, ε_0 is the permittivity of vacuum and the function χ equals unity inside the core and zero outside of it. The generalized wave equations for the transverse components follow from (7.1)

$$\begin{aligned}\partial_{tt}^2 \mathbf{E}_\perp + v^2\partial_t\mu_0\sigma\mathbf{E}_\perp - v^2\partial_{33}^2\mathbf{E}_\perp = &-v^2\partial_3\nabla_\perp E - v^2\mathbf{e}_3 \times \partial_t \nabla_\perp B \\ &-(v/c)^2\partial_{tt}^2\chi(\hat{\varepsilon}_r - \varepsilon)\mathbf{E}_\perp \\ &-v^2\partial_t\mu_0\chi(\hat{\sigma}_r - \sigma)\mathbf{E}_\perp,\end{aligned} \tag{7.2}$$

$$\begin{aligned}\partial_{tt}^2 \mathbf{B}_\perp + v^2\mu_0\sigma\partial_t\mathbf{B}_\perp - v^2\partial_{33}^2\mathbf{B}_\perp = &-v^2\partial_3\nabla_\perp B + v^2\mathbf{e}_3 \times \left(\frac{1}{c^2}\partial_t\hat{\varepsilon} + \mu_0\hat{\sigma}\right)\nabla_\perp E \\ &-(v/c)^2\partial_t\chi(\hat{\varepsilon}_r - \varepsilon)\partial_t\mathbf{B}_\perp \\ &-v^2\mu_0\chi(\hat{\sigma}_r - \sigma)\partial_t\mathbf{B}_\perp.\end{aligned} \tag{7.3}$$

Here, $\nabla = \nabla_\perp + \hat{e}_3\partial_3$, $\partial_3 = \partial/\partial z$, $\nabla_\perp = \hat{e}_1\partial_1 + \hat{e}_2\partial_2$, $\partial_t = \partial/\partial t$, $c = 1/\sqrt{\varepsilon_0\mu_0}$ is the velocity of light in a vacuum and $v = c/\sqrt{\varepsilon}$ is the phase velocity in the background. The commutants $\partial_1\hat{\varepsilon} = \hat{\varepsilon}\partial_1$ and $\partial_1\hat{\sigma} = \hat{\sigma}\partial_1$ hold true in the equations because the structure is assumed to be uniform in the longitudinal direction. It is clear from these equations that the transverse components are expressed in terms of the longitudinal ones. The longitudinal components satisfy self-consistent generalized wave equations

$$\frac{1}{c^2}\partial_{tt}^2\varepsilon E + \partial_t\mu_0\sigma E - \Delta E = -\frac{1}{c^2}\partial_{tt}^2\chi(\hat{\varepsilon}_r - \varepsilon)E - \chi\mu_0\partial_t(\hat{\sigma}_r - \sigma)E \\ + \hat{\varepsilon}^{-1}((\nabla_\perp\hat{\varepsilon}),\mathbf{E}_\perp), \tag{7.4}$$

$$\frac{1}{c^2}\partial_t\varepsilon\partial_t B + \mu_0\sigma\partial_t B - \Delta B = -\frac{1}{c^2}\partial_t\chi(\hat{\varepsilon}_r - \varepsilon)\partial_t B - \chi\mu_0(\hat{\sigma}_r - \sigma)\partial_t B. \tag{7.5}$$

The left-hand side of each of these equations has the same form as in the background, and the right-hand side is only distinct from zero in the waveguide core. Owing to the use of generalized derivatives, these equations will also include the boundary conditions for the electromagnetic fields on the waveguide walls.

It follows from Eqs. (7.2)–(7.5) that two kinds of fields can exist, as in the frequency domain: TE-fields with $\Delta_\perp E = 0$ and TM-fields with $\Delta_\perp B = 0$. In both cases, we have a scalar 3D problem in the time domain for the longitudinal components and a vector 1D problem in the time domain for the transverse ones.

7.2 Volterra Integral Equations for Non-Stationary Electromagnetic Processes in Time-Varying Dielectric Waveguides

7.2.1 Integral Equations for the Fields

To solve (7.2)–(7.5), which contain generalized derivatives and yield generalized functions as their solutions, we transform them to integral equations. At first, we consider Eqs. (7.2) and

(7.3) for the transverse components. These equations are vector ones and their solution formulate a 1D problem in the time domain. Green's function for each of these equations is defined by the formula [38]

$$G^{(1)}(t,z)=\frac{1}{2}e^{-\alpha t}I_0\left(\frac{\alpha}{v}\sqrt{(vt)^2-z^2}\right)\theta(vt-|z|), \tag{7.6}$$

where I_0 is the modified Bessel function, θ is the Heaviside unit function, and $\alpha=\frac{v^2\mu_0\sigma}{2}$ takes into account the dissipation in the cladding. This function allows us to write the integral equations for the transverse electric field as

$$\mathbf{E}_\perp=\mathbf{F}_E-\frac{v}{2}\int_{-\infty}^{t}dt'\int_{z-v(t-t')}^{z+v(t-t')}I_0\left(\frac{\alpha}{v}\sqrt{v^2(t-t')^2-(z-z')^2}\right)e^{-\alpha(t-t')}$$
$$\times\left\{\frac{1}{c^2}\partial^2_{t't'}\chi(\hat{\varepsilon}_r-\varepsilon)+\mu_0\chi\partial_{t'}(\hat{\sigma}_r-\sigma)\right\}\mathbf{E}_\perp dz' \tag{7.7}$$

and similarly for the transverse magnetic field as

$$\mathbf{B}_\perp=\mathbf{F}_B-\frac{v}{2}\int_{-\infty}^{t}dt'\int_{z-v(t-t')}^{z+v(t-t')}I_0\left(\frac{\alpha}{v}\sqrt{v^2(t-t')^2-(z-z')^2}\right)e^{-\alpha(t-t')}$$
$$\times\left\{\frac{1}{c^2}\partial_{t'}\chi(\hat{\varepsilon}_r-\varepsilon)\partial_{t'}+\mu_0\chi(\hat{\sigma}_r-\sigma)\partial_{t'}\right\}\mathbf{B}_\perp dz' \tag{7.8}$$

The free terms in these equations

$$\mathbf{F}_E=-\frac{v}{2}\int_{-\infty}^{t}dt'\int_{z-v(t-t')}^{z+v(t-t')}I_0\left(\frac{\alpha}{v}\sqrt{v^2(t-t')^2-(z-z')^2}\right)$$
$$\times e^{-\alpha(t-t')}\{\partial_{3'}\nabla_\perp E+\mathbf{e}_3\times\partial_{t'}\nabla_\perp B\}dz' \tag{7.9}$$

$$\mathbf{F}_B=-\frac{v}{2}\int_{-\infty}^{t}dt'\int_{z-v(t-t')}^{z+v(t-t')}I_0\left(\frac{\alpha}{v}\sqrt{v^2(t-t')^2-(z-z')^2}\right)$$
$$\times e^{-\alpha(t-t')}\left\{\partial_{3'}\nabla_\perp B-\mathbf{e}_3\times\left(\frac{1}{c^2}\partial_{t'}\hat{\varepsilon}+\mu_0(\hat{\sigma}-\sigma)\right)\nabla_\perp E\right\}dz' \tag{7.10}$$

are defined in both the waveguide core and cladding, and are known if the longitudinal components of the fields are known.

According to the general approach developing in this book, Eqs. (7.7) and (7.8) are Volterra integral equations for the transverse components only inside the core, where $\chi = 1$. In the cladding, the transverse components are simply equal to the free terms, $\mathbf{B}_\perp = \mathbf{F}_B$, $\mathbf{E}_\perp = \mathbf{F}_E$. In both cases, however, the problem is a vector 1D one in the time domain.

Equations (7.4) and (7.5) for the longitudinal components are scalar but 3D ones. To obtain the integral equations for these components, we have to use the appropriate Green's function [39], which in the 3D case has the form

$$G^{(3)}(t,\mathbf{r}) = \frac{1}{2\pi} e^{-\alpha t} \left[\delta(v^2t^2 - r^2)\theta(vt) + \frac{\alpha I_1\left(\frac{\alpha}{v}\sqrt{v^2t^2 - r^2}\right)}{2v\sqrt{v^2t^2 - r^2}} \theta(vt - r) \right], \tag{7.11}$$

where $I_1(t)$ is the modified Bessel function, and $\delta(t)$ is the delta function. Further, we will consider the case of the planar waveguide with width b, which allows us to reveal the main features of the transient phenomena with comparative simplicity. In this case, the problem is reduced to two dimensions and Green's function for (7.4) and (7.5) becomes simpler:

$$G^{(2)}(t, r_\perp) = \frac{1}{2\pi} e^{\alpha t} \frac{\cosh(\alpha\sqrt{(vt)^2 - r_\perp^2})}{\sqrt{(vt)^2 - r_\perp^2}} \theta(vt - r_\perp), \tag{7.12}$$

where $r_\perp = \sqrt{x^2 + z^2}$. Using this function, we obtain the Volterra integral equations for the longitudinal components in the planar waveguide

$$E = E^{(0)} - v\int_{-\infty}^{\infty} dt' \int_{-\infty}^{\infty} dz' \int_{-b/2}^{b/2} dx' G^{(2)}(t - t', |\mathbf{r}_\perp - \mathbf{r}'_\perp|) \partial_{t'}$$
$$\left[\left\{ \frac{1}{v^2} \partial_{t'} \left(\frac{\hat{\varepsilon}_r}{\varepsilon} - 1 \right) + \mu_0(\hat{\sigma}_r - \sigma) \right\} E(t', \mathbf{r}') - \hat{\varepsilon}^{-1}((\nabla_\perp \varepsilon), \mathbf{E}_\perp) \right] \tag{7.13}$$

$$B = B^{(0)} - v\int_{-\infty}^{\infty} dt' \int_{-\infty}^{\infty} dz' \int_{-b/2}^{b/2} dx' G^{(2)}(t-t', |\mathbf{r}_\perp - \mathbf{r}'_\perp|)$$

$$\left\{\frac{1}{v^2}\partial_{t'}\left(\frac{\hat{\varepsilon}_r}{\varepsilon}-1\right)+\mu_0(\hat{\sigma}_r-\sigma)\right\}\partial_{t'}B(t',\mathbf{r}') \qquad (7.14)$$

The functions $E^{(0)}$ and $B^{(0)}$ are solutions to the homogeneous Eqs. (7.4) and (7.5) or, in other words, the field that can exist in the background in the absence of the core, that is the primary or initial field. Equations (7.13) and (7.14), as well as the corresponding ones for the transverse components (7.7) and (7.8), are proper integral equations only with respect to the unknown fields in the core. Outside the core these expressions are quadrature formulas for calculating the fields in the cladding given the field in the core.

It is seen from (7.7) and (7.8) that the problem for the transverse components is a vector 1D one in the time domain while the problem for the longitudinal components (7.13), (7.14) is a scalar 2D one in the time domain. Owing to the use of generalized derivatives these equations also include the boundary conditions for the electromagnetic fields on the waveguide walls as well as the initial conditions.

7.2.2 Harmonic Waves in a Waveguide

Considering an initial TE field of the form $B^{(0)} = B^{(0)}(t, x)e^{i(\omega t - \Gamma z)}$ the problem of harmonic wave propagation in a parallel plate waveguide is easily solved. These waves are defined by Eqs. (7.13) and (7.14), which give the complete field in the core as well as in the cladding and contain the boundary conditions for the fields on the waveguide walls. Let a non-dissipative medium, $\sigma_r = 0$, in the waveguide have the permittivity $\hat{\varepsilon}_r = \varepsilon_1$. As the initial wave has a TE nature, a solution to (7.14) can be sought in the form $B = e^{i(\omega t - \Gamma z) - \alpha t}\{Me^{-i\kappa_1 x} + Ne^{i\kappa_1 x}\}$. For the typical case of dielectric waveguides $\Gamma^2 - \alpha^2/v^2 > 0$ and $\omega^2/v^2 - \Gamma^2 + \alpha^2/v^2 < 0$, so (7.14) for the field inside the core, $-b/2 \le y \le b/2$, is reduced to the system of algebraic equations

$$\begin{cases} (\kappa+\kappa_1)Me^{i\kappa_1 b/2}+(\kappa-\kappa_1)Ne^{-i\kappa_1 b/2}=\dfrac{2\nu\nu_1^2(\kappa_1^2-\kappa^2)}{\omega(\nu^2-\nu_1^2)}\sqrt{1-(\Gamma\nu/\omega)^2}e^{i\kappa b/2}B_1^{(0)} \\ (\kappa-\kappa_1)Me^{-i\kappa_1 b/2}+(\kappa+\kappa_1)Ne^{i\kappa_1 b/2}=0, \end{cases} \tag{7.15}$$

where $\kappa^2 = \omega^2/\nu^2 - \Gamma^2 + \alpha^2/\nu^2$, $\kappa = -i\bar{\kappa}$, $\bar{\kappa} = \sqrt{\Gamma^2-(\omega/\nu)^2-(\alpha/\nu)^2} > 0$, and $\kappa_1^2 = -(\alpha+i\omega)^2/\nu_1^2 - \Gamma^2$.

If the initial field $B^{(0)}$ is given, then the values of ω and Γ are fixed and these determine κ_1. In this case (7.15) yields a system for the coefficients M and N that gives the waves excited in the core by the primary field $B^{(0)}$ existing in the uniform background (i.e., without the core.) In the case of $B^{(0)} = 0$, we have an eigenvalue problem. In this case, setting the determinant of the homogeneous system (7.15) to zero leads to the well known dispersion relations for the even and odd waves in the symmetric dielectric waveguide

$$-i\tan(\kappa_1 b/2) = \kappa_1/\kappa,\; i\cot(\kappa_1 b/2) = \kappa_1/\kappa \tag{7.16}$$

The solutions of (7.16) give the values of κ_1, Γ for the guided modes for a given frequency ω. Therefore, the solutions of the integral equation are consistent with those obtained directly from the governing differential equations, consisting of forward and backward propagating waves of the known form.

If the waveguide core consists of a plasma characterized by a plasma frequency ω_e and a dielectric ε_1 and the cladding is a dissipative dielectric with the permittivity ε and conductivity σ, then the dispersion equation is

$$\tan(\kappa_1 b/2) = \frac{-\nu_1\kappa_1}{\sqrt{(1-(\nu_1/\nu)^2)\omega^2-\omega_e^2/\varepsilon_1-(\nu_1\mu_0\sigma/2)^2-(\nu_1\kappa_1)^2}}. \tag{7.17}$$

The right-hand side tends to zero when the conductivity σ tends to infinity. In this case, the equality $\tan(\kappa_1 b/2) = 0$ gives the eigen-number $\kappa_1 = 2n\pi/b$, as in the case of a metallic waveguide.

7.3 Solution for the Problem with a Time Jump Change in the Waveguide Core Permittivity

An “all-purpose” solution to the integral equation is given by a resolvent operator. Such an operator cannot be built in the general case, so we consider here a simple case, but comprising all main phenomenon features. The initial-value problem for the field in the waveguide can be solved analytically (exactly) in the 2D case when the core’s permittivity changes abruptly in time. We also simplify the problem, without loss of generality, by considering non-dissipative media in both core and cladding, i.e., $\sigma_r = \sigma = 0$. Substitution of $\left(\frac{\hat{\varepsilon}_r}{\varepsilon}-1\right)=\left(\frac{\varepsilon_1}{\varepsilon}+\frac{\varepsilon_2-\varepsilon_1}{\varepsilon}\theta(t)-1\right)$ into (7.14) gives the integral equation for the longitudinal component of the magnetic field of the TE wave

$$B = B^{(0)} - \frac{1}{v}\partial_{tt}^2 \int_{-\infty}^{0} dt' \int_{-\infty}^{\infty} dz' \int_{-b/21}^{b/2} dx' G^{(2)}(t-t',|\mathbf{r}_\perp - \mathbf{r}'_\perp|)\left(\frac{\varepsilon_1}{\varepsilon}-1\right)B^{(-)}$$

$$-\frac{1}{v}\partial_{tt}^2 \int_{0}^{\infty} dt' \int_{-\infty}^{\infty} dz' \int_{-b/2}^{b/2} dx' G^{(2)}(t-t',|\mathbf{r}_\perp - \mathbf{r}'_\perp|)\left(\frac{\varepsilon_2}{\varepsilon}-1\right)B^{(+)}, \tag{7.18}$$

where $B^{(-)}(t, x, z)$ and $B^{(+)}(t, x, z)$ are the fields before and after the change in permittivity, respectively.

As we are considering a waveguide that is uniform in the longitudinal direction, a simultaneous variation of the medium in the whole core cannot influence the longitudinal features of the field. However, the distribution of the fields across the waveguide can change because the medium is inhomogeneous in this direction. Based upon this observation, we can consider the fields to be in the form of a Fourier transform $B(t,x,z) = \int_{-\infty}^{\infty} \bar{B}(t,x)$ $e^{-i\Gamma z} d\Gamma$, which allows one to calculate the integral in (7.18) with respect to the longitudinal coordinate z'. After this the integral equation takes the form

$$\bar{B} = F + \int_{0}^{\infty} dt' \int_{-b/2}^{b/2} dx' \langle \mathbf{x} | \hat{K} | \mathbf{x}' \rangle \bar{B}^{(+)}(t',x'). \tag{7.19}$$

Here $\mathbf{x} = (t, x)$ is a time-spatial vector and the kernel is equal to

$$\left\langle \mathbf{x}\middle|\hat{K}\middle|\mathbf{x}'\right\rangle = \frac{\varepsilon-\varepsilon_2}{2v\varepsilon}\partial^2_{tt}Q(t,t',x,x'), \tag{7.20}$$

$$Q(t,t',x,x') = J_0\left(\Gamma v\sqrt{(t-t')^2-\frac{(x-x')^2}{v^2}}\right)\theta\left(t-t'-\frac{|x-x'|}{v}\right), \tag{7.21}$$

and $J_0(.)$ is the Bessel function of the first kind.

The free term of (7.19) is determined by the pre-history of the field in the waveguide and its state at the moment of the time jump of medium parameters in the core

$$F = B^{(0)}e^{i\Gamma z} - \frac{\varepsilon_1-\varepsilon}{2v\varepsilon}\partial^2_{tt}\int_{-\infty}^{0}dt'\int_{-b/2}^{b/2}dx' Q(t,t',x,x')\overline{B}^{(-)}(t',x'). \tag{7.22}$$

Equation (7.19) is a Volterra integral equation, so its solution can be given by a resolvent operator $\hat{R}$

$$\overline{B}^{(+)} = F + \hat{R}F = F + \int_0^{\infty}dt'\int_{-b/2}^{b/2}dx'\left\langle \mathbf{x}|\hat{R}|\mathbf{x}'\right\rangle F(t',x'). \tag{7.23}$$

The kernel $\left\langle \mathbf{x}|\hat{R}|\mathbf{x}'\right\rangle$ of the resolvent $|\hat{R}|$ must satisfy the equation

$$\left\langle \mathbf{x}|\hat{R}|\mathbf{x}'\right\rangle = \left\langle \mathbf{x}|\hat{K}|\mathbf{x}'\right\rangle + \int_0^{\infty}dt''\int_{-b/2}^{b/2}dy''\left\langle \mathbf{x}|\hat{K}|\mathbf{x}''\right\rangle\left\langle \mathbf{x}''|\hat{R}|\mathbf{x}'\right\rangle. \tag{7.24}$$

We find the solution to this equation by following the procedure used for the problem with a plane dielectric boundary considered in Sections 3.1 and 3.4. According to this, the resolvent form depends on the position of the observation point $M_{ij}(t, x)$ as shown in time-space diagram of Fig. 7.1a, where the two subscript indices denote different regions in which the fields are influenced in different ways by the waveguide walls after the medium parameters change.

The solution of (7.24) becomes easier by using the Laplace transform with respect to the time variable and the Fourier transform with respect to spatial coordinate. In the zone {00}, the operator $\hat{R}$ is not affected by the waveguide boundary, as shown in time-spatial diagram in Fig. 7.1b, where the integration region in (7.24) is indicated by horizontal hatching. Using the Laplace–

Fourier transform, we reduce the integral Eq. (7.24) to an algebraic one with respect to the resolvent kernel. Next, the inverse Laplace–Fourier transformation gives the kernel of the resolvent operator in a time-spatial representation which in zone {00} is equal to

$$\left\langle \mathbf{x} \,|\, \hat{R}_{00} \,|\, \mathbf{x}' \right\rangle = -\frac{\varepsilon_2 - \varepsilon}{\varepsilon_2} \frac{1}{2v_2} \hat{W}_0(t - t', |x - x'|), \tag{7.25}$$

where the $\hat{W}_0$ is defined by the expression

$$\hat{W}_0(t - t', |x - x'|) = \partial_{tt}^2 J_0\left(v_2 \Gamma \sqrt{(t - t')^2 - \frac{(x - x')^2}{v_2^2}} \right) \theta\left(t - t' - \frac{|x - x'|}{v_2} \right) \tag{7.26}$$

and $v_2 = c / \sqrt{\varepsilon_2}$.

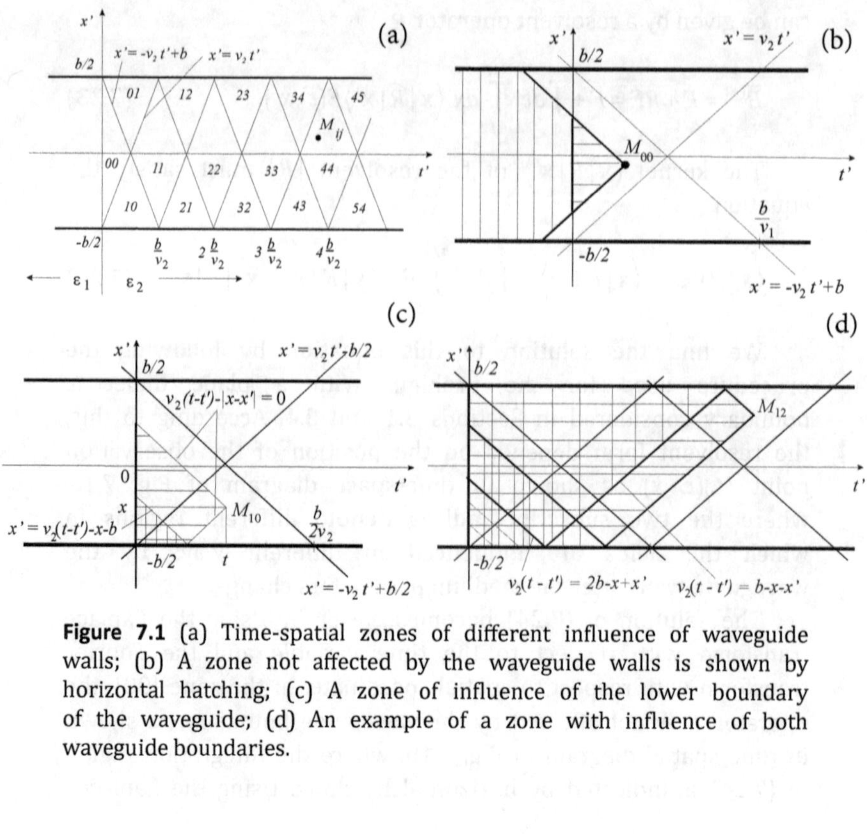

Figure 7.1 (a) Time-spatial zones of different influence of waveguide walls; (b) A zone not affected by the waveguide walls is shown by horizontal hatching; (c) A zone of influence of the lower boundary of the waveguide; (d) An example of a zone with influence of both waveguide boundaries.

In the other zones shown in Fig. 7.1a, the kernel of the integral equations is affected by the waveguide boundaries. In the zone {10} the kernel, and consequently the field, experiences an influence from the lower boundary of the waveguide, as seen in Fig. 7.1c.

The resolvent kernel for Eq. (7.19) in this zone obtained as above is equal to

$$\begin{aligned}\left\langle \mathbf{x} \,|\, \hat{R}_{10} \,|\, \mathbf{x}' \right\rangle &= \frac{\varepsilon_2 - \varepsilon}{2v_2\varepsilon_2}\,\theta(x + b/2)\theta(b/2 - x)\{-\hat{W}_0(t - t', |x - x'|) \\ &\quad + \hat{W}_1(t - t', x + x' + b)\}\theta(x' + b/2)\theta(b/2 - x'). \qquad (7.27)\end{aligned}$$

The additive function

$$\hat{W}_l(t, x) = \int_{\eta - i\infty}^{\eta + i\infty} \frac{dp}{2\pi i} p^2 \frac{R(p)^l}{\varphi_2(p)} e^{pt - \frac{\varphi_2(p)}{v_2}x}, \qquad (7.28)$$

which has index $l = 1$ in this case, is distinct from zero only in the region with vertical hatching shown in Fig. 7.1c and the influence of the lower boundary of the waveguide is present by virtue of the coefficient $R(p)$

$$R(p) = \frac{v_2\varphi(p) - v\varphi_2(p)}{v_2\varphi(p) + v\varphi_2(p)},\ \varphi_2(p) = \sqrt{p^2 + v_2^2\Gamma^2},\ \varphi(p) = \sqrt{p^2 + v^2\Gamma^2}. \qquad (7.29)$$

This coefficient plays the role of a reflectance and takes into account a single reflection of the influence line $v_2(t - t') = x - x'$ from the lower boundary of the waveguide. The line $v_2(t - t') = x - x'$ is one of the characteristics of the problem along which information about the field state propagates.

The influence of the upper boundary of the waveguide is likewise observed in the fields of zone {01} where the expression for the resolvent can be found analogously.

Successive solutions of (7.24) now give the total expression for the resolvent in the zone $\{mn\}$ as

$$\langle \mathbf{x} \,|\, \hat{R} \,|\, \mathbf{x}' \rangle = \langle \mathbf{x} \,|\, \hat{R}_0 \,|\, \mathbf{x}' \rangle + \langle \mathbf{x} \,|\, \hat{R}_{mn} \,|\, \mathbf{x}' \rangle, \qquad (7.30)$$

where

$$\langle \mathbf{x} | \hat{R}_0 | \mathbf{x}' \rangle = \frac{v^2 - v_2^2}{v^2} \frac{\theta(x + b/2)\theta(b/2 - x)}{2v_2} \langle \mathbf{x} | \hat{R}_{00} | \mathbf{x}' \rangle \theta(x' + b/2)\theta(b/2 - x') \tag{7.31}$$

and

$$\langle \mathbf{x} | \hat{R}_{mn} | \mathbf{x}' \rangle = \frac{v^2 - v_2^2}{v^2} \frac{\theta(x + b/2)\theta(b/2 - x)}{2v_2} \left\{ \sum_{l=1}^{m} \hat{W}_l(t - t', lb + x - (-1)^{i_l} x') + \sum_{l=1}^{n} \hat{W}_l(t - t', lb - x + (-1)^{i_l} x') \right\} \theta(x' + b/2)\theta(b/2 - x'), \tag{7.32}$$

where the symbol $i_l = \frac{1-(-1)^l}{2}$ is introduced and the operator-functions $\hat{W}_l$ are determined by (7.28). These functions take into account k reflections from the lower and upper boundaries, an example of which is shown in Fig. 7.1d for the case of two reflections.

Using the expression (7.28) the summation in (7.32) can be implemented explicitly, giving the resolvent in the form of the integral

$$\langle \mathbf{x} | \hat{R} | \mathbf{x}' \rangle = \frac{v^2 - v_2^2}{2v^2 v_2} \int_{\eta - i\infty}^{\eta + i\infty} \frac{dp}{2\pi i} \frac{p^2}{\varphi_2(p)} e^{p(t-t')} \left[R_0(p,x,x') + R_{mn}(p,x,x') \right] \tag{7.33}$$

The part of the resolvent that is common for all zones is determined by the expression

$$R_0(p,x,x') = -e^{-\frac{\varphi_2}{v_2}|x - x'|} \tag{7.34}$$

and each zone gives the contribution determined by

$$R_{mn}(p,x,x') = \frac{1}{1 - \tilde{R}^2} \{ C_{mn}(x) e^{-x'\varphi_2/v_2} + D_{mn}(x) e^{x'\varphi_2/v_2} \}. \tag{7.35}$$

Here,

$$C_{mn}(x) = A_m e^{-x\varphi_2/v_2} + B_n e^{x\varphi_2/v_2}, \; D_{mn}(x) = B_m e^{-x\varphi_2/v_2} + A_n e^{x\varphi_2/v_2}, \tag{7.36}$$

$$A_m = \tilde{R}(1 - \tilde{R}^{m+jm}),\ B_m = \tilde{R}^2(1 - \tilde{R}^{m-jm}),\ \tilde{R} = R(p)e^{-\varphi_2 b/v_2}. \quad (7.37)$$

7.4 Harmonic Wave Transformation Caused by a Permittivity Change in the Waveguide Core

The resolvents constructed in the previous section allow the transformation of an initial field in a waveguide with a time-varying core medium to be evaluated. Here we consider the transformation of an initial (source) field, which is a guided mode $\bar{B}^{(-)}(t, x) = e^{i\omega_0 t}$ $\cos\kappa_1 x$ that existed in the waveguide until the moment when the permittivity in the core abruptly changed. We assume that the undisturbed field $B^{(0)}$ in the free term (7.22) has such a form that the source field for $t < 0$ satisfies the equation below, following from (7.14)

$$\bar{B}^{(-)} = e^{i\omega_0 t}\cos\kappa_1 x = \bar{B}^{(0)} - \frac{v^2 - v_1^2}{2vv_1^2}\partial_{tt}^2 \int_{-\infty}^{t} dt' \int_{-b/2}^{b/2} dx' Q(t,t',x,x') e^{i\omega_0 t'}\cos\kappa_1 x', \quad (7.38)$$

where the transverse wavenumber κ_1 is a root of the dispersion equations $(\kappa + \kappa_1)^2 e^{i\kappa_1 b} - (\kappa - \kappa_1)^2 e^{-i\kappa_1 b} = 0$, $\kappa^2 - \kappa_1^2 = \omega_0^2 \frac{v_1^2 - v^2}{v_1^2 v^2}$. This field governs the free term (7.22), the calculation of which can be represented as a sum of two terms

$$F = -\frac{v^2 - v_1^2}{2vv_1^2}\partial_{tt}^2 \int_{-\infty}^{0} dt' \int_{-b/2}^{b/2} dx' Q(t,t',x,x')\bar{B}^{(-)}(t',x') = F_1 + F_2. \quad (7.39)$$

The first term in this sum

$$F_1 = \frac{v}{2v_1^2}[(v + v_1)e^{it\omega_0 v/v_1} + (v - v_1)e^{-it\omega_0 v/v_1}]\cos\kappa_1 x \quad (7.40)$$

coincides with that for the case of an homogeneous unbounded medium. The second term, written in the form of the Laplace inverse transform, is caused by the waveguide:

$$F_2 = \frac{v_1^2 - v^2}{2v_1^2} v \int_{\eta - i\infty}^{\eta + i\infty} \frac{dq}{2\pi i} \frac{q^2}{\varphi(q)} e^{qt} \frac{\Phi(\varphi(q)/v)}{(q - i\omega)[q^2 + (\omega v/v_1)^2]} \left[e^{\varphi(q)\frac{x-b/2}{v}} + e^{-\varphi(q)\frac{x+b/2}{v}} \right]. \tag{7.41}$$

Here $\Phi(a) = a\cos(\kappa_1 b/2) - \kappa_1 \sin(\kappa_1 b/2)$, and $\varphi(q) = \sqrt{q^2 + (v\Gamma)^2}$. The second term turns to zero when the waveguide walls are removed to infinity, that is when $b \to \infty$.

In a problem with a waveguide, this term is also equal to zero if the observation point satisfies the inequality $vt - b/2 < x < -vt + b/2$. As in the waveguide $v/v_1 > 1$, it means that such an observation point is located in the zone {00}.

Action of the resolvent (7.33) on the free term (7.39) gives the field in the whole waveguide at any moment $t \geq 0$

$$\bar{B} = F + \hat{R}_0 F + \hat{R}_{mn} F \tag{7.42}$$

The first part of the resolvent gives the field at any observation point (t, x)

$$\hat{R}_0 F = \langle \mathbf{x} | \hat{R}_0 F | \mathbf{x}' \rangle$$
$$= \frac{v^2 - v_2^2}{2v^2 v_2} \int_{\eta - i\infty}^{\eta + i\infty} \frac{dp}{2\pi i} \frac{p^2}{\varphi_2(p)} \int_0^{\infty} dt' \int_{-b/2}^{b/2} dx' e^{p(t-t')} R_0(p,x,x')[F_1(t'x') + F_2(t'x')] \tag{7.43}$$

and the second part gives the field in the *mn* zone of the time-space diagram

$$\hat{R}_{mn} F = \langle \mathbf{x} | \hat{R}_{mn} F | \mathbf{x}' \rangle$$
$$= \frac{v^2 - v_2^2}{2v^2 v_2} \int_{\eta - i\infty}^{\eta + i\infty} \frac{dp}{2\pi i} \frac{p^2}{\varphi_2(p)} \int_0^{\infty} dt' \int_{-b/2}^{b/2} dx' e^{p(t-t')} R_{mn}(p,x,x')$$
$$[F_1(t',x') + F_2(t',x')]. \tag{7.44}$$

7.4.1 The Early Stage of the Transient

The result of the action of the first part of the resolvent yields

$$\hat{R}_0 F = -F(t,x) + \bar{B}^{(00)}(t,x) + \bar{B}^{(\mathrm{com})}(t,x). \tag{7.45}$$

The first term on the right-hand side of (7.45) cancels the free term in (7.42), the second one is the field as in an unbounded medium

$$\overline{B}^{(00)} = \frac{v_2}{2v_1^2}[(v_1 + v_2)e^{it\omega_0 v_2/v_1} + (v_2 - v_1)e^{it\omega_0 v_2/v_1}]\cos\kappa_1 x \tag{7.46}$$

and the last term gives the transient field common for all spatial-time zones

$$\begin{aligned}\overline{B}^{(\text{com})}(t,x) = \frac{v_2}{v_1^2}\int_{\eta-i\infty}^{\eta+i\infty}\frac{dp}{2\pi i}\frac{p^2}{\varphi_2}\frac{e^{pt-\varphi_2(x+b/2)/v_2} + e^{pt+\varphi_2(x-b/2)/v_2}}{p^2 + (v\omega_0/v_1)^2} \\ \times\left[\frac{(v^2 - v_2^2)(p + i\omega_0)\Phi(\varphi_2/v_2)}{p^2 + (v_2\omega_0/v_1)^2}\right. \\ \left.-\frac{(v^2 - v_1^2)\Phi(\varphi/v)}{2v_2\varphi(p - i\omega_0)}\{(v\varphi_2 + v_2\varphi) - (v\varphi_2 - v_2\varphi)e^{-\varphi b/v}\}\right],\end{aligned} \tag{7.47}$$

where $\varphi_2 = \sqrt{p^2 + v_2^2\Gamma^2}$, $\varphi = \sqrt{p^2 + v^2\Gamma^2}$.

This term is evanescent in the unbounded space as it turns to zero when $b \to \infty$. However, this term is also equal to zero in the zone {00}, so the expression (7.46) represents the whole field at the early transient stage in the waveguide. This field consists of two waves, both with a new frequency $v_2\omega/v_1$ that propagate in opposite directions. This result agrees with the case of a change of the permittivity of a medium in a simple unbounded space considered in Chapter 2. A jump of the core permittivity does not change the transverse distribution of the field. This directly follows from the fact that in this zone, which is in the early stage of the transient, the waveguide boundaries have no effect.

7.4.2 Waves Spectra Generated by a Permittivity Time Jump

To see the structure of the transformed field in the subsequent zones where the influence of the waveguide walls comes into play, we analyze Eq. (7.47) and the equation that is the action of the second part of the resolvent giving the field in the *mn* zone

$$\overline{B}^{(mn)}(t,x)=\frac{1}{2v_1^2}\int_{\eta-i\infty}^{\eta+i\infty}\frac{dp}{2\pi i}\frac{p^2}{\varphi_2}\frac{L_m e^{pt-\varphi_2 x/v_2}+L_n e^{pt+\varphi_2 x/v_2}}{p^2+(v\omega_0/v_1)^2}$$
$$\times\left[\frac{2(v^2-v_2^2)(p+i\omega_0)\{\varphi_2\cos(\kappa_1 b/2)\sinh(\varphi_2 b/2v_2)+v_2\kappa_1\sin(\kappa_1 b/2)\cosh(\varphi_2 b/2v_2)\}}{p^2+(v_2\omega_0/v_1)^2}\right.$$
$$\left.+\frac{(v^2-v_1^2)\Phi(\varphi/v)}{\varphi(p-i\omega_0)}\{-v\varphi_2(1+e^{-\varphi b/v})\sinh(\varphi_2 b/2v_2)+v_2\varphi(1-e^{-\varphi b/v})\cosh(\varphi_2 b/2v_2)\}\right]. \tag{7.48}$$

Here, the coefficients L_l, $l = m, n$,

$$L_l=\tilde{R}\frac{1-\tilde{R}^l}{1-\tilde{R}}, \tag{7.49}$$

in which $\tilde{R}$ is as in (7.49), correspond to the mn zone. The real influence of the waveguide walls will be in the zones where at least one of the numbers m,n is non-zero.

The integrands in (7.47) and (7.48) have the same singular points: poles at the points $p_1^{\pm}=\pm iv\omega_0/v_1$, $p_2^{\pm}=\pm iv_2\omega_0/v_1$ and $p_3 = i\omega_0$ and branch points $p_{b1}^{\pm}=\pm iv\Gamma$, $p_{b2}^{\pm}=\pm iv_2\Gamma$. The integrand in (7.48) also has additional singular points where the denominator of (7.49) turns to zero.

Equations (7.47) and (7.48) show that transients in the transformed field are very significant; this can be seen clearly with this field spectra, which is obtained from the integrands (7.47) and (7.48) by substituting $p = i\omega$. The transient is accompanied by the transform of the monochromatic line spectrum of the original field into the complex spectrum consisting of the continuous spectrum and the banded one.

The spectrum of the common part of the field (Fig. 7.2) consists of the bands corresponding to the poles of the integrand. The poles $p_1^{\pm}=\pm iv\omega_0/v_1$ give two waves propagating in opposite directions with a new frequency and the transverse wavenumber $\pm\sqrt{(v\omega_0/v_1v_2)^2-\Gamma^2}$. The poles $p_2^{\pm}=\pm iv_2\omega_0/v_1$ give two waves with new frequency and the transverse wavenumber $\pm\kappa_1$ as in the original wave, and the pole $p_3 = i\omega_0$ gives the wave with the same frequency as the original one but the new transverse wavenumber $\kappa_2=\sqrt{(\omega/v_2)^2-\Gamma^2}$. These two latter waves are guided ones in the waveguide with the new core, because the pairs $\omega_0 v_2/v_1$, κ_1 and ω, κ_2 satisfy the dispersion equation corresponding to the new phase velocity in the core after its permittivity change. It is clearly confirmed by the behavior of the corresponding waves outside of the core.

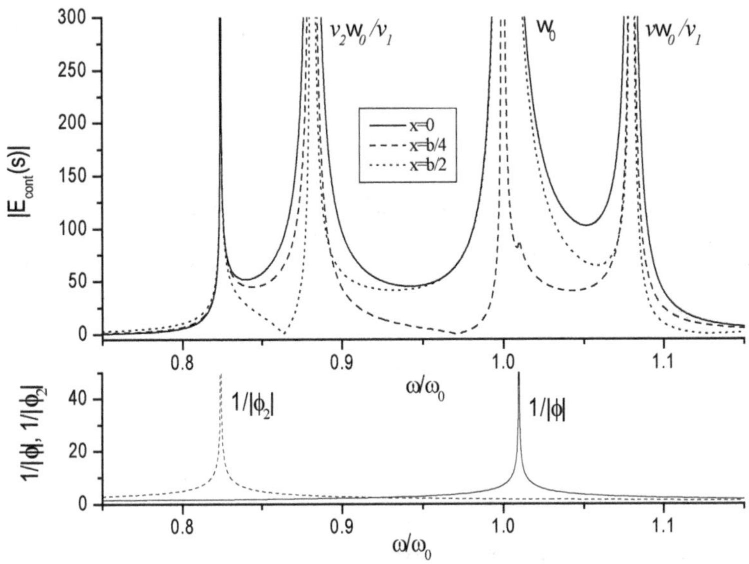

Figure 7.2 The band spectrum of the field that is common for all time-spatial zones.

This external field is easily calculated by the expression (7.19) if one considers an observation point beyond the waveguide core. In this case, the behavior of the field outside of the core is determined by the integral ($x \geq b/2$)

$$\overline{B}_{\text{out}} \sim \int_{-b/2}^{b/2} dx' \int_{0}^{t-(x-x')/v} dt' J_0(v\Gamma\sqrt{(t-t')^2-(x-x')^2/v^2})\overline{B}(t',x'). \quad (7.50)$$

Far away from the waveguide, and for a long-term time transient, this integral can be presented in the form

$$\overline{B}_{\text{out}} \approx \int_{x-b/2}^{x+b/2} dx' \int_{x'/v}^{\infty} dt' J_0(v\Gamma\sqrt{t'^2-(x'/v)^2})\overline{B}(t-t',x-x'). \quad (7.51)$$

Using this formula, we can estimate the field outside of the waveguide induced by the corresponding inner wave. If the latter has the form $\overline{B}(t,x) = e^{i\omega_s t}\cos(\kappa_s(\omega_s)x)$, $\kappa_s = \sqrt{(\omega_s/v_2)^2 - \Gamma^2}$, then the result of the integration has a different form depending on the frequency of the inner wave. If $(v\Gamma)^2 - (\omega_s)^2 \geq 0$ then

$$\bar{B}_{\text{out}} \sim \cos(\omega_s t)[A(\omega_s)e^{-(x+b/2)\sqrt{\Gamma^2-(\omega_s/v)^2}} + B(\omega_s)e^{-(x-b/2)\sqrt{\Gamma^2-(\omega_s/v)^2}}] \tag{7.52}$$

If $(\omega_s)^2 - (v\Gamma)^2 \geq 0$ then

$$\begin{aligned}\bar{B}_{\text{out}} \sim\ & A_1(\omega_s)\cos[\omega_s t-(x+b/2)(\sqrt{(\omega_s/v_2)^2-\Gamma^2}+\sqrt{(\omega_s/v)^2-\Gamma^2})] \\ & + A_2(\omega_s)\cos[\omega_s t+(x+b/2)(\sqrt{(\omega_s/v_2)^2-\Gamma^2}-\sqrt{(\omega_s/v)^2-\Gamma^2})] \\ & - A_3(\omega_s)\cos[\omega_s t-(x-b/2)(\sqrt{(\omega_s/v_2)^2-\Gamma^2}+\sqrt{(\omega_s/v)^2-\Gamma^2})] \\ & - A_4(\omega_s)\cos[\omega_s t+(x-b/2)(\sqrt{(\omega_s/v_2)^2-\Gamma^2}-\sqrt{(\omega_s/v)^2-\Gamma^2})]\end{aligned} \tag{7.53}$$

Analysis shows that $\Gamma^2 - (\omega_0 v_2/v v_1)^2 \geq 0$ and $\Gamma^2 - (\omega_0/v)^2 \geq 0$, so, the inner waves corresponding to the poles $p_2^{\pm} = \pm i v_2\omega_0/v_1$ and $p_3 = i\omega_0$ induce evanescent external waves while the inner wave corresponding to $p_1^{\pm} = \pm i v\omega_0/v_1$ gives a propagating external wave, since $\Gamma^2 - (\omega/v_1)^2 \leq 0$ and $(\omega_0 v/v_1 v_2)^2 - \Gamma^2 \geq 0$. This means that this wave, $p_1^{\pm} = \pm i v\omega_0/v_1$, is a radiated wave and it diminishes in the core with time.

The spectrum in Fig. 7.2 also has two bands determined by the branch points corresponding to zeros of the functions $\varphi(p)$, $\varphi_2(p)$ that are shown in the lower part of Fig. 7.2. But calculation of the residues in these points gives zero, so there will not be any separate waves corresponding to these bands.

The width of the bands depends on the location of the observation points in the core, as is seen clearly in Fig. 7.3 where the detailed structure of one of the bands is shown. The bandwidth has the smallest magnitude at $x \approx b/4$ and it widens towards the middle of the core as well as towards the waveguide walls.

The spectrum corresponding to the particular zone is shown in Fig. 7.4. It has peculiarities that are similar to the common spectrum of the band determined by the denominator of the coefficient (7.49).

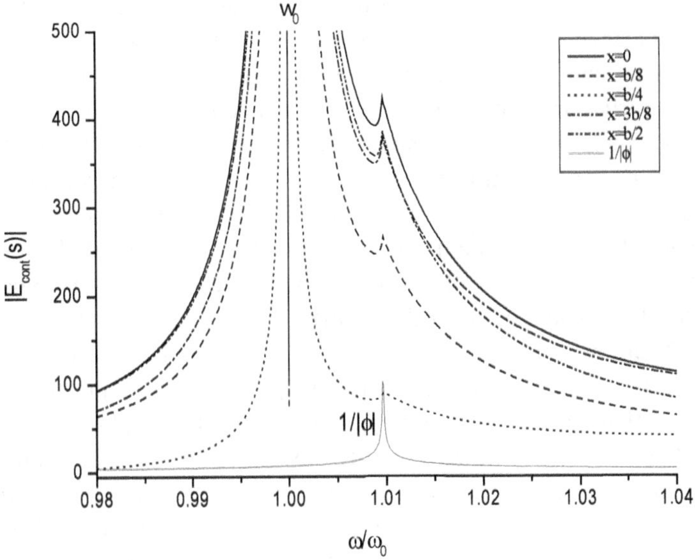

Figure 7.3 The detailed structure of the band of the common field spectrum corresponding to the frequency of the initial wave.

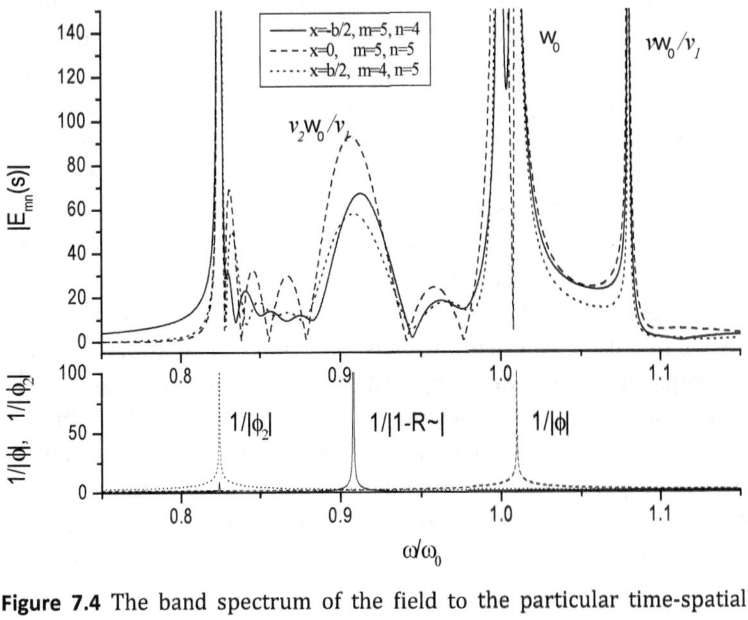

Figure 7.4 The band spectrum of the field to the particular time-spatial zone.

This spectrum changes with an increase in the zone numbers, and its asymptotic form when $m,n \to \infty$ is shown in Fig. 7.5. The bandwidth has a similar dependence on the location of the observation point as in the common spectrum (Fig. 7.6).

The joint asymptotic spectrum has the peculiarities of the common and asymptotic spectra (Fig. 7.7). Excitation of the continuous spectrum of waves is similar to the phenomenon observed above when a harmonic plane wave is obliquely incident onto a flat interface between two dielectrics one of whose permittivity changes abruptly in time.

Thereby, analysis of the transformation of an initial guided mode caused by an abrupt time change in the core permittivity shows that in the early stages of the transient it splits into two guided waves with new frequencies, which propagate in opposite directions. This result agrees with the case of an unbounded space considered in Chapter 2, namely, that a jump of the permittivity in the waveguide does not change the spatial structure of the field. A consequence of this fact is that the waveguide walls have no influence at this stage of the process. Subsequently, the influence of the waveguide walls reveals itself and analysis shows a significant complication of the structure of the transformed field. Two new kinds of guided waves appear. One of these waves has the same frequency as the initial field but has a new transverse wavenumber. This wavenumber satisfies the dispersion equation corresponding to the new phase velocity in the core after its change of permittivity. The second guided wave has a new frequency but the same transverse wavenumber as the initial wave. The field corresponding to these waves outside of the core has an evanescent form. There is also a separate wave in the core that has a radiation character because its outer field presents itself a propagating non-decaying wave. In addition to the guided waves, a continuous spectrum of waves is also excited that is similar to the phenomenon considered in Chapter 4 when a harmonic plane wave is obliquely incident onto a flat interface between two dielectrics, one of whose permittivity changes abruptly in time.

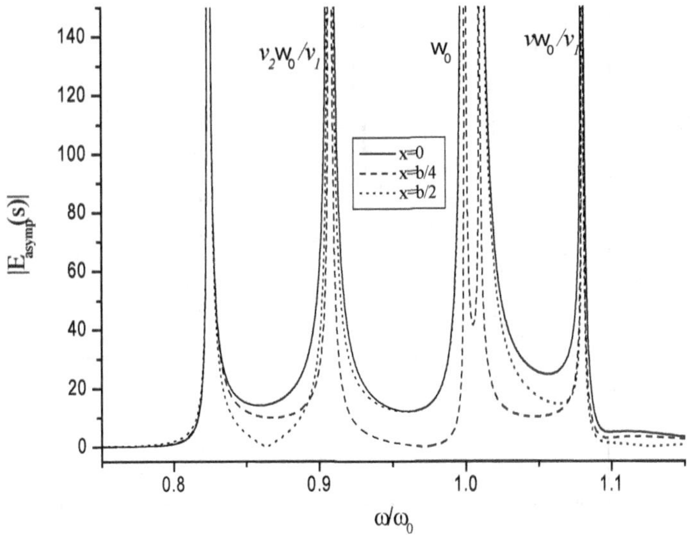

Figure 7.5 Asymptotic form of the spectrum with $m,n \to \infty$.

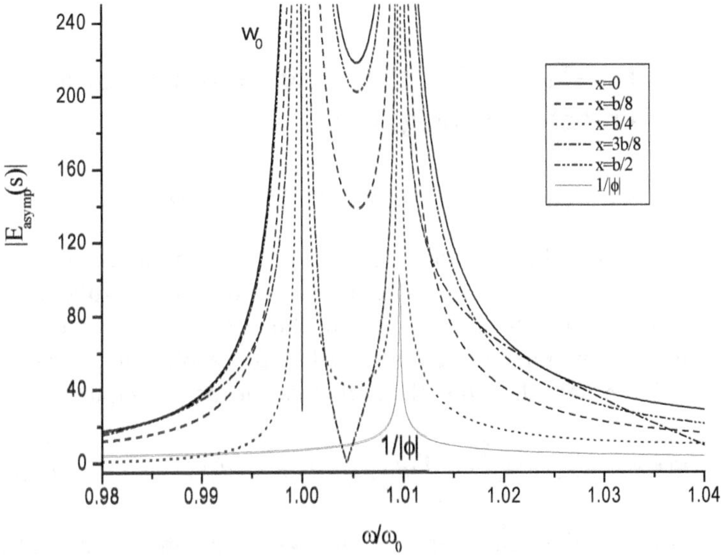

Figure 7.6 Dependence of the bandwidth on the observation point location.

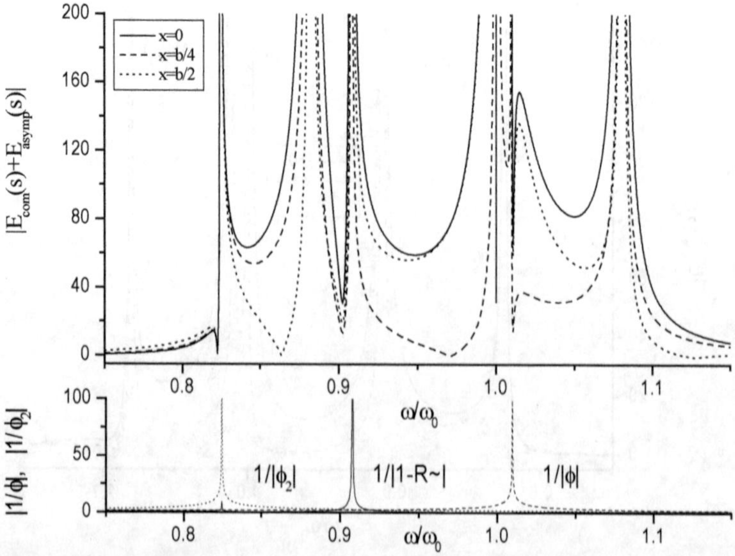

Figure 7.7 The complete spectrum taking into account the influence of the waveguide walls.

7.5 Transformation of a Wave in a Non-Linear Dielectric Waveguide

7.5.1 Step-Like Description of Field Evolution

In the case of arbitrary time variation of medium parameters, the resolvent to the integral equation cannot be found explicitly. In this case, the resolvent constructed above for an abrupt change of medium parameters can be used if the evolution process is divided into step-like intervals. To do this, the integral equation

$$\mathbf{E}(t,\mathbf{r}) = \mathbf{E}_0(t,\mathbf{r}) + \int_{-\infty}^{\infty} dt' \int_{V(t')} d\mathbf{r}' \hat{K}(t-t',\mathbf{r},\mathbf{r}',t')\mathbf{E}(t',\mathbf{r}') \qquad (7.54)$$

is written as the sum over time intervals separated by an initial moment t_n

$$\mathbf{E}(t,\mathbf{r}) = \tilde{\mathbf{F}}(t,\mathbf{r}) + \int_{t_n}^{\infty} dt' \int_{V(t')} d\mathbf{r}' \hat{K}(t-t',t',\mathbf{r},\mathbf{r}')\mathbf{E}(t',\mathbf{r}'). \qquad (7.55)$$

The free term in this equation is determined by a source field $\mathbf{E}_0$, the prehistory of the field before zero moment and by the fields in the preceding interval from zero moment until t_n.

$$\tilde{\mathbf{F}}(t,\mathbf{r}) = \mathbf{E}_0(t,\mathbf{r}) + \int_{-\infty}^{0} dt' \int_{V(t')} d\mathbf{r}' \hat{K}(t-t',\mathbf{r},\mathbf{r}',t')\mathbf{E}'_0(t',\mathbf{r}')$$

$$+\int_{0}^{t_n} dt' \int_{V(t')} d\mathbf{r}' \hat{K}(t-t',\mathbf{r},\mathbf{r}',t')\mathbf{E}(t',\mathbf{r}') \tag{7.56}$$

The kernel in this equation involves two factors, $\hat{K}(t-t',\mathbf{r},\mathbf{r}',t') = \hat{G}(t-t',|\mathbf{r}-\mathbf{r}'|)\hat{V}(t',\mathbf{r}')$, where $\hat{G}$ is a propagator, which is determined by the background medium parameters only and does not depend on the medium parameters inside the region $V(t)$, and the operator $\hat{V}$, which just describes the non-stationary medium in this region.

The solution to (7.54) can be derived explicitly only in few special cases and the simplest one of these is when the operator $\hat{V}$ does not depend on time after some moment t_n. However, a jump changing of the medium parameters at that moment is permissible. In this case, one can use the resolvent operator $\hat{R}$ to Eq. (7.55) and obtain the solution in the form

$$\mathbf{E}(t,\mathbf{r}) = \tilde{\mathbf{F}}(t,\mathbf{r}) + \int_{t_n}^{\infty} dt' \int_{V(t')} d\mathbf{r}' \hat{R}(t-t',\mathbf{r},\mathbf{r}')\mathbf{F}(t',\mathbf{r}'). \tag{7.57}$$

The dependence of the resolvent on the difference of the time arguments is very important for further consideration and can be used to solve a non-stationary problem with an arbitrary time change of the medium parameters beginning from zero moment if the time changes are approximated by a sequence of step functions. It is shown schematically for the medium operator in Fig. 7.8.

It is enough to find the resolvent operator $\hat{R}(\tau - \tau', \mathbf{r}, \mathbf{r}')$ to the equation (7.55) for the jump change of the medium parameters at some moment when $\hat{V}(t_n + \tau) = V_n = \text{const}$ for $\tau > 0$, that is to obtain the relation $\hat{G}(t-t',|\mathbf{r}-\mathbf{r}'|)\hat{V}_n(\mathbf{r}') \Rightarrow \hat{R}_n(t - t', \mathbf{r}, \mathbf{r}')$. Then one can easily construct the solution to (7.55) with the medium operator approximated by step-constant operators. The resolvents

$\hat{R}_n(t - t', \mathbf{r}, \mathbf{r}')$ for all moments t_n differ from each other by the medium parameter magnitudes only.

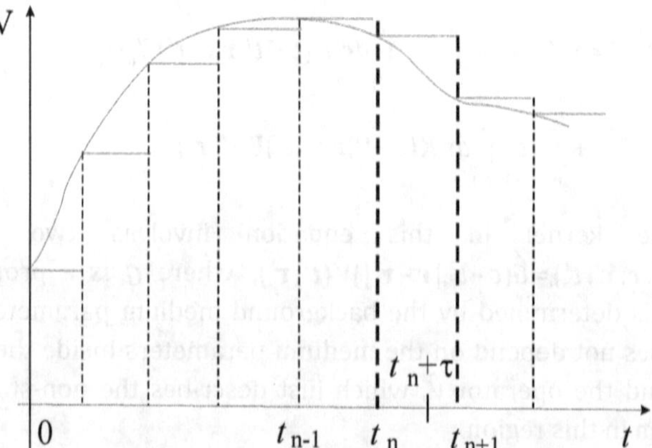

Figure 7.8 Approximation of the medium properties by step functions.

Thus, to solve a non-stationary boundary value problem it is important to know the resolvent for the initial and boundary value problem for the electromagnetic field in the medium whose parameters change in time by a jump at the initial moment. Then an arbitrary change of the parameters can be approximated by a sequence of such sharp changes. The simplest boundary is a plane and the resolvent for the problem with one such boundary has been obtained in Chapter 3, where a plane interface between two dielectric media has been considered. The resolvent for a problem with two parallel interfaces has been obtained in Section 7.3. There the existence of time-spatial zones is shown, in which the resolvents as well as electromagnetic processes described are different. Here these ideas and results are used to construct a tool for analytic-numerical investigations of electromagnetic transients in non-linear planar waveguides.

7.5.2 The Step-Resolvent Method for the Waveguide

Now we apply the step-like method to the dielectric waveguide where a TE electromagnetic signal propagates (Fig. 7.9).

The integral equation for the transverse electric field follows from (7.7) and can be written in the form

$$\mathbf{E}_{\perp} = \mathbf{F}_{\perp E} - \frac{v}{2}\partial_{tt}^{2}\int_{-\infty}^{t} dt' \int_{z-v(t-t')}^{z+v(t-t')} I_{0}\left(\frac{\alpha}{v}\sqrt{v^{2}(t-t')^{2}-(z-z')^{2}}\right) e^{\alpha(t-t')}\chi_{[b]}\hat{V}(t')\mathbf{E}_{\perp}dz', \tag{7.58}$$

where I_0 is the modified Bessel function, $\alpha = -\frac{v^2\mu_0\sigma}{2}$ takes into account the dissipation in the cladding, $c = 1/\sqrt{\varepsilon_0\mu_0}$ is the velocity of light in a vacuum, $v = c/\sqrt{\varepsilon}$ is the phase velocity in the cladding (background), the function $\chi_{[b]}$ equals unity in the core and zero outside of it, $\chi_{[b]} = 1$ if $-b/2 \le x \le b/2$ otherwise $\chi_{[b]} = 0$, and $\hat{V}$ is the medium operator, a general form of which is

$$\hat{V}(t) = \frac{1}{v^2}\left(\frac{\hat{\varepsilon}_r(t)}{\varepsilon} - 1\right) + \mu_0\int^{t} dt'(\hat{\sigma}_r(t') - \sigma), \tag{7.59}$$

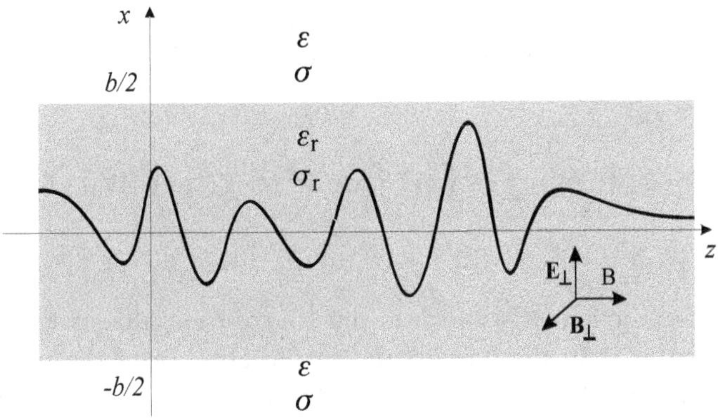

Figure 7.9 Slab waveguide media and orientation of the co-ordinate system.

The free term of Eq. (7.58) is expressed via the longitudinal component of the magnetic field

$$\mathbf{F}_{\perp E} = -\frac{v}{2}\int_{-\infty}^{t} dt' \int_{z-v(t-t')}^{z+v(t-t')} I_{0}\left(\frac{\alpha}{v}\sqrt{v^{2}(t-t')^{2}-(z-z')^{2}}\right) e^{\alpha(t-t')}[\mathbf{e}_3, \partial_{t'}\nabla_{\perp}B_3]dz', \tag{7.60}$$

and it is defined in both the waveguide core and the cladding. It is worth noting that (7.58) is the Volterra integral equation for the transverse components only inside the core, where $\chi_{[b]} \equiv 1$.

In the cladding, the transverse components are simply equal to the free terms, $\mathbf{E}_\perp = \mathbf{F}_{\perp E}$.

To calculate the free term (7.60) one has to know the longitudinal component of the magnetic field. This component is described by the self-consistent equation following from (7.14)

$$B(t,\mathbf{r}) = F_B(t,\mathbf{r}) - v\int_0^\infty dt' \int_{-\infty}^\infty dz' \int_{-b/2}^{b/2} dx' G^{(2)}(t-t',|\mathbf{r}-\mathbf{r}'|)\hat{V}(t')B(t',\mathbf{r}'), \tag{7.61}$$

where the propagator is determined by the function below, following from (7.12)

$$G^{(2)}(t-t',|\mathbf{r}-\mathbf{r}'|) = \frac{1}{2\pi}\partial_{tt}^2 e^{\alpha(t-t')} \frac{\cosh\left(a\sqrt{v^2(t-t')^2-|\mathbf{r}-\mathbf{r}'|^2}\right)}{\sqrt{v^2(t-t')^2-|\mathbf{r}-\mathbf{r}'|^2}}\theta(t-t'-|\mathbf{r}-\mathbf{r}'|/v), \tag{7.62}$$

and $\mathbf{r} = (x, z)$ is a two-dimensional vector.

The free terms are defined by the prehistory of just the longitudinal component

$$F_B(t,\mathbf{r}) = B^{(0)}(t,\mathbf{r}) - v\int_{-\infty}^0 dt' \int_{-\infty}^\infty dz' \int_{-b/2}^{b/2} dx' G^{(2)}(t-t',|\mathbf{r}-\mathbf{r}'|)\hat{V}(t')B(t',\mathbf{r}'), \tag{7.63}$$

and the function $B^{(0)}$, which is the source field, that is the field that can exist in the background in the absence of the core (in the other words the primary or initial field). Equation (7.61), as well as the corresponding one for the transverse components (7.58), is a proper Volterra integral equation only with respect to the unknown fields in the core. Outside the core the expressions (7.58) and (7.61) are quadrature formulas for calculating the fields in the cladding given the field in the core.

So, the electromagnetic field in the waveguide is governed in the time domain by Eqs. (7.58) and (7.61) that have the form similar to Eq. (7.54) and all reasoning in Sections 7.1–7.4 is applicable for them. As the transverse components are expressed in terms of the longitudinal ones, and, since the latter reveal the main features of the proposed algorithm, attention will be concentrated on them.

The initial problem with initial moment t_n for the longitudinal component B is described by the equation below that follows from (7.61) and is similar to (7.55)

$$B(t,\mathbf{r}) = \tilde{F}(t,\mathbf{r}) + \int_{t_n}^{\infty} dt' \int_{-\infty}^{\infty} dz' \int_{-b/2}^{b/2} dx' \hat{K}(t,t',|\mathbf{r}-\mathbf{r}'|)B(t',\mathbf{r}'), \quad (7.64)$$

where $\hat{K} = G^{(2)}(t-t',|\mathbf{r}-\mathbf{r}'|)\hat{V}(t')$, $\mathbf{r} = (x, z)$. The propagator in this equation is represented by the formula (7.62) and the medium operator determined by the expression (7.59) describes an arbitrary medium in the waveguide core. The free term in (7.64) is determined by the prehistory of the field before the moment t_n

$$\tilde{F}(t,\mathbf{r}) = B^{(0)}(t,\mathbf{r}) + \int_{-\infty}^{0} dt' \int d\mathbf{r}' \hat{K}(t-t',|\mathbf{r}-\mathbf{r}'|,t')B(t',\mathbf{r}')$$
$$+ \int_{0}^{t_n} dt' \int d\mathbf{r}' \hat{K}(t-t',|\mathbf{r}-\mathbf{r}'|,t')B(t',\mathbf{r}') \quad (7.65)$$

and contains the preceding intervals before the zero moment as well as from the zero moment till t_n. The arrangement of the observation point $(t, \mathbf{r})$ and the integration regions for the expression (7.65) (displayed by hatching) is shown on the time-spatial diagram of Fig. 7.10.

This statement allows one to consider various kinds of problems. For example, it is possible to investigate the case when the waveguide is created at zero moment, i.e., $\hat{\varepsilon}_r = \varepsilon$ until $t = 0$. Then $\hat{K}(t-t',|\mathbf{r}-\mathbf{r}'|,t') \equiv 0$ for $t' < 0$ and the first integral over the negative time axis in (7.65) is absent. The second example considers medium properties that are time changing under some external influence. Further, we consider the case when the non-linearity property of the medium in the waveguide core is "switched on" at zero moment. We also assume that material dissipation can be neglected, i.e., $\hat{\sigma}_r = \sigma = 0$. In this case, the medium operator is given by the expression

$$\hat{V}(t) = \frac{1}{v^2}\left(\frac{\varepsilon_d}{\varepsilon} - 1\right) + \frac{1}{c^2}\chi_{NL}^{(3)}E^2(t), \quad (7.66)$$

where ε_d is the linear part of the permittivity and $\chi_{NL}^{(3)}E^2$ is the non-linear susceptibility.

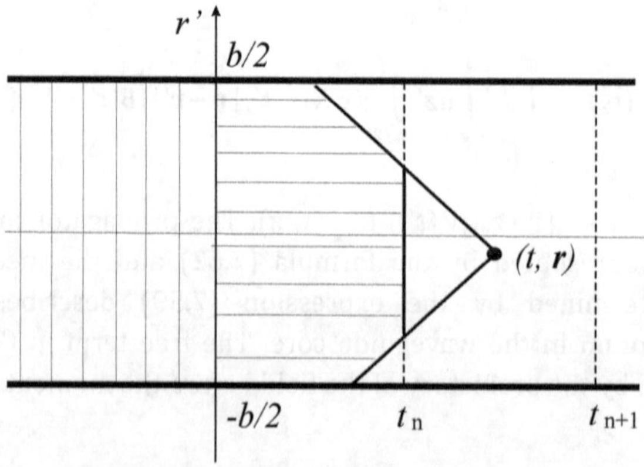

Figure 7.10 The influence (integration) regions for the free term of the integral equation.

Approximation of the time-varying properties of this medium by step functions, shown schematically in Fig. 7.11, is such that the continuous time-varying medium operator $\hat{V}(t)$ is replaced by a sequence of constant operators $\hat{V}_n = \text{const}$, each of which takes the value

$$V_n = \frac{1}{v^2}\left(\frac{\varepsilon_n}{\varepsilon} - 1\right), \tag{7.67}$$

in the nth time interval $t_n < t < t_{n+1}$. The permittivity varies in time in a step-wise manner and in the nth time interval is determined by the effective field magnitude $E_{\text{eff}\,n-1}$ at the previous interval $[t_{n-1}, t_n]$:

$$\varepsilon_n = \varepsilon_d + \chi_{NL}^{(3)} E^2_{\text{eff}\,n-1}. \tag{7.68}$$

The exact solution to Eq. (7.64) can be constructed using the corresponding resolvent operator

$$B(t,\mathbf{r}) = \tilde{F}(t,\mathbf{r}) + \int_{t_n}^{\infty} dt' \int_{-\infty}^{\infty} dz' \int_{-b/2}^{b/2} dx' \hat{R}_n(t-t',\mathbf{r},\mathbf{r}')\tilde{F}(t',\mathbf{r}') \tag{7.69}$$

To construct this operator, it is enough to know the resolvent for the initial problem with some initial moment, for example with zero initial moment $t_1 = 0$. Then for the given initial moment t_n and the corresponding equation kernel $\hat{K}_n = G^{(2)}(\tau - \tau', R)\hat{V}_n$ the resolvent $\hat{R}_n(\tau - \tau', \mathbf{r}, \mathbf{r}')$ is obtained by substituting corresponding values of medium parameters. In the frame of the approximation by step functions adopted, the formula (7.69) gives the exact solution to the problem on the transformation of the electromagnetic field in the waveguide with a non-linear material in the core. It is worth noting that Eq. (7.69) contains the known function $\tilde{F}$ under the integral instead of the unknown field, as in the initial Eq. (7.64). The formula (7.69) contains a powerful tool, the resolvent, which explicitly takes into account the medium properties change as well as influence of the waveguide walls.

A longitudinal uniformity of the waveguide allows us to use the Fourier representation with respect to the longitudinal coordinate z, $B(t,x,z) = \int_{-\infty}^{\infty} \bar{B}(t,x,\Gamma)e^{-i\Gamma z}d\Gamma$. This simplifies Eqs. (7.64) and (7.65), and facilitates seeking the resolvent. The integral equation takes the form

$$\bar{B} = F + \int_{t_n}^{\infty} dt' \int_{-b/2}^{b/2} dx' \langle \mathbf{x} | \hat{K} | \mathbf{x}' \rangle \bar{B}(t', x'), \tag{7.70}$$

where $\mathbf{x} = (t, x)$ is a time-spatial vector and the kernel in bracket notation is equal to

$$\langle \mathbf{x} | \hat{K}_n | \mathbf{x}' \rangle = \frac{\varepsilon - \varepsilon_n}{2v\varepsilon} \partial_{tt}^2 W_0(t - t', x - x'), \tag{7.71}$$

$$W_0(t,x) = J_0\left(\Gamma v \sqrt{t^2 - \frac{x^2}{v^2}}\right)\theta\left(t - \frac{|x|}{v}\right). \tag{7.72}$$

The free term in (7.70) follows from (7.65)

$$\tilde{F} = B^{(0)} - \partial_{tt}^2 \int_{-\infty}^{t_n} dt' \int_{-b/2}^{b/2} dx' \frac{\varepsilon(t') - \varepsilon}{2v\varepsilon}$$

$$J_0\left(\Gamma v \sqrt{(t - t')^2 - \frac{(x - x')^2}{v^2}}\right)\theta\left(t - t' - \frac{|x - x'|}{v}\right)\bar{B}(t', x') \tag{7.73}$$

and determines the solution to Eq. (7.70) by virtue of the resolvent operator $\hat{R}_n$

$$\bar{B} = \tilde{F} + \hat{R}_n\tilde{F} = \tilde{F} + \int_{t_n}^{\infty} dt' \int_{-b/2}^{b/2} dx' \langle \mathbf{x} | \hat{R}_n | \mathbf{x}' \rangle \tilde{F}(t', x'), \tag{7.74}$$

whose kernel must satisfy the operator equation

$$\langle \mathbf{x} | \hat{R}_n | \mathbf{x}' \rangle = \langle \mathbf{x} | \hat{K}_n | \mathbf{x}' \rangle + \int_{t_n}^{\infty} dt'' \int_{-b/2}^{b/2} dy'' \langle \mathbf{x} | \hat{K}_n | \mathbf{x}'' \rangle \langle \mathbf{x}'' | \hat{R}_n | \mathbf{x}' \rangle. \tag{7.75}$$

It was shown in Section 7.3 that the expressions for the resolvent are different in various zones of the time-spatial diagram. Thin solid lines in Fig. 7.11 separate these zones in the time interval $[t_n, t_{n+1}]$. These lines do not intersect each other if the condition $v_n \Delta t_n \leq b/2$ is fulfilled. Here $v_n = c/\sqrt{\varepsilon_n}$ is the phase velocity in the time interval $[t_n, t_{n+1}]$, $\Delta t_n = t_{n+1} - t_n$ is the duration of this interval.

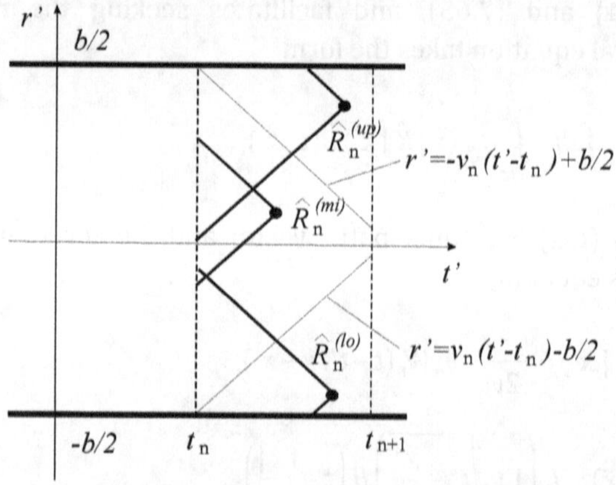

Figure 7.11 The zones of various influence of the waveguide walls in the time interval $[t_n, t_{n+1}]$. The bold points correspond to the observation point, beams coming from them restrict the integration region in (7.74).

This condition gives an estimation of the maximum duration of the n-th interval

$$\Delta t_n \leq \frac{b}{2c}\sqrt{\varepsilon_d + \chi_{NL}^{(3)} |E_{\text{eff}\,n-1}|^2} \tag{7.76}$$

that is admissible for the algorithm considered. The field E_{eff} is calculated from a formula which follows from Eq. (7.58), and which after integration with respect to x is reduced to

$$\left(1+\chi_{[b]}\frac{\varepsilon_n-\varepsilon}{\varepsilon}\right)\mathbf{E}_{\perp}(t,x,z)=-v^2\int_{-\infty}^{t}dt'\cos(v\Gamma(t-t'))[\mathbf{e}_3,\nabla_{\perp}B_3]$$
$$+v\Gamma\chi_{[b]}\frac{\varepsilon_n-\varepsilon}{\varepsilon}\int_{-\infty}^{t}dt'\sin(v\Gamma(t-t'))\mathbf{E}_{\perp}(t',x,z) \tag{7.77}$$

For jump time changing of the medium properties the solution to this equation is given by the formula

$$\mathbf{E}_{\perp}(t,x,z)=-v_n^2\int_{-\infty}^{0}dt'\cos(v\Gamma(t-t'))[\mathbf{e}_3,\nabla_{\perp}B_3]$$
$$-v_n^2\int_{0}^{t}dt'\cos(v_n\Gamma(t-t'))[\mathbf{e}_3,\nabla_{\perp}B_3] \tag{7.78}$$

In the middle zone the layer boundaries have no influence, while in the lower and upper zones only the presence of one of the walls reveals itself. The resolvent in the middle zone is equal to

$$\hat{R}_n^{(mi)}=-\frac{1-h_n^2}{2h_n}\partial_{tt}^2W_n^{(0)}(t-t',|x-x'|)\theta\left(t-t'-\frac{|x-x'|}{h_n}\right), \tag{7.79}$$

where [24]

$$\hat{W}_n^{(0)}(t,|x|)=J_0\left(\Gamma h_n\sqrt{t^2-\frac{x^2}{h_n^2}}\right),\quad h_n=\sqrt{\frac{\varepsilon}{\varepsilon_n}}, \tag{7.80}$$

$J_0(t)$ is the Bessel function, Γ is a variable of Fourier transform.

The lower and upper resolvents in the near-boundary zones get the augmentation

$$\hat{R}_n^{(\mathrm{lo,up})}=\hat{R}_n^{(mi)}+\Delta\hat{R}_n^{(\mp)} \tag{7.81}$$

$$\Delta\hat{R}_n^{(\mp)}=-\frac{1-h_n}{2h_n}\partial_{tt}^2W_n^{(nb)}(t-t',1\pm(x+x'))\theta(t-(1\pm(x+x'))/h_n), \tag{7.82}$$

where

$$W_n^{(nb)}(t,x) = (1-h_n)J_0(\Gamma h_n\sqrt{t^2-(x/h_n)^2}) + 2\Gamma h_n \int_0^{t-x/h_n} J_0(\Gamma h_n\sqrt{t^2-(h_n u + x)^2/h_n^2})U(u)du \quad (7.83)$$

with $U(u) = \sinh(\Gamma h_n u) - (1+h_n)\ \Gamma\int_0^u \cosh[\Gamma h_n(u-u')]\ \frac{J_1(\Lambda_n u')}{\Lambda_n u'}du'$, $\Lambda_n^2 = \Gamma^2(1-h_n^2)$, and $J_1(t)$ as the Bessel function.

7.5.3 Calculation Scheme for Time Steps Approximation

A numerical calculation scheme based on the resolvent algorithm described above is shown schematically in Fig. 7.12. The time interval investigated is divided into steps, whose length must not exceed

$$\Delta t_n \leq 1/2\sqrt{h_d^{-2} + \gamma E_{\text{eff}\,n-1}^{\;2}},$$

where $h_d^2 = \frac{\varepsilon}{\varepsilon_d}$, ε_d is a part of the linear permittivity of the core material and γ is a normalized non-linearity parameter $\gamma = \chi_{NL}^{(3)}|E_0|^2/\varepsilon$. Further, normalized variables are used: a time $t \to vt/b$; a spatial coordinate $x \to x/b$; and a normalized frequency $\omega \to b\omega/v$.

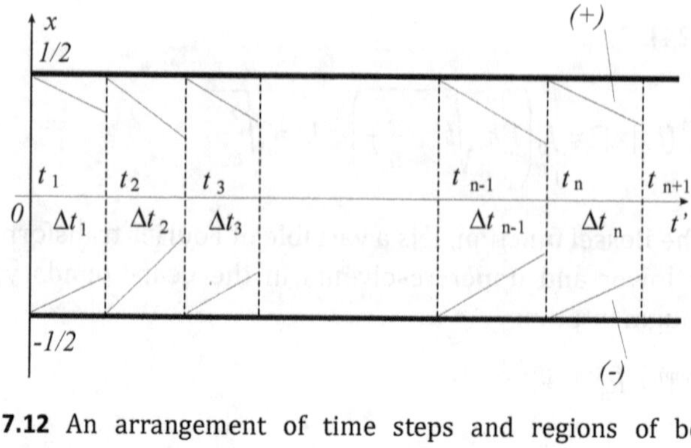

Figure 7.12 An arrangement of time steps and regions of boundary influence.

Implementation of this scheme includes the following operations: the definition of the zones of influence of the waveguide walls; calculation of the field in various regions using the proper resolvents; updating the free term of the integral equation after passing to the next time step by taking into account the field in the previous step.

This implementation uses the following formulae. The exact expression for the free term of the integral equation consists of two parts

$$F(t,x) = F_1(t,x) + F_2(t,x) \tag{7.84}$$

The first part of the free term is determined by the field pre-history until the beginning of the non-stationary process

$$F_1(t,x) = -\frac{H_0}{2}\partial_{tt}^2 \int_{-\infty}^{0} dt' \int_{x-t+t'}^{x+t-t'} W_0(t-t', x-x')\overline{B}_0(t',x')dx'. \tag{7.85}$$

The second part of the free term is determined by the field on the previous time steps after the beginning of the non-stationary process

$$F_2 = \sum_{k=1}^{n-1} \frac{H_k}{2} \int_{t_k}^{t_{k+1}} dt' \int_{x-t+t'}^{x+t-t'} W_0(t-t',x-x')\overline{B}(t',x')dx', \tag{7.86}$$

where $H_k = (1-h_k^2)/h_k^2 = \dfrac{\varepsilon_k - \varepsilon}{\varepsilon}$, $h_k^2 = \dfrac{\varepsilon}{\varepsilon_k}$.

The field in the middle zone (the main field), where the waveguide walls have no influence on the nth-time step considered, is calculated by the formula

$$\overline{B}^{(mi)}(t,x) = F - \frac{1-h_n^2}{2h_n}\partial_{tt}^2 \int_{t_n}^{t_{n+1}} dt' \int_{x-t+t'}^{x+t-t'} W_n^{(0)}(t-t',x-x')F(t',x')dx' \tag{7.87}$$

The integration region here is shown in Fig. 7.13.

The field in the lower or upper zones near the waveguide boundary consists of the main field and the augmentation of the field

$$\overline{B}^{(\text{lo,up})}(t,x) = \overline{B}^{(\text{main:lo,up})}(t,x) + \Delta\overline{B}^{(\text{lo,up})}(t,x) \tag{7.88}$$

The calculation of the main field in the near-boundary zones is made via the same formula (7.87) but taking into account a restriction of the integration region by the presence of the waveguide walls.

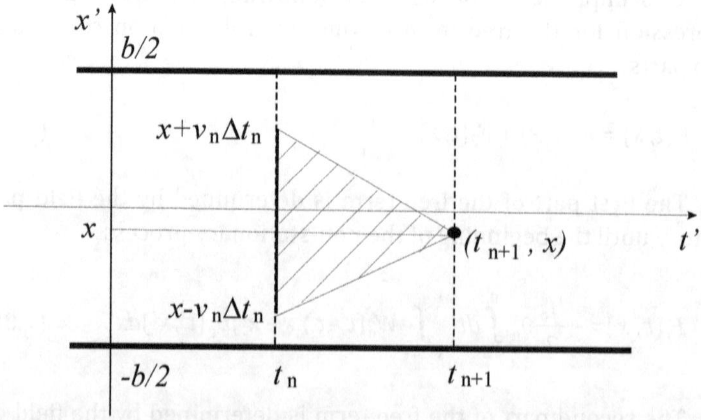

Figure 7.13 The integration interval for calculation of the field in the middle zone.

The augmentation to the field is determined via the formula

$$\Delta\bar{B}^{(\mathrm{lo,up})}(t,x) = -\frac{1-h_n}{2h_n}\partial_{tt}^2 \int_{t_n}^{t} dt' \int_{-1/2}^{1/2} W_n^{(nb)}(t-t', 1\pm(x+x'))$$

$$\theta\left(t-t'-\frac{1\pm(x+x')}{h_n}\right)F(t',x')dx', \qquad (7.89)$$

where the integration region is marked by hatching in Fig. 7.14.

To construct a direct numerical calculation scheme the problem region is discretized using a spatial discretization step Δx equal to Δt in normalized variables. Every approximation interval in time is constant: $\Delta t_n = \Delta t$. The discretization interval $\Delta = \Delta x = \Delta t$ should be quite small to represent field evolution in time and space smoothly; normally 20 intervals per wavelength is sufficient. The values of the field and the free-term are stored for every node of the defined regular quadratic mesh: $B_i^n = B(n\Delta, i\Delta)$, $F_i^n = F(n\Delta, i\Delta)$.

The second derivative in the expression (7.84) for the free-term F is first estimated analytically and then the whole expression is converted into a discretized form for numerical calculation. After this discretization the free-term could be calculated using only values of the B field from the defined mesh nodes. A similar procedure is performed with the expressions for the B field, however an additional interpolation of F is needed here, because of difference in the geometry of the integration regions. Also all nodes for the B-field along the spatial coordinate, except boundary points, belong to the middle zone (see Figs. 7.15 and 7.16) and are calculated using the formula (7.87). The boundary points are calculated using formulae (7.88) and (7.89) for the quadratic mesh defined.

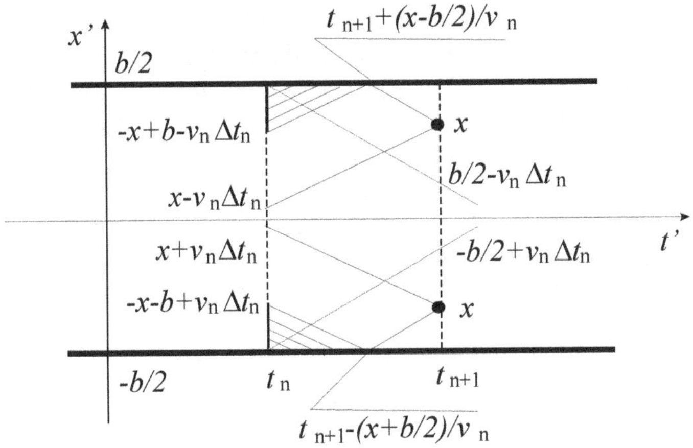

Figure 7.14 The regions for calculation of the field in the near-boundary zones.

The calculation order on some iteration n is as follows: first, the free-term F^{n+1} values together with their derivatives and coefficients for subsequent interpolation should be estimated, then the B^{n+1} field can be calculated. After the magnetic field is estimated, the medium permittivity ε^{n+1} is updated if needed (for the case of a non-linear medium).

7.5.4 Evolution of the Electromagnetic Wave after Switching of Non-Linearity in the Waveguide

Numerical calculations were performed for the transformation of an initial wave propagating in a slab dielectric waveguide, whose core medium becomes non-linear after some zero moment of time. The initial wave is a harmonic component of the guided wave $B_0(t, x) = e^{i\Omega_0 t - i\Gamma z}\cos(\kappa x)$ where $\Omega_0 = \omega b/v$ is the normalized frequency, $\kappa = bk$ is the normalized transverse eigennumber, and the propagation constants are calculated from the dispersion equations

$$\tan(\kappa/2) + \frac{\kappa}{\sqrt{\Omega_0^2(h_0^{-2}-1)-\kappa^2}} = 0,\ \Gamma = \sqrt{\Omega_0^2 h_0^{-2} - k^2}. \tag{7.90}$$

The permittivity of the cladding is $\varepsilon = 9$ and the linear part of the permittivity of the core is $\varepsilon_1 = 11$. Two values are considered for the normalized frequency of the initial wave, $\Omega_0 = 20$ and $\Omega_0 = 40$. The non-linearity of the medium is characterized by the normalized coefficient $\gamma = \chi_{NL}^{(3)} |E_0|^2/\varepsilon$, which takes the values: 0.05, 0.1, 0.3, 0.5. The linear material corresponds to $\gamma = 0$. The process is investigated for a duration τ_{max} of 3 in dimensionless quantities (corresponding to 1 ps) and the discretization step is $h = 0.01$. For 100 spatial intervals and 150 time steps the required memory is approximately 1Mb. The run-time is approximately 10 s on a PC with an Athlon 1700 XP+ processor.

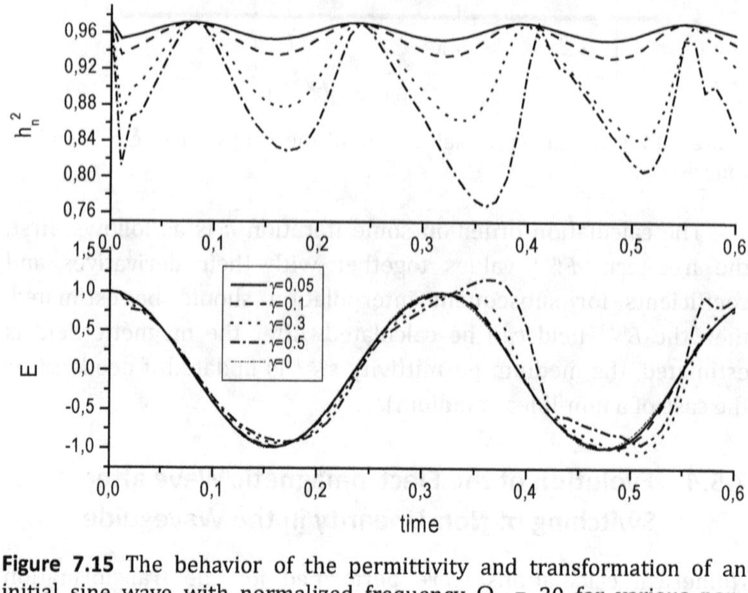

Figure 7.15 The behavior of the permittivity and transformation of an initial sine wave with normalized frequency $\Omega_0 = 20$ for various non-linearity coefficients.

A comparison of the cases with linear and non-linear dielectric waveguides, and the changes of the wave shapes, is shown in Figs. 7.18–7.20. Figure 7.15 shows the time behavior of the estimated non-linear permittivity (the upper picture) and the evolution of the wave near the waveguide wall (the lower picture) for the case where the initial wave frequency $\Omega_0 = 20$ in normalized units. As might be anticipated, increasing the material non-linearity leads to more shape distortion and signal retardation. However, high values of the non-linear coefficient also lead to instability of the field. Time-spatial diagrams for different values of the non-linear coefficient are presented in Fig. 7.16.

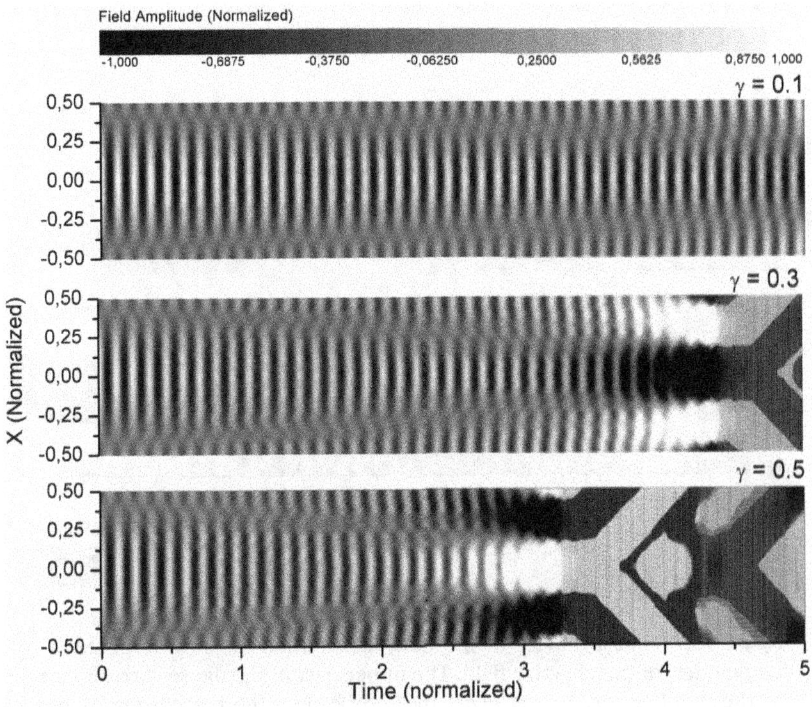

Figure 7.16 Time behavior of the field distribution in the waveguide for various non-linearity coefficients: $\gamma = 0.1$; 0.3; 0.5.

In this case, the linear part of the waveguide core permittivity is equal to 2.1904 and the cladding permittivity is 2.1784. The signal frequency $\Omega_0 = 54$ (in normalized units, this corresponds to a free space wavelength of 1.55 μm if it is assumed that the waveguide width = 9 μm). As one can see the instability of the field appears sooner for stronger non-linearities. It is the authors' opinion that the differences in the threshold for the instability is determined by the model for the effective field in the expression (7.68). This model reacts immediately to the field value changes, does not have a saturation threshold, and the effective permittivity on each new time step depends only on recent field values obtained on the previous time step. Therefore the interaction of the field and medium permittivity can lead to the appearance of stochastic oscillations.

The various models for the effective field are compared in Fig. 7.17 where the two diagrams correspond to the cases with exactly the same initial parameters as in Fig. 7.19.

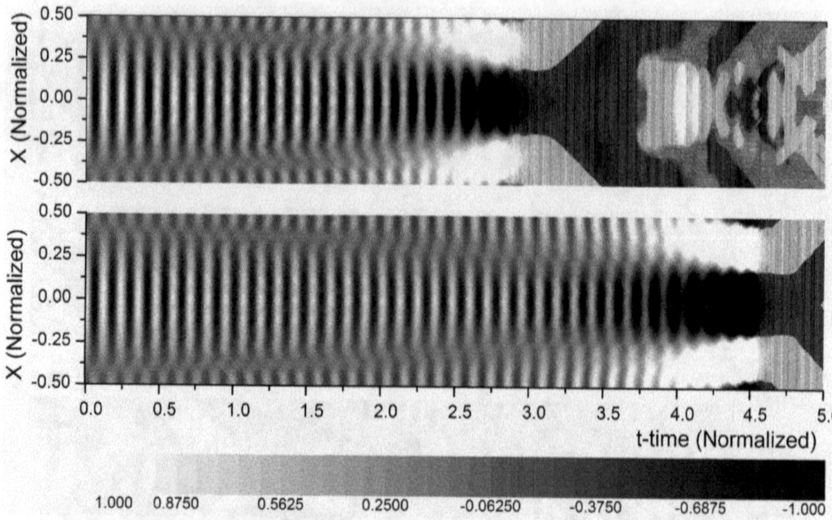

Figure 7.17 Time behavior of the field distribution in the waveguide for two models of the effective field. The upper part of the figure corresponds to the non-linearity model given by (7.68) while the bottom one is the result of the 'smoothed' simulation using (7.91).

The only difference is that the bottom diagram is the result of simulation, where the non-linearity model (7.68) is modified a little bit to "smooth" the permittivity as

$$\varepsilon_n = \varepsilon_d + \chi_{NL}^{(3)} \frac{1}{3}(E_{\text{eff}\,n-1}^2 + E_{\text{eff}\,n-2}^2 + E_{\text{eff}\,n-3}^2). \tag{7.91}$$

Even with such a trivial improvement to the model, the stability of the solution is sufficiently improved. In a practical investigation, the non-linearity model should take into account many physical properties of a particular non-linear medium of interest. It is evident that some level of saturation, reaction delay (medium inertness) and so on should be taken into account when constructing a model of non-linearity [40–43].

So, replacing a continuous changing of the non-linear dielectric core parameters of by a jump changing of the latter is the only approximation in this algorithm, excluding approximations used in the numerical implementation. Calculations show that this algorithm is operable but that the results depend on the models adopted for the non-linearity.

7.6 Two Ways for Calculating Field Evolution in a Dielectric Waveguide: Via Brillouin- or Eigen-Waves

Guided-wave dynamics can be organized around two complementary approaches, progressing and oscillatory, that are closely related to local versus global descriptions [44]. The progressing description operates with wave-fronts and it represents point-to-point propagation during which the wave fronts interact with the physical environment locally along their trajectories. The oscillatory approach operates with modes or resonances which form standing waves over extended, often global portions of the physical environment [45]. These two approaches reveal the double nature of the oscillation and wave processes in the waveguide structures. However, it is important to emphasize that the time variable has to enter in the progressive framework explicitly while the time dependence in the harmonic form is often enough for the oscillatory approach where the evolution of the electromagnetic phenomena has to be presented by the temporal Fourier series that assumes an infinite time axis. In contrast, the evolution process means the existence of a beginning of this process. This, in turn, means that the evolution of the electromagnetic phenomena has to be described as an initial and boundary value problem. The progressive approach is more suitable for such problems because it describes in a natural way those phenomena that develop from an initial time point. The progressive method is also more straightforward and is used, for example, for the calculation of transient waves reflection from multilayered media [46]. The oscillatory approach is more commonly used when the beginning of the process does not play a role in the phenomenon.

This section elucidates the conceptual distinction between these two approaches by considering an initial-boundary value electromagnetic problem in a dielectric waveguide. The beginning of the electromagnetic wave evolution in the waveguide is induced by the instant change of the core permittivity. It is a model problem for a wide range of non-stationary phenomena with many applications. We investigate here the wave transformation by virtue of the oscillatory or eigen-waves approach and the progressive or Brillouin waves approach. The calculation is made by the summation of generated oscillatory modes in the former case and by tracing the Brillouin wave fronts in the latter one. The evolution of the electromagnetic field and its spectrum is considered and compared.

7.6.1 Elastic Oscillations

To obtain deeper understanding of the two representations we first consider a simple one-dimensional problem in the form of longitudinal oscillations of an elastic rod fixed at one end and subjected to a force $F(t)$ at the other end. The oscillations are described by the one-dimensional wave equation [47]

$$\frac{\partial^2 u}{\partial t^2} - a^2 \frac{\partial^2 u}{\partial x^2} = 0, \tag{7.92}$$

where $u = u(x, t)$ is the longitudinal displacement and a^2 is a constant coefficient depending on the rod material. If the harmonic force $F = A\sin(\omega t)\theta(t)$ begins to act at zero moment of time ($\theta(t)$ is the Heaviside unit function) then the initial and boundary conditions turn into the following:

$$u\big|_{t=0} = \frac{\partial u}{\partial t}\bigg|_{t=0} = 0,\ \ u\big|_{x=0} = 0,\ \ \frac{\partial u}{\partial x}\bigg|_{x=l} = \frac{A}{E}\sin\omega t, \tag{7.93}$$

where E is the modulus of elasticity and l is the length of the rod. The Laplace transform of this problem is given by the formula

$$U(p) = \frac{B}{p(p^2+\omega^2)}\frac{\sinh(xp/a)}{\cosh(lp/a)},\ -1 \le \frac{x}{l} \le 1, \tag{7.94}$$

where p is the Laplace transform variable and $B = Aa\omega/E$. The function (7.94) has two poles $p = +i\omega$ and an infinite number of poles $p_k = i\omega_k$, $\omega_k = (2k + 1)\pi a/2l$ satisfying the equation $\cosh(lp/a) = 0$ or $e^{2lp/a} = -1$. The poles p_k represent all eigen-frequencies of the rod oscillations.

The inverse Laplace transformation can be found using the two ways described above: the oscillatory and the progressive approaches. The oscillatory approach uses all eigen-frequencies of the rod oscillations given by the poles. The calculation of all the residues at the poles gives the sum of the eigen-waves whose frequencies are determined by the rod length only. If only $2k$ poles p_k are used then the inverse Laplace transform has the form

$$\begin{aligned} u_1(t,K) &= \frac{B}{2\pi i}\int_{\eta-i\infty}^{\eta+i\infty}\frac{1}{p(p^2+\omega^2)}\frac{e^{(x+l)p/a}-e^{-(x-l)p/a}}{e^{2lp/a}+1}e^{pt}dp \\ &= \frac{B}{2\omega^2\cos(\omega l/a)}[\cos(\omega(t-x/a))\theta(t-x/a)-\cos(\omega(t+x/a))\theta(t+x/a)] \\ &+\frac{aB}{l}\sum_{k=0}^{K-1}\frac{1}{\omega_k\left[\omega_k^2-\omega^2\right]}\left\{\sin\left[\omega_k\left(t+\frac{x+l}{a}\right)\right]\theta\left(t+\frac{x+l}{a}\right)-\sin\left[\omega_k\left(t-\frac{x-l}{a}\right)\right]\theta\left(t-\frac{x-l}{a}\right)\right\} \\ &-\frac{aB}{2l}\frac{i}{\omega_K\left[\omega_K^2-\omega^2\right]}\left\{e^{i\omega_K\left(t+\frac{x+l}{a}\right)}\theta\left(t+\frac{x+l}{a}\right)-e^{i\omega_K\left(t-\frac{x-l}{a}\right)}\theta\left(t-\frac{x-l}{a}\right)\right\}, \end{aligned} \tag{7.95}$$

where $\eta > \eta_a$ and η_a is the abscissa of the absolute convergence of the Laplace transform.

The progressive approach is obtained by expanding the fraction with hyperbolic functions into the geometrical progression

$$\frac{\sinh(xp/a)}{\cosh(lp/a)} = \frac{e^{(x-l)p/a}-e^{-(x+l)p/a}}{1+e^{-2lp/a}} = \sum_{n=0}^{\infty}(-1)^n[e^{(x-(1+2n)l)p/a}-e^{(-x-(1+2n)l)p/a}] \tag{7.96}$$

In this form, all the poles of the oscillatory representation disappear and the remaining two poles of the first fraction in (7.94) give re-reflected waves at the frequency of the driving force. Finally, it gives another representation of the inverse Laplace transform

$$u_2(t,N)=\frac{B}{2\pi i}\int_{\sigma-i\infty}^{\sigma+i\infty}\frac{1}{p(p^2+\omega^2)}\frac{e^{(x-l)p/a}-e^{-(x+l)p}/a}{1+e^{-2lp/a}}e^{pt}dp$$

$$=\frac{B}{\omega^2}\sum_{n=0}^{N}(-1)^{n+1}\left\{\cos\left[\omega\left(t+\frac{x-(1+2n)l}{a}\right)\right]\theta\left(t+\frac{x+(1+2n)l}{a}\right)\right.$$

$$\left.-\cos\left[\omega\left(t-\frac{x+(1+2n)l}{a}\right)\right]\theta\left(t-\frac{x+(1+2n)l}{a}\right)\right\} \tag{7.97}$$

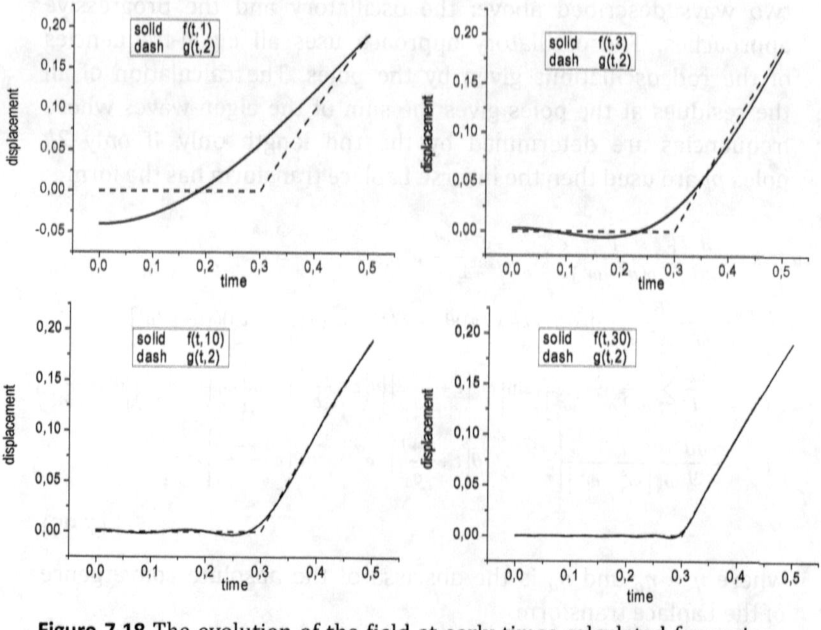

Figure 7.18 The evolution of the field at early times calculated for various number of the expansion terms in (7.95). (The parameters are: $\omega l/a = 2.7$, $x/l = 0.7$).

It is worth noting that for every given moment of time only a finite number N of terms in $u_2(t, N)$ are nonzero and this number is proportional to the time that has passed from the process beginning to the moment of observation, $N \sim (t + (x - l)a/2)$.

The difference between the two approaches is essential at the early stage of the evolution (Fig. 7.18) when the progressive approach requires only two terms in the sum while the oscillatory one requires a few tens of terms for the same accuracy. For a long interval both transforms give the same result asymptotically (Fig. 7.19) and in this case a few terms of the oscillations give a good accuracy while the number of terms in the progressive representation grows with time.

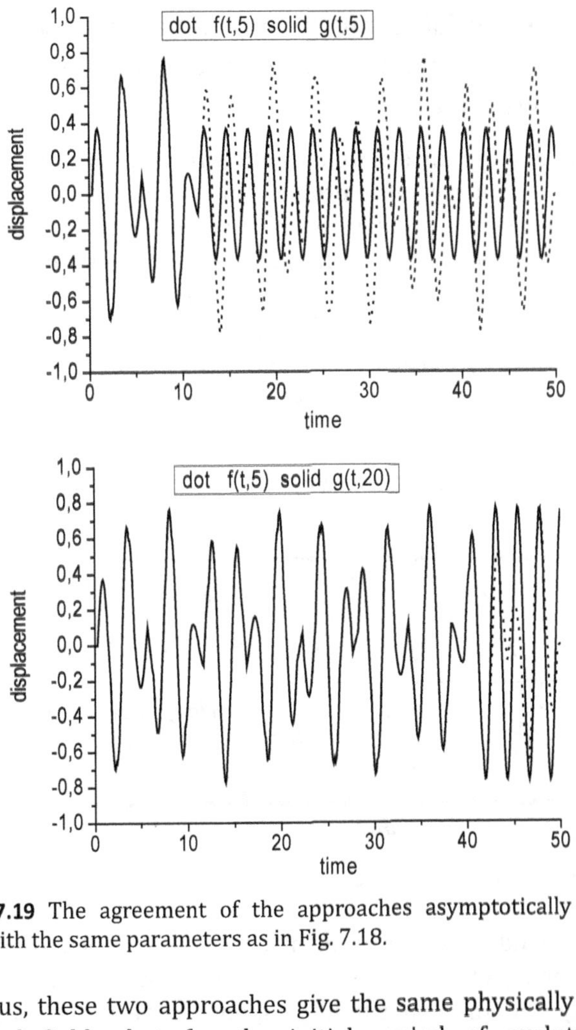

Figure 7.19 The agreement of the approaches asymptotically for long times with the same parameters as in Fig. 7.18.

Thus, these two approaches give the same physically existing observed fields, but for the initial period of evolution, the oscillatory approach requires a few tens of terms for a suitable accuracy while only two terms of the progressive approach give the exact expression for the same time interval. Concluding, one can say that the oscillatory approach gives the indefinite sum of abstract eigen-waves. Only this sum can be observed directly, not the individual waves. On the other hand, the progressive approach gives the finite sum of physically existing waves excited by the driving force and their wave fronts can be detected separately.

7.6.2 Differential Formulation of Initial and Boundary Value Electromagnetic Problem in a Dielectric Waveguide

We consider here an electromagnetic problem formulated as an initial and boundary value problem in differential statement. An example of such a problem is the two-dimensional problem of the electromagnetic field evolution in a flat dielectric waveguide after an instantaneous change of the core material parameters. The initial guided TE wave propagates along the z axis and its electric field is perpendicular to the propagation direction (along the y axis which is parallel to the waveguide walls) (Fig. 7.20).

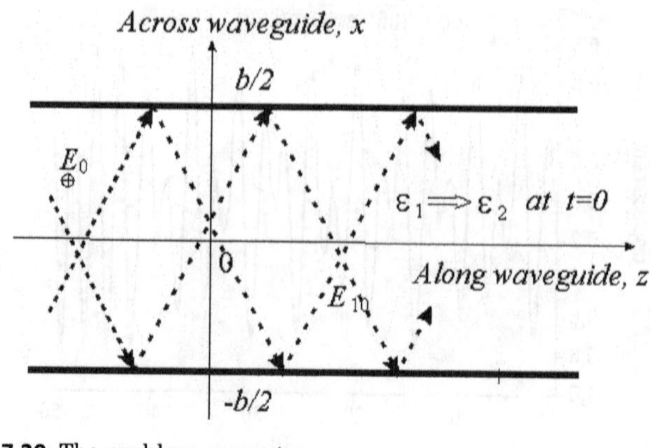

Figure 7.20 The problem geometry.

The boundary value problem considers the electromagnetic field in and out of the waveguide with the requirement of continuity for the tangential components of the electric and magnetic fields on the waveguides boundaries. The initial value problem arises from an instant change of the core permittivity at zero moment, $\varepsilon_1 \Rightarrow \varepsilon_2$. Then the field evolution is initiated by this change according to the initial conditions which can be easily obtained from Maxwell's equations:

$$B_z\big|_{t=-0} = B_z\big|_{t=+0};\ \varepsilon_1 E_y\big|_{t=-0} = \varepsilon_2 E_y\big|_{t=+0};\ \partial_t(\varepsilon_1 E_y)\big|_{t=-0} = \partial_t(\varepsilon_2 E_y)\big|_{t=+0} \quad (7.98)$$

and continuity of the tangential components of the electric and magnetic fields on the waveguides boundaries. All these determine the spatial and temporal behavior of the field. The problem, as an initial-boundary value one, can also be considered using the two approaches used in the example of the mechanical problem above: the oscillatory and the progressive ones.

We consider the initial field as the TE eigen-wave propagating along the z-axis of the dielectric waveguide:

$$E_{0y} = A_0 \sin(k_1 x)e^{i\omega t - i\Gamma z},\ B_{0z} = i(k_1/\omega)\, A_0 \cos(k_1 x)e^{i\omega t - i\Gamma z}.$$

Here, $\Gamma^2 = \omega^2/v^2 + s_1^2 = \omega^2/v_1^2 - k_1^2$ is the longitudinal wave number, $v = c/\sqrt{\varepsilon}$, $v_1 = c/\sqrt{\varepsilon_1}$ are the wave phase velocities, ε and ε_1 are the cladding and core permittivities, respectively. The transverse wave numbers κ_1 and s_1, Re $s_1 > 0$, satisfy the common dispersion equation $\tan(\kappa_1 b/2) = -\kappa_1/s_1$. The field outside the waveguide is represented by the evanescent waves $E'_{0y} = A'_0 e^{i\omega t - s_1 x - i\Gamma z}$, $B'_{0z} = -i(s_1/\omega)A'_0 e^{i\omega t - s_1 x - i\Gamma z}$.

After the moment of the core permittivity change the field inside the waveguide is controlled by the equations

$$\begin{cases} \partial_{zz}^2 E_y + \partial_{xx}^2 E_y = v_2^{-2}\partial_t^2 E_y \\ \partial_x E_y = -\partial_t B_z \end{cases} \quad (7.99)$$

Outside the waveguide the equations are similar

$$\begin{cases} \partial_{zz}^2 E_y + \partial_{xx}^2 E_y = v^{-2}\partial_t^2 E_y \\ \partial_x E_y = -\partial_t B_z \end{cases} \quad (7.100)$$

The solutions to these equations are found using Laplace transformation together with the initial and boundary conditions that gives the Laplace transform of the field inside the waveguide

$$E = \frac{v_2^2}{v_1^2} A_0 \left\{ \frac{p + i\omega}{p^2 + \omega_2^2} \sin(\kappa_1 x) - \left(\frac{v_1^2}{v_2^2} - 1 \right) \frac{v_2 \kappa_1 p^2 (p + i\omega)}{s_1 (v s_1 + \varphi)} \frac{\cos(\kappa_1 b/2)}{p^2 + \omega_2^2} 2e^{-\varphi_2 b/2v_2} \frac{\sinh(\varphi_2 x/v_2)}{D} \right\}. \tag{7.101}$$

Here, p is the Laplace transform variable, $\varphi_2 = \sqrt{p^2 + v_2^2\Gamma^2}$, $\varphi = \sqrt{p^2 + v^2\Gamma^2}$, $v_2 = c/\sqrt{\varepsilon_2}$ and b is the waveguide width. The denominator D in (7.101) is given by the formula $D = (v_2\varphi + v\varphi_2)(1 - Re^{-\varphi_2 b/v_2})$ where $R = \frac{v_2\varphi - v\varphi_2}{v_2\varphi + v\varphi_2}$ is the reflectivity of the Brillouin waves.

The frequency $\omega_2 = \omega v_2/v_1$ is a new frequency which the transformed wave acquires due to the permittivity change.

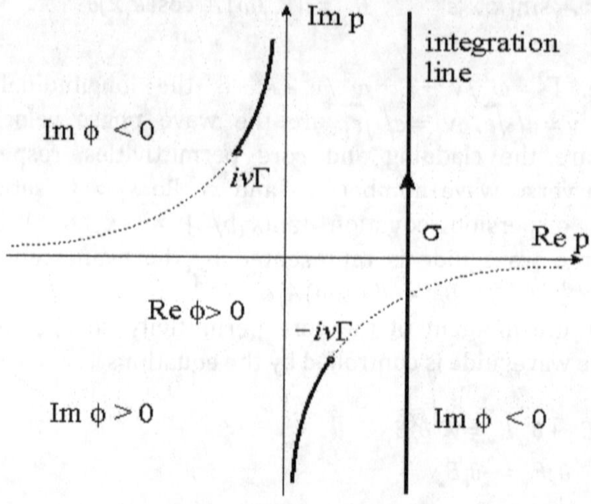

Figure 7.21 The cuts for choosing the branches of the square roots in the function φ. The Riemann surface corresponding to Re $\varphi(p) > 0$ is chosen.

The field evolution in time is given by the inverse Laplace transform which is implemented in the oscillatory and the progressive ways. The oscillatory or eigen-frequency representation of the inverse Laplace transform is given by the calculation of residues at the poles $p_{\pm} = \pm i\omega_2$ and the poles p_k which are zeros of the denominator $D = 0$ in the expression (7.101). The poles p_k are the roots of the dispersion equation

$$\tanh(\varphi_2 b/2v_2) = -(v\varphi_2/v_2\varphi) \tag{7.102}$$

and represent the eigen-waves of the waveguide.

The integration line for the calculation of the inverse Laplace transform shown in Fig. 7.21 for the function φ is passing on the right from all the poles. Analogous lines are found for the function φ_2. The Riemann surface for the functions $\varphi(p)$ and $\varphi_2(p)$ is chosen in such a way that the conditions Re $\varphi_2(p) > 0$ and Re $\varphi_2(p) > 0$ are satisfied.

The alternative, progressive representation is given by the expansion with respect to R if the corresponding fraction in (7.101) is written in the form of a geometric progression

$$\frac{\sinh\frac{\varphi_2 x}{v_2}}{D} = \frac{1}{v_2\varphi + v\varphi_2}\,\frac{e^{\frac{\varphi_2}{v_2}(x-b/2)} - e^{-\frac{\varphi_2}{v_2}(x+b/2)}}{1 - Re^{-\frac{\varphi_2 b}{v_2}}}$$
$$= \frac{1}{v_2\varphi + v\varphi_2}\sum_{m=0}^{\infty} R^m\left(e^{\frac{\varphi_2}{v_2}(x-b/2-mb)} - e^{-\frac{\varphi_2}{v_2}(x+b/2+mb)}\right) \tag{7.103}$$

7.6.3 Flat Dielectric Resonator

If $\Gamma = 0$ then the formula (7.101) for the field describes electromagnetic oscillations in a flat resonator, in which Brillouin waves propagate normally to the walls. In this case the ambiguity of this equation caused by the radicals signs disappears because $\varphi = \varphi_2 = p$; the dispersion equation and the expression for the reflectivity are simplified, $e^{pb/v_2} = -R$, $R = (v_2 - v)/(v_2 + v)$, and the Laplace transform can be inverted exactly.

In the oscillation approach all the poles and corresponding eigen-frequencies are substantially complex $p_k = \frac{v_2}{b}$ $[\ln|R| + i(2k+1)\pi]$. The field in the time domain given by the calculation of the residues at the poles is represented by the series of the mode oscillations

$$E_{\text{res}}(t,x,K) = E_{\text{com}}(t,x) + E_{\omega_2}(t,x) + B_0 \frac{b}{v_2} \sum_{k=-K}^{K} \frac{p_k}{p_k^2 + \omega_2^2}$$

$$\left[\theta\left(t + \frac{x - b/2}{v_2}\right) e^{p_k\left(t + \frac{x-b/2}{v_2}\right)} + \theta\left(t - \frac{x + b/2}{v_2}\right) e^{p_k\left(t - \frac{x+b/2}{v_2}\right)}\right], \tag{7.104}$$

where $B_0 = A_0 \frac{(v_1^2 - v_2^2)v^2}{(v+v_2)v_1^3} i\cos(\omega b/2v_1)$. The first term in (7.104) is common to both approaches

$$E_{\text{com}} = \frac{v^2}{2v_1^2} A_0 \sin(\omega x/v_1)[(v_2 + v_1)e^{i\omega_2 t} + (v_2 - v_1)e^{-i\omega_2 t}] \tag{7.105}$$

and the second one is determined by the frequency ω_2 waves

$$E_{\omega_2}(t,x) = \left\{\theta\left(t + \frac{x - b/2}{v_2}\right)\left[\cos\left(\omega_2\left(t + \frac{x - b/2}{v_2}\right)\right) - R\cos\left(\omega_2\left(t + \frac{x + b/2}{v_2}\right)\right)\right]\right.$$

$$\left. -\theta\left(t - \frac{x + b/2}{v_2}\right)\left[\cos\left(\omega_2\left(t - \frac{x + b/2}{v_2}\right)\right) - R\cos\left(\omega_2\left(t - \frac{x - b/2}{v_2}\right)\right)\right]\right\}$$

$$\frac{B_0}{1 - 2R\cos\left(\frac{\omega_2 b}{v_2}\right) + R^2} \tag{7.106}$$

The third term in (7.104) is determined by the poles $p_k = \frac{v_2}{b}$ $[\ln|R| + i(2k+1)\pi]$ and only K terms are retained in this term for calculation.

The progressive approach gives the successive reflections of the Brillouin waves all of which have the same frequency that is equal to the transformed one $\omega_2 = v_2\omega/v_1$ (as in the case of the unbounded medium):

$$E_{\text{Br}}(t,x,M) = E_{\text{com}}(t,x)$$

$$+ B_0 \sum_{m=0}^{M} R^m \left[\theta\left(t + \frac{x - (m+1/2)b}{v_2}\right)\cos\left(\omega_2 t + \omega\frac{x - (m+1/2)b}{v_1}\right)\right.$$

$$\left. -\theta\left(t - \frac{x + (m+1/2)b}{v_2}\right)\cos\left(\omega_2 t - \omega\frac{x + (m+1/2)b}{v_1}\right)\right] \tag{7.107}$$

In this case, the field is given exactly by a finite sum of terms, the number of which is determined by the moment of the observation.

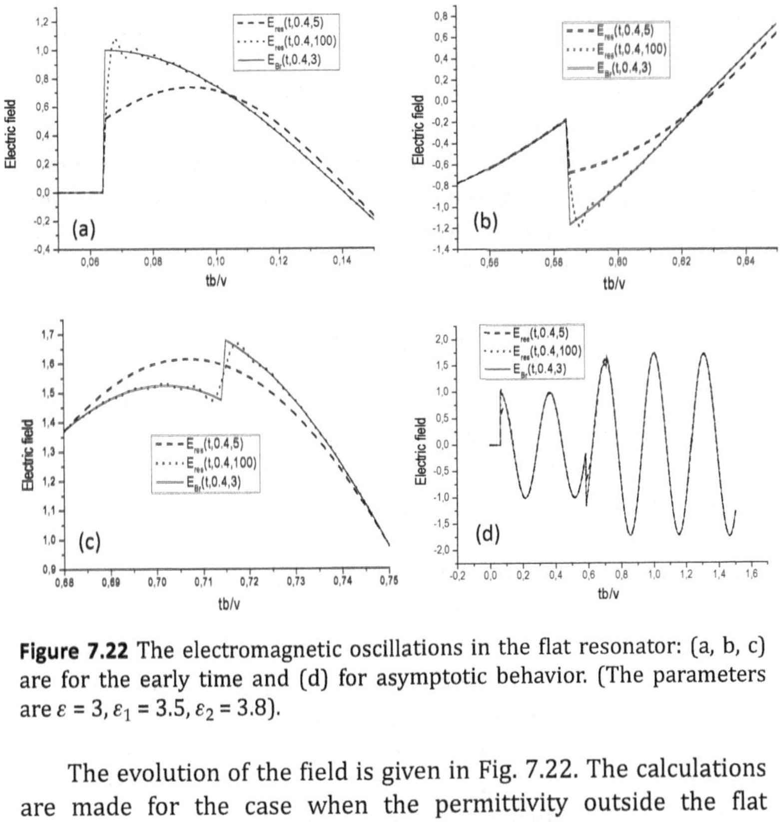

Figure 7.22 The electromagnetic oscillations in the flat resonator: (a, b, c) are for the early time and (d) for asymptotic behavior. (The parameters are $\varepsilon = 3, \varepsilon_1 = 3.5, \varepsilon_2 = 3.8$).

The evolution of the field is given in Fig. 7.22. The calculations are made for the case when the permittivity outside the flat resonator is equal to $\varepsilon = 3$. The inner permittivity changes from $\varepsilon_1 = 3.5$ to $\varepsilon_2 = 3.8$ at zero moment. The solid line in Fig. 7.22 corresponds to the approach exploiting reflections of the Brillouin waves with the reflectivity $R = (v_2 - v)/(v_2 + v)$. The dash and dot lines give the results obtained by using the eigen-frequencies. The dash line is for 5 terms of the series and the dot line is for 100 terms. It can be seen that the accuracy of the results based on the eigen-frequencies improves with the number of the terms for the early time (Fig. 7.22a,b,c) and practically does not depend on it for the asymptotic behavior (Fig. 7.22d). At the same time interval it is enough to take into account only three reflections of the Brillouin waves. Thus, the two representations of the field evolution differ at the early stage but coincide asymptotically.

As in the mechanical problem in the electromagnetic case the oscillatory approach gives an infinite sum of abstract eigen-waves. Only the sum as a whole can be observed directly not the individual waves. The progressive approach gives a finite sum of real waves initiated by the primary field and their wave fronts can be detected separately. These two approaches give the same physical observed field, but for the initial period of evolution, the oscillatory approach requires hundreds of terms for a suitable accuracy while only three terms of the progressive approach give good results for the same time interval.

7.6.4 Field Evolution in a Dielectric Waveguide

When the propagation constant is not equal to zero, $\Gamma \neq 0$, then Eq. (7.101) describes the evolution of the waves in the dielectric waveguides after the instantaneous changes of the core permittivity. In this case the ambiguity of the radical signs must be taken into account; this determines the branch points for the equation. The presence of the radicals changes considerably the essence of the dispersion Eq. (7.102) and the form of its roots. Only the contributions from the poles will be considered further as the contributions from the integrals along the cut lines drawing from the branch points are negligible. Indeed, it is enough to make the substitution $p = qt$ in the integrand of the inverse Laplace transform of (7.101) to see that the integral over the branch cut tends to zero when time t tends to infinity [32]:

$$\frac{1}{2\pi i}\int_C \frac{e^q}{t}\left\{\frac{qt+i\omega}{(qt)^2+\omega_2^2}\sin(\kappa_1 x)\right.$$
$$\left. -\left(\frac{v_1^2}{v_2^2}-1\right)\frac{\kappa_1(qt)^2(qt+i\omega)\cos(\kappa_1 b/2)}{v_2 s_1(vs_1+t\overline{\varphi})\quad (qt)^2+\omega_2^2} 2e^{-\varphi_2 b/2v_2}\frac{\sinh(\varphi_2 x/v_2)}{D}\right\}dq \tag{7.108}$$

So, the expression obtained by the eigen-frequency approach is equal to (without the integrals over the branch cuts)

$$E_{res}(t,x,K) = E_{com}(t,x) + E_{\omega_2}(t,x)$$
$$+ B_1 \frac{vv_1}{v_2^2 - v^2} \sum_{m=1}^{K} \frac{p_m(p_m + i\omega)}{(p_m^2 + \omega_2^2)} \frac{\varphi_m \varphi_{2m}(v_2\varphi_m + v\varphi_{2m})}{(vs_1 + \varphi_m)(bp_m^2\varphi_m - 2vv_2^2\Gamma^2)}$$
$$\times e^{p_m t}\left(e^{\frac{\varphi_{2m}}{v_2}(x+b/2)}\theta[t + (x - b/2)/v_2] - e^{-\frac{\varphi_{2m}}{v_2}(x-b/2)}\theta\left[t - (x + b/2)/v_2\right]\right),$$

(7.109)

where E_{com} is as in (7.105),

$$E_{\omega_2} = B_1 \frac{v_1 + v_2}{2v_1}\omega^2 \sum_{\pm}$$

$$\frac{\sin(\omega_2 t + \kappa_1 x)\theta[t + (x - b/2)/v_2] - \sin(\omega_2 t - \kappa_1 x)\theta[t - (x + b/2)/v_2]}{(vs_1 + \psi)[\psi \sin(\kappa_1 b/2) + v\kappa_1 \cos(\kappa_1 b/2)]}$$

and $B_1 = A_0 \frac{(v_1^2 - v_2^2)v\kappa_1}{v_1^3 s_1} \cos(\kappa_1 b/2)$ $\varphi_{2m} = \sqrt{p_m^2 + v_2^2\Gamma^2}$, $\varphi_m = \sqrt{p_m^2 + v^2\Gamma^2}$, $\psi = \sqrt{-\omega_2^2 + v^2\Gamma^2}$, and $\omega_2 = v_2\omega/v_1$ is the transformed frequency due to the change of the material permittivity. The eigen-frequencies p_m are the roots of the dispersion equation $D(p) = 0$ which has a finite number of real roots and an infinite number of complex ones. For the parameters considered in the above case of the flat resonator the values for the lower eigen-frequencies are given in Table 7.1. It is seen that only a few roots are imaginary and that the others have negative real parts. It means that the corresponding waves will decay fast in time, so the contribution of the complex roots to the whole field is very small and can therefore be neglected in this calculation.

Table 7.1 The lower eigen-frequencies (the first roots of the dispersion equations)

m	1	2	3	4	5	6	7	8
Re $(p_m b/v_2)$	0	0	0	−1,79	−1,91	−1,99	−2,04	−2,14
Im $(p_m b/v_2)$	5,46	10,90	16,04	34,51	40,81	47,10	53,39	72,24

The Brillouin approach gives the expression in the form of the expansion on the reflectivity of the Brillouin waves $R\varepsilon^{-j_2 b/v_2}$ =

$(v_2\varphi - v\varphi_2)e^{-\varphi_2 b/v_2}/(v_2\varphi + v\varphi_2)$. This formula gives the re-reflections of the Brillouin waves at each moment

$$E_{\mathrm{Br}}(t,M) = E_{\mathrm{com}} + B_1\frac{v_2}{2v_1}\sum_{\pm}\frac{\mp v_1 - v_2}{vs_1 + \psi}e^{\pm i\omega_2 t}\sum_{m=0}^{\infty}(R^{\pm})^m \times \left[e^{\pm i\kappa_1(x-b/2-mb)}\theta\left[t+(x-b/2-mb)/v_2\right] - e^{\mp i\kappa_1(x+b/2+mb)}\theta\left[t-(x+b/2+mb)/v_2\right]\right] \tag{7.110}$$

The formation of the transformed field is illustrated by the tracks of the Brillouin wave fronts as shown in the time-spatial diagram of Fig. 7.23. The progressive representation is a sum of a finite number of terms at each time moment. This number is equal to the number of the Brillouin wave re-reflections and is increasing with time. All re-reflected waves have the same transformed frequency $\omega_2 = \omega v_2/v_1$ as in the unbounded medium.

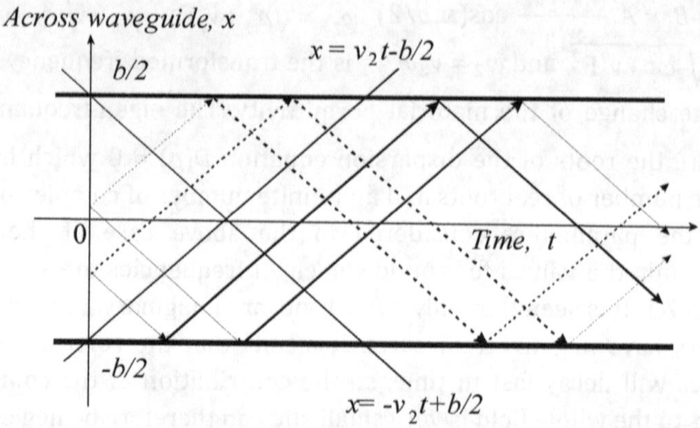

Figure 7.23 Re-reflection of the Brillouin waves in spatial-time coordinates.

The initial and transformed fields on short time intervals are given in Fig. 7.24. There the complete coincidence for both methods of calculation is shown in detail in the early stage of the process immediately after the permittivity change (Fig. 7.24a) as well as after a long period (Fig. 7.24b), which equals to 70 periods of the initial wave marked by the dotted line in this diagram. These curves are calculated for seven terms in the oscillatory approach which correspond to the pure imaginary roots as the contribution of the other terms with complex roots is very small. The number of terms in the progressive approach increases with time as a growing number of the Brillouin wave re-reflections contributes to the whole field.

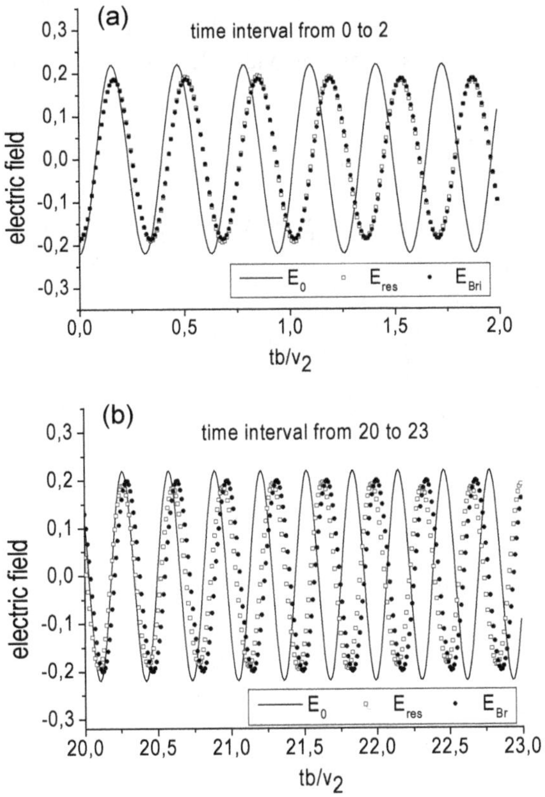

Figure 7.24 The comparison of calculations using the two approaches at different stages of the evolution.

The wave evolution on the long time interval is shown in Fig. 7.25a and the spectra presented in Fig. 7.25b show clearly the shift in the frequency.

Concluding, one can say that there are two ways for solving an initial value problem in a dielectric waveguide. These ways are applicable to the investigation of wave evolution dynamics in a waveguide (resonator) after the moment the process commenced. Each way describes one of two complementary approaches: a progressive or Brillouin waves approach and an oscillatory or eigen-frequency approach. The Brillouin wave representation is a sum of a finite number of terms at each time moment while the eigenmode representation is an infinite series. The progressing way turns out to be preferable because it does not require the calculation of the eigen-frequencies for the new modes that may be a rather difficult problem. Therewith the Brillouin approach is more intuitive and visual and also less demanding of computational resources.

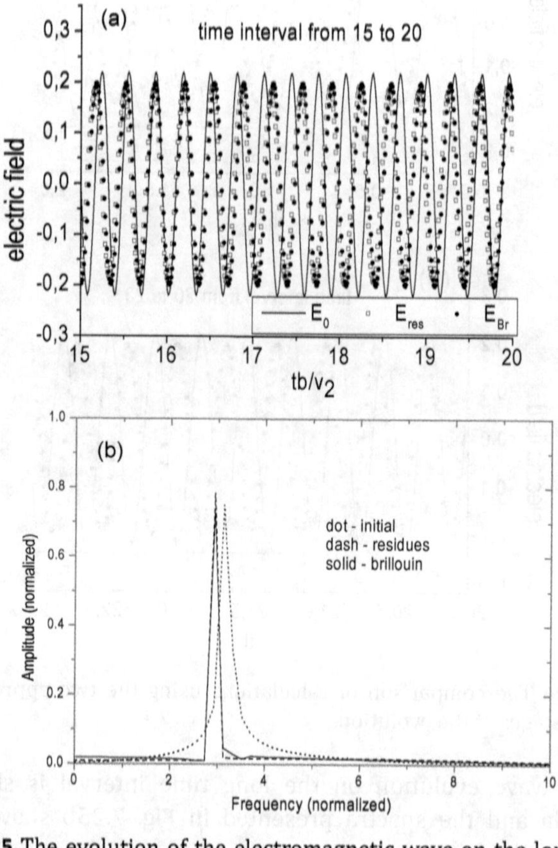

Figure 7.25 The evolution of the electromagnetic wave on the long period (a) and the change of the spectrum (b).

References

1. Cohn, G. I. (1952) Electromagnetic transients in waveguides, *Proc. Nat. Electronics Conf.*, **8**, 284–295.
2. Cotte, M. (1954) Propagation d'une impulsion sur un guide d'ondes, *Onde Elec.*, **34**, 143–146.

3. Karbowiak, A. E. (1957) Propagation of transients in waveguides, *Proc. IEE (London)*, **104**(C), 339–349.

4. Elliott, R. S. (1957) Pulse waveform degradation due to dispersion in waveguide, *IER Trans. Microwave Theory Tech.*, **MTT-5**(4), 254–257.

5. Cohen, D. A., M. Hossein-Zadeh, M., and Levi, A. F. J. (2001) High-Q microphotonic electro-optic modulator, *Solid-State Electronics*, **45**, 1577–1589.

6. Ilchenko, V. S., Savchenkov, A. A., Matsko, A. B., and Maleki, L. (2002) Sub-micro watt photonic microwave receiver, *IEEE Photonics Technol. Lett.*, **14**(11), 1602–1604.

7. Ilchenko, V. S., Savchenkov, A. A., Matsko, A. B., and Maleki, L. (2003) Whispering-gallery-mode electro-optic modulator and photonic microwave receiver, *J. Opt. Soc. Am. B*, **20**(2), 333–342.

8. Ayazi, A., Hsu, R. C. J., Houshmand, B., Steier, W. H., and Jalali, B. (2008) All-dielectric photonic-assisted wireless receiver, *Opt. Express*, **16**, 1742–1747.

9. Carter, S. G., Birkedal, V., Wang, C. S., Caldren, L. A., Maslov, A. V., and Shevrin, M. S. (2005) Quantum coherence in an optical modulator, *Science*, **310**(5748), 651–653.

10. Maslov, A. V., and Citrin, D. S. (2002) Mutual transparency of coherent laser beams through a terahertz-field-driven quantum well, *J. Opt. Soc. Am. B*, **19**(8), 1905–1909.

11. Bekker, E. V., Vukovic, A., Sewell, P., Benson, T. M., Sakhnenko, N. K., and Nerukh, A. G. (2007) An assessment of coherent coupling through radiation fields in time varying slab waveguides, *Opt. Quantum Electronics*, **39**(7), 533–551.

12. Blom, F. C., van Dijk, D. R., Hoekstra, H., Driessen, A., and Popma, T. (1997) Experimental study of integrated-optics microcavity resonators: Toward an all-optical switching device, *Appl. Phys. Lett.*, **71**(6), 747–749.

13. Savchenkov, A. A., Ilchenko, V. S., Matsko, A. B., and Maleki, L. (2005) High-order tunable filters based on a chain coupled crystalline whispering gallery-mode resonators, *IEEE Photon. Technol. Lett.*, **17**(1), 136–139.

14. Djordjiev, K., Coi, S., and Dapkus, P. (2002) Microdisk tunable resonant filters and switches, *IEEE Photon. Technol. Lett.*, **14**(6), 828–830.

15. Morichetti, F., Ferrari, C., Melloni, A. (2007) Enhanced frequency shift in optical slow-wave structures. In Proc. ICTON 2007. Rome, Italy, July, Tu.A3.3.
16. Reed, E., Soljacic, M., and Joannopoulos, J. (2003) Reversed Doppler effect in photonic crystals, *Phys. Rev Lett.*, **91**(13), 133901.
17. Reed, E., Soljacic, M., and Joannopoulos, J. (2003) Color of shock waves in photonic crystals, *Phys. Rev Lett.*, **90**(20), 203904.
18. Fan, S., Yanik, M. F., Povinelli, M. L., and Sandhu, S., (2007) Dynamic photonic crystals, *Opt. Photonic News*, **18**(3), 42–45.
19. Macdonald, K. F., Samson, Z. L., Stockman, M. I., and Zheludev, N. I., (2009) Ultrafast active plasmonics, *Nat. Photonics*, **3**, 55–58.
20. Masoudi, H. M., and Arnold, J. M. (1995) Modelling second-order nonlinear effects in optical waveguides using a parallel beam propagation method, *IEEE J. Quantum Electronics*, **31**(12), 2107–2113.
21. Masoudi, H. M., Alsunaidi, M. A., Arnold, J. M. (2000) A time-domain algorithm for the analysis of second-harmonic generation in nonlinear optical structures, *IEEE Photonics Technol. Lett.*, **12**(4), 395–397.
22. Chi, J. W. D., Chao, L., and Rao, M. K. (2001) Time-domain large-scale investigation on nonlinear interactions between an optical pulse and semiconductor waveguides, *IEEE J. Quantum Electronics*, **37**(10), 1329–1336.
23. Jeong, Y., and Lee, B. (2001) Characteristics of second-harmonic generation including third-order nonlinear interactions, *IEEE J. Quantum Electronics*, **37**(10), 1292–1300.
24. Bolomey, J., Durix, C., and Lesselier, D. (1978) Time domain integral equation approach for inhomogeneous and dispersive slab problems, *IEEE Trans. Antennas Propag.*, **AP-26**(5), 658–667.
25. Romanova, E. A., and Melnikov, L. A., (2003) Detailed analysis of spatiotemporal stability of the ultra-short optical pulses propagating in non-linear step-index optical waveguide, *Proc. 11-th Int. Workshop on Optical Waveguides Theory and Numerical Modelling, April 4–5, Prague, Czech Republic, 51.*
26. Shifman, Y., and Leviatan, Y. (2001) On the use of spatio-temporal multiresolution analysis in method of moments solutions of transient electromagnetic scattering, *IEEE Trans. Antennas Propag.*, **AP-49**(8), 1123–1129.

27. Weitzman, P. S., and Osterberg, U. (1993) A modified beam propagation method to model second harmonic generation in optical fibers, *IEEE J. Quantum Electronics,* **29**(5), 1437–1443.

28. Hsu, T.-T., and Carin, L. (1996) FDTD analysis of plane-wave diffraction from microwave devices on an infinite dielectric slab, *IEEE Microwave Guided Wave Lett.*, **6**(1), 16–18.

29. Chou, H.-F., Lin, C.-F., and Mou, S. (1999) Comparison of finite difference beam propagation methods for modelling second-order nonlinear effects, *J. Lightwave Technol.*, **17**(8), 1481–1486.

30. Stathopoulos, N. A., and Kanellopoulos, J. D. (2002) Calculation of nonlinear waves guided by optical fibers with an inhomogeneous nonlinear core, *Opt. Quantum Electronics,* **34**(9), 915–929.

31. Morgenthaler, F. R. (1958) Velocity modulation of electromagnetic waves, *IRE Trans. Microwave Theory Techn.,* **MTT-6**(2), 167–172.

32. Felsen, L. B., and Whitman, G. M. (1970) Wave propagation in time-varying media, *IEEE Trans. Antennas Propag.,* **AP-18**(2), 242–253.

33. Fante, R. L. (1971) Transmission of electromagnetic waves into time-varying media, *IEEE Trans. Antennas Propag.,* **AP-19**(3), 417–424.

34. Nerukh, A. G. (1995) Evolutionary approach in transient electrodynamics problems, *Radio Sci.,* **30**, 481–491.

35. Nerukh, A. G., Scherbatko, I. V., and Marciniak, M. (2001) *Electromagnetics of Modulated Media with Applications to Photonics,* Warsaw, National Institute of Telecommunications Publishing House.

36. Schwartz, L. (1950–1951) *Theorie Des Distributions,* vol. I-II, Paris, Hermann.

37. Gelfand, I. M., and Shilov, G. E. (1964) *Generalized Functions,* vol. 1, New York, Academic Press.

38. Morse, P. M., and Feshbach, H. (1953) *Methods of Theoretical Physics,* New York, McGraw-Hill.

39. Vladimirov, V. S. (1967) *Equations of Mathematical Physics,* "Nauka", Moscow. 437 p. (in Russian).

40. Janyani, V. (2005) Modelling of dispersive and nonlinear materials for optoelectronics using TLM, PhD thesis University of Nottingham.

41. Paul, J., Christopoulos, C., and Thomas, D. W. P. (1999) Generalized material models in TLM–Part 1: Materials with frequency dependent properties, *IEEE Trans. Antennas Propag.,* **AP-47**(10), 1528–1534.

42. Paul, J., Christopoulos, C., and Thomas, D. W. P. (1999) Generalized material models in TLM–Part 2: Materials with anisotropic properties", *IEEE Trans. Antennas Propag.*, **AP- 47**(10), 1535–1542.

43. Paul, J., Christopoulos, C., and Thomas, D. W. P. (2002) Generalized material models in TLM–Part 3: Materials with nonlinear properties", *IEEE Trans. Antennas and Propag.*, **AP-50**(7), 997–1004, 2002.

44. Felsen, L. B., and Marcuvitz, N. (1973) *Radiation and Scattering of Waves,* Englewood Cliffs, NJ, Prentice-Hall, (1973); reissue by Piscataway, NJ, IEEE Press, (1994).

45. Sevgi, L., Akleman, F., and Felsen, L. B. (2007) Groundwave propagation modeling: problem-matched analytical formulations and direct numerical techniques, *IEEE AntennasPropagation Mag.*, **49**(3), 76–94.

46. Perry, B. T., and Rothwell, E. J. (2007) Calculation of the transient plane-wave reflection from an N-layer medium by the method of subregions, *IEEE Trans. Antennas Propag*, **AP-55**(11), 3293–3299.

47. Lavrentyev, M. A., and Shabat, B. V. (1987) *Methods of Complex Variable Functions* (in Russian), "Nauka", Moscow.

Index

For Product Safety Concerns and Information please contact our EU
representative GPSR@taylorandfrancis.com
Taylor & Francis Verlag GmbH, Kaufingerstraße 24, 80331 München, Germany

www.ingramcontent.com/pod-product-compliance
Lightning Source LLC
LaVergne TN
LVHW020628300926
849161LV00004B/637
* 9 7 8 9 8 1 4 7 7 4 9 5 6 *